www.wadsworth.com

wadsworth.com is the World Wide Web site for Wadsworth and is your direct source to dozens of online resources.

At *wadsworth.com* you can find out about supplements, demonstration software, and student resources. You can also send email to many of our authors and preview new publications and exciting new technologies.

wadsworth.com
Changing the way the world learns®

Understanding Society

An Introductory Reader

MARGARET L. ANDERSEN
University of Delaware

KIM A. LOGIO
Saint Joseph's University

HOWARD F. TAYLOR
Princeton University

WADSWORTH
™
THOMSON LEARNING

Australia • Canada • Mexico • Singapore • Spain
United Kingdom • United States

WADSWORTH
™
THOMSON LEARNING

Publisher: Eve Howard
Assistant Editor: Dee Dee Zobian
Editorial Assistant: Stephanie Monzon
Marketing Manager: Matthew Wright
Signing Representative: Ron Shelly
Project Editor: Jerilyn Emori
Print Buyer: Karen Hunt

Permissions Editor: Robert Kauser
Production Service: Andrea Bednar/Shepherd Incorporated
Cover Designer: Norman Baugher
Cover Image: Hessam Abrishami
Cover Printer: Phoenix Color Corp.
Compositor: Shepherd Incorporated
Printer: R. R. Donnelley & Sons Company

Printed in the United States of America
1 2 3 4 5 6 7 04 03 02 01 00

Wadsworth/Thomson Learning
10 Davis Drive
Belmont, CA 94002-3098
USA

For more information about our products, contact us:
Thomson Learning Academic Resource Center
1-800-423-0563
http://www.wadsworth.com

International Headquarters
Thomson Learning
International Division
290 Harbor Drive, 2nd Floor
Stamford, CT 06902-7477
USA

UK/Europe/Middle East/South Africa
Thomson Learning
Berkshire House
168-173 High Holborn
London WC1V 7AA
United Kingdom

Asia
Thomson Learning
60 Albert Street, #15-01
Albert Complex
Singapore 189969

Canada
Nelson Thomson Learning
1120 Birchmount Road
Toronto, Ontario M1K 5G4
Canada

Library of Congress Cataloging-in-Publication Data

Andersen, Margaret L.
  Understanding society : an introductory reader /
Margaret L. Andersen, Kim A. Logio, Howard F. Taylor.
    p. cm.
  Includes index.
  ISBN 0-534-56666-9
    1. Sociology. 2. Sociology--Research. I. Logio, Kim A.
II. Taylor, Howard Francis, 1939- III. Title.

HM585 .A535 2000
301--dc21   00-057219

ISBN 0-534-56666-9

Contents

✦

Preface

"If you really acquire the sociological perspective, you can never be bored," writes June Jordan, a contemporary African American essayist. We agree and present these readings to help students see how fascinating the sociological perspective can be in interpreting human life. This anthology is intended for use in introductory sociology courses. Most of the students in these courses are first- or second-year students, many of whom are not majoring in sociology. We wanted to compile an anthology that would excite these students about the sociological perspective and show them what such a perspective can bring to understanding the society in which they live.

We have selected articles for this collection that would engage students and would show them what sociology can contribute to their understanding of the world. Most of the readings have been carefully excerpted and kept short for student comprehension. The collection includes articles with a variety of styles and perspectives—a global perspective is apparent throughout. Readers will find a strong focus on diversity. The book also features more current research than that found in competing texts. And, the book presents a balance of classic and contemporary readings that professors teaching the course find important. We developed this book with five themes in mind:

- **Contemporary research:** We wanted students to see examples of strong contemporary research, presented in a fashion that would be accessible to beginning undergraduates. The articles included here feature different styles of sociological research. For example, Amy Schulz's article on identity among

Navajo women uses a qualitative interview study to examine the construction and meaning of identity. In contrast, Alan Hedley's study uses a comparative analysis of seven nations wherein he examines the impact of technological changes and the information revolution on the global diffusion of culture.

- **Classical theory:** Although beginning students sometimes find classical sociological theory difficult to read, we think it is important that they learn about the contributions of classical sociological theorists. Thus, in many sections we have included selections from the classics, but we have limited the number and length of such readings to keep the book accessible to introductory students. Many of the discussion questions for these readings ask students to think about the current application of these writings. For example, Max Weber's argument about the Protestant Ethic and the spirit of capitalism is fascinating to think about in the contemporary context of increased consumerism and increased class inequality. Students might ask whether contemporary patterns of wealth and consumption no longer reflect the asceticism and moral calling about which Weber wrote. At the same time, they will see how beliefs about the worth of different groups in this stratification system continue to be shaped by the ideologies that Weber identified. Likewise, W. E. B. Du Bois's reflections on double consciousness continue to be very important in discussions of race and group perceptions. And, Georg Simmel's analysis of how group behavior shapes society can easily be used to understand social interaction in society today.

- **Diversity:** In keeping with our knowledge that the society is increasingly diverse, we have selected articles that show the range of experiences that people have by virtue of differences in race, gender, class, sexual orientation, disability, and other characteristics (like age and religion). These factors differentiate human experience in contemporary society. Numerous articles in the reader focus on African Americans, Native Americans, Latinos, Asian Americans, women, gays and lesbians, Jewish Americans, and people with disabilities, among others. Some of the selections bring a comprehensive analysis of race, class, and gender to the subject at hand, thus adding to students' understanding of how diverse groups experience the social structure of society. As an example, Yen le Espiritu's discussion of gender stereotypes of Asian American women and men shows how race and gender together construct images of social groups.

- **Global perspective:** We have also incorporated a global perspective into the reader, with many sections including articles that broaden students' worldview beyond the borders of the United States. Articles like Arlie Hochschild's essay on "The Nanny Chain" will help students see how patterns of domestic help and contemporary immigration link the experiences of those in U.S. families to women from other nations who are increasingly being employed to provide domestic help for professional workers in the United States. Similarly, Anne Fadiman's essay contrasts Western medical care with that of Hmong people, showing students how different cultures can perceive Western culture.

- **Applying sociological knowledge:** Our students commonly ask, "What can you do with a sociological perspective?" We think this is an important question and one with many different answers. Sociologists use their knowledge in a variety of ways: to influence social policy formation, to interpret current events, and to educate people about common misconceptions and stereotypes, to name a few. Because we want to show students how sociological knowledge can be used, we have included a number of readings that demonstrate how sociological analyses can be applied to specific issues. For example, Arloc Sherman and Jodi Sandfort's essay in the opening section on the sociological perspective discusses how sociologists can effectively communicate their research findings to a public increasingly influenced by media "sound bites." For a generation of students absorbed by the media, we think this piece will be especially interesting. Likewise, Jill Quadagno's selection on Social Security reform discusses the likely shortfall of funds in the Social Security system and explains some legislative measures proposed to address this problem.

In sum, with this anthology we hope to capture student interest in sociology, provide a balance of classical and contemporary research and theory, incorporate the analysis of diversity into the core of the sociological perspective, analyze the increasingly global dimensions of society, and show students how what they learn about sociology can be applied to real issues and problems. We believe this collection of readings will engage students and help them learn to think critically about social issues.

ORGANIZATION OF THE BOOK

Understanding Society can be used to accompany Margaret L. Andersen and Howard F. Taylor's text, *Sociology: Understanding a Diverse Society,* or their book, *Sociology: The Essentials,* but it can also easily be used in courses with other texts. The outline of the book begins with an introduction to the sociological perspective and sociological research, followed by sections that look at socialization, social interaction, groups and social structure, sexuality, deviance, social inequalities (by class, global stratification, race, gender, and age), social institutions (family, education, religion, work, politics, and health care), and patterns of social change (including population, urbanism and the environment; collective behavior and social movements; and social change). There are twenty-two sections in all, each section topic based on usual subjects of different chapters in introductory textbooks.

NOTE ON LANGUAGE

We have kept the language as stated in the original readings. Thus, students will see some diversity in how various authors identify particular groups (as Latino or Hispanics, for example, or African American or Black). In some of the classical articles, students may find some of the sexist language jarring, such as in the use of "man" when purportedly referring to all human beings. We ask that readers be forgiving of such sexist language, understanding it in the context of the time it was written. Indeed, it would be a good lesson in critical thinking to have students ask whether, in hindsight, the same argument would hold up had the language been more specifically inclusive.

PEDAGOGICAL FEATURES

In addition to the sociological content of this reader, we have included a number of pedagogical features to enrich student learning and to help people teach with the book. Each essay has a **brief introductory paragraph** that identifies the major themes and questions being raised in the article. We follow each article with **discussion questions** that students can use to improve their critical thinking and to reinforce their understanding of the article's major points. Many of these questions could also be used as the basis for class discussion, student papers, or research exercises and projects.

In addition, students using the book will be given passwords to the online **InfoTrac College Edition** system. We include terms at the end of each article which students will be able to use to find additional information on different subjects. These, too, can be the basis for students' papers and projects.

Unlike many anthologies, we have included a **glossary** at the end of the book that contains the definition of basic terms and concepts that students will encounter in the readings. Finally, we have also included a **subject/name index** to help students and faculty locate specific topics and authors in the book.

ACKNOWLEDGMENTS

Many people helped in a variety of ways as we were developing this anthology. We thank Sheryl Ruzek, Dave Ermann, Elizabeth Higginbotham, and Joanne Nigg for suggesting various selections. We thank Vicky Bayne, Lisa Huber, Linda Keen, Jeffrey Quirico, Rika Schmidt and Judy Watson for their help and support. Maggie especially thanks Al Camarillo, Margarita Ibarra, and Monica Wheeler of the Stanford University Center for Comparative Studies in Race and Ethnicity for their support during the time we were developing this book. We especially thank Alison Bianchi, now completing her graduate degree in Sociology at Stanford, for her expert work in compiling the glossary.

We also thank the following people who reviewed the manuscript and provided valuable feedback:

David Boden, Lake Forest College
John Bridges, Kutztown University
Kevin Early, Oakland University

Patti Guiffre, Southwest Texas State University
Jane Johnson, Southwest Texas State University
David Maines, Oakland University
Alvar Nieves, Wheaton College

We sincerely appreciate the enthusiasm and support provided by all those at Wadsworth Publishing Company who have been part of the book's development. We especially thank Dee Dee Zobian for her expert work in managing a myriad of details and Eve Howard, our editor, for all that she does to support our work. We all thank Jim, Pat, and Richard for their love and support. And, Nolan Logio Rau, born as the book was being conceived, we thank you for providing all those smiles while we worked at your kitchen table.

SUPPLEMENTS

Instructor's Manual and Testbank

This supplement offers teaching tips, lecture suggestions, student activities, and much more. It also contains test items, such as Multiple Choice and True/False, which test students' understanding of the articles.

InfoTrac College Edition

Ignite discussions or augment your lectures with the latest developments in Sociology! *InfoTrac College Edition* (available as a free option with this text) gives you and your students free access for 4 months to an easy-to-use online database of reliable, full-length articles (not abstracts) from hundreds of top academic journals and popular sources. Contact your Wadsworth/Thomson Learning representative for more information. Exclusive to Wadsworth/Thomson Learning. Available to North American college and university students only. Journals subject to change. As mentioned, included after each reading in *Understanding Society* are InfoTrac College Edition search terms that will help students and instructors find related articles.

Student Guide to InfoTrac College Edition

http://sociology.wadsworth.com/infotrac/index.html
This is an online supplement prepared by Tim Pippert, of Augsburg College, which contains exercises and suggested readings on InfoTrac College Edition. It consists of critical thinking questions for each of the following standard topics in introductory sociology: Culture, Socialization, Deviance, Social Stratification, Race and Ethnicity, Gender, Aging, Family, Economy/Work, Education, Politics/Government, Health/Medicine, Population, Social Change, and Religion. It is accessible via the web sites of Wadsworth introductory sociology texts, as well as from the Virtual Society home page.

Web Site-Virtual Society: The Wadsworth Sociology Resource Center

At **Virtual Society: Wadsworth's Sociology Resource Center** you and your students can find a career center, "surfing" lessons (tips on how to find information on the web), links to great sociology web sites, and many other selections.

Visit the Andersen and Taylor online resources that accompany each text:

Sociology: The Essentials at http://sociology.wadsworth.com/andersen_essen/

or

SOCIOLOGY at http://sociology.wadsworth.com/andersen_diverse/index.html

Features of the Andersen and Taylor online resources include:

Hypercontents: chapter-by-chapter resources available on the Internet

Chapter Quizzes: online self-quizzes for each chapter in the text

InfoTrac College Edition

Join the Forum: an online threaded discussion forum

◆

About the Editors

Margaret L. Andersen (Ph.D., University of Massachusetts, Amherst) is Professor of Sociology and Women's Studies at the University of Delaware and completed this book while a Visiting Professor at Stanford University. She is the author of *Thinking about Women: Sociological Perspectives on Sex and Gender; Race, Class and Gender: An Anthology* (with Patricia Hill Collins); *Sociology: Understanding a Diverse Society* (with Howard F. Taylor); and *Sociology: The Essentials* (also with Howard F. Taylor). She has won the University of Delaware's Excellence in Teaching Award and is a past president of the Eastern Sociological Society and former editor of *Gender & Society*.

    Kim A. Logio is an Assistant Professor of Sociology at Saint Joseph's University in Philadelphia where she teaches courses in research methods, drugs and society, and race and criminal justice. Her Ph.D. dissertation (University of Delaware, 1998) is entitled, "Here's Looking at You, Kid: Gender, Race, Body Image, and Adolescent Health." She is the coauthor of *Adventures in Criminal Justice Research*, revised edition (with George Dowdall, Earl Babbie, and Fred Halley) and is currently working on a victimization study in Philadelphia and research on the connection between drug use and body image. She is the former Managing Editor of *Gender & Society*.

    Howard F. Taylor (Ph. D., Yale University) is Professor of Sociology at Princeton University. He is the author of *Balance in Small Groups; The IQ Game; Sociology: Understanding a Diverse Society* (with Margaret L. Andersen); *Sociology: The Essentials* (also with Margaret L. Andersen); and a forthcoming book, *Race, Class, and the Bell Curve in America*. He is the winner of the DuBois-Johnson-

Frazier Award given by the American Sociological Association for distinguished research in race and ethnic relations and of Princeton University's President's Award for Distinguished Teaching. He is past president of the Eastern Sociological Society and a member of the Sociological Research Association, an honorary society for distinguished research.

1

The Sociological Imagination

C. WRIGHT MILLS

First published in 1959, C. Wright Mills' essay, taken from his book, The Sociological Imagination, *is a classic statement about the sociological perspective. A man of his times, his sexist language intrudes on his argument, but the questions he posed about the connection between history, social structure, and people's biography (or lived experiences) still resonate today. His central theme is that the task of sociology is to understand how social and historical structures impinge on the lives of different people in society.*

Nowadays men often feel that their private lives are a series of traps. They sense that within their everyday worlds, they cannot overcome their troubles, and in this feeling, they are often quite correct: What ordinary men are directly aware of and what they try to do are bounded by the private orbits in which they live; their visions and their powers are limited to the close-up scenes of job, family, neighborhood; in other milieux, they move vicariously and remain spectators. And the more aware they become, however vaguely, of ambitions and of threats which transcend their immediate locales, the more trapped they seem to feel.

Underlying this sense of being trapped are seemingly impersonal changes in the very structure of continent-wide societies. The facts of contemporary history are also facts about the success and the failure of individual men and women. When a society is industrialized, a peasant becomes a worker; a feudal lord is liquidated or becomes a businessman. When classes rise or fall, a man is employed or unemployed; when the rate of investment goes up or down, a man takes new heart or goes broke. When wars happen, an insurance salesman becomes a rocket launcher; a store clerk, a radar man; a wife lives alone; a child grows up without a father. Neither the life of an individual nor the history of a society can be understood without understanding both.

Yet men do not usually define the troubles they endure in terms of historical change and institutional contradiction. The well-being they enjoy, they do not usually impute to the big ups and downs of the societies in which they live. Seldom aware of the intricate connection between the patterns of their own lives and the course of world history, ordinary men do not usually know what this connection means for the kinds of men they are becoming and for the kinds of history-making in which they might take part. They do not possess the quality of mind essential to grasp the interplay of man and society, of biography and history,

From: C. Wright Mills. *1959. The Sociological Imagination.* New York: Oxford University Press, pp. 3–11. Reprinted with permission.

of self and world. They cannot cope with their personal troubles in such ways as to control the structural transformations that usually lie behind them. . . .

The sociological imagination enables its possessor to understand the larger historical scene in terms of its meaning for the inner life and the external career of a variety of individuals. It enables him to take into account how individuals, in the welter of their daily experience, often become falsely conscious of their social positions. Within that welter, the framework of modern society is sought, and within that framework the psychologies of a variety of men and women are formulated. By such means the personal uneasiness of individuals is focused upon explicit troubles and the indifference of publics is transformed into involvement with public issues.

The first fruit of this imagination—and the first lesson of the social science that embodies it—is the idea that the individual can understand his own experience and gauge his own fate only by locating himself within his period, that he can know his own chances in life only by becoming aware of those of all individuals in his circumstances. In many ways it is a terrible lesson; in many ways a magnificent one. We do not know the limits of man's capacities for supreme effort or willing degradation, for agony or glee, for pleasurable brutality or the sweetness of reason. But in our time we have come to know that the limits of 'human nature' are frighteningly broad. We have come to know that every individual lives, from one generation to the next, in some society; that he lives out a biography, and that he lives it out within some historical sequence. By the fact of his living he contributes, however minutely, to the shaping of this society and to the course of its history, even as he is made by society and by its historical push and shove.

The sociological imagination enables us to grasp history and biography and the relations between the two within society. That is its task and its promise. To recognize this task and this promise is the mark of the classic social analyst. . . .

No social study that does not come back to the problems of biography, of history and of their intersections within a society has completed its intellectual journey. Whatever the specific problems of the classic social analysts, however limited or however broad the features of social reality they have examined, those who have been imaginatively aware of the promise of their work have consistently asked three sorts of questions:

1. What is the structure of this particular society as a whole? What are its essential components, and how are they related to one another? How does it differ from other varieties of social order? Within it, what is the meaning of any particular feature for its continuance and for its change?

2. Where does this society stand in human history? What are the mechanics by which it is changing? What is its place within and its meaning for the development of humanity as a whole? How does any particular feature we are examining affect, and how is it affected by, the historical period in which it moves? And this period—what are its essential features? How does it differ from other periods? What are its characteristic ways of history-making?

3. What varieties of men and women now prevail in this society and in this period? And what varieties are coming to prevail? In what ways are they

selected and formed, liberated and repressed, made sensitive and blunted? What kinds of 'human nature' are revealed in the conduct and character we observe in this society in this period? And what is the meaning for 'human nature' of each and every feature of the society we are examining?

Whether the point of interest is a great power state or a minor literary mood, a family, a prison, a creed—these are the kinds of questions the best social analysts have asked. They are the intellectual pivots of classic studies of man in society—and they are the questions inevitably raised by any mind possessing the sociological imagination. For that imagination is the capacity to shift from one perspective to another—from the political to the psychological; from examination of a single family to comparative assessment of the national budgets of the world; from the theological school to the military establishment; from considerations of an oil industry to studies of contemporary poetry. It is the capacity to range from the most impersonal and remote transformations to the most intimate features of the human self—and to see the relations between the two. Back of its use there is always the urge to know the social and historical meaning of the individual in the society and in the period in which he has his quality and his being.

That, in brief, is why it is by means of the sociological imagination that men now hope to grasp what is going on in the world, and to understand what is happening in themselves as minute points of the intersections of biography and history within society. In large part, contemporary man's self-conscious view of himself as at least an outsider, if not a permanent stranger, rests upon an absorbed realization of social relativity and of the transformative power of history. The sociological imagination is the most fruitful form of this self-consciousness. By its use men whose mentalities have swept only a series of limited orbits often come to feel as if suddenly awakened in a house with which they had only supposed themselves to be familiar. Correctly or incorrectly, they often come to feel that they can now provide themselves with adequate summations, cohesive assessments, comprehensive orientations. Older decisions that once appeared sound now seem to them products of a mind unaccountably dense. Their capacity for astonishment is made lively again. They acquire a new way of thinking, they experience a transvaluation of values: in a word, by their reflection and by their sensibility, they realize the cultural meaning of the social sciences.

Perhaps the most fruitful distinction with which the sociological imagination works is between 'the personal troubles of milieu' and 'the public issues of social structure.' This distinction is an essential tool of the sociological imagination and a feature of all classic work in social science.

Troubles occur within the character of the individual and within the range of his immediate relations with others; they have to do with his self and with those limited areas of social life of which he is directly and personally aware. Accordingly, the statement and the resolution of troubles properly lie within the individual as a biographical entity and within the scope of his immediate milieu—the social setting that is directly open to his personal experience and to some extent his willful activity. A trouble is a private matter: values cherished by an individual are felt by him to be threatened.

Issues have to do with matters that transcend these local environments of the individual and the range of his inner life. They have to do with the organization of many such milieux into the institutions of an historical society as a whole, with the ways in which various milieux overlap and interpenetrate to form the larger structure of social and historical life. An issue is a public matter: some value cherished by publics is felt to be threatened. Often there is a debate about what that value really is and about what it is that really threatens it. This debate is often without focus if only because it is the very nature of an issue, unlike even widespread trouble, that it cannot very well be defined in terms of the immediate and everyday environments of ordinary men. An issue, in fact, often involves a crisis in institutional arrangements, and often too it involves what Marxists call 'contradictions' or 'antagonisms.'

In these terms, consider unemployment. When, in a city of 100,000, only one man is unemployed, that is his personal trouble, and for its relief we properly look to the character of the man, his skills, and his immediate opportunities. But when in a nation of 50 million employees, 15 million men are unemployed, that is an issue, and we may not hope to find its solution within the range of opportunities open to any one individual. The very structure of opportunities has collapsed. Both the correct statement of the problem and the range of possible solutions require us to consider the economic and political institutions of the society, and not merely the personal situation and character of a scatter of individuals.

Consider war. The personal problem of war, when it occurs, may be how to survive it or how to die in it with honor; how to make money out of it; how to climb into the higher safety of the military apparatus; or how to contribute to the war's termination. In short, according to one's values, to find a set of milieux and within it to survive the war or make one's death in it meaningful. But the structural issues of war have to do with its causes; with what types of men it throws up into command; with its effects upon economic and political, family and religious institutions, with the unorganized irresponsibility of a world of nation-states.

Consider marriage. Inside a marriage a man and a woman may experience personal troubles, but when the divorce rate during the first four years of marriage is 250 out of every 1,000 attempts, this is an indication of a structural issue having to do with the institutions of marriage and the family and other institutions that bear upon them.

Or consider the metropolis—the horrible, beautiful, ugly, magnificent sprawl of the great city. For many upper-class people, the personal solution to 'the problem of the city' is to have an apartment with private garage under it in the heart of the city, and forty miles out, a house by Henry Hill, garden by Garrett Eckbo, on a hundred acres of private land. In these two controlled environments—with a small staff at each end and a private helicopter connection—most people could solve many of the problems of personal milieux caused by the facts of the city. But all this, however splendid, does not solve the public issues that the structural fact of the city poses. What should be done with this wonderful monstrosity? Break it all up into scattered units, combining residence and work? Refurbish it as it stands? Or, after evacuation, dynamite it and build new cities according to new plans in new places? What should those plans be? And who is to decide and to accomplish whatever choice is made? These are structural issues; to confront them and to solve them requires us to consider political and economic issues that affect innumerable milieux.

In so far as an economy is so arranged that slumps occur, the problem of unemployment becomes incapable of personal solution. In so far as war is inherent in the nation-state system and in the uneven industrialization of the world, the ordinary individual in his restricted milieu will be powerless—with or without psychiatric aid—to solve the troubles this system or lack of system imposes upon him. In so far as the family as an institution turns women into darling little slaves and men into their chief providers and unweaned dependents, the problem of a satisfactory marriage remains incapable of purely private solution. In so far as the overdeveloped megalopolis and the overdeveloped automobile are built-in features of the overdeveloped society, the issues of urban living will not be solved by personal ingenuity and private wealth.

What we experience in various and specific milieux, I have noted, is often caused by structural changes. Accordingly, to understand the changes of many personal milieux we are required to look beyond them. And the number and variety of such structural changes increase as the institutions within which we live become more embracing and more intricately connected with one another. To be aware of the idea of social structure and to use it with sensibility is to be capable of tracing such linkages among a great variety of milieux. To be able to do that is to possess the sociological imagination. . . .

DISCUSSION QUESTIONS

1. Using either today's newspaper or some other source of news, identify one example of what C. Wright Mills would call an issue. How is this issue reflected in the personal troubles of people it affects? Why would Mills call it a social issue?

2. What are the major historical events that have influenced the biographies of people in your generation? in your parents' generation? What does this tell you about the influence of society and history on biography?

INFOTRAC COLLEGE EDITION

You can use your access to InfoTrac College Edition to learn more about the subjects covered in this essay. Some suggested search terms include:

divorce	urban development
sociological imagination	war
unemployment	

2

The Forest and the Trees

ALLAN G. JOHNSON

Allan Johnson uses the classic example of the forest and the trees as a metaphor to demonstrate that people in society are participating in something larger than themselves. He also argues that the strong cultural belief in individualism blunts the sociological imagination, because it makes you see only individuals, not the social structures that shape diverse group experiences.

As a form of sociological practice, I work with people in corporations, schools, and universities who are trying to deal with issues of diversity. In the simplest sense, diversity is about the variety of people in the world, the varied mix of gender, race, age, social class, ethnicity, religion, and other social characteristics. In the United States and Europe, for example, the workforce is changing as the percentages who are female or from non-European ethnic and racial backgrounds increase and the percentage who are white and male declines.

If the changing mix was all that diversity amounted to, there wouldn't be a problem since in many ways differences make life interesting and enhance creativity. Compared with homogeneous teams, for example, diverse work teams are usually better with problems that require creative solutions. To be sure, diversity brings with it difficulties to be dealt with such as language barriers and different ways of doing things that can confuse or irritate people. But we're the species with the "big brain," the adaptable ones who learn quickly, so learning to get along with people unlike ourselves shouldn't be a problem we can't handle. Like travelers in a strange land, we'd simply learn about one another and make room for differences and figure out how to make good use of them.

As most people know, however, in the world as it is, difference amounts to more than just variety. It's also used as a basis for including some and excluding others, for rewarding some more and others less, for treating some with respect and dignity and some as if they were less than fully human or not even there. Difference is used as a basis for privilege, from reserving for some the simple human dignities that everyone should have, to the extreme of deciding who lives and who dies. Since the workplace is part of the world, patterns of inequality and oppression that permeate the world also show up at work, even though people may like to think of themselves as "colleagues" or part of "the team." And just as these patterns shape people's lives in often damaging ways, they can eat away at

From: Allan G. Johnson. 1997. *The Forest and the Trees: Sociology as Life, Practice, Promise.* Philadelphia: Temple University Press, pp. 7–27. Reprinted with permission.

the core of a community or an organization, weakening it with internal division and resentment bred and fed by injustice and suffering. . . .

People tend to think of things only in terms of individuals, as if a society or a company or a university were nothing more than a collection of people living in a particular time and place. Many writers have pointed out how individualism affects social life. It isolates us from one another, promotes divisive competition, and makes it harder to sustain a sense of community, of all being "in this together." But individualism does more than affect how we participate in social life. It also affects how we *think* about social life and how we make sense of it. If we think everything begins and ends with individuals—their personalities, biographies, feelings, and behavior—then it's easy to think that social problems must come down to flaws in individual character. If we have a drug problem, it must be because individuals just can't or won't "say no." If there is racism, sexism, heterosexism, classism, and other forms of oppression, it must be because of people who for some reason have the personal "need" to behave in racist, sexist, and other oppressive ways. If evil consequences occur in social life, then it must be because of evil people and their evil ways and motives.

If we think about the world in this way—which is especially common in the United States—then it's easy to see why members of privileged groups get upset when they're asked to look at the benefits that go with belonging to that particular group and the price others pay for it. When women, for example, talk about how sexism affects them, individualistic thinking encourages men to hear this as a personal accusation: "If women are oppressed, then I'm an evil oppressor who wants to oppress them." Since no man wants to see himself as a bad person, and since most men probably don't *feel* oppressive toward women, men may feel unfairly attacked.

In the United States, individualism goes back to the nineteenth century and, beyond that, to the European Enlightenment and the certainties of modernist thinking. It was in this period that the rational mind of the individual person was recognized and elevated to a dominant position in the hierarchy of things, separated from and placed above even religion and God. The roots of individualistic thinking in the United States trace in part to the work of William James who helped pioneer the field of psychology. Later, it was deepened in Europe and the United States by Sigmund Freud's revolutionary insights into the existence of the subconscious and the inner world of individual existence. Over the course of the twentieth century, the individual life has emerged as a dominant framework for understanding the complexities and mysteries of human existence.

You can see this in bookstores and best-seller lists that abound with promises to change the world through "self-help" and individual growth and transformation. Even on the grand scale of societies—from war and politics to international economics—individualism reduces everything to the personalities and behavior of the people we perceive to be "in charge." If ordinary people in capitalist societies feel deprived and insecure, the individualistic answer is that the people who run corporations are "greedy" or the politicians are corrupt and incompetent and otherwise lacking in personal character. The same perspective argues that poverty exists because of the habits, attitudes, and skills of individual poor people, who are blamed for what they supposedly lack as people and told to change if they want anything better for themselves. To make a better world, we think we have to put the "right people" in charge or make better people by liberating human consciousness in a New Age or by changing how children are socialized or by locking up or tossing out or killing people who won't or can't be better than they are. Psychotherapy is increasingly offered as a model for changing not only the inner life of individuals, but also the world they live in. If enough people heal themselves through therapy, then the world will "heal" itself as well. The solution to collective problems such as poverty or deteriorating cities then becomes a matter not of collective solutions but of an accumulation of individual solutions. So, if we want to have less poverty in the world, the answer lies in raising people out of poverty or keeping them from becoming poor, *one person at a time.*

So, individualism is a way of thinking that encourages us to explain the world in terms of what goes on inside individuals and nothing else. We've been able to think this way because we've developed the human ability to be reflexive, which is to say, we've learned to look at ourselves *as selves* with greater awareness and insight than before. We can think about what kind of people we are and how we live in the world, and we can imagine ourselves in new ways. To do this, however, we first have to be able to believe that we exist as distinct individuals apart from the groups and communities and societies that make up our social environment. In other words, the *idea* of the "individual" has to exist before we think about ourselves as individuals, and the idea of the individual has been around for only a few centuries. Today, we've gone far beyond this by thinking of the social environment itself as just a collection of individuals: Society *is* people and people *are* society. To understand social life, all we have to do is understand what makes the individual psyche tick.

If you grow up and live in a society that's dominated by individualism, the idea that society is just people seems obvious. The problem is that this approach ignores the difference between the individual people who participate in social life and the relationships that connect them to one another and to groups and societies. It's true that you can't have a social relationship without people to participate in it and make it happen, but the people and the relationship aren't the same thing. That's why this book's title plays on the old saying about missing the forest for the trees. In one sense, a forest is simply a collection of individual trees; but it's more than that. It's also a collection of trees that exist in *a particular relation* to one another, and you can't tell what that relation is by just looking at each individual tree. Take a thousand trees and scatter them across the Great Plains of North America, and all you have are a thousand trees. But take those same trees and bring them close together and you have a forest. Same individual trees, but in one case a forest and in another case just a lot of trees.

The "empty space" that separates individual trees from one another isn't a characteristic of any one tree or the characteristics of all the individual trees somehow added together. It's something more than that, and it's crucial to understand the *relationships among* trees that make a forest what it is. Paying attention to that "something more"—whether it's a family or a corporation or an entire society— and how people are related to it is at the heart of sociological practice.

THE ONE THING

If sociology could teach everyone just one thing with the best chance to lead toward everything else we could know about social life, it would, I believe, be this: *We are always participating in something larger than ourselves, and if we want to understand social life and what happens to people in it, we have to understand what it is that we're participating in* and *how we participate in it.* In other words, the key to understanding social life isn't just the forest and it isn't just the trees. It's the forest *and* the trees and how they're related to one another. Sociology is the study of how all this happens.

The "larger" things we participate in are called social systems, and they come in all shapes and sizes. In general, the concept of a system refers to any collection of parts or elements that are connected in ways that cohere into some kind of whole. We can think of the engine in a car as a system, for example, a collection of parts arranged in ways that make the car "go." Or we could think of a language as a system, with words and punctuation and rules for how to combine them into sentences that mean something. We can also think of a family as a system—a collection of elements related to one another in a way that leads us to think of it as a unit. These include things such as the positions of mother, father, wife, husband, parent, child, daughter, son, sister, and brother. Elements also include shared ideas that tie those positions together to make relationships, such as how "good mothers" are supposed to act in relation to children or what a "family" is and what makes family members "related" to one another as kin. If we take the positions and the ideas and other elements, then we can think of what results as a whole and call it a social system.

In similar ways, we can think of corporations or societies as social systems. They differ from one another—and from families—in the kinds of elements they include and how those are arranged in relation to one another. Corporations have positions such as CEOs and stockholders, for example; but the position of "mother" isn't part of the corporate system. People who work in corporations can certainly be mothers in families, but that isn't a position that connects them to a corporation. Such differences are a key to seeing how systems work and produce different kinds of consequences. Corporations are sometimes referred to as "families," for example, but if you look at how families and corporations are actually put together as systems, it's easy to see how unrealistic such notions are. Families don't usually "lay off" their members when times are tough or to boost the bottom line, and they usually don't divide the food on the dinner table according to who's the strongest and best able to grab the lion's share for themselves. But corporations dispense with workers all the time as a way to raise dividends and the value of stock, and top managers routinely take a huge share of each year's profits even while putting other members of the corporate "family" out of work.

What social life comes down to, then, is social systems and how people participate in and relate to them. Note that people *participate* in systems without being *parts* of the systems themselves. In this sense, "father" is a position in my family, and I, Allan, am a person who actually occupies that position. It's a crucial distinction that's easy to lose sight of. It's easy to lose sight of because we're so used to thinking solely in terms of individuals. It's crucial because it means that people aren't systems, and systems aren't people, and if we forget that, we're likely to focus on the wrong thing in trying to solve our problems. . . .

INDIVIDUALISTIC MODELS DON'T WORK

Probably the most important basis for sociological practice is to realize that *the individualistic perspective that dominates current thinking about social life doesn't work.* Nothing we do or experience takes place in a vacuum; everything is always related to a context of some kind. When a wife and husband argue about who'll clean the bathroom, for example, or who'll take care of a sick child when they both work outside the home, the issue is never simply about the two of them even though it may seem that way at the time. We have to ask about the larger context in which this takes place. We might ask how this instance is related to living in a society organized in ways that privilege men over women, in part by not making men feel obliged to share equally in domestic work except when they choose to "help out." On an individual level, he may think she's being a nag; she may think he's being a jerk; but it's never as simple as that. What both may miss is that in a different kind of society, they might not be having this argument in the first place because both might feel obliged to take care of the home and children. In similar ways, when we see ourselves as a unique result of the family we came from, we overlook how each family is connected to larger patterns. The emotional problems we struggle with as individuals aren't due simply to what kind of parents we had, for their participation in social systems—at work, in the community, in society as a whole—shaped them as people, including their roles as mothers and fathers.

An individualistic model is misleading because it encourages us to explain human behavior and experience from a perspective that's so narrow it misses most of what's going on. A related problem is that *we can't understand what goes on in social systems simply by looking at individuals.* In one sense, for example, suicide is a solitary act done by an individual, typically while alone. If we ask why people kill themselves, we're likely to think first of how people feel when they do it—hopeless, depressed, guilty, lonely, or perhaps obliged by honor or duty to sacrifice themselves for someone else or some greater social good. That might explain suicides taken one at a time, but what do we have when we add up all the suicides that happen in a society for a given year? What does that number tell us, and, more importantly, about what? The suicide rate for the entire U.S. population in 1994, for example, was twelve suicides per 100,000 people. If we look inside that number, we find that the rate for males was twenty per 100,000, but the rate for females was only five per 100,000. The rate also differs dramatically by race and country and varies over time. The suicide rate for white males, for example, was 71 percent higher than for black males, and the rate for white females was more than twice that for black females. While the rate in the United States was twelve per 100,000, it was thirty-four per 100,000 in Hungary and only seven per 100,000 in Italy. So, in the United States, males and whites are far more likely than females and blacks to kill themselves; and people in the United States are almost twice as likely as Italians to commit suicide but only one third as likely as Hungarians.

If we use an individualistic model to explain such differences, we'll tend to see them as nothing more than a sum of individual suicides. If males are more likely to kill themselves, then it must be because males are more likely to feel suicidally depressed, lonely, worthless, and hopeless. In other words, the psychological factors that cause individuals to kill themselves must be more common among U.S. males than they are among U.S. females, or more common among people in the United States than among Italians. There's nothing wrong with such reasoning; it may be exactly right *as far as it goes.* But that's just the problem: It doesn't go very far because it doesn't answer the question of *why* these differences exist in the first place. Why, for example, would males be more likely to feel suicidally hopeless and depressed than females, or Hungarians more likely than Italians? Or why would Hungarians who feel suicidally depressed be more likely to go ahead and kill themselves than Italians who feel the same way? To answer such questions, we need more than an understanding of individual psychology. Among other things, we need to pay attention to the fact that words like "female," "white," and "Italian" name positions that people occupy in social systems. This draws attention to how those systems work and what it means to occupy those positions in them.

Sociologically, a suicide rate is a number that describes something about a group or a society, not the individuals who belong to it. A suicide rate of twelve per 100,000 tells us nothing about you or me or anyone else. Each of us either commits suicide during a given year or we don't, and the rate can't tell us who does what. In the same way, how individuals feel before they kill themselves isn't by itself enough to explain why some groups or societies have higher suicide rates than others. Individuals can feel depressed or lonely, but groups and societies can't feel a thing. We could consider that Italians might tend to be less depressed than people in the United States, for example, or that in the United States, people might tend to deal with feelings of depression more effectively than Hungarians. It makes no sense at all, however, to say that the United States is more depressed or lonely than Italy.

While it might work to look at what goes on in individuals as a way to explain why one person commits suicide, this can't explain *patterns* of suicide found in social systems. To do this, we have to look at how people feel and behave *in relation* to systems and how these systems work. We need to ask, for example, how societies are organized in ways that encourage people who participate in them to feel more or less depressed or to respond to such feelings in suicidal or nonsuicidal ways. We need to see how belonging to particular groups shapes people's experience as they participate in social life, and how this limits the alternatives they think they can choose from. What is it about being male or being white that can make suicide a path of least resistance? How, in other words, can we go to the heart of sociological practice to ask how people participate in something larger than themselves and see how this affects the choices they make? How can we see the relationship between people and systems that produces variations in suicide rates or, for that matter, just about everything else that we do and experience, from having sex to going to school to working to dying?

Just as we can't tell what's going on in a system just by looking at individuals, we also can't tell what's going on in individuals just by looking at systems. Something may look like one thing in the system as a whole, but something else entirely when we look at the people who participate in it. If we look at the kind of mass destruction and suffering that war typically causes, for example, an individualistic model suggests a direct link with the "kinds" of people who participate in it. If war produces cruelty, bloodshed, aggression, and conquest, then it must be that the people who participate in it are cruel, bloodthirsty, aggressive people who want to conquer and dominate others. Viewing the carnage and destruction that war typically leaves in its wake, we're likely to ask, "What kind of people could do such a thing?" Sociologically, however, this question misleads us by reducing a social phenomenon to a simple matter of "kinds of people" without looking at the systems those people participate in. Since we're always participating in one system or another, when someone drops a bomb that incinerates thousands of people, we can't explain what happened simply by figuring out "what kind of person would do such a thing." In fact, if we look at what's known about people who fight in wars, they appear fairly normal by most standards and anything but bloodthirsty and cruel. Most accounts portray men in combat, for example, as alternating between boredom and feeling scared out of their wits. They worry much less about glory than they do about not being hurt or killed and getting themselves and their friends home in one piece. For most soldiers, killing and the almost constant danger of being killed are traumatic experiences that leave them forever changed as people. They go to war not in response to some inner need to be aggressive and kill, but because they think it's their duty to go, because they'll go to prison if they dodge the draft, because they've seen war portrayed in books and movies as an adventurous way to prove they're "real men," or because they don't want to risk family and friends rejecting them for not measuring up as true patriots.

People aren't systems, and systems aren't people, which means that social life can produce horrible or wonderful consequences without necessarily meaning that the people who participate in them are horrible or wonderful. Good people participate in systems that produce bad consequences all the time. I'm often aware of this in the simplest situations, such as when I go to buy clothes or food. Many of the clothes sold in the United States are made in sweatshops in cities like Los Angeles and New York and in Third World countries, where people work under conditions that resemble slavery in many respects, and for wages that are so low they can barely live on them. A great deal of the fruit and vegetables in stores are harvested by migrant farm workers who work under conditions that aren't much better. If these workers were provided with decent working conditions and paid a living wage, the price of clothing and food would probably be a lot higher than it is. This means that I benefit directly from the daily mistreatment and exploitation of thousands of people. The fact that I benefit doesn't make me a bad person; but my participation in that system does involve me in what happens to them. . . .

DISCUSSION QUESTIONS

1. Johnson argues that there is a tendency in the United States for people to explain everything in individual terms. Using the example of suicide, why are individualistic explanations inadequate? What would sociological explanations emphasize instead?

2. Johnson opens his discussion by noting the diversity that characterizes U.S. society. How does he apply the sociological perspective to his understanding of diversity and its significance?

INFOTRAC COLLEGE EDITION

You can use your access to InfoTrac College Edition to learn more about the subjects covered in this essay. Some suggested search terms include:

diversity   suicide
Enlightenment   sweatshops
individualism

3

Learning from the Outsider Within

PATRICIA HILL COLLINS

In this classic essay, Patricia Hill Collins describes the unique contribution to sociological thought that African American women make because of their status as "outsiders" within the academy. Collins' essay rests on the idea that knowledge is constructed within a context where race, class, gender, and other differences of power influence what is known and by whom. Her essay has important implications for creating sociology that is more accurate and more inclusive of the diversity of human experiences.

From: *Social Problems* 33 (December 1986): S14–S52. Reprinted with permission.

fro-American women have long been privy to some of the most intimate secrets of white society. Countless numbers of Black women have ridden buses to their white "families," where they not only cooked, cleaned, and executed other domestic duties, but where they also nurtured their "other" children, shrewdly offered guidance to their employers, and frequently, became honorary members of their white "families." These women have seen white elites, both actual and aspiring, from perspectives largely obscured from their Black spouses and from these groups themselves.

On one level, this "insider" relationship has been satisfying to all involved. The memoirs of affluent whites often mention their love for their Black "mothers," while accounts of Black domestic workers stress the sense of self-affirmation they experienced at seeing white power demystified—of knowing that it was not the intellect, talent, or humanity of their employers that supported their superior status, but largely just the advantages of racism. But on another level, these same Black women knew they could never belong to their white "families." In spite of their involvement, they remained "outsiders."

This "outsider within" status has provided a special standpoint on self, family, and society for Afro-American women. A careful review of the emerging Black feminist literature reveals that many Black intellectuals, especially those in touch with their marginality in academic settings, tap this standpoint in producing distinctive analyses of race, class, and gender. For example, Zora Neal Hurston's 1937 novel, *Their Eyes Were Watching God,* most certainly reflects her skill at using the strengths and transcending the limitations both of her academic training and of her background in traditional Afro-American community life. Black feminist historian E. Frances White (1984) suggests that Black women's ideas have been honed at the juncture between movements for racial and sexual equality, and contends that Afro-American women have been pushed by "their marginalization in both arenas" to create Black feminism. Finally, Black feminist critic Bell Hooks captures the unique standpoint that the outsider within status can generate. In describing her small-town, Kentucky childhood, she notes, "living as we did—on the edge—we developed a particular way of seeing reality. We looked both from the outside and in from the inside out . . . we understood both" (1984:vii).

In spite of the obstacles that can confront outsiders within, such individuals can benefit from this status. Simmel's (1921) essay on the sociological significance of what he called the "stranger" offers a helpful starting point for understanding the largely unexplored area of Black female outsider within status and the usefulness of the standpoint it might produce. Some of the potential benefits of outsider within status include: (1) Simmel's definition of "objectivity" as "a peculiar composition of nearness and remoteness, concern and indifference"; (2) the tendency for people to confide in a "stranger" in ways they never would with each other; and (3) the ability of the "stranger" to see patterns that may be more difficult for those immersed in the situation to see. Mannheim (1954) labels the "strangers" in academia "marginal intellectuals" and argues that the critical posture such individuals bring to academic endeavors may be essential to the creative development of academic disciplines themselves. Finally, in assessing the potentially positive qualities of social difference, specifically marginality, Lee notes, "for a time this marginality can be a most stimulating, albeit often a painful, experience. For some, it is debilitating . . . for others, it is an excitement to creativity" (1973:64).

Sociologists might benefit greatly from serious consideration of the emerging, cross-disciplinary literature that I label Black feminist thought, precisely because, for many Afro-American female intellectuals, "marginality" has been an excitement to creativity. As outsiders within, Black feminist scholars may be one of many distinct groups of marginal intellectuals whose standpoints promise to enrich contemporary sociological discourse. Bringing this group—as well as others who share an outsider within status vis-a-vis sociology—into the center of analysis may reveal aspects of reality obscured by more orthodox approaches.

In the remainder of this essay, I examine the sociological significance of the Black feminist thought stimulated by Black women's outsider within status. First, I outline three key themes that characterize the emerging cross-disciplinary literature that I label Black feminist thought. For each theme, I summarize its content, supply examples from Black feminist and other works that illustrate its nature, and discuss its importance. Second, I explain the significance these key themes in Black feminist thought may have for sociologists by describing why Black women's outsider within status might generate a distinctive standpoint vis-*a*-vis existing sociological paradigms. Finally, I discuss one general implication of this essay for social scientists: namely, the potential usefulness of identifying and using one's own standpoint in conducting research.

THREE KEY THEMES IN BLACK FEMINIST THOUGHT

Black feminist thought consists of ideas produced by Black women that clarify a standpoint of and for Black women. Several assumptions underlie this working definition. First, the definition suggests that it is impossible to separate the structure and thematic content of thought from the historical and material conditions shaping the lives of its producers (Berger and Luckmann 1966; Mannheim 1954). Therefore, while Black feminist thought may be recorded by others, it is produced by Black women. Second, the definition assumes that Black women possess a unique standpoint on, or perspective of, their experiences and that there will be certain commonalities of perception shared by Black women as a group. Third, while living life as Black women may produce certain commonalities of outlook, the diversity of class, region, age, and sexual orientation shaping individual Black women's lives has resulted in different expressions of these common themes. Thus, universal themes included in the Black women's standpoint may be experienced and expressed differently by distinct groups of Afro-American women. Finally, the definition assumes that, while a Black women's standpoint exists, its contours may not be clear to Black women themselves. Therefore, one role for Black female intellectuals is to produce facts and theories about the Black female experience that will clarify a Black woman's standpoint for Black women. In other words, Black feminist thought contains observations and interpretations about Afro-American womanhood that describe and explain different expressions of common themes. . . .

The Meaning of Self-Definition and Self-Valuation

An affirmation of the importance of Black women's self-definition and self-valuation is the first key theme that pervades historical and contemporary statements of Black feminist thought. Self-definition involves challenging the political knowledge-validation process that has resulted in externally-defined, stereotypical images of Afro-American womanhood. In contrast, self-valuation stresses the content of Black women's self-definitions—namely, replacing externally-derived images with authentic Black female images. . . .

Black women's insistence on self-definition, self-valuation, and the necessity for a Black female-centered analysis is significant for two reasons. First, defining and valuing one's consciousness of one's own self-defined standpoint in the face of images that foster a self-definition as the objectified "other" is an important way of resisting the dehumanization essential to systems of domination. The status of being the "other" implies being "other than" or different from the assumed norm of white male behavior. In this model, powerful white males define themselves as subjects, the true actors, and classify people of color and women in terms of their position vis-a-vis this white male hub. Since Black women have been denied the authority to challenge these definitions, this model consists of images that define Black women as a negative other, the virtual antithesis of positive white male images. Moreover, as Brittan and Maynard (1984:199) point out, "domination always involves the objectification of the dominated; all forms of oppression imply the devaluation of the subjectivity of the oppressed." . . .

A second reason that Black female self-definition and self-valuation are significant concerns their value in allowing Afro-American women to reject internalized, psychological oppression (Baldwin, 1980). The potential damage of internalized control to Afro-American women's self-esteem can be great, even to the prepared. Enduring the frequent assaults of controlling images requires considerable inner strength. Nancy White, . . . also points out how debilitating being treated as less than human can be if Black women are not self-defined. She notes, "Now, you know that no woman is a dog or a mule, but if folks keep making you feel that way, if you don't have a mind of your own, you can start letting them tell you what you are" (Gwaltney, 1980:152). Seen in this light, self-definition and self-valuation are not luxuries—they are necessary for Black female survival.

The Interlocking Nature of Oppression

Attention to the interlocking nature of race, gender, and class oppression is a second recurring theme in the works of Black feminists (Beale, 1970; Davis, 1981;

Dill, 1983; Hooks, 1981; Lewis, 1977; Murray, 1970; Steady, 1981). While different socio-historical periods may have increased the saliency of one or another type of oppression, the thesis of the linked nature of oppression has long pervaded Black feminist thought. . . .

The Black feminist attention to the interlocking nature of oppression is significant for two reasons. First, this viewpoint shifts the entire focus of investigation from one aimed at explicating elements of race or gender or class oppression to one whose goal is to determine what the links are among these systems. The first approach typically prioritizes one form of oppression as being primary, then handles remaining types of oppression as variables within what is seen as the most important system. For example, the efforts to insert race and gender into Marxist theory exemplify this effort. In contrast, the more holistic approach implied in Black feminist thought treats the interaction among multiple systems as the object of study. Rather than adding to existing theories by inserting previously excluded variables, Black feminists aim to develop new theoretical interpretations of the interaction itself. . . .

Second, Black feminist attention to the interlocking nature of oppression is significant in that, implicit in this view, is an alternative humanist vision of societal organization. . . .

Black feminists who see the simultaneity of oppression affecting Black women appear to be more sensitive to how these same oppressive systems affect Afro-American men, people of color, women, and the dominant group itself. Thus, while Black feminist activists may work on behalf of Black women, they rarely project separatist solutions to Black female oppression. Rather, the vision is one that . . . takes its "stand on the solidarity of humanity."

The Importance of Afro-American Women's Culture

A third key theme characterizing Black feminist thought involves efforts to redefine and explain the importance of Black women's culture. In doing so, Black feminists have not only uncovered previously unexplored areas of the Black female experience, but they have also identified concrete areas of social relations where Afro-American women create and pass on self-definitions and self-valuations essential to coping with the simultaneity of oppression they experience. . . .

THE SOCIOLOGICAL SIGNIFICANCE
OF BLACK FEMINIST THOUGHT

Taken together, the three key themes in Black feminist thought—the meaning of self-definition and self-valuation, the interlocking nature of oppression, and the importance of redefining culture—have made significant contributions to the task of clarifying a Black women's standpoint of and for Black women. While this accomplishment is important in and of itself, Black feminist thought has potential contributions to make to the diverse disciplines housing its practitioners. . . .

TOWARD SYNTHESIS: OUTSIDERS WITHIN SOCIOLOGY

Black women are not the only outsiders within sociology. As an extreme case of outsiders moving into a community that historically excluded them, Black women's experiences highlight the tension experienced by any group of less powerful outsiders encountering the paradigmatic thought of a more powerful insider community. In this sense, a variety of individuals can learn from Black women's experiences as outsiders within: Black men, working-class individuals, white women, other people of color, religious and sexual minorities, and all individuals who, while from social strata that provided them with the benefits of white male insiderism, have never felt comfortable with its taken-for-granted assumptions.

Outsider within status is bound to generate tension, for people who become outsiders within are forever changed by their new status. Learning the subject matter of sociology stimulates a reexamination of one's own personal and cultural experiences; and, yet, these same experiences paradoxically help to illuminate sociology's anomalies. Outsiders within occupy a special place—they become different people, and their difference sensitizes them to patterns that may be more difficult for established sociological insiders to see. Some outsiders within try to resolve the tension generated by their new status by leaving sociology and remaining sociological outsiders. Others choose to suppress their difference by striving to become bonafide, "thinking as usual" sociological insiders. Both choices rob sociology of diversity and ultimately weaken the discipline.

A third alternative is to conserve the creative tension of outsider within status by encouraging and institutionalizing outsider within ways of seeing. This alternative has merit not only for actual outsiders within, but also for other sociologists as well. The approach suggested by the experiences of outsiders within is one where intellectuals learn to trust their own personal and cultural biographies as significant sources of knowledge. In contrast to approaches that require submerging these dimensions of self in the process of becoming an allegedly unbiased, objective social scientist, outsiders within bring these ways of knowing back into the research process. At its best, outsider within status seems to offer its occupants a powerful balance between the strengths of their sociological training and the offerings of their personal and cultural experiences. Neither is subordinated to the other. Rather, experienced reality is used as a valid source of knowledge for critiquing sociological facts and theories, while sociological thought offers new ways of seeing that experienced reality.

What many Black feminists appear to be doing is embracing the creative potential of their outsider within status and using it wisely. In doing so, they move themselves and their disciplines closer to the humanist vision implicit in their work—namely, the freedom both to be different and part of the solidarity of humanity.

REFERENCES

Baldwin, Joseph A. 1980. "The psychology of oppression." Pp. 95–110 in Molefi Kete Asante and Abdulai S. Vandi (eds.), *Contemporary Black Thought.* Beverly Hills, CA: Sage.

Beale, Frances. 1970. "Double jeopardy: To be Black and female." Pp. 99–110 in Toni Cade (ed.). *The Black Woman.* New York: Signet.

Berger, Peter L. and Thomas Luckmann. 1966. *The Social Construction of Reality.* New York: Doubleday.

Brittan, Arthur and Mary Maynard. 1984. *Sexism, Racism, and Oppression.* New York: Basil Blackwell.

Davis, Angela. 1981. *Women, Race and Class.* New York: Random House.

Dill, Bonnie Thornton. 1983. "Together and in harness': Women's traditions in the sanctified church." *Signs* 10: 678–699.

Gwaltney, John Langston. 1980. *Drylongso, A Self-Portrait of Black America.* New York: Vintage.

hooks, bell. 1981. *Ain't I a Woman: Black Women and Feminism.* Boston: South End Press.

Lee, Alfred McClung. 1973. *Toward Humanist Sociology.* Englewood Cliffs, NJ: Prentice-Hall.

Lewis, Diane. 1977. "A response to inequality: Black women, racism, and sexism." *Signs* 3: 339–361.

Mannheim, Karl. 1954. *Ideology and Utopia: An Introduction to the Sociology of Knowledge.* New York: Harcourt, Brace & Co.

Murray, Pauli, 1970. "The liberation of Black women." Pp. 87–102. in Mary Lou Thompson (ed.), *Voices of the New Feminism.* Boston: Beacon Press.

Simmel, Georg. 1921. "The sociological significance of the 'stranger'" Pp. 322–27 in Robert E. Park and Ernest W. Burgess (eds.). *Introduction to the Science of Sociology.* Chicago: University of Chicago Press.

Steady, Filomina Chioma. 1981. "The Black woman cross-culturally: an overview." Pp. 7–42 in Filomina Chioma Steady (ed.). *The Black Woman Cross-culturally.* Cambridge, MA: Schenkman.

White, E. Frances. 1984. "Listening to the voices of Black feminism." *Radical America* 18:7–25.

DISCUSSION QUESTIONS

1. What does it mean to say that Black women are "outsiders within?" How does Patricia Hill Collins see this social status as influencing Black women's contributions to sociological thought?

2. What three themes does Collins identify as characterizing Black feminist thought and how are these important to the study of sociology?

INFOTRAC COLLEGE EDITION

You can use your access to InfoTrac College Edition to learn more about the subjects covered in this essay. Some suggested search terms include:

Black feminist thought
marginality
outsiders

paradigm
sociology of knowledge
standpoint theory

4

Fighting Child Poverty in America: How Research Can Help

ARLOC SHERMAN AND JODI SANDFORT

Arloc Sherman and Jodi Sandfort see an important role for sociologists in the formation of social policy, in this case, to address the problem of child poverty. They review some of the current research that is pertinent to understanding child poverty and policies meant to alleviate it, while also reminding sociologists of the importance of presenting their research in a style that will be accessible to policymakers and the lay public.

The passage of the 1996 federal welfare law left many social scientists who cared about children's poverty feeling both discouraged by the policy changes and ignored by policy makers. National policy debates had defined the problem as welfare rather than poverty and focused narrowly on ending the existing welfare system rather than on how to replace it with something better. Compared with prior waves of welfare reform, social science research played little role in the debate. Some researchers complained that policy makers seemed "immune to evidence," and the language of the final legislation did little to contradict this complaint. Revealing the distance between researchers and lawmakers, Congress declared in the welfare law's preamble that "the increase in the number of children receiving public assistance is closely related to the increase in births to unmarried women" and clarified that the new law was "intended to address the crisis." Only two years earlier, a strongly worded statement that "welfare has not played a major role in the rise in out-of-wedlock childbearing" had been signed by 76 top scholars . . . and circulated widely on Capitol Hill (Center on Budget and Policy Priorities, 1994).

The political and economic landscape has since changed in several ways. First, the old cash assistance program (Aid to Families with Dependent Children) is gone, taking with it much of the public's focus on welfare as the cause of social problems. Second, in every state except Hawaii, caseloads have fallen in the new Temporary Assistance for Needy Families (TANF) block grant that replaced AFDC. Under the frozen block grant funding structure, this decline frees up a windfall of federal funds that states can use to expand services and payments,

From: *Contemporary Sociology* 27 (November 1998): 555–561. Reprinted with permission.

either for current recipients or for other needy families with children. Third, many governors campaigned on their ability to spearhead successful welfare reform, and they now feel some pressure to demonstrate that low-income families are finding employment and children are not being harmed. Finally, several major policy changes made outside the welfare system— including additional funding for children's health insurance, increases in federal and state earned income tax credits, higher federal minimum wage rates, and more federal dollars for child care—have made it easier for families to go to work (and provide an indication of the public's new focus on helping families who are struggling with low- and moderate-wage jobs).

Alongside these reasons for optimism, there are certainly reasons to be concerned about the direction of welfare-to-work and antipoverty policy. Many states have been exercising their new authority to limit needy families' access to TANF assistance, either by imposing shorter time limits or by imposing stricter requirements for applying and receiving aid, or both. States are rewarded for limiting access to TANF not only by the publicity gained from falling welfare caseloads (which thus far has been overwhelmingly favorable) but also by strong fiscal incentives in TANF that increase a state's windfall with every family dropped from the rolls. Even greater troubles loom on the horizon: Over the next few years, economic and policy pressures on low-income families will grow enormously because of the implementation of the federal five-year time limits on TANF receipt, the accelerating work requirements for those recipients who remain on TANF, and the inevitable downturns in the economy that will drive up the number of needy families while decreasing state revenue. Under the framework of the new welfare law, the amount of federal TANF funds available to help families will remain largely frozen in the face of these rising needs (Greenberg and Savner 1997; Savner and Greenberg 1997).

This new policy landscape, though complex, appears to offer major opportunities for assisting poor children by expanding pro-work, "pro-family" antipoverty policies. In the short term, at least, some of the most dramatic opportunities lie in expanding the supports that states provide to working poor families, including to poor two-parent families frequently excluded by the old welfare system. Key supports include child care, transportation, and other work-related services; new child health insurance programs; and a variety of cash supplements for working parents with low wages, such as state earned income tax credits and earnings disregards in cash assistance programs. In addition, advocates must work to ease harsh welfare "reform" practices that deny families cash assistance, food stamps, and medical care, without regard to their ability to find a job or earn adequate wages. In the longer term, the focus for innovation may to some degree shift back to the federal level. New federal funding and changes in federal policy may become necessary with the onset of an economic recession or the documentation of large numbers of families hitting cash assistance time limits with limited opportunities for work. (One opportunity for change will occur when block grant funding expires after FY 2001.) Such federal changes might include targeted public job creation granting to states greater discretion in instituting time limits and other rules, or an overall expansion in funding. In this later phase, the potential for positive policy changes—rather than simply more backlash against the welfare system—will continue to depend on having current research that is made accessible to the public, the media, and policy makers.

MAKING RESEARCH MORE USEFUL FOR ADVOCACY

Social scientists are already the fundamental source from which to build the program knowledge and public understanding needed for reducing child poverty. Advocates for poor children, including organizations such as the Children's Defense Fund (CDF), depend on social science research to make informed judgments and present credible arguments. . . . The following types of findings are particularly compelling in the public policy sphere:

- cross-national differences in children's poverty status, and policy reasons why an American child is two or more times as likely to be poor as a British child, a French child, or a German child (e.g., Cornia and Danziger 1997);

- racial and class disparities in children's poverty experiences and neighborhood environment. More than 50 percent of black children—but only 3 percent of white children—suffer the double jeopardy of both living in a high-poverty neighborhood and experiencing a year or more of poverty during a six-year period, according to our calculations from data in Brooks-Gunn, Duncan, and Aber (1997);

- poor children's lost learning and education (Duncan and Brooks-Gunn 1997), including such telling details as the chain of connections among poverty, parental stress, family conflict, harsh parenting, and children's lost self-esteem, which together account for most of poverty's measured effects on school grades among tenth-graders in one Iowa study (Conger et al. 1997); . . .

- the insight that the societal trend toward single-parenthood—which is frequently blamed for the bulk of child poverty— itself stems in part from economic insecurity. For instance, about 30 percent of the increase in one-parent families from 1968 to 1988 is attributable to the higher rate of marital breakup observed during recession years than nonrecession years, according to the Census Bureau's Donald J. Hernandez (1993, 1997).

- Duncan and Brooks-Gunn's *Consequences of Growing Up Poor* (1997) also contributes to public understanding by showing that—contrary to widespread belief among policy makers—child poverty's negative effects on learning, school completion, and young adult earnings are large, are consistently separate from the effects of living in a single-parent family, and are generally stronger than single-parent effects.

Why are these studies so useful? Partly, they are able succinctly to illustrate their conclusions with clear nontechnical sentences, straightfoward cross-tabulations, or both. In the policy world, where conclusions longer than a one-line "sound bite" are rarely heeded, these studies clearly state their "bottom line."

Multivariate findings are particularly hard to explain to an audience of voters or legislators unless rendered in quotable English, as in this relatively intelligible finding from a team of federal researchers: "Even after we [accounted for differences in] both race and family structure, poor children were almost 3 times more likely to be in fair or poor health in 1989 through 1991. . . . Poverty . . . has the strongest effect on child health, and is not explained by the other variables" (Montgomery et al. 1996). Using simple language doesn't mean making exaggerated claims. But it does mean making an effort to describe both your caveats and your findings in accessible terms.

Most importantly, useful studies engage the questions, prejudices, and assumptions held by policy makers and the public. These are sometimes different from the questions that matter most to academics. Although social scientists must explore academic theories in order to advance the state of knowledge, they also have a role to play in identifying, articulating, and testing the assumptions underpinning public debate. Public policy too often is made from rudimentary hypotheses and misconceptions that researchers would be almost embarrassed to explore. Much that is "known" about the plight of poor children and their families is never transmitted into the public arena.

In communicating your findings with a policy audience, be especially aware of the power of language. Metaphors and examples are good because they make an abstract point concrete and clear, but they also can be dangerously misconstrued. As an illustration, one scholarly book recently stated on its dust jacket that "parental characteristics that employers value and are willing to pay for, such as skills, honesty, good health, and reliability, also improve children's life chances." Journalists took this to mean that poorer parents have bad character and that is why their children fail. The researcher meant something else—that there is a category of attributes that the researcher could not identify but that appear to influence both child success and family income. The book could as accurately have listed well-connected relatives, a good neighborhood, or freedom from domestic violence. But her speculative examples were taken as findings. While it is impossible to avoid being misconstrued now and then, it is important to think ahead about how words will be heard in a policy climate that is often hostile to the poor.

To be effective, of course, research must be disseminated. Scholars interested in having a larger impact on public policy should explore a few tools for disseminating their research more broadly. For one, the Internet is changing the way information is shared. Research centers, such as the Institute for Research on Poverty (www.ssc.wisc.edu/irp), the Joint Center for Poverty Research (www.jcpr.org), and the National Bureau of Economic Research (www.nber.org) post working papers on-line to facilitate easy access. Similarly, the National Center on Children in Poverty (www.researchforum.org) and the Welfare Information Network (www-welfareinfo.org) maintain lists of current welfare research projects, while the Institute for Women's Policy Research (www.iwpr.org) maintains a listserve for welfare researchers to share information with each other and with advocates. Many advocates and policy organizations check these sites or subscribe to the printed versions. Second, several national advocacy and research organizations (such as the Children's Defense Fund, the Food Research and Action Center, the Center for Law and Social Policy, and the National Center on Children in Poverty) issue newsletters to state and local advocates around the country that summarize new research. Finally, some researchers have developed mailing lists of national advocates, policy researchers, and state and local child advocates to whom they regularly distribute copies of their papers. The National Association of Child Advocates and Kids Count organizations have contacts in every state who use research on child poverty. If you send your findings to a policy organization, remember that including multiple copies of the article and a cover letter identifying how the findings are relevant to public policy will improve the chances that your study will land on the right desk. . . .

Children's poverty can be reduced dramatically. Political and economic opportunities exist to do so in the near future. Most states are not yet focusing on child poverty as a major policy concern, in welfare reform or any other context. But if researchers, policy advocates, community groups, direct service organizations, and others can work together to keep the focus on children's needs, the United States could yet move toward the kind of new antipoverty system that children have always needed.

REFERENCES

Brooks-Gunn, Jeanne, Greg J. Duncan, and J. Lawrence Aber. 1997. *Neighborhood Poverty, Vol. 1: Context and Consequences for Children.* New York: Russell Sage Foundation.

Center on Budget and Policy Priorities. 1994. "Welfare and Out of Wedlock Births: A Research Summary." Washington, DC.

Conger, Rand D., Katherine Jewsbury Conger, and Glen H. Elder, Jr. 1997. "Family Economic Hardship and Adolescent Adjustment: Mediating and Moderating Process." *In Consequences of Growing Up Poor,* edited by Greg J. Duncan and Jeanne Brooks-Gunn. New York: Russell Sage Foundation.

Cornia, Giovanni Andrea, and Sheldon Danziger (eds.). 1997. *Child Poverty and Deprivation in the Industrialized Countries, 1945–1995.* New York: Oxford University Press.

Duncan, Greg J. and Jeanne Brooks-Gunn, eds. 1997. *Consequences of Growing Up Poor.* New York: Russell Sage Foundation.

Greenberg, Mark, and Steve Savner. 1997. *A Brief Summary of Key Provisions of the Temporary Assistance for Needy Families Block Grant.* Washington, DC: Center for Law and Social Policy, revised May.

Hernandez, Donald J. 1993. *America's Children: Resources from Family, Government and the Economy.* New York: Russell Sage Foundation.

———. 1997. "Poverty Trends." Pp. 18–34 in *Consequences of Growing Up Poor,* edited by Greg J. Duncan and Jeanne Brooks-Gunn. New York: Russell Sage Foundation.

Montgomery, Laura E., Lohn L. Kiely, and Gregory Pappas. 1996. "The Effects of Poverty, Race, and Family Structure on US Children's Health: Data from the NHIS, 1978 through 1980 and 1989 through 1991." *American Journal of Public Health* 86 (October).

Savner, Steve and Mark Greenberg. 1997. *The New Framework: Alternative State Funding Choices Under TANF.* Washington, DC: Center for Law and Social Policy.

DISCUSSION QUESTIONS

1. What are some of the major findings about child poverty that can be found in current sociological research?
2. What advice do Sherman and Sandfort offer to sociologists who want their research to inform the development of social policy?

INFOTRAC COLLEGE EDITION

You can use your access to InfoTrac College Edition to learn more about the subjects covered in this essay. Some suggested search terms include:

Anna Casey Foundation

child advocacy groups

child health

child poverty

Children's Defense Fund

Temporary Assistance for Needy
Families (TANF)

5

Science as a Vocation

MAX WEBER

Max Weber, one of the major classical theorists in sociology, saw a particular role for university teaching. Weber taught in the early twentieth century in Germany when faculty were expected to support government interests through what they taught. Weber responded by insisting that faculty should help students analyze facts and values, even when scientific analyses reveals certain "inconvenient facts" (i.e., those that are counter to dominant views). His essay raises important questions about the role of values in sociological analysis.

Today one usually speaks of science as 'free from presuppositions.' Is there such a thing? It depends upon what one understands thereby. All scientific work presupposes that the rules of logic and method are valid; these are the general foundations of our orientation in the world; and, at least for our special question, these presuppositions are the least problematic aspect of science. Science further presupposes that what is yielded by scientific work is important in the sense that it is 'worth being known.' In this, obviously, are contained all our problems. For this presupposition cannot be proved by scientific means. It can only be *interpreted* with reference to its ultimate meaning, which we must reject or accept according to our ultimate position towards life.

Furthermore, the nature of the relationship of scientific work and its presuppositions varies widely according to their structure. The natural sciences, for instance, physics, chemistry, and astronomy, presuppose as self-evident that it is worth while to know the ultimate laws of cosmic events as far as science can construe them. This is the case not only because with such knowledge one can attain technical results but for its own sake, if the quest for such knowledge is to be a 'vocation.' Yet this presupposition can by no means be proved. And still less can it be proved that the existence of the world which these sciences describe is worth while, that it has any 'meaning,' or that it makes sense to live in such a world. Science does not ask for the answers to such questions.

. . . .Consider the historical and cultural sciences. They teach us how to understand and interpret political, artistic, literary, and social phenomena in terms of their origins. But they give us no answer to the question, whether the existence of these cultural phenomena have been and are *worth while*. And they do not answer the further question, whether it is worth the effort required to know them. They presuppose that there is an interest in partaking, through this procedure, of the community of 'civilized men.' But they cannot prove 'scientifically' that this is

From: Hans Gerth and C. Wright Mills. 1958. *From Max Weber: Essays in Sociology.* New York: Oxford University Press, pp. 143–156.

the case; and that they presuppose this interest by no means proves that it goes without saying. In fact it is not at all self-evident.

. . . Let us consider the disciplines close to me: sociology, history, economics, political science, and those types of cultural philosophy that make it their task to interpret these sciences. It is said, and I agree, that politics is out of place in the lecture-room. It does not belong there on the part of the students. If, for instance, in the lecture-room of my former colleague Dietrich Schäfer in Berlin, pacifist students were to surround his desk and make an uproar, I should deplore it just as much as I should deplore the uproar which anti-pacifist students are said to have made against Professor Förster, whose views in many ways are as remote as could be from mine. Neither does politics, however, belong in the lecture-room on the part of the docents, and when the docent is scientifically concerned with politics, it belongs there least of all.

To take a practical political stand is one thing, and to analyze political structures and party positions is another. When speaking in a political meeting about democracy, one does not hide one's personal standpoint; indeed, to come out clearly and take a stand is one's damned duty. The words one uses in such a meeting are not means of scientific analysis but means of canvassing votes and winning over others. They are not plow shares to loosen the soil of contemplative thought; they are swords against the enemies: such words are weapons. It would be an outrage, however, to use words in this fashion in a lecture or in the lecture-room. If, for instance, 'democracy' is under discussion, one considers its various forms, analyzes them in the way they function, determines what results for the conditions of life the one form has as compared with the other. Then one confronts the forms of democracy with non-democratic forms of political order and endeavors to come to a position where the student may find the point from which, in terms of his ultimate ideals, he can take a stand. But the true teacher will beware of imposing from the platform any political position upon the student, whether it is expressed or suggested. 'To let the facts speak for themselves' is the most unfair way of putting over a political position to the student.

Why should we abstain from doing this? I state in advance that some highly esteemed colleagues are of the opinion that it is not possible to carry through this self-restraint and that, even if it were possible, it would be a whim to avoid declaring oneself. Now one cannot demonstrate scientifically what the duty of an academic teacher is. One can only demand of the teacher that he have the intellectual integrity to see that it is one thing to state facts, to determine mathematical or logical relations or the internal structure of cultural values, while it is another thing to answer questions of the *value* of culture and its individual contents and the question of how one should act in the cultural community and in political associations. These are quite heterogeneous problems. If he asks further why he should not deal with both types of problems in the lecture-room, the answer is: because the prophet and the demagogue do not belong on the academic platform.

To the prophet and the demagogue, it is said: 'Go your ways out into the streets and speak openly to the world,' that is, speak where criticism is possible. In the lecture-room we stand opposite our audience, and it has to remain silent. I deem it irresponsible to exploit the circumstance that for the sake of their career the students have to attend a teacher's course while there is nobody present to oppose him with criticism. The task of the teacher is to serve the students with his knowledge and scientific experience and not to imprint upon them his personal political views. It is certainly possible that the individual teacher will not entirely succeed in eliminating his personal sympathies. He is then exposed to the sharpest criticism in the forum of his own conscience. And this deficiency does not prove anything; other errors are also possible, for instance, erroneous statements of fact, and yet they prove nothing against the duty of searching for the truth. I also reject this in the very interest of science. I am ready to prove from the works of our historians that whenever the man of science introduces his personal value judgment, a full understanding of the facts *ceases*. . . .

The primary task of a useful teacher is to teach his students to recognize 'inconvenient' facts—I mean facts that are inconvenient for their party opinions. And for every party opinion there are facts that are extremely inconvenient, for my own opinion no less than for others. I believe the teacher accomplishes more than a mere intellectual task if he compels his audience to accustom itself to the existence of such facts. I would be so immodest as even to apply the expression 'moral achievement,' though perhaps this may sound too grandiose for something that should go without saying. . . .

Those of our youth are in error who react to all this by saying, 'Yes, but we happen to come to lectures in order to experience something more than mere analyses and statements of fact.' The error is that they seek in the professor something different from what stands before them. They crave a leader and not a teacher. But we are placed upon the platform solely as teachers. And these are two different things, as one can readily see. . . .

Science today is a 'vocation' organized in special disciplines in the service of self-clarification and knowledge of interrelated facts. It is not the gift of grace of seers and prophets dispensing sacred values and revelations, nor does it partake of the contemplation of sages and philosophers about the meaning of the universe. This, to be sure, is the inescapable condition of our historical situation. We cannot evade it so long as we remain true to ourselves. And if Tolstoi's question recurs to you: as science does not, who is to answer the question: 'What shall we do, and, how shall we arrange our lives?' or, in the words used here tonight: 'Which of the warring gods should we serve? Or should we serve perhaps an entirely different god, and who is he?' then one can say that only a prophet or a savior can give the answers. If there is no such man, or if his message is no longer believed in, then you will certainly not compel him to appear on this earth by having thousands of professors, as privileged hirelings of the state, attempt as petty prophets in their lecture-rooms to take over his role. All they will accomplish is to show that they are unaware of the decisive state of affairs: the prophet for whom so many of our younger generation yearn simply does not exist. But this knowledge in its forceful significance has never become vital for them. The inward interest of a truly religiously 'musical' man can never be served by veiling to him and to others the fundamental fact that he is destined to live in a godless and prophetless time by giving him the *ersatz* of armchair prophecy. The integrity of his religious organ, it seems to me, must rebel against this. . . .

DISCUSSION QUESTIONS

1. Does Weber think that sociology is value-free? Do you think this is possible?

2. Weber worked in a context where the government was insisting that faculty teach views that supported the government's interest. Did Weber conclude that there is no place for politics in the classroom? What does he mean by suggesting that the teacher's primary task is to teach students to recognize inconvenient facts? How would you illustrate this with a contemporary example?

INFOTRAC COLLEGE EDITION

You can use your access to InfoTrac College Edition to learn more about the subjects covered in this essay. Some suggested search terms include:

Max Weber   value-free sociology
politics and education   verstehen

6

The Practice of Social Research

EARL BABBIE

In this essay, Earl Babbie shows the underlying assumptions that guide different sociological paradigms. In doing so, he shows the value of diverse theoretical arguments in sociology and helps us see how you might use different theoretical paradigms to study particular questions. As he shows, sociological research is guided by the theoretical paradigms that sociologists use.

Because theories organize our observations and make sense of them, there is usually more than one way to make sense of things. Different points of view usually yield different explanations. This is true in daily life: Liberals and conservatives, for example, often explain the same phenomenon quite differently; so do atheists and fundamentalists.

We begin our examination, then, with some of the major points of view social scientists have taken in the search for meaning. Thomas Kuhn (1970) refers to

From: Earl Babbie. 1998. *The Practice of Social Research (8ᵗʰ* ed.). Belmont, CA: Wadsworth, pp. 42–51. Reprinted with permission.

the fundamental points of view characterizing a science as its *paradigms*. In the history of the natural sciences, major paradigms include Newtonian mechanics, Einsteinian relativism, Darwin's evolutionary theory, and Copernicus's heliocentric theory of heavenly motion, to name a few.

While we sometimes think of science as developing gradually over time, marked by important discoveries and inventions, Kuhn says it was typical for one paradigm to become entrenched, resisting any substantial change. Eventually, however, as the shortcomings of that paradigm became obvious, a new paradigm would emerge and supplant the old one. Thus, the view that the sum revolves around the earth was supplanted by the view that the earth revolves around the sun. Kuhn's classic book on this subject is titled, appropriately enough, *The Structure of Scientific Revolutions*.

Social scientists have developed several paradigms for understanding social behavior. The fate of supplanted paradigms in the social sciences has differed from what Kuhn has observed in the natural sciences, however. Natural scientists generally believe that the succession from one paradigm to another represents progress from a false view to a true one. No modern astronomer believes that the sun revolves around the earth, for example.

In the social sciences, on the other hand, theoretical paradigms may gain or lose popularity, but they are seldom discarded. As you'll see shortly, the paradigms of the social sciences offer a variety of views, each of which offers insights the others lack—but ignores aspects of social life that the others reveal.

Thus, each of the paradigms we are about to examine offers a different way of looking at human social life. Each makes certain assumptions about the nature of social reality. I advise you to examine each as to how it might open up new understandings for you, rather than to try to decide which is true and which false. Ultimately, paradigms cannot be true or false; as ways of looking, they can only be more or less useful. Try to find ways these paradigms might be useful to you. . . .

EARLY POSITIVISM

When the French philosopher Auguste Comte (1798–1857) coined the term *sociologie* in 1822, he launched an intellectual adventure that is still unfolding today. Most important, Comte identified society as a phenomenon that can be studied scientifically. (Initially, he wanted to label his enterprise "social physics," but that term was coopted by another scholar.)

Prior to Comte's time, society simply *was*. To the extent that people recognized different kinds of societies or changes in society over time, religious paradigms generally predominated in explanations of the differences. The state of social affairs was often seen as a reflection of God's will. Alternatively, people were challenged to create a "City of God" on earth to replace sin and godlessness.

Comte separated his inquiry from religion. He felt that society could be studied scientifically, that religious belief could be replaced with scientific objectivity. His "positive philosophy" postulated three stages of history. A "theological stage" predominated throughout the world until about 1300. During the next five hundred years, a "metaphysical stage" replaced God with ideas such as "nature" and "natural law."

Finally, Comte felt he was launching the third stage of history, in which science would replace religion and metaphysics—basing knowledge on observations through the five senses rather than on belief. Comte felt that society could be studied and understood logically and rationally, that sociology could be as scientific as biology or physics.

Comte's view came to form the foundation for subsequent development of the social sciences. In his optimism for the future, he coined the term *positivism* to describe this scientific approach—in contrast to what he regarded as negative elements in the Enlightenment. Only in recent decades has the idea of positivism come under serious challenge, as you'll see later in this discussion.

SOCIAL DARWINISM

Comte's major work on his positivist philosophy was published between 1830 and 1842. One year after the publication of the first volume in that series, a young British naturalist set sail on the HMS *Beagle,* beginning a cruise that would profoundly affect the way we have come to think of ourselves and our place in the world.

In 1858, when Charles Darwin published his *Origin of the Species,* he set forth the idea of *evolution* through the process of *natural selection.* Simply put, the theory states that as a species coped with its environment, those individuals most suited to success would be the most likely to survive long enough to reproduce. Those less well suited would perish; therefore the former group would come to dominate the species.

As scholars began to study society analytically, it was perhaps inevitable that Darwin's ideas would be applied to the changes that had taken place in the structure of human affairs. The journey from simple hunting-and-gathering tribes to large, industrial civilizations was easily seen as the evolution of progressively "fitter" forms of society.

Herbert Spencer (1820–1903) was one who concluded that society was getting better and better. Indeed, his native England had profited greatly from the development of industrial capitalism, and Spencer favored a system of free competition, which he felt would insure continued progress and improvement. Spencer may even have coined the phrase, "the survival of the fittest." In any event, he believed it was a primary force in shaping the nature of society. *Social Darwinism* or *social evolution* was a popular view in Spencer's time, although it was not universally accepted. . . .

CONFLICT PARADIGM

One of Spencer's contemporaries took a sharply different view of the evolution of capitalism. Karl Marx (1818–1883) suggested that social behavior could best be seen as the process of conflict: the attempt to dominate others and to avoid being dominated. Marx primarily focused on the struggle among economic classes. Specifically, he examined the way capitalism produced the oppression of workers by the owners of industry. As you know, Marx's interest in this topic did not end with analytical study: He was also ideologically committed to restructuring economic relations to end the oppression he observed.

The contrast in the views set forth by Spencer and Marx indicates the influence that paradigms have on research. These fundamental viewpoints shape the kinds of observations we are likely to make, the facts we will seek to discover, and the conclusions we draw from those facts. Whereas economic classes were essential to Marx's analysis, for example, Spencer was more interested in the relationship between individuals and society—particularly, the amount of freedom individuals had to surrender for society to work.

The conflict paradigm is not limited to economic analyses. Georg Simmel (1858–1918) was particularly interested in small-scale conflict, in contrast to the class struggle that interested Marx. Simmel noted, for example, that conflicts among members of a tightly knit group tended to be more intense than those among people who did not share feelings of belonging and intimacy.

Where it is perhaps natural to see conflict as a threat to organized society, Lewis Coser (1956) pointed out that conflict can sometimes promote social solidarity. Conflict between two groups tends to increase cohesion within each. Or, the expression of conflict within a group can often serve the function of "letting off steam" before stresses become too great to be resolved.

These few examples should illustrate some of the ways you might view social life if you were taking your lead from the conflict paradigm. To explore the applicability of this paradigm, you might take a minute to skim through a daily newspaper or news magazine and identify events that you could interpret in terms of individuals and groups attempting to dominate each other and avoid being dominated. The theoretical concepts and premises of the conflict paradigm might help you make sense out of these events.

SYMBOLIC INTERACTIONISM

In his overall focus, Georg Simmel differed from both Spencer and Marx. Whereas they were chiefly concerned with macrotheoretical issues—large institutions and whole societies in their evolution through the course of history—Simmel, by contrast, was more interested in the ways in which individuals interacted with one another. He began by examining dyads (two people) and triads (three people), for example. Similarly, he wrote about "the web of group affiliations."

Simmel was one of the first European sociologists to influence the development of U.S. sociology. His focus on the nature of interactions particularly influenced George Herbert Mead (1863–1931), Charles Horton Cooley (1864–1929), and others who took up the cause and developed it into a powerful paradigm for research.

Cooley, for example, introduced the idea of the "primary group," those intimate associates with whom we share a sense of belonging, such as our family, friendship cliques, and so forth. Cooley also wrote of the "looking-glass self" we form by looking into the reactions of people around us. If everyone treats us as beautiful, for example, we conclude that we are. See how fundamentally this paradigm differs from the society-level concerns of Spencer and Marx.

Mead emphasized the importance of our human ability to "take the role of the other," imagining how others feel and how they might behave in certain circumstances. As we gain an idea of how people in general see things, we develop a sense of what Mead called the "generalized other." See how this relates to Cooley's "looking-glass self."

Mead also had a special interest in the role of communications in human affairs. Most interactions, he felt, revolved around the process of individuals reaching common understanding through the use of language and other symbol systems, hence the term *symbolic* interactionism.

Here's one way you might apply this paradigm to an examination of your own life. The next time you meet someone new, pay attention to how you get to know each other. To begin, what assumptions do you make about the other person based merely on how they look, how they talk, and the circumstances under which you've met. ("What's someone like you doing in a place like this?") Then watch how your knowledge of each other unfolds through the process of interaction. Notice also any attempts you make to manage the image you are creating in the other person's mind.

ROLE THEORY

. . . The insights of Mead, Cooley, and others regarding the nature of interaction yielded some fundamental social scientific concepts. Ralph Linton (1895–1953) specified the ideas of status and role. Status is a position we occupy in society (such as daughter, lawyer, Republican, American), and a *role* is a set of expected behaviors. When a professor and a student meet to discuss a term paper assignment, there are numerous expectations about the way each will behave in the interaction: The student will plead for an extension, the professor will heartlessly refuse, and so on.

Some social scientists have focused their analysis on how people deal with their various role expectations. This approach yields a variety of new concepts, such as "role strain" (too many expectations for one person to manage) and "role conflict" (my role as professor calls for one thing, my role as the student's son, as it so happens, calls for another).

To get a better idea of this paradigm, you might pay special attention to the ways you and others present yourselves in different situations. Do you speak differently to your professors than to your friends? When you need to deal with the college administration, what role do you assume? How about the roles you assume at worship services, in a supermarket, at a sports event, at a party? After reflecting on your repertoire of roles, you might want to consider how they relate to who you "really" are. What would *you* be like if you weren't acting out any social role?

ETHNOMETHODOLOGY

While some social scientific paradigms emphasize the impact of social structure (such as norms, values, control agents) on human behavior, others do not. Thus, while our social statuses establish expectations for our behavior, everyone deals with these expectations somewhat differently.

Harold Garfinkel, a contemporary sociologist takes the point of view that people are continually creating social structure through their actions and interactions—that they are, in fact, creating their realities. Thus, when you and I meet to discuss your term paper, even though there are myriad expectations about how we should act, our conversation will be somewhat different from any of those that have occurred before, and how we act will somewhat modify our expectations in the future. That is, discussing your term paper will impact our interactions with other professors/students in the future.

Given the tentativeness of reality in this view, Garfinkel suggests that people are continuously trying to make sense of the life they experience. In a sense, he suggests that everyone is acting like a social scientist: hence the term *ethnomethodology,* or "methodology of the people."

How would you go about learning about people's expectations and how they make sense out of their world? One technique ethnomethodologists use is to *break the rules,* to violate people's expectations. Thus, if you try to talk to me about your term paper and I keep talking about football, that might reveal the expectations you had for my behavior. We might also see how you make sense out of my behavior. ("Maybe he's using football as an analogy for understanding social systems theory.") . . .

There is no end to the opportunities you have for trying on the ethnomethodological paradigm. For instance, the next time you get on an elevator, spend your ride facing the rear of the elevator. Don't face front and watch the floor numbers whip by (that's the norm). Just stand quietly facing the rear. See how others react to this behavior. Just as important, notice how *you* feel about it. If you do this experiment a few times, you should begin to develop a feel for the ethnomethodological paradigm.

STRUCTURAL FUNCTIONALISM

Structural functionalism, sometimes also known as "social systems theory," grows out of a notion introduced by Comte and Spencer: that a social entity, such as an organization or a whole society, can be viewed as an *organism*. Like other organisms, a social system is made up of parts, each of which contributes to the functioning of the whole.

By analogy, consider the human body. Each component—such as the heart, lungs, kidneys, skin, and brain—has a particular job to do. The body as a whole cannot survive unless each of these parts does its job, and none of the parts can survive except as a part of the whole body. Or consider an automobile. It is composed of the tires, the steering wheel, the gas tank, the spark plugs, and so forth. Each of the parts serves a function for the whole; taken together, that system can get us across town. None of the individual parts would be of much use to us by itself, however.

The view of society as a social system, then, looks for the "functions" served by its various components. We might consider a football team as a social system— one in which the quarterback, running backs, offensive linemen, and others each have their jobs to do for the team as a whole. Or, we could look at a symphony orchestra and examine the functions served by the conductor, the first violinist, and the other musicians.

Social scientists using the structural functional paradigm might note that the function of the police, for example, is to exercise social control—encouraging people to abide by the norms of society and bringing to justice those who do not. We could just as reasonably ask what functions criminals serve in society, however. Within the functionalist paradigm, we'd see that criminals serve as job security for the police. In a related observation, Emile Durkhelm (1858–1917) suggested that crimes and their punishment provided an opportunity for the reaffirmation of a society's values. By catching and punishing a thief, we reaffirm our collective respect for private property.

To get a sense of the structural-functional paradigm, you might thumb through your college or university catalog and begin assembling a list of the administrators (such as president, deans, registrar, campus security, maintenance personnel). Figure out what each of them does. To what extent do these roles relate to the chief functions of your college or university, such as teaching or research? Suppose you were studying some other kind of organization. How many of the school administrators' functions would also be needed in, say, an insurance company?

FEMINIST PARADIGMS

When Ralph Linton concluded his anthropological classic, *The Study of Man,* (1937:490), speaking of "a store of knowledge that promises to give man a better life than any he has known," no one complained that he had left *women* out. Linton was using the linguistic conventions of his time; he implicitly included women in all his references to men. Or did he?

When feminists (of both genders) first began questioning the use of the third-person masculine whenever gender was ambiguous, their concerns were often viewed as petty, even silly. At most, many felt the issue was one of women having their feelings hurt, their egos bruised.

In fact, feminism has established important theoretical paradigms for social research. In part it has focused on gender differences and how they relate to the rest of social organization. These paradigms have drawn attention to the oppression of women in a great many societies, which has in turn shed light on general oppression.

Because men and women have had very different social experiences throughout history, they have come to see things differently, with the result that their conclusions about social life differ in many ways. In perhaps the most general example, feminist paradigms have challenged the prevailing notions concerning consensus in society. Most descriptions of the predominant beliefs, values, and norms of a society are written by people representing only portions of society. In the United States, for example, such analyses have typically been written by middle-class white men—not surprisingly, they have written about the beliefs, values, and norms they themselves share. Though George Herbert Mead spoke of the "generalized other" that each of us becomes aware of and can "take the role of," feminist paradigms question whether such a *generalized* other even exists.

Where Mead used the example of learning to play baseball to illustrate how we learn about the generalized other, Janet Lever's research suggests that understanding the experience of boys may tell us little about girls.

> Girls' play and games are very different. They are mostly spontaneous, imaginative, and free of structure or rules. Turn-taking activities like jumprope may be played without setting explicit goals. Girls have far less experience with interpersonal competition. The style of their competition is indirect, rather than face to face, individual rather than team affiliated. Leadership roles are either missing or randomly filled.
>
> (LEVER 1986:86)

. . . To try out feminist paradigms, you might want to look into the possibility of discrimination against women at your college or university. Are the top administrative positions held equally by men and women? How about secretarial and clerical positions? Are men's and women's sports supported equally? Read through the official history of your school; is it a history that includes men and women equally? (If you attend an all-male or all-female school, of course, some of these questions won't apply.)

THE EXCHANGE PARADIGM

George Homans, a contemporary social scientist, has suggested that all human behavior can be seen to reflect people's calculations of costs and benefits. More than any of the previously mentioned theorists, Homans consciously states his views as explicit propositions, such as these:

> The more often a particular action of a person is rewarded, the more likely the person is to perform that action.
>
> (HOMANS 1974:16)
>
> The more valuable to a person is the result of his action, the more likely he is to perform the action.
>
> (HOMANS 1974:25)
>
> The more often in the recent past a person has received a particular reward, the less valuable any further unit of that reward becomes for him.
>
> (HOMANS 1974:29)

Notice that Homans, a sociologist, approaches *social* relations from a fundamentally *psychological* starting point, grounded in how individuals see things, how they reason, and how they draw conclusions. Other theorists, such as Peter Blau

(1975), have shown that the basic notion of *exchange,* however, can also be applied to groups and organizations: companies, departments within companies, or countries dealing with one another.

To get a sense of this paradigm, you might review your decision to attend the college or university you are now attending. Did you have alternative schools to choose from? At the very least, you had the choice not to attend college. What were the pluses and minuses of each possible choice? Can you see any ways your deliberations conformed with Homans's propositions?

Notice that the exchange paradigm contains the implicit assumption that human beings are basically rational in their life choices. For this reason, the exchange paradigm is often included within a larger paradigm, labeled *rational-choice paradigm.* Some contemporary scholars, however, question the links between rationality and social science.

RATIONAL OBJECTIVITY RECONSIDERED

We began with Comte's assertion that we can study society rationally and objectively. Since his time, the growth of science, the relative decline of superstition, and the rise of bureaucratic structures all seem to put rationality more and more in the center of social life. As fundamental as rationality is to most of us, however, some contemporary scholars have raised questions about it.

For example, positivistic social scientists have sometimes erred in assuming that human will always act rationally. I'm sure your own experience offers ample evidence to the contrary. And yet, many modern economic models fundamentally assume that people will make rational choices in the economic sector: They will choose the highest-paying job, pay the lowest price, and so forth. However, this assumption ignores the power of such matters as tradition, loyalty, and image that compete with reason in the determination of human behavior. . . .

More radically, we can question whether social life abides by rational principles at all. In the physical sciences, developments such as *chaos theory, fuzzy logic* and *complexity* have suggested that we may need to rethink fundamentally the orderliness of events on our planet.

The contemporary challenge to positivism, however, goes beyond the question of whether people behave rationally. In part, the criticism of positivism challenges the idea that scientists can be as objective as the scientific ideal assumes. Most scientists would agree that personal feelings can and do influence the problems scientists choose to study, what they choose to observe, and the conclusions they draw from their observations.

As with rationality, there is a more radical critique of objectivity. Whereas scientific objectivity has long stood as an unquestionable ideal, some contemporary researchers suggest that subjectivity might actually be preferred in some situations, as we glimpsed in the discussions of feminism and ethnomethodology.

From the seventeenth century through the middle of the twentieth, the belief in an objective reality that people could see ever more clearly predominated in science. For the most part, it was not simply held as a useful paradigm but as *The*

Truth. The term *positivism* has generally represented the belief in a logically ordered, objective reality that we can come to know. This is the view challenged today by the postmodernists and others.

Some say that the ideal of objectivity conceals as much as it reveals. As we saw earlier, much of what was agreed on as scientific objectivity in years past was actually an agreement primarily among white, middle-class European men. Subjective experiences common to women, to ethnic minorities, or to the poor, for example, were not necessarily represented in that reality.

The early anthropologists are now criticized for often making modern, Westernized "sense" out of the beliefs and practices of nonliterate tribes around the world—sometimes portraying their subjects as superstitious savages. We often call nonliterate tribal beliefs about the distant past "creation myth," whereas we speak of our own beliefs as "history." Increasingly today, there is a demand to find the native logic by which various peoples make sense out of life.

Ultimately, we'll never know whether there is an objective reality that we experience subjectively or whether our concepts of an objective reality are illusory. So desperate is our need to know just what is going on, however, that both the positivists and the postmodernists are sometimes drawn into the belief that their view is real and true. There is a dual irony in this. On the one hand, the positivist's belief in the reality of the objective world must ultimately be based on faith; it cannot be proven by "objective" science, since that's precisely what's at issue. And the postmodernists who say nothing is objectively so, do at least feel the absence of objective reality is really the way things are.

Rather than align yourself with either of these approaches as a religion, I encourage you to treat them as two distinct arrows in your quiver. Each approach brings special strengths, and each conpensates for the weaknesses of the other. Why choose, therefore? Work both sides of the street.

These brief remarks on the critique of positivism are intended to illustrate the rich variety of theoretical perspectives that can be brought to bear on the study of human social life. While the attempt to establish formal theories of society has been closely associated with the belief in a discoverable, objective reality, the issues involved in theory construction are of interest and use to all social researchers, from the positivists to the postmodernists—and all those in between. . . .

REFERENCES

Blau, Peter M., ed. 1975. *Approaches to the Study of Social Structure.* New York: Free Press.

Coser, Lewis. 1956. *The Functions of Social Conflict.* New York: Free Press.

Homans, George. 1974. *Social Behavior: Its Elementary Forms.* New York: Harcourt, Brace, Jovanovich.

Kuhn, Thomas. 1970. *The Structure of Scientific Revolutions.* Chicago: University of Chicago Press.

Lever, Janet. 1986. "Sex Differences in the Complexity of Children's Play and Games," Pp. 74–89 in *Structure and Process,* edited by Richard J. Peterson and Charlotte A. Vaughan. Belmont, CA: Wadsworth.

Linton, Ralph. 1936. *The Study of Man.* New York: D. Appleton and Century.

DISCUSSION QUESTIONS

1. At the end of his discussion of each paradigm, Babbie gives several examples of how you can explore the assumptions these different paradigms make through examination of a contemporary issue. Take one of his suggested exercises and describe how the example at hand is illuminated by one of the paradigms he describes.

2. Take two of the paradigms that Babbie presents and compare and contrast the assumptions that each makes about social behavior.

INFOTRAC COLLEGE EDITION

You can use your access to InfoTrac College Edition to learn more about the subjects covered in this essay. Some suggested search terms include:

conflict theory
ethnomethodology
exchange theory
feminist theory

paradigm shift
positivism
Social Darwinism

7

Not Our Kind of Girl

ELAINE BELL KAPLAN

Elaine Bell Kaplan's research on African American teen mothers debunks a number of myths about teenage pregnancy and about African American mothers. She conducted her research using the method of participant observation. Here she describes how she did her research, including how her position as both an insider and an outsider influenced her research project. Her project and its results are presented in her book, Not Our Kind of Girl: Unraveling the Myths of Black Teenage Motherhood *(1995).*

"If we want to solve the problems of the Black community, we have to do something about illegitimate babies born to teenage mothers." The caller, who identified himself as Black, was responding to a radio talk show discussion about the social and economic problems of the Black community. According to

From: Elaine Bell Kaplan. 1995. *Not Our Kind of Girl: Unraveling the Myths of Black Teenage Motherhood.* Berkeley: University of California Press, pp. xviii–xxiii, 10–26.

this caller's view, Black teen mothers' children grow up in fatherless households with mothers who have few moral values and little control over their offspring. The boys join gangs; the girls stand a good chance of becoming teen mothers themselves. The caller's perspective captures the popular view of many Americans: that marital status and age-appropriate sexual behavior ensure the well-being of the family and the community. . . .

According to mainstream ideology, men who through hard work have moved up the career ladder and provide their families with decent food on the dinner table, clothes on their backs, and an occasional family vacation have achieved the American Dream. Women's achievements are measured by their marriage and child rearing, done in proper order and at an appropriate age. Teenage girls are expected to replicate these values by refraining from sexual relations before adulthood and marriage.

Certainly, such traditional ideas held sway over the Black community I knew. Two decades ago unmarried teenagers with babies were a rare and unwelcome presence in my Harlem community. These few girls would be subjected to gossip about their lack of morals and stigmatized if they were on welfare. But by the 1980s so many young Black girls were pushing strollers around inner-city neighborhoods that they became an integral part of both the reality and the myth concerning the sexuality of Black underclass culture and Black family values. These Black teenage mothers did not fit in with the American ethic of hard work and strong moral character. . . .

If this conservative ideology is extended to teen mothers, their situation can be explained only as a result of aberrant moral character. If Black adolescent girls fail to achieve, something in their nature prevents them from doing so. As president, Ronald Reagan often urged teenage mothers to "just say no" so that taxpayers would no longer be forced to pay for their sexual behavior. The "Just Say No" slogan invoked by the Reagan and Bush administrations in the 1980s was utilized in the 1990s by both Black and White conservatives in the attempt "to change welfare as we know it." If these politicians have their way, teenage mothers will be shunned, hidden, and ignored.

As I made my way through East Oakland and downtown Richmond to interview teen mothers,* I witnessed a different scenario from the one devised by politicians. Teenage mothers are housed in threatening, drug-infested environments, schooled in jail-like institutions, and obstructed from achieving the American Dream. In our ostensibly open society, teenage mothers are disqualified from full participation and are marked as deviant. Black teenage girls aged fourteen, fifteen, and sixteen—many of them just beginning to show an adolescent interest in wearing makeup, dressing in the latest fashions, and reading teen magazines— are stigmatized. These teen mothers attempt to cope as best they can by redefining their situation in terms that involve the least damage to their self-respect.

Are Black teenage mothers responsible for the socioeconomic problems besetting the Black community, as the radio show caller would have us believe? Do

---

* Except when necessary for clarification, all teenage mothers and older women who were previously teenage mothers will for the sake of brevity be referred to as teen mothers—a term they use. When appropriate the teen mothers' own mothers will be referred to as adult mothers. All names and places have been changed to protect confidentiality.

Black teenage mothers have different moral values than most Americans? Do they have babies in order to collect welfare, as politicians suggest? Do the families of Black teenage mothers condone their deviant behavior, as the popular view contends? Or, as William J. Wilson's economic theory suggests, is Black teenage motherhood simply a response to the economic problems of the Black community? Black teenage girls confront a world in which gender norms, poverty, and racism are intertwined. Accordingly, to answer questions about these young mothers, we must sort out a host of complex economic and social problems that pervade their lives. I hope the questions I have asked and the answers given will provide portraits of real teenage mothers involved in real experiences.

The reality of these teenage mothers is that they have had to adopt strategies for survival that seem to them to make sense within their social environment but are as inadequate for them as they were for teenage mothers in the past.

These ethnographic pictures illuminate the way structural contradictions act on psychological well-being and the way people construct and reconstruct their lives in order to cope on a daily basis. One issue that comes through quite clearly in this study, and one that is often overlooked by politicians and various studies on Black teenage mothers, is that these teenagers know what constitutes a successful life. Black teenage mothers . . . struggle against being considered morally deviant, underclass, and unworthy.

If we are to understand the stories of these teenage mothers and generalize from their experiences in any significant way, we must place them within the current theoretical and political discussions concerning Black teenage mothers. As T.S. Eliot noted long ago, reality is often more troubling than myth.

What is begging for our attention is the fact that adolescence is a time when Black girls, striving for maturity, lose the support of others in three significant ways. First, they are abandoned by the educational system; second, they become mere sexual accompanists for boys and men; third, these problems create a split between the girls and their families and significant others. What is needed to understand the losses, the stresses, and the large and small violences that render such teenage girls incapable of successfully completing their adolescent tasks is a gender, race, and class analysis. When early motherhood is added to these challenges, they become insurmountable. The adolescent mothers I saw were deprived of every resource needed for any human being to function well in our society: education, jobs, food, medical care, a secure place to live, love and respect, the ability to securely connect with others. In addition, these girls were silenced by the insidious and insistent stereotyping of them as promiscuous and aberrant teenage girls. . . .

TALKING TO TEEN MOTHERS

I began my search to understand the rise in Black motherhood by interviewing two teen mothers referred to me by friends. They came to my house early one Saturday morning and stayed for three hours. Although I had prepared a series of general questions, the young women had so much more to say that I was compelled to create a more extensive set. Next, the director of a local family planning center let me attend a teen parent meeting, where I left a letter of introduction inviting those who were interested in my project to contact me. These teen mothers referred me to others. Eventually, I created a list of fifteen teen mothers.

The director of the family planning center also introduced me to Mary Higgins, the director of the Alternative Center in East Oakland. The Center operated with a grant from a large charity organization that allowed it to develop outreach programs geared to the needs of the local teenage population. These programs included an alternative school, day care, self-esteem development, parenting skills training, and personal counseling. Mary in turn introduced me to Ann Getty, a counselor at the center. Through Ann I met Claudia Wilson, a counselor for the Richmond Youth Counseling program. A short time after that meeting, I began to work as a volunteer consultant for the Alternative Center and to attend meetings with counselors and others who visited the center.

Through my contacts at the center, in the autumn of 1985 I met De Vonya Smalls and twenty of the sample of thirty-two teen mothers who participated in this study. The rest of my sample was drawn from other contacts I made in a network of community workers at the Richmond Youth Service Agency and through my work as a volunteer consultant there. The youth agency's counselors introduced me to teenage mothers who lived in the downtown Richmond area. As a consultant, I was able to talk extensively with the adolescents who took part in teen mother programs.

After several months of making contacts, losing some, and making new ones, I was able to pull together the sample of thirty-two teenage mothers. Of this sample, I "hung out" with a core group of seven teen mothers for a period of seven months, including sixteen-year-old De Vonya Smalls. The other six teen mothers who participated were sixteen-year-old Susan Carter, a mother of a two-month-old baby, who was living with her mother and sister in East Oakland; seventeen-year-old Shana Leeds, a mother with a nine-month-old baby, who was living with a family friend in downtown Richmond; and eighteen-year-old Terry Parks, a mother of a two-year-old, who was sharing her East Oakland apartment with twenty-year-old Dana Little and her five-year-old son. The group also included twenty-year-old Diane Harris, who had become pregnant at seventeen and within months had exchanged a middle-class lifestyle for that of a welfare mother and was now living in a run-down apartment in East Oakland; Lois Patterson, a twenty-seven-year-old mother of two and long-term welfare recipient, who was living with her extended family in a small, crowded house in East Oakland; and Evie Jenkins, a forty-three-year-old mother of two, who was living on monthly disability insurance in a housing project near downtown Richmond. Like Diane Harris, Evie lost her middle-class status when she became a teenage welfare mother at age seventeen.

I accompanied these women to the Alternative Center, to the welfare office, and to visits with their mothers. Some of the teen mothers could not find private places to talk, so we talked in the back seat of my car, over lunch or dinner in coffee shops, in a shopping mall, at teenage program meetings, or while moving boxes to a new apartment—in other words, anywhere they would let me join them.

Interviewing the teen mothers on a regular basis was difficult: they frequently moved, appointments were missed, telephones were disconnected. One day I tried to call five mothers about planned participant observation sessions only to find all their telephones disconnected. A few mothers were willing to be interviewed because they thought they would benefit in some way. One mother let me interview her because she thought I had access to housing and could get her an apartment. Another thought I would be able to get her into a teen parent program. A few mothers did not bother returning my telephone calls once they discovered I could not pay them.

I did not pay the teen mothers or the others for taking part in these interviews. In exchange for their information, I told the teen mothers about my own family, gave out information about welfare assistance and teen parent programs, and drove them to various stores. I helped De Vonya Smalls move into her first apartment. I went out with the teen mothers to eat Chinese food, shared takeout dinners, and bought potato chips and sodas for, so it seemed, everyone's sisters, brothers, and cousins. I was in some homes so often that the families began to treat me like a friend.

I found myself caught up in the teen mothers' lives more than I had planned. I was able to capture changes in their lives. I watched a teen mother break up with her baby's father. I witnessed De Vonya Smalls and Shana Leeds move in and out of three different homes. I saw Shana Leeds go through the process of applying for AFDC. I sat through long afternoons with Diane Harris discussing her baby's "womanizing" father, only to attend their wedding a few months later.

I also talked to everyone else I could, including the teen mothers' mothers, Black and White teenage girls who were not mothers, teachers, counselors, directors of teen programs, social workers, and Planned Parenthood counselors. Many have definite views about teenage mothers, some representing a more conservative voice than we usually hear in the Black community.

Sadly, most of the teen mothers' fathers and their babies' fathers were not involved in their lives in any significant way. The teen mothers' lack of knowledge about the babies' fathers' whereabouts made it impossible for me to interview the men. The few men who were still involved with the teen mothers refused to be interviewed. The best I could do was to observe some of the dynamics between two teen fathers and mothers.

PERSONAL HISTORIES

The teen mothers' ages ranged from fourteen to forty-three. Seventeen of them were currently teen mothers (aged fourteen to nineteen), and fifteen were older women who had previously been teen mothers (aged twenty to forty-three). The presence of the two age groups enabled me to appreciate the dynamic quality and long-term effects of teenage pregnancy on the mothers. The current teen mothers brought to the study a "here and now" aspect: I witnessed some of the family drama as it unfolded. The older women brought a sense of history and their reflective skills; the problems of being a teenage mother did not disappear when the teenage mothers became adults. The older women's stories served two goals for this book: to show that the black community has a history of not condoning teenage motherhood, and to locate emerging problems within the structural changes of our society that have affected everyone in recent years. . . .

As a group, the teen mothers' personal histories reveal both common and not so common patterns among teenage mothers. The youngest teen mother was fourteen and the oldest was eighteen at the time of their first pregnancies. Seventeen teen mothers were currently receiving welfare aid. But contrary to the commonly held assumption that welfare mothers beget welfare mothers, only five teen mothers reported that their families had been on welfare for longer than five years. Twenty-four of the teen mothers had grown up in families headed by a single mother—a common pattern among teenage mothers. Thirteen reported that their mothers had been teenage mothers. Unlike other studies that focus on poor teenage mothers, this study also included five middle-class and three working-class teenage mothers whose parents were teachers, civil service managers, or nursing assistants. Nine of the teen mothers were attending high school (of whom six were attending alternative high school). Several had taken college courses, and two had managed to obtain a college degree.

Along with capturing an ethnographic snapshot of the seven teen mothers, I conducted semistructured interviews in which I asked all the teen mothers specific questions about their experiences before, during, and after their pregnancies. I asked questions about various common perceptions: the idea of passive and promiscuous teenage girls, the role of men in their lives, the notion of strong cultural support for their pregnancies, the concept of extended family support networks, and the idea that teenage mothers have babies in order to receive welfare aid. Each teenage mother was interviewed for two to two and one-half hours. I audiotaped and transcribed all of the interviews.

I transcribed the material verbatim except for names and other identifying markers, which were changed during the transcription. I coded each teen mother on background variables and patterns. I read and reread my fieldnotes, supporting documents, and relevant literature. For this book I chose those quotations that would best represent typical responses, overall categories, and major themes. I used quotations from the core sample of seven as well as from the larger sample of thirty-two to include a wide range of responses.

Whenever possible I have tried to capture the teen mothers' emotional responses to the questions or issues. Often a teen mother would express through a sigh or a laugh feelings about some issue that contradicted her verbal response. For instance, when Terry Parks laughed as she described her feelings about being on welfare, I added a note about her laughter because it indicated to me that she was embarrassed about the subject. Without that notation, I would not have been able to communicate the emotional intensity with which she said the word "welfare" as she talked about her welfare experiences.

THROUGH THE ETHNOGRAPHIC LENS

I use an ethnographic approach to provide an intricate picture of how gender and poverty dictate the lives of these young teenage mothers and how societal gender, race, and class struggles are played out at the personal level. An ethnographic approach can bridge the gap between the sociological discussion of field research and the actual field experience. Studying these women through the lens of ethnography helped me move the teen mothers' personal stories to an objective level of analysis. The ethnographic method allowed the teen mothers to express to me personal information that was close to the heart. The method also allowed me to bring these Black teenage mothers into sociology's purview, to better understand them as persons, to make their voices heard, and to make their lives important to the larger society. The interviews and observations show that Black teenage girls' experiences are structural and troublesome. At all times I have attempted to make these teen mothers' stories real and visible by presenting the teen mothers' own words with as little editing as possible and by revealing their own insights into the interlocking structures of gender, race, and class.

THE INSIDER INTERVIEWER

I could not walk easily into some teen mothers' lives. Being close to the people being interviewed made me both pleased and tense. Being an insider—someone sharing the culture, community, ethnicity, or gender background of the study participants—has its advantages and disadvantages. When the interviewer can identify with the class and ethnic background of the person being interviewed, there is a greater chance of establishing rapport. The person will express a greater range of attitudes and opinions, especially when the opinions to be expressed are somewhat opposed to general public opinion. The situation is more complex when interviewees are asked to reveal information that may serve the researcher's interest but not that of the group involved. "Don't wash dirty linen in public," they remind the researcher.

The most difficult questions I faced, as do most insider interviewers, had to do with the politics of doing interviews in my own community. As an insider I had to decide whether making certain issues public would benefit the group at the same time that it served my research goals. I imagine that these interviews will raise questions. How will the White community perceive Black families if I discuss the conflicts between teen mothers and their mothers, or fathers who refuse to support their children, or the heavy negative sanctioning of these teen mothers by some in the Black community? My work would be taken out of context, several people warned me.

Every Black researcher who works on issues pertaining to her or his community grapples with these questions. We think about the possibility that our findings may contradict what the Black community wants outsiders to know. Some researchers select nonthreatening topics. Others romanticize Black life despite the evidence that life is hard for those on the bottom. And others simply adopt a code of silence, taking a position similar to that of the Black college teacher who in another context made the point to me, "I'm socialized to bear my pain in silence and not go blabbing about my problems to White folks, let alone strangers."

Being an insider did not help me gain the confidence of the teen mothers and others immediately. Most were suspicious of researchers. I lost a chance to interview one group of teen mothers involved in a special school project because the counselors who worked with them did not like the way a White male researcher had treated the teen mothers previously. Indeed, these teen mothers had the right to be suspicious. What these girls and women say about their lives can be used against them by public policy makers, since the Black community is often blamed for its own social and economic situations.

But overall, being a Black woman was helpful, because eventually the teen mothers, realizing we had much in common, stopped being suspicious of me and began to talk candidly of their lives. Occasionally I could not find a babysitter and had to bring my little boy along. I found my son's presence helped reduce the aloofness of my role as researcher and the powerlessness of the teens' position as interview subjects. I was surprised at how helpful my son was in breaking through the first awkward moments. We made him the topic of discussion—mothers can always compare child-care problems. His presence also helped me counter some of the teenagers' tendencies to deny problems. When I talked to De Vonya Smalls about my son's effects on my own schedule, like having to get up at five in the morning instead of at seven, she relaxed and told me about her efforts to study for a test while her baby cried for attention. She also admitted to doing poorly in school.

I decided to study these teenage mothers because Black teenage mothers are not going away, no matter how much we ignore, romanticize, or remain silent about their lives. I strongly disagree with approaches that let the group's code of silence supersede the need to understand the problems and issues of Black teenage mothers. That kind of false ideology only perpetuates the myths about Black teenage motherhood and causes researchers to neglect larger sociological issues or fail to ask pertinent questions about the lives of these mothers. In the name of racial pride, then, we essentially overlook how the larger society shares a great deal of responsibility for these problems. The only way to reduce the number of teenage pregnancies or to improve the lives of teenage mothers is to understand the societal causes by examining the realities of these girls' lives. The time had arrived, as Nate Hare put it, for an end to the unrealistic view of Black lives.

DISCUSSION QUESTIONS

1. How did Elaine Bell Kaplan's status as a Black woman give her an insider's view of Black teenage mothers? In what ways did she remain an outsider and how does this affect her research?

2. Why is participant observation a particularly good research method for investigating the questions that Kaplan was asking about Black teen mothers?

INFOTRAC COLLEGE EDITION

You can use your access to InfoTrac College Edition to learn more about the
subjects covered in this essay. Some suggested search terms include:

Black adolescents   teen mothers
contraceptives   teen pregnancy
participant observation   teenage sex

8

Barbie Doll Culture
and the American Waistland

KAMY CUNNINGHAM

Cultural norms establish expectations about beauty and body image that, as Kamy Cunningham points out, can be unrealistic to attain. Yet, many women (young and old) try to meet these standards, often harming their bodies and their self-concepts. Yet, the profits generated from cultural icons like Barbie ring enormous benefit to those who produce these cultural artifacts.

A Waistland is a land where, if you're a woman, you have to have a tiny waist in order to not feel like something the cat drug out of the garbage bin. I remember at age ten gazing at my first Barbie, the sloe-eyed version with painted toenails in its zebra-striped suit, and deciding, well, I guess this pneumatic creature (I already had a pretty sophisticated vocabulary back then) with the long, horsey legs and Scarlett O'Hara waist was what I was supposed to grow up to look like.

Doesn't every little, and big, American girl want to look like Barbie? And doesn't she want to *be* Barbie, wholesome and popular and perky? In short, a plastic doll.

Barbie beckons us little, and big, girls, but toward what? And if it's toward beauty, what sort of beauty is this, with its tiny waist?

During a moment of epiphany in the middle of a television commercial the other night (most people just go to the bathroom), I speculated that it might be the beauty of the Heartland of America, cholesterol free and patriotically waving tubs of margarine called Promise.

The show between the commercials was the Ms. Teenage America Pageant followed, a couple of days later, by a grown-up beauty contest, the Supermodel of the Year. The teenage hopefuls were all ruffles and tans and soufflés of clichés, each determined to be herself and not succumb to peer pressure and to work with the handicapped, the learning disabled, the old, and the terminally ill because, of course, the most wonderful thing in life is to help others and be the best you can be.

Their supermodel counterparts were slinky, and slid along the stage like skinny eels, in that funny model posture, pelvis jutting forward, small bosoms re-

From: Kamy Cunningham. "Barbie Doll Culture and the American Waistland" *Symbolic Interaction* 16, no. 1 (1993). Copyright © 1993 by JAI Press.

ceding onto the terrace of the breastbone, that makes a woman look like a limp piece of spaghetti about to fall over—backwards.

Barbie combines the prototypes of the two pageants—she's all ruffles and cuteness *and* all experienced slinkiness. A recent version, the Fashion Play Barbie, is a good illustration of what I mean. Clad in a Frederick's of Hollywood wisp of lingerie and topped by luxuriant platinum tresses, the doll has lavender eyes, both willing and innocent, that look out of a face cutely dimpled and empty of feminine guile, yet somehow eerily seductive.

In her own plastic person, Barbie carries the Virgin/Whore paradox to an even more tensile extreme than does, say, a Marilyn Monroe, or a Madonna. Marilyn combined helpless, yielding child with voluptuous, knowing woman in a caricature of the two that was almost obscene. Madonna—shrewd, ruthless, experienced, slightly perverse—is a walking, strutting contradiction to the name of the Virgin she has appropriated.

Slip off that wisp of lingerie, barely clinging to those fulsome curves, and a naked Barbie doll is a sexy thing. Pouty bosom, that tiny waist so oft spoke of, flared hips, lissome legs. Squeeze her and knead her and she has a rubbery life of her own. Cup her, King Kong fashion, and feel the points of the breasts press into your palm. Run your hand down the full 11½" and experience cool, clean silk feel of plastic. Wholesome and seductive.

Is this beauty? Egyptian woman painted their eyelids a heavy charcoal black. Medieval paintings show that women with small tulip breasts and big, oven-rounded bellies were desirable. Rubens and Titian and Ingres thought that women layered like lily-white, hothouse marshmallows were best. In some cultures the male is the heavily painted and artificial one. Like an obscene, opalescent peacock or an aquamarine bird of paradise with blue dragons curling around his arms, in sinuous indigo, he gyrates in front of the womenfolk, hoping (and hopping) to be "pretty" enough to be picked.

The question is not really one of beauty, of course, but of the oppressive equation of beauty, however we define it, with worth. Surface so dominates essence in America that the equation has gotten out of hand. The reason is obvious. We are bombarded by images of Barbie doll women. On a recent *Smithsonian World,* a popular culture critic called advertising "one of the predominate art forms of our time." Advertising is so dominate, the show goes on to say, that "its messages are the only ones being heard." "America is about selling" and "we accept the marketplace as the arbiter of values."

Ads create the symbols of our culture; they suggest that the Johnson's make-me-your-baby-powder woman is the only acceptable version of the feminine.

When I was ten years old, I didn't know that I was longing for a Rubenesque or a Titianesque, rather than a Barbiesque, visual model. I didn't know at the time that she was influencing and reinforcing impossible cultural norms of physical beauty, norms that I would never be able to even approximate. I didn't know that most men want Barbie doll women, the ones with long blonde hair, innocent baby-blue (or baby-lavender) eyes, substantial Cosmocover melons, tiny waists, flat tummies, taut bottoms, and long graceful legs. (And absurdly small feet: Barbie doesn't even have to wear heels in order to be "hobbled"—it's built into those ridiculous concubine feet.) I didn't know that to be considered desirable, I would have to be a centerfold, zipped into my nakedness like a shrimp in its casing.

If I had known all of this, I would probably have thrown myself off Hoover Dam and never reached age eleven.

There are some other things wrong with Barbie too. She's a simulacrum of a human being, a sad grotesquerie: her creators gave her breasts but no nipples, flared hips but no womb, seductively spread legs but no vagina. No milk, no sucklings, no procreation. A twilight zone creature, as strange as her life-sized counterpart—the department store mannequin with the sterility of a lavender sheen on its cadaverous, blue-grey cheeks—she is an emblem of frustration and unfulfillment.

In Las Vegas, at Caesars Palace, an enormous figurehead of Cleopatra juts out over the casino. With her huge bronze breasts that dangle above your head and her ample but shapely girth, she looks as if she could have mothered the whole human race. Instead, taut in every disappointed muscle, she strains out into nothing, gazing at this sterile indoor cosmos of star-spangled chandeliers.

Las Vegas showgirls look manufactured—identical lanky clones carrying ten pounds of feathers above eyes so mascared no eyes are there. Ads across the country misrepresent them as voluptuous; actually, by some ironic twist, they're all tiny-bosomed because big breasts bounce around on the stage with the least step or jiggle. All the girls would look like cumbersome, milkheavy cows. Go to a Las Vegas show and the sensation is eerie: two-hundred identical breasts with tiny peppermint nipples point your way, like the pink noses of puppies. Beneath the nipples, identical Rockettes' legs. Large breasts might be an improvement: the hilarity would relieve the manufactured look of the women.

The "simply irresistible" clone women, of Pepsi commercial and MTV fame, produce a similar shuddering sensation. Painted over with that lavender sheen of the mannequin, with big, hard, dark eyes, and starved cheeks, their faces look like those of boxed dolls—identical and inexpressive. Only their bodies are alive, in a mechanical way, as they move. Their eyes are dead. Some even wear goggles, blinders that make them look like horses in harness. They have been zapped of all their vitality, by being turned into mindless doll-like clones. Dead dolls, vampire women. Plastic. Manufactured. Artificial. Unreal. No room for the appealing flaws and living warmth of "real" women, those whose Rubenesque curves might spread a little and whose Titianesque arms might have a bit of the soft sway of the basset hound. Warm arms, motherly arms.

Barbie, with all of her accessories (thousands of little outfits, and dozens of pieces of pink plastic furniture, and Hollywood hot tubs and sleek racing cars) brings in three quarters of a billion dollars a year for Mattel. Every two seconds someone somewhere in the world buys a Barbie. Numerically, there are 2.5 Barbies for every household in America.

She is obviously a powerful cultural icon, but what is her iconography? What text is she illustrating? The text of woman as manufactured cadaver? Woman robbed of any insides because she has to be all outside?

Barbie's living clone, Vanna White, Goddess of the Empty Woman, seems to be illustrating a depressing blankness (note the name *White*). Vanna's message is that if you look like her and dress beautifully and smile warmly and turn letters with great skill and remain forever, mentally and emotionally, on the level of an untroubled child, then you will be valued and given lots of money. Turning letters counts for far more, apparently, than turning a phrase. Rarely, in the history of womankind mankind, has so little been so richly rewarded.

One night, on *Wheel of Fortunate,* she gushed, in a see-Spot-run vocabulary, over an "island paradise" vacation she'd just taken that was "simply wonderful" and "so great." (Her narrative, childishly adjectival and without a story line, had not quite reached the level of sophistication of the "cow jumped over the moon.") I feel resentful that Vanna and her ilk (all those manufactured mannequins and cadaverous clones) can pile up fortunes by selling their bodies and that I can't make anything by selling my mind.

But, to temper my tirade (a bit), I feel a little sorry for her (and them). And happy for me, a little. In a world where you have to sell something to survive, maybe it's better to have a mind to market than a transient body.

Perhaps, decades from now, I may be able to entertain myself with books and thoughts after all the centerfolds have sagged and the Vannas have died away from their own untroubled boredom.

DISCUSSION QUESTIONS

1. When you were growing up, what toys did you play with? How do you think these might have influenced your concepts of good-looking women and men? Has this established any norms for your own appearance now?

2. What beauty images do you see as most frequently represented in various forms of popular culture now? What cultural expectations do these images establish based on gender? age? race? class?

INFOTRAC COLLEGE EDITION

You can use your access to InfoTrac College Edition to learn more about the subjects covered in this essay. Some suggested search terms include:

anorexia (nervosa)   eating disorders
beauty myths   gender and self-esteem
body image

9

The Shameless World of Phil, Sally and Oprah:

Television Talk Shows and the Deconstructing of Society

VICKI ABT AND MEL SEESHOLTZ

Television talk shows are a major source of entertainment for millions of viewers. Abt and Seesholtz argue that the cultural scripts such shows promote distort views of human relationships and break down concepts of privacy and deviance in such a way that people begin to accept preposterous behavior as an ordinary feature of life. Their research also shows the enormous power of the communications industry to influence the cultural orientation of a society.

To experience the virtual realities of television talk shows is to confront a crisis in the social construction of reality. Television talk shows create audiences by breaking cultural rules, by managed shocks, by shifting our conceptions of what is acceptable, by transforming our ideas about what is possible, by undermining the bases for cultural judgment, by redefining deviance and appropriate reactions to it, by eroding social barriers, inhibitions and cultural distinctions.

> Social order is not part of the "nature of things," and it cannot be derived from the "laws of nature." Social order exists only as a product of human activity . . . Man himself provide[s] a stable environment for his conduct. (Berger and Luckmann 52)

If social order is not a given, if it is not encoded in our DNA, then mechanisms must evolve to accommodate the construction of society and social identities. To this extent, we are always in the process of producing "virtual realities," some more functional than others for our survival (Rheingold; Seesholtz 14–20). Habits, routines, institutionalization are the patterns that create the "world taken for granted." Knowledge of how to behave in social situations is contained in cultural scripts or blueprints that are themselves products of human interaction

---

From: *Journal of Popular Culture* 28 (Summer 1994): 171–191. Reprinted with permission.

and symbolic communication about the nature of "reality." "Morality," norms, values, judgmental expressions of group conventions, create limits on social behavior. These limits are maintained either by "internalizing the scripts" or by reliance on external threats and punishments. Society is a result, then, of its boundaries, of what it will and won't allow. Shame, guilt, embarrassment are controlling feelings that arise from "speaking the unspeakable" and from violating cultural taboos.

To the extent that individual and social order exist as products of human cultural judgments and definitions, then incessant nonjudgmental focus on "real life," increasingly bizarre narratives of deviance and pathology threatens to normalize, routinize and trivialize these behaviors and situations. Like addicts developing "tolerance," it has been suggested we become desensitized and need increasing stimuli to get excited or outraged (Boorstin; Slater). Historian Hannah Arendt's phrase "the banality of evil," written about the holocaust, referred to "normal" people's ability to accept unspeakable atrocities as ordinary, everyday reality. . . .

As we watch, listen and are entertained, television is rewriting our cultural scripts, altering our perceptions, our social relationships and our relationships to the natural world. While all modes of communication, including speech, are devices for fixing perception and organizing experience about the world, television is a radical new way of organizing experience. . . .

In their competition for audience share, ratings and profits, television talk shows co-opt deviant subcultures, break taboos and eventually, through repeated, nonjudgmental exposure, make it all seem banal and ordinary. The addictive nature of this lies in the fact that increasingly bizarre stories are constructed to maintain audience share. Television talk shows offer us an anomic world of blurred boundaries and at best normative ambiguity. Cultural distinctions between public and private, credible and incredible witnesses, truth and falseness, good and evil, sickness and irresponsibility, normal and abnormal, therapy and exploitation, intimate and stranger, fragmentation and community are manipulated and erased for our distraction and entertainment. Nothing makes conventional sense in this deconstructed society. . . .

TV TALK SHOWS AND THE COMFORT OF STRANGERS?

The evolution of daytime talk shows over the past 25 years seems to coincide with parallel changes in American society. Earlier versions of television talk shows were very different from the three syndicated contemporary daytime talk shows—*The Donahue Show, The Sally Jessy Raphael Show, The Oprah Winfrey Show*—all named after their hosts. These three talk shows are the primary focus of this paper. The original 1960's version of *The Phil Donahue Show* focused, as did Johnny Carson's *Tonight Show,* on celebrity interviews with those interested in self-promotion of their books, movies, television programs or political agendas. Today's

Donahue Show is a relentless display of deviants, conflict and personal stories of real-life private people trying to "fix themselves" through therapy (Taylor 26–34; Zoglin 79). However, neither the earlier nor the contemporary versions of talk shows deal with social activism attempting to change the too complex, impersonal, bureaucratic world, and neither are a realistic sampling of human daily life or lifestyles. But contemporary talk shows further blur the boundaries between the real and the fictive and the acceptable and the deviant on these "real life" soap operas (Abt, "Appeal" 108–18). The structures of modern society are largely ignored or reduced into personal stories, like the recent focus on presidential candidates' sex lives rather than on the complexities of the national debt. As a visual medium television cannot cope easily with these abstractions. (To this extent, McLuhan was right: the medium *is* the message.) Television talk shows privatize our social concerns while collapsing boundaries between public and private spheres. Everything that is private becomes public, and everything that is public gets reduced to private stories; ordinary life rhythms and routines are ignored. Attention is paid to "therapy" not social change.

The 12-step therapeutic model permeates the popular literature on addiction "treatment" (Abt and McGurrin 657–70). It has become the ideology of the new television talk show. The underlying assumption that most social pathology is the result of a medical problem beyond the control of the so-called "victim" encourages, at least indirectly, people to come on to these shows confessing outrageous stories of anti-social behavior to millions of strangers in the global village. This certainly changes the notion of shame, privacy, appropriateness and guilt. . . .

. . . Rather than being mortified, ashamed or trying to hide their stigma, "guests" willingly and eagerly discuss their child molesting, sexual quirks and criminal records in a effort to seek "understanding" for their particular disease. After all, according to the talk show ideology, they are all "victims" rather than possibly being irresponsible, weak people. Of course the fact is that their bad behavior is their pass "to get on television"—the reason they're invited to participate is the provocative nature of their problem. These people remain caricatures, plucked out of the context of their lives, unimportant except for their entertaining problem. Therapy as entertainment is the appeal of these shows. The so-called hosts rely on the cynical use of the therapeutic model as psychological sound bites, and the need to educate and inform the audience is voiced rational for getting the so-called guests to give ever more titillating details of their misdeeds, or of the misdeeds done to them by family or friends (often not on the show). Frequently, minor children of "guests" either appear on stage or in the studio audience, and we watch as they ashamedly listen to their parents' most intimate secrets. (At least they know enough to be ashamed, before the therapeutic model takes hold to tell them they need not be ashamed.)

Traditional boundaries between very private matters and public discussions are continuously breached; i.e., boundaries that would exist between therapist/patient and the outside *world,* or between confessor/priest and the outside world or even the anonymity granted those in a "self-help group" modeled after the original 12-step program of Alcoholics Anonymous are obliterated. This electronic confessional encourages "show and tell" performances on the part of real people who couldn't publicly act out their stories anywhere else in real time and space without being sanctioned. It's as if television is a "third place," where people have no sense of place (Oldenburg; Meyrowitz). . . .

MANAGED OUTRAGE AND OTHER CYNICAL DISTORTIONS

Reluctant guests are frequently kept from declining to reveal too much by the coaxing facilitator—the host—who reminds them, after all, that "this is a talk show—so talk." Pseudo-intimacy is encouraged by the use of first names—Phil, Sally, Oprah. Polite guests who make the mistake of saying "Mr. Donahue" or "Miss Winfrey" are teasingly chided for this breach of the talk show's of etiquette of "*instimacy.*" It is also possible that use of the words "host" and "guests" further obscures the commercial nature of the show. After all, we have been taught that "hosts" will protect and assist "guests" in their attempts at "impression management," and that in return the good guest reciprocally tries to please the host. . . .

The audience at various points in the hour shows has a chance to get on television too. The camera follows the host into the studio audience with his/her mike. S/he holds the mike so that a member of the audience can ask questions of the guests or give "advice." Their questions are often rude by conventional standards and reinforce the host's requests for more potentially entertaining details. Sometimes audience members even tell their own stories of pain and pathology so as to reinforce the guests on stage. Their advice ranges from merely simplistic, under the circumstances, to misleading and erroneous. For example in a recent *Sally Jessy Raphael Show* entitled "When Your Best Friend is Sleeping with Your Father," the daughters on stage were advised to "just love them both and accept the situation." The new couple were let off the hook by the audience when they said they weren't to blame because they didn't plan to fall in love—"it just happened." Similarly in a recent *Donahue Show* about a woman who married her son's 14-year-old friend, the audience failed to comment on the boy/husband's dropping out of school, the couple's dependence on welfare, the boy/husband's threatening of his parents with violence if they didn't sign the permission, as well as the boy/husband's wearing of a swastika on a torn T-shirt. At one point the boy/husband did say he "liked the way it looked," otherwise it had no meaning for him.

Most problematic part of this is the generally nonjudgmental tenor of the dialogue between the "guests" and the audience. Society's conventions are flouted with impunity, and the hidden message is that the way to get on television is to be as outrageous and antisocial as possible. Many talk shows devote themselves to the theme of family dysfunction and also to hate groups. In a show featuring "skinheads," members of KKK, there was no shame displayed indeed, the guests wore their uniforms proudly on stage, and mouthed anti-Semitic and racist erroneous information. No one seriously attempted to correct their inaccurate rendering of historical events, or challenge their reliance on pseudo-biology during their diatribes against Jews and other, unrelated minority groups. Our argument is not against First Amendment rights, but rather the unequal access afforded extremist views. An *Oprah Winfrey Show* on the LA riots had the audience saying, unchallenged, that the Korean shopkeepers shouldn't be in black neighborhoods. No one pointed out the parallel when whites want to exclude blacks from their communities, nor did anybody discuss the "false consciousness" of competing minorities within a complex capitalistic system.

When Phil, Sally and Oprah make statements in a show about rape that "Rape is never about sex, but always about control," or "You're not a bad person," or "Thank you for sharing your story," or when they simply urge guests to "get treatment," they are not offering scientific evidence for the efficacy of these homilies, but are simply mouthing mantras of pop-therapy. Moreover, the host and occasional guest "experts" fail to refute negative or misleading information offered by the guests or studio audience members. The studio audience participation keeps the action going and also contributes to the narrative that all opinions count equally in this democratic global village, and that we're all "plain folks" interested in helping our fellow man. In other words, all "hierarchies of credibility" that exist in society are obliterated. The facsimile of a "town meeting" or "the people's right to know" and the fraudulent egalitarian theme of the global village are part of the dynamics of the television talk show.

No one ever says, at least in the 60 contiguous shows we viewed over a period of a month (as well as additional shows since), "it's none of our business" (the audience) or "it's none of your business" (guests), followed by "I'm leaving." It must be reported that, however rarely, some audience members do ask the guests why they are telling these secrets on television, but there is no follow-up to the answer, and they always seem satisfied by answers such as "I want my story to help others," or "I'm sick of secrets," or "I want to force my mother [father, brother, sister, etc.] to confront what they did to me." In real life someone just might question the benefits of publicly confessing to people who really don't care about you or don't have the expertise to give advice. The absurdity of the situation is analogous to dialing a number randomly and proceeding to explain the most intimate details of your life to whoever answered the phone. . . .

Talk *is* cheap. These shows don't have to pay for the expensive celebrity guests who appeared on earlier forms of talk shows, they don't have to employ expensive screenwriters or actors, they can count on an endless supply of people wanting to get on *Phil, Sally or Oprah*. Some may want to push strange new religions (a female witch whose religion demands that she have sex with all fellow worshippers, but who says she's really an ordinary housewife, "most of the time") or lifestyles (nudist families who swap mates to demonstrate openness and sharing). Some may be exhibitionists (grandmothers who strip for fun, fat women who are proud of their fat and wear bikinis on the show to prove it, daughters who strip for a living and who have brought their fathers on the show and do their act for them, as we watch). Some may be pushing business ventures (men who own family restaurants where nude waitresses serve hamburgers). These people are clearly exploiting the medium which exploits them. But many of the shows are devoted to serious abuse and pathology. These people may really be seeking help or understanding. Appropriate reactions seem virtually impossible under the circumstances. Despite the topic, the talk show follows the same routine format. Cheerful theme music (slowed down for "serious" discussion) opens the show as the host's name is boldly flashed on the screen, along with shots of her/his smiling face. Guests are introduced at the beginning of the show and their narratives are highlighted by subtitles that summarize their stories: "Susan hasn't spoken to her mother in four years." Feuding families are seated side-by-side on the stage as they are encouraged to scream at each other. Abused children sit next to their parents. Hosts, despite differences in style, are all adept at managing outrage, encouraging the telling of secrets, cooling off the proceedings if they threaten the continuity of the show, shutting off boring guests, putting people on the spot, summing up with clichés and platitudes complex situations, making the audience feel comfortable witnessing private matters. Hosts use their first names to create the illusion of intimacy and equality. Hosts often refer to their guests' relatives as "Mother" or "Dad" rather than "your mother" or "your dad" in order to create a feeling of family for the show. No one, despite the seriousness of his/her problem, receives more than a few minutes to tell his/her story. The audience must not be allowed to get bored of restless. A show on child abuse may bring four or five separate cases on stage to tell their stories so that if one story isn't interesting, others can take over. And all of this happens with the 45 minutes of actual show (minus commercials and trailers).

Phil, Sally and Oprah always know best. They take the roles of caring parents, understanding friend, knowing therapist. They may not have professional credentials to give advice, but they do so freely. Of course, they probably make considerably more money than therapists who still bother to get professional training, credentials and licenses. It was recently reported in *The Philadelphia Inquirer* (April 21, 1992: B2) that Oprah made $80 million last year. Her syndicated talk show, which she owns, gets the same audience share as the perennially popular primetime show *60 Minutes*. Despite differences, it is no accident that both shows are exposés. Inexorably, they focus on the pathological and bizarre. One can only imagine what this constant attention to the fringes of society, to those who break rules, is doing to our society's ability to define and constrain deviance. One thing seems fairly certain: law-abiding, privacy-loving, ordinary people who have had reasonable happy childhoods and are satisfied with their lives, probably won't get to tell their stories to Phil, Sally or Oprah. But if they did get on a television talk show, they would have to highlight the problematic aspects of their lives. Television talk shows are not interested in adequately reflecting or representing social reality, but in highlighting and trivializing its underside for fun and profit. . . .

REFERENCES

Abt, Vicki. "The Appeal of the Soap Opera Heroine: Role, Ritual and Romance." Eds. Ray B. Browne and Marshall

Fishwick. *The Hero in Transition*. Bowling Green, OH: Bowling Green State University Popular Press, 1983.

Abt, Vicki, and Martin McGurrin. "The Politics of Problem Gambling: Issues in the Professionalization of Addiction Counseling." *Gambling and Public Policy.* Eds. William Eadington and Judy Cornelius. Reno, NV: Institute for the Study of Commercial Gambling and Commercial Gaming, U of Nevada Press, 1991.

Arendt, Hannah. *Eichmann in Jerusalem: A Report on the Banality of Evil.* Rev. ed. New York: Viking-Penguin, 1977.

Berger, Peter, and Thomas Luckmann. *The Social Construction of Reality: A Treatise in the Sociology of Knowledge.* New York: Doubleday, 1967.

Boorstin, Daniel. *The Image: A Guide to Pseudo Events in America.* New York: Atheneum, 1961.

Meyrowitz, Joshua. *No Sense of Place: The Impact of Electronic Media on Social Behavior.* New York: Oxford, 1985.

Oldenburg, Ray. *The Great Good Place.* New York: Paragon House, 1989.

Rheingold, Howard. *Virtual Reality.* New York: Summit, 1991.

Seesholtz, Mel. "ExoTechnologies: *Human Efforts to Evolve Beyond Human Being." Thinking Robots, An Aware Internet, and Cyberpunk Librarians: Proceedings and Background Essays of the 1992 LITA President's Program.* Eds. R. Bruce Miller and Milton Wolf. Chicago: Library and Information Technology, 1992.

Slater, Philip. *The Pursuit of Loneliness: American Culture at the Breaking Point.* Boston: Beacon, 1970.

Taylor, John. "Don't Blame Me: The Culture of Victimization." *New York* 3 June, 1991.

Zoglin, Richard. "Running Off at the Mouth: Mother-in-Law from Hell and Other Lunacies Fuel the Proliferating Talk Shows." *Time* 14 Oct. 1991.

DISCUSSION QUESTIONS

1. According to Abt and Seesholtz, how do television talk shows promote individualistic solutions to social issues? Why do they see this as problematic?

2. One outcome of the popularity of television talk shows, according to Abt and Seesholtz's argument, is that they numb people from concerns about real human suffering. How does this occur and who benefits from it?

INFOTRAC COLLEGE EDITION

You can use your access to InfoTrac College Edition to learn more about the subjects covered in this essay. Some suggested search terms include:

mass media
media images
popular culture

social construction of reality
television talk shows

10

Family Rituals and the Construction of Reality

SCOTT COLTRANE

Sociologists see rituals as patterned activities that help groups define their beliefs and social roles. As such, various kinds of rituals are an important element of the culture of a society. Here, Scott Coltrane analyzes a common family ritual—a Thanksgiving dinner—to reveal the family relationships and gender roles that this ritual reaffirms.

HOME FOR THE HOLIDAYS

In the Jodie Foster film *Home for the Holidays,* Claudia (played by Holly Hunter) is a single mother who takes a trip to her parents' house for Thanksgiving. After getting fired from her job at a Chicago museum, Claudia gets a ride to the airport from her teenage daughter. As Claudia grabs her bags and leans over to say good-bye, her daughter tells her that she is planning to have sex with her boyfriend over the weekend. Thus begins a humorous look at one of the most popular family rituals in America—the Thanksgiving holiday feast.

This movie, like television sitcoms, shows what is expected of a "typical" American family at holiday time and simultaneously illustrates some departures from the "normal." Suffering from a cold, lamenting her job loss, worrying about her daughter, and fretting about the upcoming visit, Claudia picks up the airplane telephone and calls her brother in Boston. When he doesn't answer, she frantically recounts her tales of woe for the answering machine and pleads with him to join her at their parents' house for the holiday. (This little performance leaves her feeling even more stupid and foolish than she felt before.)

Arriving at her parents' house, Claudia is forced to answer probing questions from her mother (played by Anne Bancroft) about why her daughter did not come with her. Her mother immediately intuits that Claudia was fired from her job, despite Claudia's saying only that she is thinking about a change, whereon her mother closes the door and admonishes her not to tell her father. When Claudia's gay brother, Tommy (played by Robert Downey Jr.), arrives with a handsome friend in tow, Claudia mistakenly assumes that he is her brother's new lover. Later in the movie, we learn that Tommy and his longtime lover, Jack, had a

From: Scott Coltrane. 1997. *Gender and Families.* Thousand Oaks, CA: Pine Forge Press, pp. 13–22. Used with permission.

marriage ceremony on a Massachusetts beach some months before. Among the immediate family, only Claudia seems ready to accept Tommy's being gay, and in the beginning, only she accepts Jack as an important part of Tommy's life.

The plot thickens as Claudia and Tommy's high-strung sister shows up with her uptight husband and two bratty children. For some unstated reason, and with much tension, the sister brings in her own turkey and several other Thanksgiving dishes. The women (and daughter) migrate to the kitchen to finish the meal preparations, while the men (and son) sit in the living room. As the feast begins, the women recruit men to carve the turkeys, with obnoxious Tommy reluctantly hacking away at his mother's bird. Unintentionally, but with little remorse, Tommy ends up dumping the entire greasy turkey onto the lap of his uptight sister, who freaks out and flies into a rage. Aunt Gladie, eccentric and senile, has too much to drink and tells everyone how wonderful it was to kiss her brother-in-law some 30 years before. With the tension high and decorum broken, Claudia asks Tommy to field a question about how *her* life is going. To her surprise and amusement, he reveals that she was fired from her job, kissed her 60-year-old boss, and is expecting her daughter to have sex with her boyfriend that evening (which doesn't actually happen after all). Obviously flustered and angry, the mother storms out of the room, retreating to the pantry for a cigarette. Claudia catches up to her and tries to console her, followed by Tommy, who ends up eating with Claudia in the kitchen while the others finish the feast in the dining room.

After the meal, the women go to the kitchen to clean up, and the men go outside to play football. Tommy gets into a fistfight with his sister's husband, and the father, who is washing the sister's car, turns the hose on the "boys," creating even more havoc and leading to a swift departure of the daughter's family. Claudia cannot patch things up with the sister, has both harsh and tender words with her mother, shares a beer watching football with her father, resists having sex with Tommy's amorous friend, and in the end is surprised and delighted to find him joining her on the airplane for a romantic flight back to Chicago.

MAKING SENSE OF THE THANKSGIVING RITUAL

This movie version of a Thanksgiving holiday ritual is a good place to begin exploring how families and gender are socially constructed. Many of today's families look different from this one, and holidays are celebrated differently, depending on family composition, geographical location, ethnicity, income, and family traditions. But we can use this Eastern, white, suburban middle-class family Thanksgiving, with all its quirks and comic relief, to see how family rituals work and why they are important.

Family rituals such as Thanksgiving appear timeless to many people and invoke nostalgic images of "the good old days" and "old-fashioned family values." As historians point out, however, most family rituals looked different in the past, and most holiday rituals such as Thanksgiving and Christmas were not the isolated family-centered emotional events that they have become during the 20th century. Prior to the 1900s, civic festivals and Fourth of July parades in America were much more important occasions for celebration and strong emotion than family holidays. Only in the 20th century did the family come to be the center of festive attention and emotional intensity (Coontz, 1992, p. 17; Skolnick, 1991).

Celebrating family holidays such as Thanksgiving has become more common during the past century, and holidays have increasingly taken on special emotional significance. Why do many American families go through some version of this feast year after year, and what purpose does it serve? Sociologists and anthropologists suggest that celebrating holidays such as Thanksgiving and sharing special meals is one way that families create and reaffirm a sense of themselves. As noted above, families do not have an automatic meaning or definition. That is, they tend to have fuzzy boundaries and need some shared activities to give them shape (Gubrium & Holstein, 1990). Although our culture provides us with an overarching sense of what a family should be, we need to learn what it means to be in a particular family through direct experience and learning, and we need to re-create a sense of belonging over and over again. As described below, periodic family rituals such as birthdays and holiday feasts, along with other activities, actually construct family boundaries and teach us who we are as family members. Who is in, who is out, and what it means to be part of a particular family are literally created and re-created through these routine ceremonial events (see, for example Berger & Kellner, 1964; Gubrium & Holstein, 1990; Imber-Black, Roberts, & Whiting, 1988). Routine rituals help create family scripts—mental representations of ordered events that guide people's actions within and across family settings (Byng-Hall, 1988; Stack & Burton, 1994).

. . . These ceremonial occasions—along with countless other routine family practices—combine to create a sense of ourselves as gendered beings: as mothers or fathers, wives or husbands, daughters or sons, women or men, boys or girls. Ideals from the larger society contribute to these gender ideals, but routine family practices give personal meaning to the larger social definitions and provide us with interpersonal scripts to follow. Ritual family events such as Thanksgiving reinforce conventional expectations about what it is to be a man or a woman, and these messages are then generalized to other social settings.

In Claudia's family's Thanksgiving, as in most American families, the women orchestrated the ritual event. They prepared and served the meal, as well as cleaned up afterward, and we can assume that they also planned the menus, bought the food, and began preparing it well in advance of Thanksgiving Day. The holiday was organized around the feast, and it was the women who made it happen. This was a common pattern in the past and is still typical in many contemporary American families, although things are changing in some families (Coltrane, 1996; DeVault, 1991; diLeonardo, 1987; Fenstermaker-Berk, 1985; Luxton, 1980; Thompson & Walker, 1989). Why aren't men more involved in meal preparation? Why don't boys get asked to set the table or help out with cooking as much as girls? This gender-based division of labor has important implications for the development of different feelings of competence and entitlement in girls and boys, men and women (Hochschild, 1989; Pyke & Coltrane, 1996).

Throughout history, family work—such as cooking and preparing food—and productive work—such as growing and harvesting food—have been closely tied together. Only relatively recently have housework and paid work seemed like separate things, although they are still closely intertwined. Without being fed, clothed, and housed, workers could not stay on the job. Without the resources that come from jobs, people could not maintain households. Recent research shows how paid and unpaid work continue to be linked in modern American families. Studies by economists and sociologists demonstrate that men's ability to get higher-paying jobs allows them to avoid housework (Becker, 1981; Coltrane, 1996; Delphy, 1984; Hartmann, 1981; Shelton, 1992). Because most employers have assumed that wives will perform family work, women's wages and chances for job promotion have been limited (Baxter, 1993; Hochschild, 1989; Luxton, 1980). Opportunities in the job market thus affect decisions about family work at the same time that assumptions about family work affect the structure of the job market (Reskin & Padavic, 1994). What happens in individual families has profound effects on our assumptions about which type of work is appropriate for men and which is appropriate for women (Coltrane, 1989; West & Fenstermaker, 1993).

What makes something women's work or men's work? . . . The tasks performed by one gender or the other are subject to change as historical circumstances change and people's needs are evaluated differently. It seems as if men are supposed to carve turkeys only because people have been exposed to it year after year in family rituals such as Thanksgiving. In *Home for the Holidays,* the father is physically unable to carve the turkey (or at least his wife insists that he is), so Tommy is recruited to perform the task, even though he makes a mockery of it and drops the greasy bird in his horrified sister's lap. Why didn't Claudia or one of the other women assume the task of carving the turkey?

A look back across different civilizations shows that divisions of labor are not the same everywhere. In one society, men might be the only ones to set up the dwellings, whereas in another, this might be a woman's job. In the United States, men are rarely asked to cook but are expected to barbecue hamburgers and carve turkeys. Why should it be "feminine" to stuff and bake the turkey but "masculine" to cut it up? Why are the top professional bakers and chefs virtually always men, although women do most of these tasks in the vast majority of American homes? These questions cannot be answered without looking at issues of power and control in families, topics discussed throughout this book (see also Blumberg & Coleman, 1989; Ferree, 1957; Komter, 1989; Thompson & Walker, 1989; Thorne, 1992; Vannoy-Hiller, 1984).

As shown in *Home for the Holidays,* watching or playing football is also a Thanksgiving tradition in some American families—or more accurately, it is a tradition for some men. While women cook or clean up, men congregate in the living room to watch football on television or go outside to throw the ball around. Although not often acknowledged, this ritual teaches boys that they are entitled to special privileges. They get to play games while the girls are expected to help with the meal preparation, cleanup, and child care. Of course, this does not happen in every family, but for many American men, Thanksgiving is a re-laxing event. For most women, it entails work. This does not mean that women do not also find such events to be fun and rewarding. It is mostly women who initiate, plan, and conduct them year after year. But it is clear that ceremonial meals are primarily downtime for men, whereas for women they include work as well as pleasure. This is another example of what sociologist Arlie Hochschild (1989) calls the "leisure gap"—men's greater opportunities for relaxation at home. The leisure gap is narrowing in some families, but it is still likely that at Thanksgiving gatherings in homes across America, the women are waiting on the men.

Another gender division in many families revolves around caring for others and talking about emotions. In *Home for the Holidays,* Claudia and her mother check in with other family members to see how they are doing and initiate con-versations about their feelings. They show their concern by talking. Tommy, in contrast, is joking most of the time, and although he is expressive and playful, his teasing is often at others' expense. His father shows his affection for his wife by saying "come 'ere, gorgeous" and dancing with her, but when things get tense around the dinner table, he gets quiet. He demonstrates his concern for his daughter by going outside and washing her car.

In most American families, women consider it their duty to worry about fam-ily members and to take care of their everyday needs. They derive great satisfac-tion from doing so. Often, mothers (and sometimes fathers) focus on making sure that everyone is well fed. Preparing and serving food are thus more than just work because they represent love and care that are given to family members. Women also stereotypically show their love by talking about feelings and trying to help everyone get along. In *Home for the Holidays,* as the women cooked and served, they chatted about the food but also about people and relationships. When the conversation turned argumentative and offensive at the dinner table, Claudia's mother tried, in vain, to make it "nicer," and Claudia eventually took responsibil-ity for letting things get out of hand. The men, in contrast, talked less and focused their conversation on things: work, investments, and football. The tone and style of talk differed as well. The men argued and teased one another, talked louder, changed topics more frequently, and interrupted the women and each other. These conversational patterns reflect gender differences that are common in many American families (Pearson, West, & Turner, 1995; West & Zimmerman, 1987; Wood, 1996).

RITUALS REAFFIRM FAMILY TIES

Home for the Holidays showed an atypical Thanksgiving feast, insofar as the usual tensions and disagreements were contentiously and comically revealed, rather than staying submerged. More typically, people at family gatherings get along a little better, downplay their disagreements and hostilities, and pretend that everything is OK. Acting as if everything is OK (even when it isn't) is one way to normalize the situation and maintain a sense of family unity and continuity. Ritual celebrations such as Thanksgiving allow families to do this on a regular basis.

In a more general way, rituals help construct group identity and create a shared sense of reality. Historically, most rituals started out as community affairs, whereby people got together on special occasions to reaffirm their commitment to some common purpose—an alliance with another clan, a shared religion, a new community settlement, or allegiance to a king or other ruler. These ceremonial gatherings brought people together in face-to-face interaction; focused their attention on some common symbols; heightened their emotions through a group activity such as singing, chanting, and dancing; and linked those emotions to the symbols and to the common purposes of the group (Collins, 1988; Durkheim, 1915/1957; Goffman, 1967). These rituals served many purposes but, most important, gave people a sense of belonging to the group and reaffirmed everyone's commitment to its purposes and symbols.

Although times have changed, we still have many rituals, and they still give people a sense of belonging. Increasingly, modern rituals—at least the face-to-face ones—are centered on family activities. In the past, many public rituals and celebrations were also family based because wealthy families sponsored them, and they tended to solidify alliances between families. Today, family alliances are still important, but they are no longer the central basis of marriages, politics, business, and warfare as they once were. As public displays of family alliances have become less important, family rituals have become more focused on the personal relationships within them. Like earlier rituals, modern family celebrations and activities continue to provide family members with an important sense of place and reinforce feelings of joint membership. These rituals are sometimes linked to national holidays, such as Thanksgiving, but everyday routines, such as eating meals together and watching television together, can serve the ritual function of solidifying family bonds.

Various types of rituals create families, insofar as the family has no fixed definition, and ritualized practices are needed to define and reinforce the meaning of family. If we did not have family rituals such as weddings, anniversaries, and birthdays and holiday get-togethers, we would have a weaker sense of what it means to be a family member. If we did not have more mundane family rituals such as eating together, going on outings, and sharing inside jokes, family would seem less real. Periodic holidays and everyday rituals thus combine to create a shared sense of the family and reinforce our connections to other family members.

In the past, families had many more connections that gave a strong sense of belonging. Most people lived on farms, and family members and relatives contributed on a regular basis to the needs of everyday life. Young adults were dependent on parents and other relatives to learn skills and get jobs. Most people had frequent contact with their parents and grandparents (if they were still alive) because they relied on them for a place to live and for their means of survival. In those days, changes in family membership shaped one's access to resources, and a family member's marriage or death could significantly change the allocation of wealth in the family. The ritual of a wedding symbolically joined the fates of two families, and the ritual of a funeral reaffirmed the living family members' commitments to each other and to the larger family. . . .

REFERENCES

Baxter, J. (1993). *Work at home: The domestic division of labor.* St. Lucia, Australia: Queensland University Press.

Becker, G. (1981). *A treatise on the family.* Cambridge, MA: Harvard University Press.

Berger, P., & Kellner, H. (1964). "Marriage and the construction of reality." *Diogenes, 46,* 1–23.

Blumberg, R. L. (1984). "A general theory of gender stratification." In R. Collins (Ed.), *Sociological Theory 1984.* San Francisco: Jossey-Bass.

Byng-Hall, J. (1988). "Scripts and legends in families and family therapy." *Family Process, 27,* 167–179.

Collins, R. (1988). *Theoretical sociology.* San Diego, CA: Harcourt Brace Jovanovich.

Coltrane, S. (1996). *Family man: Fatherhood, housework, and gender equity.* New York: Oxford University Press.

Coontz, S. (1992). *The way we never were.* New York: Basic Books.

Delphy, C. (1984). *Close to home: A materialist analysis of women's oppression* (D. Leonard, Trans.). London: Hutchison.

DeVault, M. (1991). *Feeding the family: The social construction of caring as gendered work.* Chicago: University of Chicago Press.

diLeonardo, M. (1987). "The female world of cards and holidays: Women, families, and the work of kinship." *Signs, 12,* 440–453.

Durkheim, E. (1957). *The elementary forms of the religious life.* New York: Free Press. (original work published 1915)

Fenstermaker-Berk S. (1985). *The gender factory.* New York: Plenum.

Ferree, M. (1987). "She works hard for a living." In B. Hess & M. Ferree (Eds.), *Analyzing gender* (pp. 322–347). Newbury Park, CA: Sage.

Goffman, E. (1967). *Interaction ritual.* New York: Doubleday.

Gubrium, J., & Holstein, J. (1990). *What is family?* Mountain View, CA: Mayfield.

Hartmann, H. (1981). "The family as the locus of gender, class, and political struggle: The example of housework." *Signs, 6,* 366–394.

Hochschild, A. (with Manning, A.). (1989). *The second shift.* New York: Viking.

Imber-Black, E., Roberts, J., & Whiting, R. (Eds.). (1988). *Rituals in families and family therapy.* New York: Norton.

Komter, A. (1989). "Hidden power in marriage." *Gender & Society, 3,* 187–216.

Luxton, M. (1980). *More than a labor of love: Three generations of women's work in the home.* Toronto, Ontario, Canada: Women's Press.

Pearson, J., West, R., & Turner, L. (1995). *Gender and communications.* Dubuque, IA: Brown & Benchmark.

Pyke, K., & Coltrane, S. (1996). "Entitlement, gratitude, and obligation in family work." *Journal of Family Issues, 17,* 60–82.

Reskin, B. F., & Padavic, I. (1994). *Women and men at work.* Thousand Oaks, CA: Pine Forge.

Shelton, B. A. (1992). *Women, men, time.* New York: Greenwood.

Skolnick, A. (1991). *Embattled paradise: The American family in an age of uncertainty.* New York: Basic Books.

Stack, C. B., & Burton, L. M. (1994). "Kinscripts: Reflections of family, generation, and culture." In E. N. Glenn, G. Chang, & L. R. Forcey (Eds.), *Mothering: Ideology, experience, and agency* (pp. 33–44). New York: Routledge.

Thompson, L., & Walker, A. J. (1989). "Gender in families: Women and men in marriage, work, and parenthood." *Journal of Marriage and the Family, 51,* 845–871.

Thorne, B. (with Yalom, M., Eds.). (1992). *Rethinking the family: some feminist questions* (Rev. ed.). Boston: Northeastern University Press.

Vannoy-Hiller, D. (1984). "Power dependence and division of family work." *Sex Roles,* 1003–1019.

West, C., and Fenstermaker, S. (1993). "Power and the accomplishment of gender: An ethnomethological perspective." In P. England (Ed.), *Theory on gender/feminism on theory* (pp. 151–174). New York: Aldine de Gruyter.

West, C., & Zimmerman, D. (1987). "Doing gender." *Gender & Society, 1,* 125–151.

Wood, J.T. (1996). "She says/he says: Communication, caring, and conflict in heterosexual relationships." In J.T. Wood (Ed.), *Gendered relationships* (pp. 149–162). Mountain View, CA: Mayfield.

DISCUSSION QUESTIONS

1. Identify a family ritual other than Thanksgiving and discuss the particular ways that this ritual reaffirms family relationships. What evidence of gender relationships does observation of this ritual reveal?

2. By using a popular film to analyze family rituals, Coltrane shows the influence that popular culture has on the social construction of cultural values. Choose another example from popular culture (such as a TV sit-com or a contemporary movie) and describe the family ideals that this artifact of culture represents.

INFOTRAC COLLEGE EDITION

You can use your access to InfoTrac College Edition to learn more about the subjects covered in this essay. Some suggested search terms include:

cultural symbols
family ideal

family rituals
rites of passage

11

Geisha

LIZA CRIHFIELD DALBY

When looking at other cultures, people often impose their own cultural framework on the cultures of others. Stereotypically, Americans have tended to see the Japanese geisha as a sex symbol, but, in doing so, they have misunderstood geisha society. Lisa Crihfield Dalby, an American social scientist, became a geisha as part of her research on Japanese geisha culture. In this selection she shows how the meaning of sisterhood is used in the context of geisha society.

From: Liza Crihfield Dalby. 1983. *Geisha.* Berkeley: University of California Press, pp. 39–41. Reprinted with permission.

Sisterhood is basic to Kyoto geisha society. What does it mean to be sisters? First of all, one is never just *a* sister, but specifically an older sister or a younger sister. Far from having the overtones of equality implied by the English word "sisterhood," in Japan this relationship primarily indicates hierarchy. A new geisha becomes the younger sister of a more experienced geisha in this web of relationships expressed in terms of the family. She and her older sister form a pair, but an unequal one.

Not only do geisha take on the capacity of sisters to one another, but they call the women who run teahouses "mother." In Japan, the parent-child relationship is also hierarchical: an unequal pair, with rightful expectations on the one side, obligations on the other. The use of kinship terms does not necessarily call into play the sentimental notions about family that we of Western European cultural heritage have come to presume are natural. Instead, the terms older sister and younger sister, mother and daughter, define the unequal but complementary sets of categories that are the basis of geisha society.

Geisha are by no means the only Japanese who live and work in a social group defined by kin terms, but this phenomenon does appear most explicitly in traditional occupations: among carpenters, miners, sumo wrestlers, and gangsters, for example. The geisha sisterhood, however, differs from all these other groups in a distinctive way.

In the more usual form of these so-called ritual kinship groups, the *oyabun,* or the "one in the role of parent," is the linchpin of the entire organization. He (in almost every case it is a he) has a following of several *kobun,* "those in the role of child." Such groups are tightly knit, hierarchical, and to some degree authoritarian. Here, the roles of ritual brotherhood are merely extensions of the main parent-child tie.

The difference for geisha is precisely the primacy of sisterhood. Although the many "mothers," as the mistresses of teahouses, are powerful figures in the day-to-day work of the geisha world, they are not equivalent to the single, all-powerful position of the oyabun. Geisha communities have nothing resembling the pyramid of authority seen in other ritual kinship groups. Geisha mothers, daughters, and sisters participate in separate relationships, each created on its own terms, rather than as parts of one overarching whole. As with Ichiume and myself, a pair of sisters often have different mothers.

Although the key element in the relationship of sisterhood is hierarchy, geisha nowadays feel that there should be empathy, loyalty, and camaraderie between sisters. An onēsan expects deference from her younger sister, but tyranny is not supposed to be her style. Ideally, the older sister is at once mentor and friend. One has no say in who one's blood relatives are, but there is an *en,* an affinity, between two geisha who choose one another as sisters. Whatever else may enter into the considerations of this choice, compatibility between the two women is essential today.

A young woman embarking on the career of a geisha is similar to a bride leaving her natal family for her husband's home. Obviously the comparison does not hold on every point, but in general the "older sister" is similar to the bridegroom and the "younger sister" to the bride. The new geisha leaves her home to live in a place where she calls the proprietress "mother." She enters a subordinate relationship with a previously unrelated person who then becomes kin, her "older sister."

Finally, she is expected to put her old family behind her in her devotion to the new group.

The similarities between new geisha and bride are not just incidental. They are quite explicit in the ceremony that creates the bond of sisterhood.

The traditional Japanese marriage ceremony reaches its culmination when the bride and groom take three sips of sake from each of three lacquered cups. This exchange of nuptial cups is called *sansan-kudo*, "thrice three, nine times," and the phrase can be used to mean a wedding. Most people think of sansan-kudo only in connection with the marriage of a man and a woman, but its meaning and use are broader than that. The sharing of ritual sake creates a deep and solemn bond between two people whom we and the Japanese ordinarily consider unrelated. Thereafter, they are kin. This is basically what marriage is about, of course, but the tie can also bind the older and younger sisters of the geisha world.

In Kyoto, when a new maiko or geisha joins the ranks, she and her older sister-to-be enter sisterhood by performing the ritual of sansan-kudo. "The tying together of destinies" signifies marriage in Japan, and geisha use the same term, *en musubi*, to talk about their special sister ties. An en is a connection between people, usually a created connection rather than a "natural" one. The Buddhist meaning of en, karma, lies behind the notion of human connections; but in everyday usage, when Japanese say that two people have an en, they are not necessarily thinking of the metaphysical reasons for it. If you have an en with someone, there is some sort of special affinity between you. One step further is the "tying of en" (en musubi), which creates a bond not easily loosened. In fact, if such a bound pair should separate, the tie cannot simply be undone, but is said to be "cut" (*en o kiru*).

Because this bond is not to be trifled with, Ichiume and I did not actually go through the sisterhood ceremony. My own reasons for becoming a geisha were clear to everyone, but it was also clear that I was not making a long-term commitment to the geisha life. For practical purposes Ichiume was my older sister, but for us to have exchanged the ritual cups of sake would have been a sham. . . .

DISCUSSION QUESTIONS

1. What are the cultural expectations associated with sisterhood in the United States? How would you compare and contrast the cultural meaning of sisterhood as it is used in different contexts in this culture—such as in families, in sororities, and in the women's movement?

2. Like "sisterhood," "brotherhood" connotes specific social and cultural expectations. What relationships are implied by the word "brotherhood" and how does this compare to the concept of "sisterhood"?

INFOTRAC COLLEGE EDITION

You can use your access to InfoTrac College Edition to learn more about the
subjects covered in this essay. Some suggested search terms include:

geisha society   tea ceremony
Japanese marriage   teahouses
sisterhood

12

The Self

GEORGE HERBERT MEAD

In his classic writing, first published in 1934, George Herbert Mead discusses the origins of the conscious self. Our understanding of the process of socialization originates with Mead's argument that the self arises through social interaction and the ability to take the role of the "generalized other." He outlines a child's various stages of development (the play and the game stage) in learning to take the role of the other. He also distinguishes between the "I" and the "me" in social interaction. Mead's arguments continue to serve as the basis for contemporary sociological studies of the individual in society.

The language process is essential for the development of the self. The self has a character which is different from that of the physiological organism proper. The self is something which has a development; it is not initially there at birth but arises in the process of social experience and activity, that is, develops in the given individual as a result of his relations to that process as a whole and to other individuals within that process. The intelligence of the lower forms of animal life, like a great deal of human intelligence, does not involve a self. In our habitual actions, for example, in our moving about in a world that is simply there and to which we are so adjusted that no thinking is involved, there is a certain amount of sensuous experience such as persons have when they are just waking up, a bare "thereness" of the world. Such characters about us may exist in experience without taking their place in relationship to the self. One must, of course, under those conditions, distinguish between the experience that immediately takes place and our own organization of it into the experience of the self. One says upon analysis that a certain item had its place in his experience, in the experience of his self. We inevitably do tend at a certain level of sophistication to organize all experience into that of a self. We do so intimately identify our experiences, especially our affective experiences, with the self that it takes a moment's abstraction to realize that pain and pleasure can be there without being the experience of the self. Similarly, we normally organize our memories upon the string of our self. When we date things we always date them from the point of view of our past experiences. We frequently have memories that we cannot date, that we cannot place. A picture comes before us suddenly, and we are at a loss to explain when that experience originally took place. We remember perfectly distinctly the picture, but we do not have it definitely placed, and until we can place it in terms of our past experience we are not satisfied. Nevertheless, I think it is obvious,

From: Anselm Strauss. 1934. *George Herbert Mead On Social Psychology.* Chicago: University of Chicago Press, pp. 199–233. Reprinted with permission.

when one comes to consider it, that the self is not necessarily involved in the life of the organism, nor involved in what we term our sensuous experience, that is, experience in a world about us for which we have habitual reactions. . . .

The individual experiences himself as such, not directly, but only indirectly, from the particular standpoints of other individual members of the same social group or from the generalized standpoint of the social group as a whole to which he belongs. For he enters his own experience as a self or individual, not directly or immediately, not by becoming a subject to himself, but only insofar as he first becomes an object to himself just as other individuals are objects to him or are in his experience; and he becomes an object to himself only by taking the attitudes of other individuals toward himself within a social environment or context of experience and behavior in which both he and they are involved.

The importance of what we term "communication" lies in the fact that it provides a form of behavior in which the organism or the individual may become an object to himself. It is that sort of communication which we have been discussing—not communication in the sense of the cluck of the hen to the chickens, or the bark of a wolf to the pack, or the lowing of a cow but communication in the sense of significant symbols, communication which is directed not only to others but also to the individual himself. So far as that type of communication is a part of behavior, it at least introduces a self. Of course, one may hear without listening; one may see things that he does not realize; do things that he is not really aware of. But it is when one does respond to that which he addresses to another and when that response of his own becomes a part of his conduct, when he not only hears himself but responds to himself, talks and replies to himself as truly as the other person replies to him, that we have behavior in which the individuals become objects to themselves.

The self, as that which can be an object to itself, is essentially a social structure, and it arises in social experience. After a self has arisen, it in a certain sense provides for itself its social experiences, and so we can conceive of an absolutely solitary self. But it is impossible to conceive of a self arising outside of social experience. When it has arisen, we can think of a person in solitary confinement for the rest of his life, but who still has himself as a companion and is able to think and to converse with himself as he had communicated with others. That process to which I have just referred, of responding to one's self as another responds to it, taking part in one's own conversation with others, being aware of what one is saying and using that awareness of what one is saying to determine what one is going to say thereafter—that is a process with which we are all familiar. We are continually following up our own address to other persons by an understanding of what we are saying and using that understanding in the direction of our continued speech. We are finding out what we are going to say, what we are going to do, by saying and doing, and in the process we are continually controlling the process itself. In the conversation of gestures what we say calls out a certain response in another and that in turn changes our own action, so that we shift from what we started to do because of the reply the other makes. The conversation of gestures is the beginning of communication. The individual comes to carry on a conversation of gestures with himself. He says something and that calls out a certain reply in himself which makes him change what he was going to say. . . .

This process of abstraction cannot be carried on indefinitely. One inevitably seeks an audience, has to pour himself out to somebody. In reflective intelligence one thinks to act, and to act solely so that this action remains a part of a social process. Thinking becomes preparatory to social action. The very process of thinking is, of course, simply an inner conversation that goes on, but it is a conversation of gestures which in its completion implies the expression of that which one thinks to an audience. One separates the significance of what he is saying to others from the actual speech and gets it ready before saying it. He thinks it out and perhaps writes it in the form of a book; but it is still a part of social intercourse in which one is addressing other persons and at the same time addressing one's self, and in which one controls the address to other persons by the response made to one's own gesture. That the person should be responding to himself is necessary to the self, and it is this sort of social conduct which provides behavior within which that self appears. I know of no other form of behavior than the linguistic in which the individual is an object to himself, and, so far as I can see, the individual is not a self in the reflective sense unless he is an object to himself. It is this fact that gives a critical importance to communication, since this is a type of behavior in which the individual does so respond to himself.

We realize in everyday conduct and experience that an individual does not mean a great deal of what he is doing and saying. We frequently say that such an individual is not himself. We come away from an interview with a realization that we have left out important things, that there are parts of the self that did not get into what was said: What determines the amount of the self that gets into communication is the social experience itself. Of course, a good deal of the self does not need to get expression. We carry on a whole series of different relationships to different people. We are one thing to one man and another thing to another. There are parts of the self which exist only for the self in relationship to itself. We divide ourselves up in all sorts of different selves with reference to our acquaintances. We discuss politics with one and religion with another. There are all sorts of different selves answering to all sorts of different social reactions. It is the social process itself that is responsible for the appearance of the self; it is not there as a self apart from this type of experience. . . .

. . . One set of background factors in the genesis of the self is represented in the activities of play and the game.

We find in children the invisible, imaginary companions which a good many children produce in their own experience. They organize in this way the responses which they call out in other persons and call out also in themselves. Of course, this playing with an imaginary companion is only a peculiarly interesting phase of ordinary play. Play in this sense, especially the stage which precedes the organized games, is a play at something. A child plays at being a mother, at being a teacher, at being a policeman; that is, it is taking different roles, as we say. We have something that suggests this in what we call the play of animal; a cat will play with her kittens and dogs play with each other. Two dogs playing with each other will attack and defend, in a process which, if carried through, would amount to an actual fight. There is a combination of responses which checks the depth of the bite. But we do not have in such a situation the dogs taking a definite role in the sense that a child deliberately takes the role of another. This tendency on the part of the children is what we are working with in the kindergarten where the roles which the children assume are made the basis for training. When a child does assume a role he has in himself the stimuli which call out that particular response or group of responses. He may, of course, run away when he is chased, as the dog does, or he may turn around and strike back just as the dog does in his play. But that is not the same as playing at something. Children get together to "play Indian." This means that the child has a certain set of stimuli which call out in itself the responses that they would call out in others and which answer to an Indian. In the play period the child utilizes his own responses to these stimuli which he makes use of in building a self. The response which he has a tendency to make to these stimuli organizes them. He plays that he is, for instance, offering himself something, and he buys it; he gives a letter to himself and takes it away; he addresses himself as a parent or as a teacher; he arrests himself as a policeman. He has a set of stimuli which call out in himself the sort of responses they call out in others. He takes this group of responses and organizes them into a certain whole. Such is the simplest form of being another to one's self. It involves a temporal situation. The child says something in one character and responds in another character, and then his responding in another character is a stimulus to himself in the first character, and so the conversation goes on. A certain organized structure arises in him and in his other which replies to it, and these carry on the conversation of gestures between themselves.

When we contrast play with the situation in an organized game, we note the essential difference that the child who plays in a game must be ready to take the attitude of everyone else involved in that game and that these different roles must have a definite relationship to each other. Take a very simple game such as hide-and-seek. Everyone with the exception of the one who is hunting is a person who is hiding. A child does not require more than the person who is hunted and the one who is hunting. When a child is playing in the first sense he just goes on playing, but there is no basic organization gained. In that early stage he passes from one role to another just as the whim takes him. But in a game when a number of individuals are involved, the child taking one role must be ready to take the role of everyone else. When he gets in a baseball game, he must have the responses of each position involved in his own position. He must know what everyone else is going to do in order to carry out his own play. He has to take all of these roles. They do not all have to be present in consciousness at the same time, but at some moments he has to have three or four individuals present in his own attitude, such as the one who is going to throw the ball, the one who is going to catch it, and so on. These responses must be, in some degree, present in his own make-up. In the game, then, there is a set of responses of such others so organized that the attitude of one calls out the appropriate attitudes of the other.

This organization is put in the form of the rules of the game. Children take a great interest in rules. They make rules on the spot in order to help themselves out of difficulties. Part of the enjoyment of the game is to get these rules. Now, the rules are the set of responses which a particular attitude calls out. You can demand a certain response in others if you take a certain attitude. These responses are all in yourself as well. There you get an organized set of such responses as that to which I have referred, which is something more elaborate than the roles found at play. Here there is just a set of responses that follow on each other indefinitely. At such a stage we speak of a child as not yet having a fully developed self. The child responds in a fairly intelligent fashion to the immediate stimuli that come to him, but they are not organized. He does not organize his life as we would like to have him do, namely, as a whole. There is just a set of responses of the type of play. The child reacts to a certain stimulus, and the reaction is in himself that is called out in others, but he is not a whole self. In his game he has to have an organization of these roles; otherwise he cannot play the game. The game represents the passage in the life of the child from taking the role of others in play to the organized part that is essential to self-consciousness in the full sense of the term. . . .

PLAY, THE GAME, AND THE GENERALIZED OTHER

The fundamental difference between the game and play is that in the former the child must have the attitude of all the others involved in that game. The attitudes of the other players which the participant assumes organize into a sort of unit, and it is that organization which controls the response of the individual. The illustration used was of a person playing baseball. Each one of his own acts is determined by his assumption of the action of the others who are playing the game. What he does is controlled by his being everyone else on that team, at least insofar as those attitudes affect his own particular response. We get then an "other" which is an organization of the attitudes of those involved in the same process.

The organized community or social group which gives to the individual his unity of self can be called "the generalized other." The attitude of the generalized other is the attitude of the whole community. Thus, for example, in the case of such a social group as a ball team, the team is the generalized other insofar as it enters—as an organized process on social activity—into the experience of any one of the individual members. . . .

It is in the form of the generalized other that the social process influences the behavior of the individuals involved in it and carrying it on, that is, that the community exercises control over the conduct of its individual members; for it is in this form that the social process or community enters as a determining factor into the individual's thinking. In abstract thought the individual takes the attitude of the generalized other toward himself, without reference to its expression in any particular other individuals; and in concrete thought, he takes that attitude insofar as it is expressed in the attitudes toward his behavior of those other individuals with whom he is involved in the given social situation or act. But only by taking the attitude of the generalized other toward himself, in one or another of these ways, can he think at all; for only thus can thinking—or the internalized conversation of gestures which constitutes thinking—occur. And only through the taking by individuals of the attitude or attitudes of the generalized other toward themselves is the existence of a universe of disclosure, as that system of common or social meanings which thinking presupposes at its context, rendered possible. . . .

I have pointed out, then, that there are two general stages in the full development of the self. At the first of these stages, the individual's self is constituted simply by an organization of the particular attitudes of other individuals toward himself and toward one another in the specific social acts in which he participates with them. But at the second stage in the full development of the individual's self, that self is constituted not only by an organization of the social attitudes of the generalized other or the social group as a whole to which he belongs. . . .

What makes the organized self is the organization of the attitudes which are common to the group. A person is a personality because he belongs to a community, because he takes over the institutions of that community into his own conduct. He takes its language as a medium by which he gets his personality, and then through a process of taking the different roles that all the others furnish, he comes to get the attitude of the members of the community. Such, in a certain sense, is the structure of a man's personality. There are certain common responses which each individual has toward certain common things, and insofar as those common responses are awakened in the individual when he is affecting other persons, he arouses his own self. The structure, then, on which the self is built is this response which is common to all for one has to be a member of a community to be a self. Such responses are abstract attitudes, but they constitute just what we term a man's character. They give him what we term his principles, the acknowledged attitudes of all members of the community toward the values of that community. He is putting himself in the place of the generalized other, which represents the organized responses of all the members of the group. It is that which guides conduct controlled by principles, and a person who has such an organized group of responses is a man whom we say has character, in the moral sense. . . .

THE "I" AND THE "ME"

We have discussed at length the social foundations of the self and hinted that the self does not consist simply in the bare organization of social attitudes. We may now explicitly raise the question as to the nature of the "I" which is aware of the social "me." I do not mean to raise the metaphysical question of how a person can be both "I" and "me" but to ask for the significance of this distinction from the point of view of conduct itself. Where in conduct does the "I" come in as over against the "me"? When one determines what his position is in society and feels himself as having a certain function and privilege, these are all defined with reference to an "I," but the "I" is not a "me" and cannot become a "me." We may have a better self and a worse self, but that again is not the "I" as against the "me," because they are both selves. We approve of one and disapprove of the other, but when we bring up one or the other, they are there for such approval as "me's." The "I" does not get into the limelight; we talk to ourselves, but do not see ourselves. The "I" reacts to the self which arises through the taking of the attitudes of others. Through taking those attitudes, we have introduced the "me" and we react to it as an "I."

The simplest way of handling the problem would be in terms of memory. I talk to myself, and I remember what I said and perhaps the emotional content that went with it. The "I" of this moment is present in the "me" of the next moment. There again I cannot turn around quick enough to catch myself. I become a "me" insofar as I remember what I said. The "I" can be given, however, this functional relationship. It is because of the "I" that we say we are never fully aware of what we are, that we surprise ourselves by our own action. It is as we act that we are aware of ourselves. It is in memory that the "I" is constantly present in experience. We can go back directly a few moments in our experience, and then we are dependent upon memory images for the rest. So that the "I" in memory is there as the spokesman of the self of the second, or minute, or day ago. As given, it is a "me," but it is a "me" which was the "I" at the earlier time. If you ask, then, where directly in your own experience the "I" comes in, the answer is that it comes in as a historical figure. It is what you were a second ago that is the "I" of the "me." It is another "me" that has to take that role. You cannot get the immediate response or the "I" in the process. The "I" is in a certain sense that with which we do identify ourselves. The getting of it into experience constitutes one of the problems of most of our conscious experience; it is not directly given in experience.

The "I" is the response of the organism to the attitudes of the others; the "me" is the organized set of attitudes of others which one himself assumes. The attitudes of the others constitute the organized "me," and then one reacts toward that as an "I." . . .

The two are separated in the process, but they belong together in the sense of being parts of a whole. They are separated and yet they belong together. The separation of the "I" and the "me" is not fictitious. They are not identical, for, as I have said, the "I" is something that is never entirely calculable. The "me" does call for a certain sort of an "I" insofar as we meet the obligations that are given in conduct itself, but the "I" is always something different from what the situation itself calls for. So there is always that distinction, if you like, between the "I" and the "me." The "I" both calls out the "me" and responds to it. Taken together they constitute a personality as it appears in social experience. The self is essentially a social process going on with these two distinguishable phases. If it did not have these two phases, there could not be conscious responsibility and there would be nothing novel in experience. . . .

DISCUSSION QUESTIONS

1. Think about the first time you were able to play an organized game with others. Do you remember learning to "take turns" or to work as part of a team? How does childhood play translate into adult participation in society?

2. How do you identify yourself with relation to others? Try naming all the roles you play that make up part of your self identity (for example, son/daughter; friend; sister/brother).

INFOTRAC COLLEGE EDITION

You can use your access to InfoTrac College Edition to learn more about the subjects covered in this essay. Some suggested search terms include:

game   play
generalized other   self
language/communication

13

Navajo Women and the Politics of Identity

AMY J. SCHULZ

In this original research, Schulz examines the construction of American Indian identity among three generations of Navajo women. Her interviews reveal how these women balance multiple identities in response to historical and social situations. Against the backdrop of American Indian history, Schulz concludes that Navajo women construct and negotiate their identities as both Indian and Navajo.

The resurgence—or continuation—of ethnic identification and conflict in the United States belies the American myth of the "melting pot" and calls into question the inevitability of the melting process. The continued salience of group identities invites attention to the processes and structures that

From: *Social Problems* 45 (August 1997): 336–353. Reprinted with permission.

influence their construction, deconstruction, and reconstruction over time (Anderson 1983; Barth 1969; Nagel 1996; Waters 1990). How are collective identities created and transformed? How do changes in social, political, economic, and cultural contexts influence the problematics as well as the possibilities of identities? Identities both reflect and potentially disrupt or recreate social and political relationships within and between groups. Examining the construction and reconstruction of identities over time contributes to our understanding of social and political processes through which individuals and groups locate themselves in relation to others, understand themselves, and define their possibilities.

This analysis addresses these questions by examining intergenerational changes in the construction of collective identities and the meanings associated with those identities. The analysis focuses on two of many dimensions of identity available and salient to the women who participated in the study: "Indian" as a supratribal identity; and "Navajo" or tribal identity. I examine differences between younger and older Navajo women in the social, economic, political, and cultural contexts they experienced, the problems associated with Indian and Navajo identities, and the resources available to them to negotiate those identities. Drawing on in-depth interviews conducted between 1990 and 1992, I explore these dimensions of identity as they are constructed in the women's narratives and in their interactions with me, an Anglo researcher. The approach used in the analysis emphasizes the situated nature of identities, and the creative—but also socially structured—actions of individuals in constructing collective and personal identities. Furthermore, as collective identities are inextricably connected to efforts to construct a coherent sense of self, my analysis explores the use of multiple, sometimes fragmented and conflicting identities as women define themselves in relationship to others (Calhoun 1994; Lemert 1994; Woodward 1997). . . .

CONCEPTUAL FRAMEWORK

Identities are not unilateral or constant. Rather, there are many different dimensions or layers of identity, including nationality, ethnicity, gender, family, social class, and sexuality (Barth 1969; Hall 1990; Taylor 1989; Woodward 1997). The salience of these identities may vary with situational and political factors (Cohen 1985; Cornell 1988; Nagel 1996). Furthermore, the identities themselves may be contradictory, conflicting, or fragmented, creating tension within the individual as well as within and between groups as identities and their meanings are negotiated (Calhoun 1994; Waters 1990; Wiley 1994; Woodward 1997). The politics of identity involve the disruption or reconstruction of identities, or the meanings that have been associated with them, and are associated with efforts to shift power relations within or between groups (Calhoun 1994).

The identities examined in the following pages are arenas in which social and political relationships between Anglos and American Indians are negotiated and contested. Before contact with Europeans, those indigenous to North America organized their political, social, and economic relationships in a wide variety of ways, from hierarchical to loosely-knit networks (Cornell 1988). Kinship and clan networks, bands, and broader social groups came together to create tribes, as they searched for ways to respond to the social and political disruptions created by contact with European-Americans (Cornell 1988). The multiple levels of group identity available to contemporary American Indians, including supratribal or pan-Indian, tribal, and subtribal (clan, kinship), were shaped by these processes and reflect historical and contemporary relationships between Anglos and Indians. In this study I focus on self and collective identities as key conceptual frames through which individuals define or locate themselves, and through which I examine political and social relationships. Excerpts from women's narratives emphasize the creative actions of individuals as they negotiate identities within particular social, economic, political, and cultural contexts (Nagel 1996; Waters 1990; Woodward 1997). Furthermore, these narrative accounts speak to the important role played by women's day-to-day actions in resisting assimilation into dominant social and cultural systems (Ward 1993).

METHODS

The analysis presented here used primary and secondary sources to examine United States Indian policies as records of the political context, and of Anglo representations of—and beliefs about—American Indians between the late 1800s and the late 1900s. In addition, I present findings from an inductive analysis of thirty-one in-depth interviews conducted between 1990 and 1992 with women living on the Navajo Nation, ranging in age from 15 to 76 at the time of the interviews, to discuss the implications of these policies and beliefs for Navajo women. My purpose was to understand women's experience of, and their strategies for managing, pressures to be incorporated into Western political, economic, and cultural systems.

The interviews consisted of accounts of personal life histories, guided by questions to encourage elaboration about experiences with the schools, families, and work. In addition, participants were asked to describe important challenges or conflicts they had experienced and their responses to those challenges. Finally, the interviews included questions about the salience and meanings of their identities as Indian, Navajo, and women:[1] for example, "what has it meant to you to be Navajo?"; and "what does it mean to be 'Indian'?"[2] Each participant received a transcript of her interview and was invited to comment or to discuss it with me. In addition, I shared preliminary analyses with several respondents as a means of member validation and to encourage further discussion of the material (Erikson 1976; Mbilinyi 1989; Oakley 1981). This process led to many rich discussions with participants that were subsequently incorporated into the analysis.

I located participants through snowball sampling, beginning with interviews with women who were friends or acquaintances, and who then helped identify other potential study participants. Criteria for participation in the study were broadly defined to include women who; currently lived on the Navajo Nation;

spoke English; and were interested in and willing to talk about their experiences of social and cultural change. This selection process favored women with relatively more exposure to United States educational and labor systems than was the norm for women living on the Navajo Nation in 1990. For example, 68 percent of the participants in this study had completed high school and 12 percent had completed college, as compared to 42 percent and 3 percent, respectively, of all women living on the Navajo Nation in 1990.[3] Similarly, the women who participated in this study were more likely than the average woman on the Navajo Nation to participate in wage work, with 57 percent of participants engaged in the labor force as compared to 39 percent of all women on the Navajo Nation in 1990.[4] Most of my respondents live within extended family networks on the Navajo Nation; however, they have had relatively more exposure to, and day-to-day experience with, United States educational and labor systems than the average woman living on the Nation. Thus, the women who participated in this study live on the boundaries between the Navajo and Anglo worlds.

All of the women interviewed for this study spoke Navajo more or less fluently, in addition to English. Women in the oldest cohort and most of those in the middle cohort spoke Navajo as their first language, and had learned English in school. For most of the youngest cohort, English was their first language, and most had struggled to learn Navajo in school with varying degrees of success. Nearly all of the respondents had lived for some period of their life away from the reservation: at boarding school; doing migrant farm work; at college; or living or working in an urban community. The timing of these experiences varied, as some had left during childhood, some during adolescence, and others during early adulthood. Most had returned to the reservation by middle adulthood, although a few of the older women had retired from wage work in nearby communities and then returned to their home communities on the Navajo Nation.

The analysis presented here draws on the thirty-one interviews described above, divided into three cohorts: thirteen women born prior to 1946 (aged 46 to 76 at the time of the interviews); nine women born between 1946 and 1960 (aged 32–44 at the time of the interviews); and nine women born between 1961 and 1976 (aged 15–27 at the time of the interviews). These three cohorts were chosen because of their relationship to changes in United States Indian policies, these policies influenced the social conditions and formal educational institutions that women experienced as they came to adulthood.

I argue in the following pages that Navajo women's experiences are socially and historically patterned, and their responses to social and historical circumstances are shaped by the resources available to them at various historical moments. On the average, the women in this study have more exposure to educational institutions and labor force participation than is the norm for women living on the Navajo Nation and more day-to-day exposure to extended kinship networks, spiritual practices, and Navajo language than many Navajo who live in urban communities. Their particular social location, as members of extended family networks on the Navajo Nation and as participants in United States educational and labor force systems, mean that they experience tensions related to the construction of "Indian" and "Navajo" identities in their daily lives. This social location, along with their historical location as members of the cohorts described above, shaped the issues they confronted and the resources at their disposal as they negotiated these identities. The women in this study, in the language of grounded theory methodology, represent "particularly rich cases" (Strauss and Corbin 1990) whose experiences offer insights into the construction and reconstruction of identity group boundaries and meanings precisely because of the salience of these negotiations in their lives. . . .

NEGOTIATING INDIANNESS

. . . The different historical periods in which women came of age influenced their personal histories and the specific issues, concerns, resources, and constraints that they encountered as they negotiated their identities. Despite their varied experiences, as women talked about what it meant to "be Indian" a central theme that cut across cohorts was the idea of difference. For women of all ages to be Indian meant to be set apart, to be distinguished from others (Schulz 1994). For example, Nora, born in 1937, left her Tewa family of origin to live with an Anglo foster family in Oklahoma to attend junior and senior high school in the 1950s. She described what it had meant to her to be identified as Indian at that time:

> Sometimes, like when I went away to school, I felt among all the Anglo that I was different. Then inside, I knew that I wasn't all that different from them. I'm alive just like them, my body is just like them, only the skin is different. But I have a mind just like them and think like them. I guess the only thing is the Indian culture is different from yours. You have a culture too, but being an Indian—I'm proud that I am an Indian. . . . It was not really how they . . . they treated me well. It's just . . . I always thought that Anglos knew more than I did. Like when you watch movies, in the wild west kind, they make the Indians lower than themselves and I guess I had that feeling in myself.
>
> (Nora, age 54)

In this excerpt, Nora struggled with an externally imposed identity that defined her as not only "different," but "lower than" Anglos. The salience of Indianness was highlighted as she lived among, and interacted with, non-Indians. Living in a predominantly White community in post World War II America, she encountered systems of beliefs that constructed her as a member of a group that was not only different, but devalued. The "antipodal categories" articulated by Cornell (1988) . . . were salient in the 1950s, juxtaposing Anglos as good and right and powerful against Indians as "pagan, savage, nearly dumb." These dichotomies served to devalue Indians and to legitimize their political, social, and economic marginalization (Berkhofer 1979; Burt 1986).

As Nora pulled apart the strands of her experience, she noted that her White foster family "treated me well" but they were unable to protect her from the effects of these deeply rooted ideologies of racial hierarchy. Looking back on this experience some 40 years later, Nora noted that while she *knew* intellectually that she was not different or inferior to Anglos—that both had skin, minds, and cultures—she still *felt* that she was different and believed that she was inferior to Anglos.

As she described her experience as a child, Nora was unable to construct an alternative identity that would disrupt or challenge the representations of Indianness she encountered in her everyday life. As an adult, she addressed these representations by pointing out to me that I too had a culture even though it differed from her own. Emphasizing our sameness, with the exception of the color of our skins and our cultures, she challenged the differentiation underlying the devaluation of Indians as a group and of herself as a member of that group.

By the later 1940s, more elementary schools began to be located on the Navajo Nation and children increasingly lived with family members while they attended elementary school. Adele, born in 1945, attended day school through her elementary years, where she encountered strictly enforced policies that forbade the use of Navajo language or other "Indian" practices within the school. As she spoke of her school experiences, Adele also described her parents' efforts to confront school personnel and bring the attention of other tribal members to bear on the harsh treatment of children in the schools. In addition to modeling active resistance, her parents also provided her with an alternative sense of herself, grounded in her Navajo and Apache heritage. Adele's connection to those tribal identities was apparent as she spoke about what it meant to be Indian:

> I'm considered as an Indian but within Indian society there are different tribes that have different languages and cultures. There are differences among the Indian people. I think it's unique that I'm from two tribes—from Navajo and Apache. I'm able to speak and understand two languages. I was raised up by two cultures and . . . I understand two languages and two cultures too.
>
> (Adele, age 46)

Adele's choice of words in this excerpt accentuated the externally imposed nature of Indian identity—her use of passive voice emphasized that she was considered *by others* (non-Indians) to be Indian, but that she did not claim that identity. Adele's response interrupted stereotypic and homogeneous images of Indians by highlighting tribal differences. Finally, she used herself as an example to drive home the point that "Indians" are actually many different communities with different languages, cultures, and identities. In so doing, she actively negotiated the meanings associated with Indian identity, resisting externally imposed and reductionist ideas and promoting instead a more complex notion of what it meant to be Indian.

Born in 1952, Ursula also lived on the Navajo Nation within an extended family as she attended school. Like Adele, she spoke of incidents in school in which her Navajo spiritual beliefs and practices were explicitly devalued by school personnel, describing the impact of those experiences as she said ". . . and the more and more messages I got like that, I had really bad feelings about myself." However, Ursula also described support for the development of a positive sense of herself, noting that her father "used to encourage us to speak up, he'd teach us

that it was important for our voice to be heard and our opinion was important."
Thus, like Adele, Ursula had support from family members that provided alternatives to the sense of self derived from representations of Indians constructed by
outsiders. When I asked what it meant to her to be Indian, Ursula replied:

> I guess when I think about myself as an Indian I get in touch with my hu-
> manity. Because to me, the way I was taught, being Indian is a label. And who
> you are is *ashdlaa'o*[6]—man with five fingers. You exist as part of the human
> race. As an Indian you have skin of color—*bitsj' yishtlizh*—they call it—the
> color of the earth. Your skin is like the color of the earth, so therefore you are
> an earth person. So this is the Navajo coming in, for the way they teach us
> is—first you're a man with five fingers, that puts you in touch with the
> human race. When you get in touch with your humanness, then you have
> respect for others that are like you . . . whether they have different colors or
> not, they're still humans. They teach you to have respect for that.
>
> (Ursula, age 44)

Ursula's response highlighted multiple dimensions of identity and the ways in
which they may be intertwined or in conflict. She began by defining Indian as a
label imposed by outsiders, not an identity that she would claim. Like Nora, she
emphasized similarities among human beings, challenging the construction of an
Indian "other" and representations of Indians as less than human, or as entities to
be civilized, assimilated, or annihilated. Her words challenged the meanings that
outsiders have associated with the term Indian. . . .

Each of these women constructed her response to questions about identity
that I—an Anglo/outsider—had asked them. Their responses cannot be separated
from the context in which they were elicited, shaped by the salience of my outsider status and what that represented in our interactions (Cooley 1902). Their
responses were also shaped by a history in which Indians have been defined as
"different" by people like me, and that difference has been used as justification
for their political and economic marginalization as well as the destruction of their
unique identities as Navajo. The women drew upon more complex identities as
Navajo, human beings, and in Adele's case, Apache. They used the Navajo language and knowledge of their own and others' histories, to disrupt and challenge
stereotypic images and to construct alternative images of what it meant to "be
Indian."

The responses of women born after 1960 were qualitatively different from
those of the older cohorts when asked what it meant to be Indian. Most women
in this youngest cohort embraced an Indian identity at the same time that they
recognized its stigmatizing and marginalizing potential. For example, Reyna, born
in 1974, said:

> For me, being an Indian is the greatest thing that ever happened to me. Even
> though . . . some [in] White society put us down. But we're still there, and
> we have all these cultures and traditions, different values that we carry are
> really sacred and spiritual. There's no way no one can take that away from us.
> Because we've come through that and it's helped us along the way . . . it's

our own values and our traditions and we should keep it. It makes us stay
together—it holds us together with a lot of strength.

(Reyna, age 17)

Unlike the women in the previous cohort, Reyna did not deny or deconstruct
the meanings associated with the term Indian. However, as she embraced it, she
also recognized conflicts associated with that identity—in particular, the denigra-
tion and marginalization of those identified as Indian. She noted that culture, tradition, and values have been central to the ability of the group to stay together as
a coherent entity, and to overcome struggles for survival. She did not, however,
explicitly recognize *different* cultures, traditions, and values that are specific to different groups included within the term "Indian"—a distinction made repeatedly
by the women in the previous cohort.

Reyna's response reflects the general tendency of women in the youngest cohort not to deny an Indian identity. In contrast to the women in the older cohorts, whose encounters with negative labels associated with Indianness were
direct and explicit, Reyna's experience attending a community-run school on
the Navajo reservation had buffered her somewhat from similar experiences. Furthermore, growing up following the social movements of the 1960s and 1970s,
Reyna had access to, and social support for articulating more positive constructions of pan-Indian identity.

Women in this youngest cohort tended to use different strategies from the older
women to resist the marginalizing and stigmatizing potential of Indian identification.
For example, Sandra noted that "I know that it's really hard to be an Indian off the
reservation. There's a lot of times when the stereotypes can really bring you down."
She went on to describe her efforts to confront her college classmates' stereotypic
portrayal of "all Indians (as) drunks, . . . or not smart, . . . or not responsible . . .":

I always want to show them that those things are not true. . . . (I tell them)
that my mom was a personnel manager of a school, and it's not like that at
all. There are some people that are that way, and there are some people that
aren't.

(Sandra, age 27). . .

The education and other life experiences of this youngest cohort were shaped
by the shift to self-determination that emerged from the activism of the 1960s
and 1970s (Senese 1991; Szasz 1973). As noted earlier, most women in this age
range had attended elementary and often junior and senior high schools located
on the Navajo Nation, and several had attended community-run schools with an
explicit emphasis on bilingual education and Navajo history and culture. Many of
these young women had grown up in homes where their parents, products of the
boarding schools of the 1940s and 1950s, spoke English fluently and where they
learned English as their first language. As a result, while their parents had struggled to learn the English language in school, many of these young women struggled to learn to speak, read, and write the Navajo language. Their sense of
themselves as Navajo was developed in part through their interactions at school
and the intentional efforts of teachers and parents to teach them what it meant to

be Navajo. This exposure to Navajo language and identity differed from that of the older cohorts, who developed a Navajo identity through interactions with extended family members and encountered explicit efforts to eradicate it in the schools.

These results illuminate changes in the ways that women living on the Navajo Nation negotiated the meanings of Indianness, and the influence of social, economic, and political contexts on those negotiations. Women born prior to the social movements of the 1960s were more likely than younger women to distance themselves from an Indian identity, emphasizing or making salient their tribal affiliations and using them to disrupt dominant group images of Indians. Those who came of age after 1960 were much more likely to actively embrace an Indian identity. Women in both groups, however, sought to claim particular meanings associated with Indianness (e.g., citizens, human beings) and to disrupt others (e.g., drunken, illiterate, warriors), positioning themselves as different from Anglos but resisting the association of difference with deficiency. Thus, while the particular strategies and forms through which Indianness was contested changed, the contestation itself continued. . . .

CONCLUSION

The processes that shape ethnic group identities are complex and multifaceted, constructed against a backdrop of history, inter- and intra-group relationships. The analysis presented in the preceding pages emphasizes the interplay among multiple identities, and the historically as well as socially situated nature of those identities (Taylor 1989; Woodward 1997). As women describe their efforts to negotiate Indian and Navajo identities, the complexities of constructing dynamic and multidimensional personal identities within particular political and social contexts become visible.

The salience of different layers of identity varies situationally and contextually. Across the three cohorts, Indian identity was salient in relation to Anglos and reflected politicized relationships between the groups. As women spoke about what it meant to them to be Indian, their examples referenced interactions with Anglo foster families, teachers, school mates, tourists, and researchers, as well as their encounters with negative and stereotypic representations of Indians in film and literature. Throughout the interviews women attempted to interrupt power relations between Indians and Anglos by disrupting and reconstructing Indian identities and their associated meanings (Calhoun 1994; Cohen 1985; Cornell 1988; Nagel 1996).

The consistency of this resistance across all the cohorts in this study speaks to the importance of women's day-to-day resistance to the assimilationist efforts of the dominant group (Ward 1993; Yuval-Davis 1997). Women struggle to construct identities that affirm their cultural distinctiveness, but that do not accept the pervasive negative constructions of that difference; they assert and seek to maintain distinct, valued identities. The strategies they use toward this end vary according to the resources, both material and conceptual, available at different historical moments (Cornell 1988; Ortner 1989; Swidler 1986). These resources include language, history, and alternative identities, and access to them is influenced by historical patterns and events that shape the social and political contexts within which women's personal histories unfold.

Women who came of age prior to the civil rights movements of the 1960s, in general, encountered a social and political context in which American Indians were denigrated and efforts to promote economic and cultural assimilation were explicit. Those who had grown up in extended families used their Navajo identities as alternative frames to distance themselves from constructions of Indians as inferior, savage, or uncivilized. Arguing that they were labeled *by others* as Indian, women rejected both the label and the meanings associated with this externally imposed identity. Their claim to particular tribal identities subverted the homogenizing and marginalizing potential of Indianness.

For these women, the politics of identity involved not simply naming or labeling, but defining the meanings associated with the name and, perhaps most importantly, establishing who has claim to define those meanings (Barth 1969; Cohen 1985; Nagel 1996). The subjective experience of themselves as Navajo, Navajo language, and Navajo teachings or belief systems were resources upon which they drew to define themselves actively *against* the definitions of outsiders. They used Navajo identity to locate themselves within a history that they claimed, positioning themselves in relation to the Navajo, rather than in relation to Anglos. The themes of strength, resilience, and survival that were central to Navajo identity offered an alternative to the portrayal of Indians in Western history. As others have noted, access to such oppositional frames supports the ability to claim a separate and positive group identity that resists and disrupts pervasive negative group constructions (Groch 1994; Hall 1990; hooks 1990; Mulling 1992). As such, the ability to name and claim one's identity becomes an explicitly political act.

However, disruptions in patterns of practice that were central to Navajo identity occurred as post World War II era policy initiatives were implemented; it interrupted access to these frames for many women. Women like Nora were exposed to dominant group representations of Indians without access to the cultural resources with which to develop and articulate alternate constructions of identity. They described the impact of those representations on their sense of themselves. This does not mean that Navajo women who grew up away from the reservation did not resist representations of Indianness. They were, however, less likely to draw upon Navajo identities as they did so—in some cases perhaps because of the contested nature of those tribal identities. This suggests that social and historical patterns that distance Navajo women from language, patterns of practice, spiritual practices, and other dimensions of Navajo identity, as in relocation to urban communities, may interrupt the use of Navajo identity as a means to disrupt outsider notions of Indianness. Collective actions to maintain or recreate cultural practices are one strategy for countering these processes; for example, bilingual schools emerged on the Navajo Nation in the 1970s with an explicit agenda to teach Navajo language, culture, and history.

In contrast to women who grew up in the post-war era, those who came of age following the self-determination and civil rights movements of the 1960s actively embraced Indian identities. Like the women who grew up in the 1940s, these younger women also sought to redefine the content and meanings associated with Indian identity, claiming particular meanings associated with Indianness and disrupting others. Unlike the older women who grew up in extended families, and more like the older women who grew up among Anglos, the younger cohort did not draw upon their particular tribal identities to disrupt constructions of Indians. Rather, they sought to reconstruct Indianness by claiming the more pluralist constructions made available through the political struggles of the 1960s, and by directly challenging negative representations. These women drew upon frameworks that claimed Indian identity as one that distinguished them from Anglos, and actively worked to challenge prevailing constructions of that difference as deficiency.

. . . While women's specific strategies varied with the historical moment and also with individual life trajectories, women across all cohorts developed strategies to resist cultural annihilation. They illustrate both the constancy of the pressure toward incorporation and women's creativity as they draw upon available resources to maintain distinct—but not static—identities as Indian and as Navajo. The construction of identities that disrupt dominant group representations of Indians emerge as everyday acts of political resistance that challenge the continued pressures toward incorporation and the loss of a distinct and valued group identity.

NOTES

1. The gendered nature of ethnic or tribal identities is examined in detail elsewhere (Schulz 1994).

2. The specific wording of questions varied according to the particulars of the interview.

3. Data on Navajo women's education is from Table NN10: Social Characteristics of the Navajo Nation 1990. 1990 Census: Population and Housing Characteristics of the Navajo Nation (The Printing Company: Scottsdale, Arizona, 1993). See Schulz (1994) for a more complete description of the characteristics of study respondents as compared to all women living on the Navajo Nation in 1990.

4. Data on Navajo women's labor force participation is from Table NN11: Labor Force and Commuting Characteristics of the Navajo Nation: 1990. 1990 Census: Population and Housing Characteristics of the Navajo Nation. (The Printing Company: Scottsdale, Arizona, 1993).

5. See Schultz (1994) for a more detailed description of the analysis process.

6. Navajo spellings are from Garth A. Wilson's (1989) *Conversational Navajo Dictionary.*

REFERENCES

Anderson, Benedict. 1983. *Imagined Communities: Reflections on the Origins and Spread of Nationalism.* London: Verso Press.

Barth, Frederik. 1969. *Ethnic Groups and Boundaries: The Social Organization of Culture Difference.* Boston, Massachusetts: Little, Brown and Company.

Berkhofer, Robert F., Jr. 1979. *The White Man's Indian: Images of the American Indian from Columbus to the Present.* New York: Vintage Books.

Burt, Larry C. 1986. *Tribalism in Crisis: Federal Indian Policy 1953–1961.* Albuquerque, New Mexico: University of New Mexico Press.

Calhoun, Craig. 1994. "Social theory and the politics of identity." In *Social Theory and the Politics of Identity,* ed. C. Calhoun, 9–36. Cambridge, MA: Blackwell.

Cohen, Anthony P. 1985. *The Symbolic Construction of Community.* London: Routledge.

Cooley, Charles H. 1902. *The Looking Glass Self: Human Nature and the Social Order.* New York: Charles Scribner Publishers.

Cornell, Stephen. 1988. *The Return of the Native: American Indian Political Resurgence.* New York: Oxford University Press.

Erikson, Kai. 1976. *Everything in its Path: Destruction of Community in the Buffalo Creek Flood.* New York: Simon and Schuster.

Groch, Sharon A. 1994. "Oppositional consciousness: Its manifestation and development: The case of people with disabilities." *Sociological Inquiry* 64:369–395.

Hall, Stuart. 1990. "Cultural identity and the diaspora." In *Identity: Community, Culture, Difference,* ed. J. Rutherford, 222–237. London: Lawrence and Wishart.

hooks, bell. 1990. *Yearning: Race, Gender and Cultural Politics.* Boston, MA: South End Press.

Karp, David. 1996. *Speaking of Sadness:* New York: Oxford University Press.

Lemert, Charles. 1994. "Dark thoughts about the self." In *Social Theory and the Politics of Identity,* ed. C. Calhoun, 100–130. Cambridge, MA: Blackwell.

Mbilinyi, Marjorie. 1989. "I'd have been a man: Politics and the labor process in producing personal narratives." In *Interpreting Women's Lives: Feminist Theory and Personal Narratives,* eds. The Personal Narratives Group, 204–227. Bloomington, IN: Indiana University Press.

Mead, George H. 1934. *Mind, Self and Society from the Standpoint of a Social Behaviorist.* Chicago: University of Chicago Press.

Mulling, Leith. *Race, Class and Gender: Representations and Reality.* Center for Research on Women, Memphis, Tennessee: Memphis State University.

Nagel, Joane. 1996. *American Indian Ethnic Renewal: Red Power and the Resurgence of Identity and Power.* New York: Oxford University Press.

Navajo Government Publication. 1993. *Population and Housing Characteristics of the Navajo Nation, 1990.* Scottsdale, AZ: The Printing Company.

Oakley, Anne. 1981. "Interviewing women: A contradiction in terms. In *Doing Feminist Research,* ed. H. Roberts, 30–61. London: Routledge Kegan Paul.

Ortner, Sherry. 1989. *High Religion: A Cultural and Political History of Sherna Buddhism,* Princeton, NJ: Princeton University Press.

Schulz, Amy. 1994. "I raised my children to speak Navajo. . . . My Grandkids are all English speaking people': Identity, resistance and transformation among Navajo women." Doctoral Dissertation. (University of Michigan, Ann Arbor)

————. 1995. "I didn't want a life like that': Constructing a life between cultures." In *Women Creating Lives: Identities, Resilience and Resistance,* eds. Carole Franz and Abigail Stewart, 127–140. Boulder, CO: Westview Press.

————. 1998. "How would you write about that?: Language, identity and the knowing self." In *Outside the Master Narrative: Women's Untold Stories,* eds. Mary Romero and Abigail J. Stewart, New York: Routledge.

Senese, Guy B. 1991 "Self determination and the social education of Native Americans." *Journal of Thought.* New York: Praeger Press.

Sewell, William. 1989. "Towards a theory of structure: Duality, agency and transformation." CSST Working Paper #29. The University of Michigan, Ann Arbor.

Strauss, Anselm, and Juliet Corbin. 1990. *Basics of Qualitative Research: Grounded Theory Procedures and Techniques.* Newbury Park, CA: Sage.

Swidler, Anne. 1986. "Culture in action: Symbols and strategies." *American Sociological Review* 51:273–286.

Szasz, Margaret. 1973. *Education and the American Indian: On the Road to Self Determination, 1928–1973,* Albuquerque, NM: University of New Mexico Press.

Taylor, Charles. 1989. *Sources of the Self: The Making of Modern Identity.* Cambridge, MA: Harvard University Press.

Ward, Kathryn B. 1993. "Reconceptualizing world systems theory to include women." In *Theory of Gender/Feminism on Theory,* ed. Paula England, 43–68. New York: Aldine de Gruyter.

Waters, Mary C. 1990. *Ethnic Options: Choosing Identities in America.* Berkeley, CA: University of California Press.

White, Richard. 1983. *The Roots of Dependency: Subsistence, Environment, and Social Change among the Choctaws, Pawnees, and Navajos.* Lincoln, NE: University of Nebraska Press.

Wiley, Norbert. 1994. "The politics of identity in American history." In *Social Theory and the Politics of Identity.* ed. C. Calhoun, 131–149. Cambridge, MA: Blackwell.

Wilson, Garth A. 1989. *Conversational Navajo Dictionary.* Blanding, UT: Conversational Navajo Publications.

Woodward, Kay. 1997. *Identity and Difference.* London: Sage.

Yuval-Davis, Nira. 1997. *Gender and Nation.* Thousand Oaks, CA: Sage.

DISCUSSION QUESTIONS

1. How does the history of these Navajo women specifically influence the development of their identity? How does gender influence the identities of these Navajo women?

2. How could you replicate this study among a different racial-ethnic group? What historical information is important to understanding the construction of identity among another racial-ethnic group?

INFOTRAC COLLEGE EDITION

You can use your access to InfoTrac College Edition to learn more about the subjects covered in this essay. Some suggested search terms include:

activism	life histories
constructing identity	Navajo women
ethnicity	socialization
identity	traditions

14

Catching Sense:

Learning from Our Mothers to Be Black and Female

SUZANNE C. CAROTHERS

This research uncovers the importance of family relations in the socialization process. Specifically, Suzanne Carothers interviews Black women about the lessons passed down from their mothers. She outlines both the substance of these teachings and how they were taught. The women in this study develop identities as Black and as women through interaction with other women family members.

Black parents are required to prepare their children to understand and live in two cultures—Black American culture and standard American culture. To confront the bicultural nature of their world, these parents must respond in distinctive ways. In the following essay, I show how this can be seen in the practices and beliefs of several generations of Black women through their descriptions of seemingly ordinary and commonplace activities.

The first setting in which people usually experience role negotiations is the home. Boys and girls will draw from important lessons learned at home during childhood to negotiate their future roles as viable members of society. Distant though the lessons may seem from the perspective of an adult, they were taught directly and indirectly in the context of day-to-day family life. Of the many dyads occurring within families, the interactions between mothers and daugh-

From: Faye Ginsburg and Anna Lowenhaupt Tsing, eds. 1990. *Uncertain Terms: Negotiating Gender in American Culture.* Boston: Beacon Press, pp. 232–247. Reprinted with permission.

ters are a critical source of information on how women perceive what it means to be female. I have been particularly interested in these perceptions among Black mothers and daughters because of the unique socio-economic, political circumstances in which these women find themselves in American culture. During the Fall, Winter, and Spring of 1980–81, I returned to my home town of Hemington, fictitiously named, to collect data for my study. It is the contradiction that emerged between my experience of having grown up in the Black community of Hemington and my graduate school reading of the social science literature on Black family life and mother-daughter relationships that led me to engage in this research. . . .

BACKGROUND

The implications of Black women's strength have not been explored fully in the literature on American mother-daughter relationships. Black women have traditionally combined mothering and working roles, while white middle-class women in the United States until recently have not. In Western cultures, mothering is regarded as a role that directly conflicts with women's other societal roles. In response to this condition, many theorists of female status consider the mothering role to be the root cause of female dependence on and subordination to men.[1] Yet, this has not been the experience of Black mothers. As others have argued, "Women have been making culture, political decisions, and babies simultaneously and without structural conflicts in all parts of the world."[2]

During the 1970s, a recurring theme in the United States literature on mother-daughter relationships was the ambivalence and conflict existing between mothers and daughters. The literature describes competition and rivalry and suggests a negative cycle of influences passed from mothers to their daughters. For example, Judith Arcana suggests,

> The oppression of women created a breach among us, especially between mothers and daughters. Women cannot respect their mothers in a society which degrades them; women cannot respect themselves. Mothers socialize their daughters into the narrow role of wife-mother; in frustration and guilt, daughters reject their mothers for their duplicity and incapacity—so the alienation grows in the turning of the generations.[3]

The above quote is generally inapplicable to the relationship between Black mothers and their daughters. The Black cultural tradition assumes women to be working mothers, models of community strength, and skilled women whose competence moves beyond emotional sensitivity. It is through this tradition of a dual role that Black women acquire their identity, develop support systems (networks), and are surrounded by examples of female initiative, support, and mutual respect.

Black mothers do not raise their children in isolation. In contrast, Nancy Chodorow argues,

> The household with children has become an exclusively parent and child realm; infant and child care has become the exclusive domain of biological mothers who are increasingly isolated from other kin, with fewer social contacts and little routine assistance during their parenting time. . . .[4]

The families to whom Chodorow refers above are child-centered. In the arrangement she describes, the needs of the domestic unit are shaped and determined primarily by those of the children. Scholars of Black family life offer evidence of other arrangements.[5] According to them, Black women raise their children in the context of extended families in which social and domestic relations, as well as kinship and residence structures offer a great deal of social interaction among adults that includes children. In addition, these researchers have shown that child rearing is only one of many obligations to be performed within Black family households. They agree that child rearing cannot be evaluated in the singular context of an individual but rather in the plural context of the household. The process through which daughters learn from their mothers in Black families, therefore, contradicts the wave of literature on mother-daughter relationships.

In order to appreciate the contrast in orientation of Black mothers, it is necessary to consider the wider social context of Black parenting. Black parents in American society have a unique responsibility. They must prepare their children to understand and live in two cultures—Black American culture and standard American culture.[6] Or as Wade W. Nobles[7] has suggested, Black families must prepare their children to live near and be among white people without becoming white. This phenomenon has been referred to as *biculturality* by Ulf Hannerz and by Charles Valentine, an idea derived from W.E.B. Du Bois who wrote in the early 1900s about the idea of double consciousness: "that Blacks have to guard their sense of blackness while accepting the rules of the games and cultural consciousness of the dominant white culture."[8] Because Black parents recognize that their children must learn to deal with institutional racism and personal discrimination, Black children are encouraged to test absolute rules and absolute authority.[9] It is therefore critical to the socialization of Black children that their parents provide them with ample experiences dealing with procedures of interpersonal interaction rather than rules of conduct. The children are socialized to be part of a Black community rather than just Black families or "a fixed set of consanguinal and affixed members."[10] Beginning early in childhood, the wider social context in which Black children are raised usually involves not only their mothers, but also many adults—all performing a variety of roles in relation to the child, the domestic unit, and the larger community. Furthermore, the transmission of knowledge and skills in Black family life is not limited to domestic life but occurs in public life arenas in which Black women are expected to participate. Working outside their homes to contribute to the economic resources of the family has been only one of the many roles of the majority of Black women. As members of the labor market, Black mothers simultaneously manage their personal lives, raise their children, organize their households, participate in community and civic organizations, and create networks to help each other cope with seemingly insurmountable adversities.

Participation in work and community activities broadens the concept and practice of mothering for Black women. How do the women learn these roles? What must they pass on to their female children if they are to one day perform these roles? An exploration of the social interactions between Black working mothers and their daughters, as well as the cultural context and content can extend our knowledge of the cultural variation in mothering roles and mother-daughter relationships and the processes by which mothers shape female identity.

THE STUDY

Several reasons prompted my decision to return to Hemington for this research. Typically, studies of Black family life have been conducted in urban ghetto communities in the north and mid-west, some in the deep rural South and most with lower class or poor people.[11] In developing my research, I wanted to avoid choosing a location where there were vast differences between the Black and white standard of living that so often characterize the research settings in which Blacks are studied. In these settings, the usual distinguishing features for the Black population are poverty, unemployment, under employment, low wages, and inadequate housing. The white communities are more varied economically, ranging from relatively wealthy managerial elites to welfare recipients living in housing projects. Yet, there are many middle-sized and large southern cities with large, varied and stable Black communities that researchers have not adequately explored. Hemington is such an example.[12]

Prior to 1960 and urban redevelopment, the Black community of Hemington was primarily concentrated in the area closest to the main business district of the city. At the time of the study, most Blacks still lived in predominantly all Black neighborhoods located on the west side of town, where a broad range of housing suggests an economically varied Black community. Blacks in Hemington live in public housing projects, low income housing (both private homes and apartments), modern apartment complexes, and privately owned homes, ranging from modest to lavish.

Although I had not lived in Hemington for more than ten years, my kinship ties to the community meant that I had access to people, situations, and information that an outsider might not have or would need a considerably longer time to acquire. My experiences of growing up there and then studying and working in educational institutions in northern white society made me sensitive to differences in cultural patterns and more eager to analyze them.

Forty-two women and nine girls between the ages of 11 and 86 from twenty families agreed to participate in the study. I asked them to help me understand the meaning of mothering and working in their lives and how this meaning was passed on from mother to daughter—generation to generation. The stories that the women and girls shared with me about the very ordinary day-to-day activities of their lives became a rich source for understanding how women found and created meaning in the less-than-perfect world in which they lived.[13]

The study was of women whom Alice Walker would call the anonymous Black mothers whose art goes unsigned and whose names are known only by their families.[14] Many of the women in the study have known me all my life. They are great-grandmothers, who have lived to see their grandchildren's children born and grandmothers who have raised their children. I grew up with some of the mothers who are raising their children. Still others, the young girls, I remember from the time they were born.[15]

Seventy-five years separate the births of the oldest from the youngest participants. . . .

These women perceive themselves as middle-class. All are or have been employed. They are people who share a common system of values, attitudes, sentiments, and beliefs which indicate that an important measure of "class" for these Black Americans is the range of resources available to the extended family unit. This system, then—based on extensive inter-household sharing—is not synonymous with traditional criteria for social class structure which includes wealth, prestige, and power.[16]

WHAT BLACK MOTHERS TEACH: CONCRETE LEARNINGS AND CRITICAL UNDERSTANDINGS

In order to understand the teaching and learning process taking place between Black mothers and daughters, I observed the women and asked them questions about their seemingly ordinary and commonplace daily activities. When the participants in the study were asked from whom they learned, their answers included their mothers, fathers, stepfathers, stepmothers, grandfathers, grandmothers, great-grandmothers, aunts, the lady next door, an older sister, a brother, a teacher—in short, their community. When asked *what* they had learned from these people, their responses touched a range of possibilities, which can be grouped into two broad categories. Cooking, sewing, cleaning, and ironing are examples of activities that are associated with the daily routines of households that I call "concrete learnings." The regular performance of these leads to what I refer to as "critical understandings," which include such things as achieving independence, taking on responsibility, feeling confident, getting along with others, or being trustworthy. The acquisition of these is not easy to pinpoint and define. They are not taught as directly as the concrete learnings, but they are consistently expected. Their outcomes are not as immediately measurable because they usually take a longer time to develop.

What do mothers do to pass on to their daughters the understandings that are considered critical to a daughter's well being, and the skills, the learned power of doing a thing competently? Mothers teach their daughters what to take into account in order to figure out how to perform various tasks, recognizing that the individual tasks that they and their daughters are required to perform change over time. Therefore, the preparation that mothers provide includes familiarity with the task itself, as well as a total comprehension of the working of the home or other situations within which the task is being done. The women's interviews reveal that mothers teach by the way that they live their own lives ("example"), by pointing out critical understandings they feel their daughters need ("showing"), and by instructing their daughters how to do a task competently. Their teaching is both verbal and nonverbal, direct and indirect. Daughters learn not only from their mothers, but also from other members of family and the community.

The data indicate that concrete learnings teach—in ways that verbal expression alone does not—a route toward mastery and pride that integrates the child into the family and community. Daughters learn competency through a sense of aesthetics, an appreciation for work done beautifully. The women described this notion as follows: "You don't see pretty clothes hanging on the line like you used to;" "Mama could do a beautiful piece of ironing;" "You always iron the back of the collar first. The wrinkles get on the back and it makes the front of the collar smooth;" or "Now if I got in the kitchen and say I saw these pretty biscuits, I might say Mama how did you get your biscuits to look this pretty . . ." This aesthetic quality becomes one of the measures of competently done work as judged by the women themselves and by other members of their community.

As each generation encountered technological changes in household work, mothers became less rigid about teaching their daughters concrete learnings. However, like previous generations, mothers still teach their daughters responsibility through chores, which gives them opportunities to practice and get better at doing them, both alone and with others. These activities are not contrived but rather they constitute real work and contribute to the daily needs of their households. Participation in these activities encourages mastery of them. . . .

Each woman described a certain kind of pride in herself for having learned and accomplished a task well. Such mastery reinforced and established the woman's confidence in her ability to perform well. Having chores to do was the important link bridging concrete learnings to critical understandings germane to a daughter's well-being.

DEMANDS OF DOUBLE CONSCIOUSNESS
OR LESSONS OF RACISM

While mothers teach critical understandings through example, maxims, and practical lessons, they use what I am calling "dramatic enactments" as a powerful tool to teach their daughters ways to deal with white people in a racist society. Thus, daughters learn critical understandings that are specific to the Black experience.

When mothers teach by example, they enact before their daughters the particular skills necessary to achieve the task at hand. By contrast, dramatic enactments expose children to conflict or crisis and are often reserved for complex learning situations. "Learning to deal with white people," for example, was viewed by some women in the study as important to their survival, and dramatic enactment was identified as being a powerful technique for acquiring this skill. One thirty-one-year-old woman . . . explained how she learned this critical understanding through dramatic enactments from her grandmother, who did domestic work.

> My sister and I were somewhat awed of white people because when we were growing up, we did not have to deal with them in our little environment. I mean you just didn't have to because we went to an all-Black school, an all-Black church, and lived in an all-Black neighborhood. We just didn't deal with them. If you did, it was a clerk in a store.
>
> Grandmother was dealing with them. And little by little she showed us how. First, [she taught us that] you do not fear them. I'll always remember that. Just because their color may be different and they may think differently, they are just people.
>
> The way she did it was by taking us back and forth downtown with her. Here she is, a lady who cleans up peoples' kitchens. She comes into a store to spend her money. She could cause complete havoc if she felt she wasn't being treated properly. She'd say things like, "If you don't have it in the store, order it." It was like she had $500,000 to spend. We'd just be standing there and watching. But what she was trying to say [to us] was, they will ignore you if you let them. If you walk in there to spend your 15 cents, and you're not getting proper service, raise hell, carry on, call the manager but don't let them ignore you.

Preparing their daughters to deal with encounters in the world beyond home was a persistent theme in the stories offered by the women in this study. By introducing their daughters and granddaughters to such potentially explosive situations and showing the growing girls how older women could handle the problems spurred by racism, mothers and grandmothers taught the lessons needed for survival, culturally defined as coping with the wider world. . . .

THE COMMUNITY CONTEXT

The concept of Black family units working in concert to achieve the common goals they value, arises out of the inherent expectations of helping and assuming responsibility for each other as part of a conscious model of social exchange.[17] Thus, giving and receiving are the understood premises for participating in community life. Different from guilt, this system has been fueled by the racial and economic oppressions that have plagued Black families since their introduction into American society. Women in these family units traditionally assume a critical role in meeting these responsibilities. This does not end when children reach maturity, nor is it hierarchical, from mother to child; it is part of the larger community value that the women believe in and sustain. . . .

Given the difficult conditions that racism and economic discrimination have imposed on the Black community (including, of course, the middle-class Black community), it is important for children to know that their parents can survive the difficult situations they encounter. Children need to trust that the world is sufficiently stable to give meaning to what they are learning. *Dependability*—based on elements of character such as hard work, faith, and the belief that their children can live a better life—provides the context for that trust and the daughters' sense of their mothers' competence to deal with the world.

Despite the conflicts that sometimes arose, daughters generally acknowledged the ongoing lessons their mothers had to teach them and that the process of *lifelong learning* was central to their relationship, as Mrs. Washington's quote illustrates,

> Kitty and James, either one haven't got to the place today where I couldn't tell them if they were doing something wrong. And people gets on me for that. And I say, well you never get too old to learn. I say, if I know it's right why can't I correct them? They say, "When a child gets up on his own, you ought to let 'em alone." Then I say, well I'm going to bother mine as long as I live if I see 'em doing something wrong. I'm going to speak to them and if they don't do what I say, at least they don't tell me that they aren't. They just go someplace else and do it.

This sense of teaching and learning as an ongoing part of the mother-child relationship adds an impetus to the daughters' frequently expressed belief that they fulfilled their mothers' dreams and in a sense justified their mothers' lives through gaining an education.

> This is what Mama was working toward [my going to college]. This was her ambition. It was just like she was going to college herself. The first summer I finished high school and every summer after that, Mama took sleep-in jobs up in the mountains to earn extra money for my college education . . . . She very much wanted us to take advantage of all the things she never had. This is what she worked for. . . .

Black daughters learn their mothers' histories by seeing their mothers in the roles of mamas who nurture as friends, who become confidantes and companions; as teachers, who facilitate and encourage their learning about the world; and as advisors who counsel. For these women, the role of mother is not seen as "a person without further identity, one who can find her chief gratification in being all day with small children, living at a pace tuned to theirs.[18] From early on, the women in the present study see their mothers as complex beings. Knowing her mother's history intensifies the bond between mother and daughter, and helps daughters understand more about the limits under which their mothers have operated.

Getting along however, is not necessarily dependent on the women always reaching agreement, or daughters following the advice of their mothers. It would not be unusual for a mother and daughter to fall out about an issue one day and speak to each other the next. Such interactions provide continuing opportunities for daughters to practice developing and defending their own points of view, a skill useful in the outside world. These interactions insure the back and forth between Black mothers and daughters, which promotes the teaching and learning process and increases Black daughters' respect for what mothers have done and who they are.

Although respect remains a key value in mother daughter relationships and helps to foster the teaching and learning process, generational differences between mothers and daughters lead to tensions that threaten the teaching and learning context. Daughters need to *balance* their loyalty to their mothers with their own needs to grow up in accord with the terms of their own generation: realities. Mothers, on the other hand, need to balance their need for their daughters' allegiance with the knowledge that the daughters require a high degree of independence to survive and achieve in the world.

Loyalty is the unspoken but clear message in the words spoken by the mothers and daughters in this study. They describe it in terms of faithfulness and continuing emotional attachment. As loyalty defines what Black mothers and daughters expect from each other, it also is part of the conflict between them— when Black women are unable to separate the tangled threads that bind them so closely. Their obligation to each other and the deep understanding of the plight they share sometimes nurtures a desire to protect, rather than commit what would be seen as an act of desertion. . . .

CONCLUSION

Women in this study routinely have confronted very early on the contradictions between the world in which they were born and raised and the one away from their homes. The result is that the women are not thrown by that which is different or contradictory to their home practices; rather, they can accept, understand, negotiate and deal with the differences in reasonable ways.

A high degree of mutual respect and camaraderie characterize the teaching and learning processes taking place between Black mothers and daughters. The community value of mutual responsibility makes this possible. Because of the multiple roles that these mothers play, the interactions between Black mothers and daughters require that mothers balance these roles and determine which one is appropriate in different situations. It also requires that daughters actively consider the context and purpose of the interaction and the mood of her mother to determine the appropriate response. Thus, the issue of authority that is often a major concern and obstacle in school learning for both teacher and students, shifts to mutuality in the teaching and learning process occurring between Black working mothers and their daughters.

The daughters have learned from their mothers by being exposed to the complications, complexities, and contradictions that as working women, their mothers faced in a society which has traditionally viewed working and mothering as incompatible roles. The recognition of this difference requires that Black women, as a condition of their daily existence, constantly negotiate an alternative understanding of female identity that challenges the dominant gender paradigm in American culture.

NOTES

1. Nancy Chodorow, *The Reproduction of Mothering: Psychoanalysis and the Sociology of Gender* (Berkeley: University of California Press, 1978).

2. Karen Sacks, *Sisters and Wives: The Past and Future of Sexual Equity* (Westport, CT: Greenwood Press, 1979).

3. Judith Arcana, *Our Mothers' Daughters* (Berkeley: Shameless Hussy Press, 1979).

4. Chodorow, p. 5.

5. Joyce Aschenbrenner, *Lifelines: Black Families in Chicago* (New York: Holt, Rinehart & Winston, 1975). Cynthia Epstein, "Positive Effects of the Multiple Negatives: Explaining the Success of Black Professional Women," in *Changing Women in a Changing Society,* ed. J. Huber (Chicago: University of Chicago Press, 1973). T. R. Kennedy, *You Gotta Deal With It: Black Family Relations in a Southern Community* (New York: Oxford University Press, 1980). E. P. Martin & J. M. Martin, *The Black Extended Family* (Chicago: University Press, 1978). Karen Sacks, *Sisters and Wives* (Westport, CT: Greenwood Press, 1979). V. H. Young, "Family and Childhood in a Southern Negro Community," in *American Anthropologist* 72 (1970), 269–88.

6. T. Morgan, "The World Ahead: Black Parents Prepare Their Children for Pride and Prejudice," in *The New York Times Magazine* (1985, October 27), 32. V. H. Young, "A Black American Socialization Pattern," in *American Ethnologist* 1 (1974), 405–513.

7. See Nobles in J. E. Hale, *Black Children: Their Roots, Culture and Learning Styles* (Provo, UT: Brigham Young University Press, 1982).

8. Du Bois, W. E. B. *The Gift of Black Folk: The Negroes in the Making of America* (New York: Washington Square Press, 1970), xii. For "biculturality" see Ulf Hannerz, *Soulside: Inquiries Into Ghetto Culture a Community*

(New York: Columbia University Press, 1969), and Charles Valentine, "Deficit, Difference, and Bicultural Models of Afro American Behavior," in *Harvard Educational Review* 41, no. 2 (1971).

9. Young (1974), 405–513.

10. Kennedy, 223.

11. Kennedy, 226.

12. According to the 1980 census approximately one-third of Hemington's population was Black. The population of Hemington County was 404,270. Blacks were 27% of this population. In the City of Hemington Blacks were 32% of the population.

13. The research consisted of 51 taped interviews of two, three and four generations of mothers and their daughters. In addition, a questionnaire was given to each of the participants on a day other than the interview.

14. Alice Walker, *In Search of our Mothers' Gardens* (New York; Harcourt Brace and Jovanovich, 1983), 231–243.

15. They represent five different sets of consanguineous generations including: 1) seven grandmothers and mothers; 2) four great grandmothers, grandmothers, and mothers; 3) two great-grandmothers, grandmothers, mothers, and unmarried daughters; 4) three grandmothers, mothers, and unmarried daughters; and 5) four mothers and teenage daughters. The method of selecting the participants was primarily through a snowball sample technique using personal contacts of women in my mother's network of friends, neighbors, and co-workers. The initial source of participants was an older subdivision call Fenbrook Park. Names of participants, when used, have been changed.

16. This study employs the definition of social class as discussed by John F. Cuber and William F. Kenkel in *Social Stratification in the*

United States (New York: Appleton-Century-Crofts, 1954). They suggest that "*Social class* has been defined in so many different ways that a systematic treatment would be both time consuming and of doubtful utility. One central core of meaning, however, runs throughout the varied usages, namely, the notion that the hierarchies of differential statuses and of privilege and disprivilege fall into certain clearly distinguishable categories set off from one another. Historically, this conception seems to have much better factual justification than it does in contemporary America . . . Radical differences, to be sure, do exist in wealth, privilege, and possessions; but the differences *seem to range along a continuum with imperceptible gradation from one person to another,* so that no one can objectively draw 'the line' between the 'haves' and 'the have nots,' the 'privileged' and 'underprivileged,' or for that matter, say who is in the 'working class,' who is 'the common man,' or who is a 'capitalist.' The differences are not categorical, but continuous" (p. 12). For an in-depth discussion of class see Rayna Rapp's article, "Family and Class in Contemporary America: Notes Toward an Understanding of Ideology," *Science and Society* 42(3): 278–300.

17. See I. G. Joseph and J. Lewis, *Common Differences: Conflicts in Black and White Feminist Perspectives* (Garden City, NY: Anchor Books/Doubleday, 1981), 76–126.

18. A. Rich, *Of Woman Born* (New York: N. W. Norton, 1976), 3.

DISCUSSION QUESTIONS

1. Think of the things you learned from your parents/guardians. What are some examples of how you were socialized to behave in ways consistent with your parents' expectations? Do your values and beliefs reflect or challenge your parents/guardians?

2. The women in Carother's article talk about learning how to interact with white people. How has your interaction with members of a different racial-ethnic group been influenced by your parents/guardians? Can you think of times when someone from a different racial-ethnic group reacted to you in a particular way because of your race?

INFOTRAC COLLEGE EDITION

You can use your InfoTrac College Edition access to learn more about the subjects covered in this essay. Some suggested search terms include:

African-American mothers' roles
biculturality
black family

family structure
kinship

15

Alternative Masculinity
and Its Effects on Gender
Relations in the Subculture
of Skateboarding

BECKY BEAL

Becky Beal's research examines the construction of masculinity in the subculture of skate-boarding. She argues that skateboarding encourages a non-traditional form of masculinity, while still reproducing masculine dominance in gender relations. Her research exemplifies how female and male skateboarders are socialized into gender specific roles within the subculture.

With the recent movement in men's studies there has been a growing popularity of investigating different forms of masculinity and their consequences for men, their relationships with each other, and their relationships with women. According to Clatterbaugh (1990), there have been several avenues of the men's movement including a conservative, profeminist, socialist, and gay and black perspective. Each avenue carries with it different social agendas with priorities for addressing social problems. For example, a conservative approach can be correlated with a pro-family values orientation which asserts traditional gender norms and family relations as a way of reestablishing social order. On the other hand, a profeminist approach sees traditional masculinity as the root of women's oppression, and therefore seeks to change traditional gender roles as a way of promoting a more democratic society. The black and gay perspectives have demonstrated that there is not just one form of masculinity from which all men equally benefit. Gay men must deal with homophobia and blacks must deal with racism, and to more of an extent than whites, poverty. Both these circumstances show that minority men have a different experience of masculinity.

Carrigan, Connell, and Lee (1987) effectively summarized the history of research on masculinity as moving from a "sex roles," (or an assumption of "natural" differences determining social behavior) approach to an emphasis on the social construction of gender. The latter approach has emphasized power relations associated with different genders. Carrigan et al. (1987) clarified that power relations are not only between masculinity and femininity, but among different

From: Becky Beal, (1996) *Journal of Sport Behavior,* 9 (August 1996): 204–217.

forms of masculinity as well (e.g., gay & black perspective). The most powerful form of masculinity is called hegemonic masculinity. They stated "what emerges from this line of argument is the very important concept of hegemonic masculinity, not as "the male role," but as a particular variety of masculinity to which others—among them young and effeminate as well as homosexual men—are subordinated" (p. 174).

This paper will combine a profeminist and critical perspective to describe how one group of young males created a non-hegemonic or alternative form of masculinity. The subculture of skateboarders I investigated chose not to live completely by the traditional and hegemonic forms of masculinity. In doing so, they created an alternative masculinity, one which explicitly critiqued the more traditional form. This paper will not only describe how they distinguished their subculture from traditional sport and hegemonic masculinity, but also investigate the resulting gender relations within the subculture, particularly how the males maintained the privilege of masculinity by differentiating and elevating themselves from females and femininity. . . .

HEGEMONIC MASCULINITY IN SPORT

Sport is one of the most significant institutions of male bonding and male initiation rites. . . .

As noted by several sport sociologists (e.g., Curry, 1991; Kidd, 1987; Messner, 1992a; Sabo, 1989) the paradigm of hegemonic masculinity which abounds in mainstream sport includes physical domination, aggression, competition, sexism, and homophobia. These are often seen as the ideal of manliness which affects male athletes' views about themselves, their relationships with other men, and their attitudes about women.

ALTERNATIVE MASCULINITY

For pro-feminists an alternative masculinity gives hope that alternative values can be promoted which will decrease violence, sexism, and homophobia. In other words, alternative values which could improve the quality of life for a variety of men as well as women. It is from this perspective that I became interested in how one subculture of sport, skateboarding, redefined and lived an alternative masculinity. The following will describe how these skateboarders differentiated themselves from hegemonic sport and hegemonic masculinity, what they called the "jock" image. These skateboarders resisted many of the characteristics of the rituals of masculinity that Sabo and Panepinto (1990) described. In particular, skaters challenged a deference to adult male authority and the conformity and control fostered by that authority.

METHODOLOGY

I used qualitative methods of observation, participant-observation, and semi-structured in-depth interviews to investigate the subculture of skateboarding in northeastern Colorado. My research began in June 1989, when I started observing skateboarders in Jamestown and Welton, Colorado at local hangouts, skateboard shops, and even a locally sponsored skateboard exhibition.

Most of my participants I met by stopping them while they were skateboarding on the streets, and asking if I could talk with them. (They call themselves "skaters," and the act of skateboarding they call "skating.") I met other skaters through interactions I had with their parents. In addition, I met one female skater (a rarity) through mutual membership in a local feminist group. These initial contacts snowballed to many others. Over a two year period (1990, 1992) I talked with 41 skaters, 2 skateboard shop owners, and several parents and siblings.

Thirty-seven of the 41 participants were male and 4 were female. In addition, all were Anglo except two who were Hispanic males. The average age of those participating was 16, but ranged from 10 to 25 years. The participants had skateboarded for an average of four years, but the range of their participation was from one to 15 years. Of the 41 skateboarders, 24 I interviewed more than once, and 6 of whom I had on-going communication which gave me a vital source of feedback and helped to refine my conclusions and questions. In addition, I spent over 100 hours observing skateboarders, many of whom I had not interviewed (they were observed in public spaces).

After I finished gathering data, the information and analysis of their subculture was presented to approximately one third of the participants. Their comments served to confirm and fine tune my conclusions. They especially wanted me to note that although they shared many norms and values, they did not share all values and, therefore, just because they were all skateboarders did not mean that they were all good friends. This was evident in the variety of friendship groups within skateboarding such as "hippies," "punks," "Skinheads," and "old-timers.". . .

CONFORMITY AND SELF-EXPRESSION

The popular practice of skateboarding lacked a strict formal structure. In fact, the appeal for many skaters was precisely to use skateboarding as a means of self-expression and of challenging their own physical limits. Generally, a skating "session" (the time spent skateboarding) involves creating and practicing certain techniques, finding fun places to skate, and trying new tricks on the obstacles found. For example, a favorite spot of skaters in Welton was a loading dock located on the backside of a grocery store. The space used for the trucks to dock was lower than the parking area which created an U-shaped ramp where the skaters did a variety of tricks. The flexibility of skateboarding was one of the main attractions. Many skaters commented on their attraction to a sport in which there are not rules, referees, set plays, nor coaches. For example, Paul claimed that to skateboard "[you] don't need uniforms, no coach to tell you what to do and how to do it." Philip added, "I quit football because I didn't like taking orders." Skaters' general disregard for conforming to an authority carried over into a criticism of those who did. Many of the skaters distinguished themselves by considering themselves more reflective than their average peers. Philip and Jeff discussed the issue: "We might look at everything twice whereas everybody else will just go 'oh ya.' The whole war (Persian Gulf) thing, I don't know, it seems a little too, very convenient . . . We're not saying skating doesn't have any conformity, but it's more by your own choice." Jeff:

> It's not conformist conformity . . . . I think skaters are more aware of conformity than jocks, I think jocks just seem to deal with it and say "OK, well that's just the way it is," but skaters go, "geez, why do I have to do that, man. I don't want to buy these shoes, I don't want to have to buy 100 dollar shoes just to fit in," you know, that kind of thing. . . .

Many athletes as well as sports sociologists would claim that one can express themselves through organized sport, but the point is that these skateboarders felt that they had more freedom to be creative in skateboarding than traditional organized sport. The lack of a formal structure controlled by adult males is an essential element in this subculture which is reflected in the lack of standardized criteria to judge performance. As Craig commented, "There is no such thing as a perfect '10' for a trick." What particularly stood out was the skater's preference not to be judged by adult authority figures. . . .

This subculture resisted many of the tenets of organized sport and of hegemonic masculinity with particular regard to the deference to adult male authority in formal structures. Another alternative behavior encouraged in the subculture is the emphasis on participation and cooperation as opposed to elite competition found in traditional sport.

COMPETITION AND PARTICIPATION

. . . Although skateboarding can be practiced as a highly competitive sport, the vast majority of skaters I interviewed described their sport as different from competition. Jeff stated: "I don't know if I would classify it (skating) as a sport. I suppose I just find sport as competition; unless you are on the pro or amateur circuit you're not really competing against anybody." Pamela, an 18 year old skater made this comparison:

> Soccer is a lot of pressure . . . you have to be as good if not better than everybody else, you have to be otherwise you don't play at all. Skating you can't do that, you just push yourself harder and harder . . . swimming is just sort of there, you get timed, now for me you go against the clock. Now when you skate you don't go against anything, you just skate. That's what it is. . . .

Although there is a status hierarchy within the subculture it is not determined through competition with others. The criteria for status is twofold: One must be highly skilled and creative, and one must not use that skill to belittle others. . . .

From my two years of observing skaters, I found that their emphasis was placed on cooperation and encouragement as opposed to a cutthroat form of competition. For example, I had observed skaters I did not formally know in public spaces, and the following incident characterized the type of interaction I commonly observed. Outside a local video store was one popular hang out in Welton because there were several parking blocks on which skaters would slide their skateboards. (That skill is termed a "rail slide.") Two teenaged skaters were practicing their rail slides on these parking blocks. Although they were on different blocks, they would check on each other by asking how the other one was doing. The interaction consisted of congratulations for achieving some goal, or encouragement to keep trying if one missed a trick. This is in marked contrast to other observations I made of more mainstream sport. . . .

CONTRADICTIONS IN THE ALTERNATIVE SUBCULTURE

The subculture of skateboarding is not solely or purely an alternative form of masculinity. It differs from traditional sport in that it demotes competition and rule-bound behavior while it promotes self-expression. Yet, (to my initial surprise) it also serves as an alternative conduit for promoting an ideology of male superiority and of patriarchal relations within the subculture. This contradiction resulted because the subculture of skateboarding provided an informal structure for its participants to create their own rules, yet it simultaneously provided an avenue for the participants to create gender stratification.

As stated previously, organized sport has historically been a realm in which males have bonded and created and reinforced a hegemonic masculinity which has demoted femininity, females, and homosexuals. The subculture of skateboarding I investigated had similar elements. It was male dominated (90% of the participants were male) and promoted a separation and stratification of males and females and of masculinity and femininity.

One of my formal questions addressed female participation. The responses generally reflected the dominant ideology that males and females "naturally" have different social roles, and that sport, and by extension skateboarding, is a male role.

SEX-SEGREGATION

While talking with the skateboarders, I commented on the lack of female participation and asked their opinion about why it occurred. Most males were taken back and they spent time reflecting on it (as if they had not given it much thought before), and their explanations ranged from describing "natural" differences to social preferences of males and females. All the females discussed the issue directly and with depth, and it is my interpretation that they thought about this often. For both males and females the sex-segregation of skateboarding was typically justified as a reflection of feminine and masculine behaviors. In their explanations they did not distinguish between sex (as biological behaviors) and gender (as socially expected behaviors). It appeared that the dominant ideology of "natural" differences between males and females was a fundamental assumption of these skaters. . . .

Female Appearance

One common explanation for the low number of female participants was that skate-boarding does not promote the traditional feminine appearance of the immaculately groomed, petite female. The skaters' assumption is that women want to appear traditionally feminine in all realms of their lives. Part of a feminine appearance is frailness and purity. Most males could not reconcile the physical risk-taking nature of skateboarding with female behavior. For example, Craig stated that skating "is a rough sport where people get scarred, and girls don't want to have scars on their shins, it wouldn't look good." He also added that girls would get tired of wearing tennis shoes all the time. In a separate interview another skater, Stuart, a 21 year old male stated, "girls probably don't skate because they don't look good with bruises.". . .

The bruising of one's body demonstrates a traditional masculine characteristic of risking bodily injury. Most males flaunted their bruises, and often proudly told stories of past injuries. Overall, the skaters did not associate courageous injury as a feminine (and therefore, in their assumptions a female) attribute. It appeared that these males thought that bruises did not look good or "appropriate" on females which reflected their expectations of females as much (if not more) as females expectations of themselves.

Female Social Roles

Most of the skaters presumed that males and females have different social roles. Doug replied to my question of what is a cool skater by stating, "Someone who is not ashamed of it. They don't hide it in the closet around their girlfriends." The statement reflects the assumption of different social realms for males and females; skaters are male (assuming heterosexuality), and females are not typically exposed to skating. I then asked directly about female participation, and Doug responded, "there's not nearly as many, it's too bad." He seemed sincere, so I commented on the idea that it appeared to be an open sport, and he replied: "Ya, but it's also pretty aggressive, kinda, I mean, there's that end of it, it kind of looks aggressive maybe, and women don't get into it." This switch in mid-sentence from a natural difference ("but, it's also pretty aggressive") to a matter of choice ("it kind of looks aggressive, and women don't get into that") was a typical response. I interpreted him as saying: it's not that women can't be aggressive, but it's that they chose not to. Either way, woman are relegated to a different social role; they could choose differently but it's not in their "nature" to do so. Males did not expect masculine behavior from women, and therefore did not interact with females in such a way as to encourage it. . . .

Many skaters saw no physical or tangible barriers to females' involvement, and therefore assumed that females freely chose not to be involved. Other skaters were aware of social forces that may hinder females from skating such as lack of other female participants and lack of peer support. Rarely did the male skaters ever consider their behavior as a reason why females did not participate more regularly. Through my interviews I became aware that males thought of the female skater as an exception. More often, they commented about females as playing a marginalized role in the subculture of skateboarding. "Skate Betties" is the name given to most females associated with skating. Skate Betties are female groupies whose intentions (according to males) are instrumental: to meet cute guys and associate with an alternative crowd. Females are not perceived as expressive or fully engaged in the values of the subculture. . . .

SUMMARY AND CONCLUDING REMARKS

This study of skateboarding is also an example of how masculinity is not "naturally predetermined or universal, but instead a creation of the participants which varies according to the social context." The emphasis on participant control, self expression, and open participation differ greatly from the hegemonic values of adult authority, conformity, and elite competition.

An interesting contradiction arises within this subculture. Even though the participants' challenged mainstream masculinity, they defined skateboarding as primarily a male activity. This subculture of skateboarding illustrates some of the incongruities that arise when people negotiate new social relations. For on one level skateboarding displayed resistance by redefining masculine behavior, yet on another level it reproduced patriarchal relations similar to Young's findings within the rugby subculture. What is essential for the maintenance of patriarchy is creating different social roles for males and females, and marginalizing the female role. Skaters did this by redefining masculinity which preserved skateboarding as a male realm.

Because many of these male skateboarders did not participate in mainstream athletics (either by choice or size/ability), it is my contention that they created an alternative sport which met some of their specific needs, such as participant control and a de-emphasis on elite competition, and skateboarding also served to meet social needs that traditional athletics have met for other males—a place where boys create friendships and differentiate themselves from girls and that which is labeled feminine.

Although the weight of this paper is on the males' attitudes and behaviors, it is evident that both the males and the females have internalized the dominant ideology of sport as a male social role. This affects how females negotiate their position within the subculture. Some of the responses are typical of females in male dominated settings such as feeling a need to constantly demonstrate one's ability as well as fitting into the dominant culture by being "one of the boys." (Theberge, 1993). And as Theberge (1993) noted, as long as females are judged by a standard of masculinity in a patriarchal society they will always be marginalized.

REFERENCES

Carrigan, T., Connell, B., & Lee, J. (1987). Hard and heavy: Toward a new theory of masculinity. In M. Kaufman (Ed.), *Beyond patriarchy: Essays by men on pleasure, power, and change.* New York: Oxford University.

Clatterbaugh, K. (1990). *Contemporary perspectives on masculinity: Men, women, and politics in modern society.* Boulder: Westview.

Curry, T.J. (1991). Fraternal bonding in the locker room: A profeminist analysis of talk about competition and women. *Sociology of Sport Journal, 8,* 119–135.

Kidd, B. (1987). Sports and masculinity. In M. Kaufman (Ed.), *Beyond patriarchy: Essays by men on pleasure, power, and change.* New York: Oxford University.

Messner, M. (1992a). *Power at play: Sports and the problem of masculinity.* Boston: Beacon.

Messner, M. (1992b). *Like family: Power, intimacy, and sexuality in male athletes' friendships.* In P. Nardi (Ed.), *Men's Friendships.* Newbury Park, CA: Sage.

Sabo, D. (1989). Pigskin, patriarchy, and pain. In Kimmel & Messner (Eds.), *Men's lives.* New York: Macmillan.

Sabo, D., & Panepinto, J. (1990). Football ritual and the social reproduction of masculinity. In M. Messner & D. Sabo (Eds.), *Sport, men and the gender order: Critical feminist perspectives.* Champaign, IL: Human Kinetic Books.

Theberge, N. (1993). The construction of gender in sport: Women, coaching, and the naturalization of difference. *Social Problems, 40,* 301–313.

DISCUSSION QUESTIONS

1. Examine another sport or subculture. Within this sport or subculture, are men's and women's roles different from or similar to roles in dominant society? What are some examples of how women and/or men are socialized within this sport or subculture?

2. Can you think of other examples when women act in more masculine ways to fit into the particular setting? What about women in traditionally male professions?

INFOTRAC COLLEGE EDITION

You can use your access to InfoTrac College Edition to learn more about the subjects covered in this essay. Some suggested search terms include:

gender norms   patriarchal
hegemonic masculinity   sociology of sport
men's movement

16

The Problem of Sociology

GEORG SIMMEL

Georg Simmel, an early social theorist, sees society as composed from the many social inter-
actions that people have with each other. A phenomenon becomes "social" when there is a
reciprocal influence among people in these interactions. In this essay, Simmel is analyzing
the connection between the abstraction that sociologists call society and the observable social
interaction that people have.

Society exists where a number of individuals enter into interaction. This interaction always arises on the basis of certain drives or for the sake of certain
purposes. Erotic, religious, or merely associative impulses; and purposes of
defense, attack, play, gain, aid, or instruction—these and countless others cause
man to live with other men, to act for them, with them, against them, and thus to
correlate his condition with theirs. In brief, he influences and is influenced by
them. The significance of these interactions among men lies in the fact that it is
because of them that the individuals, in whom these driving impulses and purposes are lodged, form a unity, that is, a society. For unity in the empirical sense
of the word is nothing but the interaction of elements. An organic body is a unity
because its organs maintain a more intimate exchange of their energies with each
other than with any other organism; a state is a unity because its citizens show
similar mutual effects. In fact, the whole world could not be called one if each of
its parts did not somehow influence every other part, or, if at any one point the
reciprocity of effects, however indirect it may be, were cut off.

This unity, or sociation, may be of very different degrees, according to the
kind and the intimacy of the interaction which obtains. Sociation ranges all the
way from the momentary getting together for a walk to the founding of a family,
from relations maintained "until further notice" to membership in a state, from
the temporary aggregation of hotel guests to the intimate bond of medieval guild.
I designate as the content—the materials, so to speak—of sociation everything
that is present in individuals (the immediately concrete loci of all historical reality)—drive, interest, purpose, inclination, psychic state, movement—everything
that is present in them in such a way as to engender or mediate effects upon others or to receive such effects. In themselves, these materials which fill life, these

From: "The Problem of Sociology," translated by Kurt H. Wolff, in *Georg Simmel,*
1858–1918: A Collection of Essays, with Translations and a Bibliography, edited by Kurt H.
Wolff. Copyright 1959 by the Ohio State University Press, pp. 23–35. All rights reserved.
Originally published in German as "Das Problem der Soziologie," in *Soziologie* (Munich
and Leipzig: Duncker & Humblot, 1908).

motivations which propel it, are not social. Strictly speaking, neither hunger nor love, work nor religiosity, technology nor the functions and results of intelligence, are social. They are factors in sociation only when they transform the mere aggregation of isolated individuals into specific forms of being with and for one another, forms that are subsumed under the general concept of interaction. Sociation is the form (realized in its numerably different ways) in which individuals grow together in a unity and within which their interests are realized. And it is the basis of their interests—sensuous or ideal, momentary or lasting, conscious or unconscious, causal or teleological—that individuals form such unities.

In any given social phenomenon, content and societal form constitute one reality. A social form severed from all content can no more attain existence than a spatial form can exist without a material whose form it is. Any social phenomenon or process is composed of two elements which in reality are inseparable: on the other hand, an interest, a purpose, or a motive; on the other, a form of mode of interaction among individuals through which, or in the shape of which, that content attains social reality.

It is evident that that which constitutes society in every current sense of the term is identical with the kinds of interaction discussed. A collection of human beings does not become a society because each of them has an objectively determined or subjective impelling life-content. It becomes a society only when the vitality of these contents attains the form of reciprocal influence; only when one individual has an effect, immediate or mediate, upon another, is mere spatial aggregation or temporal succession transformed into society. If, therefore, there is to be a science whose subject matter is society and nothing else, it must exclusively investigate these interactions, these kinds and forms of sociation. For everything else found within "society" and realized through and within its framework is not itself society. It is merely a content that develops or is developed by this form of coexistence, and it produces the real phenomenon called "society" in the broader and more customary sense of the term only in conjunction with this form. To separate, by scientific abstraction, these two factors of form and content which are in reality inseparably united; to detach by analysis the forms of interaction or sociation from their contents (through which alone these forms become social forms); and to bring them together systematically under a consistent scientific viewpoint—this seems to me the basis for the only as well as the entire, possibility of a special science of society such. Only such a science can actually treat the facts that go under the name of sociohistorical reality upon the plane of the entirely social. . . .

A given number of individuals may be a society to a greater or a smaller degree. With each formation of parties, with each joining for common tasks or in a common feeling or way of thinking, with each articulation of the distribution of positions of submission and domination, with each common meal, with each self-adornment for others—with every growth of new synthesizing phenomena such as these, the same group becomes "more society" than it was before. There is no such thing as society "as such"; that is, there is no society in the sense that it is the condition for the emergence of all these particular phenomena. For there is no such thing as interaction "as such"—there are only specific kinds of interaction. And it is with their emergence that society too emerges, for they are neither the cause nor the consequence of society but are, themselves, society. The fact that an extraordinary multitude and variety of interactions operate at any one moment has given a seemingly autonomous historical reality to the general concept of society. Perhaps it is this hypostatization of a mere abstraction that is the reason for the peculiar vagueness and uncertainty involved in the concept of society and in the customary treatises in general sociology. We are here reminded of the fact that not much headway was made in formulating a concept of "life" as long as it was conceived of as an immediately real and homogeneous phenomenon. The science of life did not establish itself on a firm basis until it investigated specific processes within organisms—processes whose sum or web life is; not until, in other words, it recognized that life consists of these particular processes. . . .

DISCUSSION QUESTIONS

1. In what way does Simmel liken the study of society to how organisms are studied?
2. What does Simmel mean when he says that a given number of individuals may be a society to a greater or lesser degree?

INFOTRAC COLLEGE EDITION

You can use your access to InfoTrac College Edition to learn more about the subjects covered in this essay. Some suggested search terms are:

Georg Simmel   organic metaphor
group formation   social interaction

17

The Presentation of Self in Everyday Life

ERVING GOFFMAN

Erving Goffman likens social interaction to a "con game," in which we are consistently trying to put forward a certain impression or "self" in order to get something from others. Although many will not see human behavior so cynically, Goffman's analysis sheds light on how people try to manage the impression that others have of them.

From: Erving Goffman. 1959. *The Presentation of the Self in Everyday Life*. Garden City, NY: Anchor Doubleday, pp. 17–27.

When an individual plays a part he implicitly requests his observers to take seriously the impression that is fostered before them. They are asked to believe that the character they see actually possesses the attributes he appears to possess, that the task he performs will have the consequences that are implicitly claimed for it, and that, in general, matters are what they appear to be. In line with this, there is the popular view that the individual offers his performance and puts on his show "for the benefit of other people." It will be convenient to begin a consideration of performances by turning the question around and looking at the individual's own belief in the impression of reality that he attempts to engender in those among whom he finds himself.

At one extreme, one finds that the performer can be fully taken in by his own act; he can be sincerely convinced that the impression of reality which he stages is the real reality. When his audience is also convinced in this way about the show he puts on—and this seems to be the typical case—then for the moment at least, only the sociologist or the socially disgruntled will have any doubts about the "realness" of what is presented.

At the other extreme, we find that the performer may not be taken in at all by his own routine. This possibility is understandable, since no one is in quite as good an observational position to see through the act as the person who puts it on. Coupled with this, the performer may be moved to guide the conviction of his audience only as a means to other ends, having no ultimate concern in the conception that they have of him or of the situation. When the individual has no belief in his own act and no ultimate concern with the beliefs of his audience, we may call him cynical, reserving the term "sincere" for individuals who believe in the impression fostered by their own performance. It should be understood that the cynic, with all his professional disinvolvement, may obtain unprofessional pleasures from his masquerade, experiencing a kind of gleeful spiritual aggression from the fact that he can toy at will with something his audience must take seriously.

It is not assumed, of course, that all cynical performers are interested in deluding their audiences for purposes of what is called "self-interest" or private gain. A cynical individual may delude his audience for what he considers to be their own good, or for the good of the community, etc. For illustrations of this we need not appeal to sadly enlightened showmen such as Marcus Aurelius or Hsun Tzŭ. We know that in service occupations practitioners who may otherwise be sincere are sometimes forced to delude their customers because their customers show such a heartfelt demand for it. Doctors who are led into giving placebos, filling station attendants who resignedly check and recheck tire pressures for anxious women motorists, shoe clerks who sell a shoe that fits but tell the customer it is the size she wants to hear—these are cynical performers whose audiences will not allow them to be sincere. Similarly, it seems that sympathetic patients in mental wards will sometimes feign bizarre symptoms so that student nurses will not be subjected to a disappointingly sane performance. So also, when inferiors extend their most lavish reception for visiting superiors, the selfish desire to win favor may not be the chief motive; the inferior may be tactfully attempting to put the superior at ease by simulating the kind of world the superior is thought to take for granted.

I have suggested two extremes: an individual may be taken in by his own act or be cynical about it. These extremes are something a little more than just the ends of a continuum. Each provides the individual with a position which has its own particular securities and defenses, so there will be a tendency for those who have traveled close to one of these poles to complete the voyage. Starting with lack of inward belief in one's role, the individual may follow the natural movement described by Park:

> It is probably no mere historical accident that the word person, in its first meaning, is a mask. It is rather a recognition of the fact that everyone is always and everywhere, more or less consciously, playing a role . . . It is in these roles that we know each other; it is in these roles that we know ourselves.[1]

In a sense, and in so far as this mask represents the conception we have formed of ourselves—the role we are striving to live up to—this mask is our truer self, the self we would like to be. In the end, our conception of our role becomes second nature and an integral part of our personality. We come into the world as individuals, achieve character, and become persons.[2] . . .

Front, then, is the expressive equipment of a standard kind intentionally or unwittingly employed by the individual during his performance. For preliminary purposes, it will be convenient to distinguish and label what seem to be the standard parts of front.

First, there is the "setting," involving furniture, décor, physical layout, and other background items which supply the scenery and stage props for the spate of human action played out before, within, or upon it. . . .

If we take the term "setting" to refer to the scenic parts of expressive equipment, one may take the term "personal front" to refer to the other items of expressive equipment, the items that we most intimately identify with the performer himself and that we naturally expect will follow the performer wherever he goes. As part of personal front we may include: insignia of office or rank; clothing; sex, age, and racial characteristics; size and looks; posture; speech patterns; facial expressions; bodily gestures; and the like. Some of these vehicles for conveying signs, such as racial characteristics, are relatively fixed and over a span of time do not vary for the individual from one situation to another. On the other hand, some of these sign vehicles are relatively mobile or transitory, such as facial expression, and can vary during a performance from one moment to the next. . . .

In addition to the fact that different routines may employ the same front, it is to be noted that a given social front tends to become institutionalized in terms of the abstract stereotyped expectations to which it gives rise, and tends to take on a meaning and stability apart from the specific tasks which happen at the time to be performed in its name. The front becomes a "collective representation" and a fact in its own right.

When an actor takes on an established social role, usually he finds that a particular front has already been established for it. Whether his acquisition of the role was primarily motivated by a desire to perform the given task or by a desire to maintain the corresponding front, the actor will find that he must do both.

Further, if the individual takes on a task that is not only new to him but also unestablished in the society, or if he attempts to change the light in which his task is viewed, he is likely to find that there are already several well-established fronts among which he must choose. Thus, when a task is given a new front we seldom find that the front it is given is itself new.

NOTES

1. Robert Ezra Park, *Race and Culture* (Glencoe, IL: The Free Press, 1950), p. 249.

2. *Ibid.,* p. 250.

DISCUSSION QUESTIONS

1. How many "selves" do you think you could "play" or "do" in order to accomplish something with another person? Discuss two such selves and try to get them to be quite different from each other.

2. "All the world's a stage," wrote William Shakespeare. So might Goffman have said this. How does his analysis of the presentation of self in everyday life suggest that life is a drama where we all play our parts?

INFOTRAC COLLEGE EDITION

You can use your access to InfoTrac College Edition to learn more about the subjects covered in this essay. Some suggested search terms include:

definition of the situation
dramatic metaphor

impression management
presentation of self

18

Code of the Street

ELIJAH ANDERSON

Elijah Anderson's study of interaction on the street shows the vast array of implicit "codes" of behavior or rules that guide street interaction. His analysis helps explain the complexity of street interaction and provides a sociological explanation of street violence.

In some of the most economically depressed and drug- and crime-ridden pockets of the city, the rules of civil law have been severely weakened, and in their stead a "code of the street" often holds sway. At the heart of this code is a set of prescriptions and proscriptions, or informal rules, of behavior organized around a desperate search for respect that governs public social relations, especially violence, among so many residents, particularly young men and women. Possession of respect—and the credible threat of vengeance—is highly valued for shielding the ordinary person from the interpersonal violence of the street. In this social context of persistent poverty and deprivation, alienation from broader society's institutions, notably that of criminal justice, is widespread. The code of the street emerges where the influence of the police ends and personal responsibility for one's safety is felt to begin, resulting in a kind of "people's law," based on "street justice." This code involves a quite primitive form of social exchange that holds would-be perpetrators accountable by promising an "eye for an eye," or a certain "payback" for transgressions. In service to this ethic, repeated displays of "nerve" and "heart" build or reinforce a credible reputation for vengeance that works to deter aggression and disrespect, which are sources of great anxiety on the inner-city street. . . .

In approaching the goal of painting an ethnographic picture of these phenomena, I engaged in participant-observation, including direct observation, and conducted in-depth interviews. Impressionistic materials were drawn from various social settings around the city, from some of the wealthiest to some of the most economically depressed, including carryouts, "stop and go" establishments, Laundromats, taverns, playgrounds, public schools, the Center City indoor mall known as the Gallery, jails, and public street corners. In these settings I encountered a wide variety of people—adolescent boys and young women (some incarcerated, some not), older men, teenage mothers, grandmothers, and male and female schoolteachers, black and white, drug dealers, and common criminals. To protect the privacy and confidentiality of my subjects, names and certain details have been disguised. . . .

From: Elijah Anderson. 1999. *Code of the Street.* New York: W.W. Norton, pp. 9–11, 32–34, 312–317. Reprinted with permission.

Of all the problems besetting the poor inner-city black community, none is more pressing than that of interpersonal violence and aggression. This phenomenon wreaks havoc daily on the lives of community residents and increasingly spills over into downtown and residential middle-class areas. Muggings, burglaries, carjackings, and drug-related shootings, all of which may leave their victims or innocent bystanders dead, are now common enough to concern all urban and many suburban residents.

The inclination to violence springs from the circumstances of life among the ghetto poor—the lack of jobs that pay a living wage, limited basic public services (police response in emergencies, building maintenance, trash pickup, lighting, and other services that middle-class neighborhoods take for granted), the stigma of race, the fallout from rampant drug use and drug trafficking, and the resulting alienation and absence of hope for the future. Simply living in such an environment places young people at special risk of falling victim to aggressive behavior. Although there are often forces in the community that can counteract the negative influences—by far the most powerful is a strong, loving, "decent" (as inner-city residents put it) family that is committed to middle-class values—the despair is pervasive enough to have spawned an oppositional culture, that of "the street," whose norms are often consciously opposed to those of mainstream society. These two orientations—decent and street—organize the community socially, and the way they coexist and interact has important consequences for its residents, particularly for children growing up in the inner city. Above all, this environment means that even youngsters whose home lives reflect mainstream values—and most of the homes in the community do—must be able to handle themselves in a street-oriented environment.

This is because the street culture has evolved a "code of the street," which amounts to a set of informal rules governing interpersonal public behavior, particularly violence. The rules prescribe both proper comportment and the proper way to respond if challenged. They regulate the use of violence and so supply a rationale allowing those who are inclined to aggression to precipitate violent encounters in an approved way. The rules have been established and are enforced mainly by the street-oriented; but on the streets the distinction between street and decent is often irrelevant. Everybody knows that if the rules are violated, there are penalties. Knowledge of the code is thus largely defensive, and it is literally necessary for operating in public. Therefore, though families with a decency orientation are usually opposed to the values of the code, they often reluctantly encourage their children's familiarity with it in order to enable them to negotiate the inner-city environment.

At the heart of the code is the issue of respect—loosely defined as being treated "right" or being granted one's "props" (or proper due) or the deference one deserves. However, in the troublesome public environment of the inner city, as people increasingly feel buffeted by forces beyond their control, what one deserves in the way of respect becomes ever more problematic and uncertain. This situation in turn further opens up the issue of respect to sometimes intense interpersonal negotiation, at times resulting in altercations. In the street culture, especially among young people, respect is viewed as almost an external entity, one that is hard-won but easily lost—and so must constantly be guarded. The rules of the code in fact provide a framework for negotiating respect. With the right amount of respect, individuals can avoid being bothered in public. This security is important, for if they *are* bothered, not only may they face physical danger, but they will have been disgraced or "dissed" (disrespected). Many of the forms dissing can take may seem petty to middle-class people (maintaining eye contact for too long, for example), but to those invested in the street code, these actions, a virtual slap in the face, become serious indications of the other person's intentions. Consequently, such people become very sensitive to advances and slights, which could well serve as a warning of imminent physical attack or confrontation.

The hard reality of the world of the street can be traced to the profound sense of alienation from mainstream society and its institutions felt by many poor inner-city black people, particularly the young. The code of the street is actually a cultural adaptation to a profound lack of faith in the police and the judicial system—and in others who would champion one's personal security. The police, for instance, are most often viewed as representing the dominant white society and as not caring to protect inner-city residents. When called, they may not respond, which is one reason many residents feel they must be prepared to take extraordinary measures to defend themselves and their loved ones against those who are inclined to aggrression. Lack of police accountability has in fact been incorporated into the local status system: the person who is believed capable of "taking care of himself" is accorded a certain deference and regard, which translates into a sense of physical and psychological control. The code of the street thus emerges where the influence of the police ends and where personal responsibility for one's safety is felt to begin. Exacerbated by the proliferation of drugs and easy access to guns, this volatile situation results in the ability of the street-oriented minority (or those who effectively "go for bad") to dominate the public spaces. . . .

The attitudes and actions of the wider society are deeply implicated in the code of the street. Most people residing in inner-city communities are not totally invested in the code; it is the significant minority of hard-core street youth who maintain the code in order to establish reputations that are integral to the extant social order. Because of the grinding poverty of the communities these people inhabit, many have—or feel they have—few other options for expressing themselves. For them the standards and rules of the street code are the only game in town.

And as was indicated above, the decent people may find themselves caught up in problematic situations simply by being at the wrong place at the wrong time, which is why a primary survival strategy of residents here is to "see but don't see." The extent to which some children—particularly those who through upbringing have become most alienated and those who lack strong and conventional social support—experience, feel, and internalize racist rejection and contempt from mainstream society may strongly encourage them to express contempt for the society in turn. In dealing with this contempt and rejection, some youngsters consciously invest themselves and their considerable mental resources in what amounts to an oppositional culture, a part of which is the code of the street. They do so to preserve themselves and their own self-respect. Once they do, any respect they might be able to garner in the wider system pales in comparison with the respect available in the local system; thus they often lose interest in even attempting to negotiate the mainstream system.

At the same time, many less alienated young people have assumed a street-oriented demeanor as way of expressing their blackness while really embracing a much more moderate way of life; they, too, want a nonviolent setting in which to live and one day possibly raise a family. These decent people are trying hard to be part of the mainstream culture, but the racism, real and perceived, that they encounter helps legitimate the oppositional culture and, by extension, the code of the street. On occasion they adopt street behavior; in fact, depending on the demands of the situation, many people attempt to codeswitch, moving back and forth between decent and street behavior. . . .

In addition, the community is composed of working-class and very poor people since those with the means to move away have done so, and there has also been a proliferation of single-parent households in which increasing numbers of kids are being raised on welfare. The result of all this is that the inner-city community has become a kind of urban village, apart from the wider society and limited in terms of resources and human capital. Young people growing up here often receive only the truncated version of mainstream society that comes from television and the perceptions of their peers. . . .

According to the code, the white man is a mysterious entity, a part of an enormous monolithic mass of arbitrary power, in whose view black people are insignificant. In this system and in the local social context, the black man has very little clout; to salvage something of value, he must outwit, deceive, oppose, and ultimately "end-run" the system.

Moreover, he cannot rely on this system to protect him; the responsibility is his, and he is on his own. If someone rolls on him, he has to put his body, and often his life, on the line. The physicality of manhood thus becomes extremely important. And urban brinksmanship is observed and learned as a matter of course. . . .

Urban areas have experienced profound structural economic changes, as deindustrialization—the movement from manufacturing to service and high-tech—and the growth of the global economy have created new economic conditions. Job opportunities increasingly go abroad to Singapore, Taiwan, India, and Mexico, and to nonmetropolitan America, to satellite cities like King of Prussia, Pennsylvania. Over the last fifteen years, for example, Philadelphia has lost 102,500 jobs, and its manufacturing employment has declined by 53 percent. Large numbers of inner-city people, in particular, are not adjusting effectively to the new economic reality. Whereas low-wage jobs—especially unskilled and low-skill factory jobs—used to exist simultaneously with poverty and there was hope for the future, now jobs simply do not exist, the present economic boom notwithstanding. These dislocations have left many inner-city people unable to earn a decent living. More must be done by both government and business to connect inner-city people with jobs.

The condition of these communities was produced not by moral turpitude but by economic forces that have undermined black, urban, working-class life and by a neglect of their consequences on the part of the public. Although it is true that persistent welfare dependency, teenage pregnancy, drug abuse, drug dealing, violence, and crime reinforce economic marginality, many of these behavioral problems originated in frustrations and the inability to thrive under conditions of economic dislocation. This in turn leads to a weakening of social and family structure, so children are increasingly not being socialized into mainstream values and behavior. In this context, people develop profound alienation and may not know what to do about an opportunity even when it presents itself. In other words, the social ills that the companies moving out of these neighborhoods today sometimes use to justify their exodus are the same ones that their corporate predecessors, by leaving, helped to create.

Any effort to place the blame solely on individuals in urban ghetos is seriously misguided. The focus should be on the socioeconomic structure, because it was structural change that caused jobs to decline and joblessness to increase in many of these communities. But the focus also belongs on the public policy that has radically threatened the well-being of many citizens. Moreover, residents of these communities lack good education, job training, and job networks, or connections with those who could help them get jobs. They need enlightened employers able to understand their predicament and willing to give them a chance. Government, which should be assisting people to adjust to the changed economy, is instead cutting what little help it does provide. . . .

The emergence of an underclass isolated in urban ghettos with high rates of joblessness can be traced to the interaction of race prejudice, discrimination, and the effects of the global economy. These factors have contributed to the profound social isolation and impoverishment of broad segments of the inner-city black population. Even though the wider society and economy have been experiencing accelerated prosperity for almost a decade, the fruits of it often miss the truly disadvantaged isolated in urban poverty pockets.

In their social isolation an oppositional culture, a subset of which is the code of the street, has been allowed to emerge, grow, and develop. This culture is essentially one of accommodation with the wider society, but different from past efforts to accommodate the system. A larger segment of people are now not simply isolated but ever more profoundly alienated from the wider society and its institutions. For instance, in conducting the fieldwork for this book, I visited numerous inner-city schools, including elementary, middle, and high schools, located in areas of concentrated poverty. In every one, the so-called oppositional culture was well entrenched. In one elementary school, I learned from interviewing kindergarten, first-grade, second-grade, and fourth-grade teachers that through the first grade, about a fifth of the students were invested in the code of the street; the rest are interested in the subject matter and eager to take instruction from the teachers—in effect, well disciplined. By the fourth grade, though, about three-quarters of the students have bought into the code of the street or the oppositional culture.

As I have indicated throughout this work, the code emerges from the school's impoverished neighborhood, including overwhelming numbers of single-parent homes, where the fathers, uncles, and older brothers are frequently incarcerated—so frequently, in fact, that the word "incarcerated" is a prominent part of the young child's spoken vocabulary. In such communities there is not only a high rate of crime but also a generalized diminution of respect for law. As the residents go about meeting the exigencies of public life, a kind of people's law results, . . . Typically, the local streets are, as was we saw, tough and dangerous places where people often feel very much on their own, where they themselves must be personally responsible for their own security, and where in order to be safe and to travel the public spaces unmolested, they must be able to show others that they are familiar with the code—that physical transgressions will be met in kind.

In these circumstances the dominant legal codes are not the first thing on one's mind; rather, personal security for self, family, and loved ones is. Adults, dividing themselves into categories of street and decent, often encourage their children in this adaptation to their situation, but at what price to the children and at what price to wider values of civility and decency? As the fortunes of the inner city continue to decline, the situation becomes ever more dismal and intractable. . . .

DISCUSSION QUESTIONS

1. List several ways that subtle or nonverbal behavior becomes important "on the street."
2. What specific ways does Anderson see street behavior as stemming from social structural conditions for African Americans?

INFOTRAC COLLEGE EDITION

You can use your access to InfoTrac College Edition to learn more about the subjects covered in this essay. Some suggested search terms include:

inner-city violence   street violence
oppositional culture   streetwise
police harassment

19

Identity in the Age of the Internet

SHERRY TURKLE

In a world increasingly influenced by technology and computing, human interaction itself may change. Sherry Turkle argues that technology permits us to put forth multiple selves as we interact in cyberspace, creating selves in new interactive contexts.

We come to see ourselves differently as we catch sight of our images in the mirror of the machine. A decade ago, when I first called the computer a second self, these identity-transforming relationships were almost always one-on-one, a person alone with a machine. This is no longer the case. A rapidly expanding system of networks, collectively known as the Internet, links millions of people in new spaces that are changing the way we think, the nature of our sexuality, the form of our communities, our very identities.

At one level, the computer is a tool. It helps us write, keep track of our accounts, and communicate with others. Beyond this, the computer offers us both new models of mind and a new medium on which to project our ideas and fantasies. Most recently, the computer has become even more than tool and mirror: We are able to step through the looking glass. We are learning to live in virtual worlds. We may find ourselves alone as we navigate virtual oceans, unravel virtual mysteries, and engineer virtual skyscrapers. But increasingly, when we step through the looking glass, other people are there as well.

The use of the term "cyberspace" to describe virtual worlds grew out of science fiction, but for many of us, cyberspace is now part of the routines of everyday life. When we read our electronic mail or send postings to an electronic bulletin board or make an airline reservation over a computer network, we are in cyberspace. In cyberspace, we can talk, exchange ideas, and assume personae of our own creation. We have the opportunity to build new kinds of communities, virtual communities, in which we participate with people from all over the world, people with whom we converse daily, people with whom we may have fairly intimate relationships but whom we may never physically meet.

This [essay] describes how a nascent culture of simulation is affecting our ideas about mind, body, self, and machine. We shall encounter virtual sex and

From: Sherry Turkle. 1997. *Life on the Screen*. New York: Schuster, pp. 9–26. Reprinted with permission.

cyberspace marriage, computer psychotherapists, robot insects, and researchers who are trying to build artificial two-year-olds. Biological children, too, are in the story as their play with computer toys leads them to speculate about whether computers are smart and what it is to be alive. Indeed, in much of this, it is our children who are leading the way, and adults who are anxiously trailing behind.

In the story of constructing identity in the culture of simulation, experiences on the Internet figure prominently, but these experiences can only be understood as part of a larger cultural context. That context is the story of the eroding boundaries between the real and the virtual, the animate and the inanimate, the unitary and the multiple self, which is occurring both in advanced scientific fields of research and in the patterns of everyday life. From scientists trying to create artificial life to children "morphing" through a series of virtual personae, we shall see evidence of fundamental shifts in the way we create and experience human identity. But it is on the Internet that our confrontations with technology as it collides with our sense of human identity are fresh, even raw. In the real-time communities of cyberspace, we are dwellers on the threshold between the real and virtual, unsure of our footing, inventing ourselves as we go along.

In an interactive, text-based computer game designed to represent a world inspired by the television series *Star Trek: The Next Generation,* thousands of players spend up to eighty hours a week participating in intergalactic exploration and wars. Through typed descriptions and typed commands, they create characters who have casual and romantic sexual encounters, hold jobs and collect paychecks, attend rituals and celebrations, fall in love and get married. To the participants, such goings-on can be gripping; "This is more real than my real life," says a character who turns out to be a man playing a woman who is pretending to be a man. In this game the self is constructed and the rules of social interaction are built, not received.

In another text-based game, each of nearly ten thousand players creates a character or several characters, specifying their genders and other physical and psychological attributes. The characters need not be human and there are more than two genders. Players are invited to help build the computer world itself. Using a relatively simple programming language, they can create a room in the game space where they are able to set the stage and define the rules. They can fill the room with objects and specify how they work; they can, for instance, create a virtual dog that barks if one types the command "bark Rover." An eleven-year-old player built a room she calls the condo. It is beautifully furnished. She has created magical jewelry and makeup for her dressing table. When she visits the condo, she invites her cyberfriends to join her there, she chats, orders a virtual pizza, and flirts. . . .

FROM A CULTURE OF CALCULATION TOWARD A CULTURE OF SIMULATION

Most people over thirty years old (and even many younger ones) have had an introduction to computers similar to the one I received in [a] programming course. But from today's perspective, the fundamental lessons of computing that I was taught are wrong. First of all, programming is no longer cut and dried. Indeed, even its dimensions have become elusive. Are you programming when you customize your word processing software? When you design "organisms" to populate a simulation of Darwinian evolution in a computer game called Sim-Life? Or when you build a room in a MUD so that opening a door to it will cause "Happy UnBirthday" to ring out on all but one day of the year? In a sense, these activities are forms of programming, but that sense is radically different from the one presented in my 1978 computer course.

The lessons of computing today have little to do with calculation and rules; instead they concern simulation, navigation, and interaction. The very image of the computer as a giant calculator has become quaint and dated. Of course, there is still "calculation" going on within the computer, but it is no longer the important or interesting level to think about or interact with. Fifteen years ago, most computer users were limited to typing commands. Today they use off-the-shelf products to manipulate simulated desktops, draw with simulated paints and brushes, and fly in simulated airplane cockpits. The computer culture's center of gravity has shifted decisively to people who do not think of themselves as programmers. The computer science research community as well as industry pundits maintain that in the near future we can expect to interact with computers by communicating with simulated people on our screens, agents who will help organize our personal and professional lives. . . .

The meaning of the computer presence in people's lives is very different from what most expected in the late 1970s. One way to describe what has happened is to say that we are moving from a modernist culture of calculation toward a postmodernist culture of simulation.

The culture of simulation is emerging in many domains. It is affecting our understanding of our minds and our bodies. For example, fifteen years ago, the computational models of mind that dominated academic psychology were modernist in spirit: Nearly all tried to describe the mind in terms of centralized structures and programmed rules. In contrast, today's models often embrace a postmodern aesthetic of complexity and decentering. Mainstream computer researchers no longer aspire to program intelligence into computers but expect intelligence to emerge from the interactions of small subprograms. If these emergent simulations are "opaque," that is, too complex to be completely analyzed, this is not necessarily a problem. After all, these theorists say, our brains are opaque to us, but this has never prevented them from functioning perfectly well as minds.

Fifteen years ago in popular culture, people were just getting used to the idea that computers could project and extend a person's intellect. Today people are embracing the notion that computers may extend an individual's physical presence. Some people use computers to extend their physical presence via real-time video links and shared virtual conference rooms. Some use computer-mediated screen communication for sexual encounters. An Internet list of "Frequently Asked Questions" describes the latter activity—known as netsex, cybersex, and (in MUDS) TinySex—as people typing messages with erotic content to each other, "sometimes with one hand on the keyset, sometimes with two." . . .

Sexual encounters in cyberspace are only one (albeit well-publicized) element of our new lives on the screen. Virtual communities ranging from MUDs to computer bulletin boards allow people to generate experiences, relationships, identities, and living spaces that arise only through interaction with technology. In the many thousands of hours that Mike, a college freshman in Kansas, has been logged on to his favorite MUD, he has created an apartment with rooms, furniture, books, desk, and even a small computer. Its interior is exquisitely detailed, even though it exists only in textual description. A hearth, an easy chair, and a mahogany desk warm his cyberspace. "It's where I live," Mike says. "More than I do in my dingy dorm room. There's no place like home."

As human beings become increasingly intertwined with the technology and with each other via the technology, old distinctions between what is specifically human and specifically technological become more complex. Are we living life *on* the screen or life *in* the screen? Our new technologically enmeshed relationships oblige us to ask to what extent we ourselves have become cyborgs, transgressive mixtures of biology, technology, and code. The traditional distance between people and machines has become harder to maintain. . . .

. . . Along with the movement from a culture of calculation toward a culture of simulation have come changes in what computers do *for* us and in what they do *to* us—to our relationships and our ways of thinking about ourselves. . . .

It is computer screens where we project ourselves into our own dramas, dramas in which we are producer, director, and star. Some of these dramas are private, but increasingly we are able to draw in other people. Computer screens are the new location for our fantasies, both erotic and intellectual. We are using life on computer screens to become comfortable with new ways of thinking about evolution, relationships, sexuality, politics, and identity.

DISCUSSION QUESTIONS

1. What does Turkle mean that computers can create identity-transforming relationships?
2. What is the *culture of simulation* and how does it affect the interaction humans have with each other? With technology?

INFOTRAC COLLEGE EDITION

You can use your access to InfoTrac College Edition to learn more about the subjects covered in this essay. Some suggested search terms include:

automation	cybernation
culture of calculation	cyberspace communities
culture of simulation	cyberspace interaction

20

Primary Groups

CHARLES HORTEN COOLEY

Groups differ in a number of ways. Yet, all groups involve people interacting with each other, although in greatly different contexts. Informal primary groups, such as families or peer groups, allow us to accomplish certain things as the classical theorist Charles Horton Cooley shows. He contrasts these with secondary groups with more formal modes of interaction and less personalized connections.

By primary groups I mean those characterized by intimate face-to-face association and coöperation. They are primary in several senses, but chiefly in that they are fundamental in forming the social nature and ideals of the individual. The result of intimate association, psychologically, is a certain fusion of individualities in a common whole, so that one's very self, for many purposes at least, is the common life and purpose of the group. Perhaps the simplest way of describing this wholeness is by saying that it is a "we"; it involves the sort of sympathy and mutual identification for which "we" is the natural expression. One lives in the feeling of the whole and finds the chief aims of his will in that feeling.

It is not to be supposed that the unity of the primary group is one of mere harmony and love. It is always a differentiated and usually a competitive unity, admitting of self-assertion and various appropriative passions; but these passions are socialized by sympathy, and come, or tend to come, under the discipline of a common spirit. The individual will be ambitious, but the chief object of his ambition will be some desired place in the thought of the others, and he will feel allegiance to common standards of service and fair play. So the boy will dispute with his fellows a place on the team, but above such disputes will place the common glory of his class and school.

The most important spheres of this intimate association and coöperation—though by no means the only ones—are the family, the play-group of children, and the neighborhood or community group of elders. These are practically universal, belonging to all times and all stages of development; and are accordingly a chief basis of what is universal in human nature and human ideals. The best comparative studies of the family . . . show it to us as not only a universal institution, but as more alike the world over than the exaggeration of exceptional customs by an earlier school had led us to suppose. Nor can any one doubt the general prevalence of play-groups among children or of informal assemblies of

From: Charles Horten Cooley, 1962 [1909]. *Social Organization: A Study of the Larger Mind.*
New York: Schocken Books, pp. 23–31.

various kinds among their elders. Such association is clearly the nursery of human nature in the world about us, and there is no apparent reason to suppose that the case has anywhere or at any time been essentially different.

As regards play, I might, were it not a matter of common observation, multiply illustrations of the universality and spontaneity of the group discussion and coöperation to which it gives rise. The general fact is that children, especially boys after about their twelfth year, live in fellowships in which their sympathy, ambition and honor are engaged even more, often, than they are in the family. Most of us can recall examples of the endurance by boys of injustice and even cruelty, rather than appeal from their fellows to parents or teachers—as, for instance, in the hazing so prevalent at schools, and so difficult, for this very reason, to repress. And how elaborate the discussion, how cogent the public opinion, how hot the ambitions in these fellowships.

Nor is this facility of juvenile association, as is sometimes supposed, a trait peculiar to English and American boys; since experience among our immigrant population seems to show that the offspring of the more restrictive civilizations of the continent of Europe form self-governing play-groups with almost equal readiness. Thus Miss Jane Addams, after pointing out that the "gang" is almost universal, speaks of the interminable discussion which every detail of the gang's activity receives, remarking that "in these social folk-motes, so to speak, the young citizen learns to act upon his own determination."

Of the neighborhood group it may be said, in general, that from the time men formed permanent settlements upon the land, down, at least, to the rise of modern industrial cities, it has played a main part in the primary, heart-to-heart life of the people. Among our Teutonic forefathers the village community was apparently the chief sphere of sympathy and mutual aid for the commons all through the "dark" and middle ages, and for many purposes it remains so in rural districts at the present day. In some countries we still find it with all its ancient vitality, notably in Russia, where the mir, or self-governing village group, is the main theatre of life, along with the family, for perhaps fifty millions of peasants.

In our own life the intimacy of the neighborhood has been broken up by the growth of an intricate mesh of wider contacts which leaves us strangers to people who live in the same house. And even in the country the same principle is at work, though less obviously, diminishing our economic and spiritual community with our neighbors. How far this change is a healthy development, and how far a disease, is perhaps still uncertain.

Besides these almost universal kinds of primary association, there are many others whose form depends upon the particular state of civilization; the only essential thing, as I have said, being a certain intimacy and fusion of personalities. In our own society, being little bound by place, people easily form clubs, fraternal societies and the like, based on congeniality, which may give rise to real intimacy. Many such relations are formed at school and college, and among men and women brought together in the first instance by their occupations—as workmen in the same trade, or the like. Where there is a little common interest and activity, kindness grows like weeds by the roadside.

But the fact that the family and neighborhood groups are ascendant in the open and plastic time of childhood makes them even now incomparably more influential than all the rest.

Primary groups are primary in the sense that they give the individual his earliest and completest experience of social unity, and also in the sense that they do not change in the same degree as more elaborate relations, but form a comparatively permanent source out of which the latter are ever springing. Of course they are not independent of the larger society, but to some extent reflect its spirit. . . .

The view here maintained is that human nature is not something existing separately in the individual, but a *group-nature or primary phase of society,* a relatively simple and general condition of the social mind. It is something more, on the one hand, than the mere instinct that is born in us—though that enters into it—and something less, on the other, than the more elaborate development of ideas and sentiments that makes up institutions. It is the nature which is developed and expressed in those simple, face-to-face groups that are somewhat alike in all societies; groups of the family, the playground, and the neighborhood. In the essential similarity of these is to be found the basis, in experience, for similar ideas and sentiments in the human mind. In these, everywhere, human nature comes into existence. Man does not have it at birth; he cannot acquire it except through fellowship, and it decays in isolation.

If this view does not recommend itself to common sense I do not know that elaboration will be of much avail. It simply means the application at this point of the idea that society and individuals are inseparable phases of a common whole, so that wherever we find an individual fact we may look for a social fact to go with it. If there is a universal nature in persons there must be something universal in association to correspond to it.

What else can human nature be than a trait of primary groups? Surely not an attribute of the separate individual—supposing there were any such thing—since its typical characteristics, such as affection, ambition, vanity, and resentment, are inconceivable apart from society. If it belongs, then, to man in association, what kind or degree of association is required to develop it? Evidently nothing elaborate, because elaborate phases of society are transient and diverse, while human nature is comparatively stable and universal. In short the family and neighborhood life is essential to its genesis and nothing more is.

Here as everywhere in the study of society we must learn to see mankind in psychical wholes, rather than in artificial separation. We must see and feel the communal life of family and local groups as immediate facts, not as combinations of something else. And perhaps we shall do this best by recalling our own experience and extending it through sympathetic observation. What, in our life, is the family and the fellowship; what do we know of the we-feeling? Thoughts of this kind may help us to get a concrete perception of that primary group-nature of which everything social is the outgrowth.

DISCUSSION QUESTIONS

1. What needs of yours do primary groups meet? How would these needs not be met by a secondary group?
2. What role do primary groups play in modern society?

INFOTRAC COLLEGE EDITION

You can use your access to InfoTrac College Edition to learn more about the subjects covered in this essay. Some suggested search terms include:

Charles Horton Cooley
Chicago School of Sociology

face-to-face interaction
intimacy

21

Bureaucracy

MAX WEBER

Large formal organizations are a form of group behavior based on impersonal ties and bound by rules and regulations. Max Weber, an early social theorist, provides the classic statement on the organization and functioning of bureaucracies, one of the most formalized kinds of groups in society.

Modern officialdom functions in the following specific manner:

I. There is the principle of fixed and official jurisdictional areas, which are generally ordered by rules, that is, by laws or administrative regulations.

1. The regular activities required for the purposes of the bureaucratically governed structure are distributed in a fixed way as official duties.
2. The authority to give the commands required for the discharge of these duties is distributed in a stable way and is strictly delimited by rules concerning the coercive means, physical, sacerdotal, or otherwise, which may be placed at the disposal of officials.

From: H. H. Gerth and C. Wright Mills. 1946. *Max Weber: Essays in Sociology,* New York: Oxford University Press, pp. 196–244.

3. Methodical provision is made for the regular and continuous fulfillment of these duties and for the execution of the corresponding rights; only persons who have the generally regulated qualifications to serve are employed.

In public and lawful government these three elements constitute 'bureaucratic authority.' In private economic domination, they constitute bureaucratic 'management.' Bureaucracy, thus understood, is fully developed in political and ecclesiastical communities only in the modern state, and, in the private economy, only in the most advanced institutions of capitalism. Permanent and public office authority, with fixed jurisdiction, is not the historical rule but rather the exception. This is so even in large political structures such as those of the ancient Orient, the Germanic and Mongolian empires of conquest, or of many feudal structures of state. In all these cases, the ruler executes the most important measures through personal trustees, table-companions, or court-servants. Their commissions and authority are not precisely delimited and are temporarily called into being for each case.

II. The principles of office hierarchy and of levels of graded authority mean a firmly ordered system of super- and subordination in which there is a supervision of the lower offices by the higher ones. Such a system offers the governed the possibility of appealing the decision of a lower office to its higher authority, in a definitely regulated manner. With the full development of the bureaucratic type, the office hierarchy is monocratically organized. The principle of hierarchical office authority is found in all bureaucratic structures: in state and ecclesiastical structures as well as in large party organizations and private enterprises. It does not matter for the character of bureaucracy whether its authority is called 'private' or 'public.'

When the principle of jurisdictional 'competency' is fully carried through, hierarchical subordination—at least in public office—does not mean that the 'higher' authority is simply authorized to take over the business of the 'lower.' Indeed, the opposite is the rule. Once established and having fulfilled its task, an office tends to continue in existence and be held by another incumbent.

III. The management of the modern office is based upon written documents ('the files'), which are preserved in their original or draught form. There is, therefore, a staff of subaltern officials and scribes of all sorts. The body of officials actively engaged in a 'public' office, along with the respective apparatus of material implements and the files, make up a 'bureau.' In private enterprise, 'the bureau' is often called 'the office.' . . .

IV. Office management, at least all specialized office management—and such management is distinctly modern—usually presupposes thorough and expert training. This increasingly holds for the modern executive and employee of private enterprises, in the same manner as it holds for the state official.

V. When the office is fully developed, official activity demands the full working capacity of the official, irrespective of the fact that his obligatory time in the bureau may be firmly delimited. In the normal case, this is only the product of a long development, in the public as well as in the private office. Formerly, in all cases, the normal state of affairs was reversed: official business was discharged as a secondary activity.

VI. The management of the office follows general rules, which are more or less stable, more or less exhaustive, and which can be learned. Knowledge of these rules represents a special technical learning which the officials possess. It involves jurisprudence, or administrative or business management.

The reduction of modern office management to rules is deeply embedded in its very nature. The theory of modern public administration, for instance, assumes that the authority to order certain matters by decree—which has been legally granted to public authorities—does not entitle the bureau to regulate the matter by commands given for each case, but only to regulate the matter abstractly. This stands in extreme contrast to the regulation of all relationships through individual privileges and bestowals of favor, which is absolutely dominant in patrimonialism, at least in so far as such relationships are not fixed by sacred tradition. . . .

DISCUSSION QUESTIONS

1. Describe one of your recent encounters in a large bureaucracy. How does this encounter illustrate the characteristics of bureaucracy that Weber analyzes?
2. What social needs does a bureaucracy fill? How does this differ from other kinds of groups?

INFOTRAC COLLEGE EDITION

You can use your access to InfoTrac College Edition to learn more about the subjects covered in this essay. Some suggested search terms include:

bureaucratization
mass society and bureaucracy

organizational behavior
Taylorism

22

Enchanting a Disenchanted World

GEORGE RITZER

George Ritzer sees modern-day extensions and manifestations in the proliferation of "cathedrals of consumption" that he finds in society. These locations offer evidence of the increasing rationalization of social behavior and the strong place of consumerism in the character of contemporary social institutions.

As a consumer, you can do many things today that you could not do a few decades ago, including

- Shopping in an immense, brightly lit, colorful mall with several hundred shops many of them part of well-known chains and with literally millions of goods and services from which to choose.

- Spending a day or more in an even bigger and more dazzling megamall that encompasses not only shops but also an amusement park.

- Gambling the day (and night!) away in a casino that not only is an enormous hotel but also includes a shopping mall and an amusement park on the grounds of the complex.

- Whiling away a week on a deluxe, 100,000-ton cruise ship that offers the expanse of the sea and the beauty of tropical islands and also hotel-like facilities, a casino, a mall, a health spa, amusements, and many other places to spend money.

- Eating in a "themed" restaurant, part of an upscale chain, where the setting, the accoutrements, the staff, and the food bring to mind a tropical rain forest or the world of rock music.

- Vacationing at a theme park where the restaurants and everything else—attractions, customers, employees' utterances—portray such themes as life in the future, life in other parts of the world, or animals of the world.

. . .A revolutionary change has occurred in the places in which we consume goods and services, and it has had a profound effect not only on the nature of consumption but also on social life.

From: George Ritzer. 1999. *Enchanting a Disenchanting World: Revolutionizing The Means of Consumption*. Thousand Oaks, CA: Pine Forge Press, pp. ix–xi, 2–11.

One of the concepts used to describe the settings of concern in this [essay] is *means of consumption.* These settings, as means, allow us to consume a wide range of goods and services. But as you will see, these places do more than simply permit us to consume things; they are structured to lead and even coerce us into consumption.

Another important concept . . . is *cathedrals of consumption,* which points up the quasi-religious, "enchanted" nature of these new settings. They have become locales to which we make "pilgrimages" in order to practice our consumer religion.

These new means of consumption have helped to entice us to consume far more than we ever did in the past; we have been led in the direction of hyper-consumption. We are also being led to consume differently than we did in the past. For instance, we are now more likely to consume alone, to purchase many different kinds of goods and services in one locale, and to buy many of the same things and consume in many of the same kinds of places as most other people. In fact, because our cathedrals of consumption are actively being exported to the rest of the world, people in many other countries consume as actively as we do and obtain many of the same goods and services.

The new means of consumption have been so successful with the public that other social settings—ballparks, universities, hospitals, museums, churches, and the like—have begun to look quite dated and stodgy. Thus, those who control these other types of settings are rushing to emulate the cathedrals of consumption. For example, universities increasingly have "theme" dorms, fast-food restaurants, souvenir shops, and video arcades. Students and parents are more likely than ever to be treated like consumers who need to be lured to a particular university and "enchanted" so they will stay there for the duration of the college years. Similarly, a modern baseball stadium offers far more than just a baseball game. Sometimes the game seems like a minor distraction from the other attractions on the stadium grounds: theme restaurants, food courts, swimming pools, hotels, exploding computerized scoreboards, and numerous souvenir shops. We are more likely to be treated as consumers than as fans in such a setting; in turn, we are more likely to view baseball stadiums in the same way we view shopping malls and theme parks.

The result of these changes is that consumption pervades our lives; we are increasingly consumed by consumption. We frequently find ourselves in settings devoted to consumption. Even our homes have become means of consumption, penetrated by telemarketing, junk mail, catalogs, home shopping television, and cybershops. We get little respite from the pressure to consume.

Of course, most of us are enthusiastically a part of the consumer society and are eager to visit the cathedrals of consumption and to obtain what they have to offer. We frequent those settings that offer the greatest spectacle. But we easily grow bored. Thus consumption settings, from fast-food restaurants to Las Vegas casino-hotels, compete to outdo one another and to see which one can put on the greatest show. The result is increasingly spectacular displays and a continual escalation of efforts to lure consumers.

Ironically, when these efforts at enchantment are successful, they draw large numbers of consumers necessitating rationalization and bureaucratization. Thus the cathedrals of consumption grow increasingly disenchanting to the jaded consumers they have been designed to attract. . . .

It should come as no surprise to you that this new world of consumption exists. Many of you spend a great deal of time in such settings. However, you may not have recognized how closely consumer enchantment and disenchantment are related, how the principles used to design cathedrals of consumption are being employed in locales that we do not often think of as devoted to consumption, and how these principles are helping to create a social world increasingly dominated by consumption.

. . . Consumption plays an ever-expanding role in the lives of individuals around the world. To some, consumption defines contemporary American society, as well as much of the rest of the developed world. We consume many obvious things—fast food, t-shirts, a day at Walt Disney World®—and many others that are not so obvious—a lecture, medical service, a day at the ballpark. We consume many goods and services that we must have in order to live and many more that we simply have come to want. Often we must go to particular settings to obtain these goods and services (although, as we will see, more and more of them are coming to us). This [essay] is concerned with those settings: shopping malls, cybermalls, fast food restaurants, theme parks, and cruise ships, to name a few. We will be concerned with the issue of why we go to some places rather than others, and we will also deal with the ways in which these settings contribute to the high and increasing level of consumption that characterizes society today.

Unlike many treatments of the subject of consumption, this is *not* . . . about the consumer or the increasing profusion of goods and services. Rather, it is about the almost dizzying proliferation of settings that allow, encourage, and even compel us to consume so many of those goods and services. The settings of interest will be termed the *new means of consumption*. These are, in the main, settings that have come into existence or taken new forms since the end of World War II and that, building on but going beyond earlier settings, have dramatically transformed the nature of consumption. Because of important continuities, it is not always easy to clearly distinguish between new and older means of consumption, but it is the more contemporary versions, singly and collectively, that will concern us. . . .

THE NEW MEANS OF CONSUMPTION

Disney World is of interest to us because it represents a model of a new *means of consumption,* or in other words the settings or structures that enable us to consume all sorts of things. As a new means of consumption, Disney World has many continuities with older settings, as do many of the other new means of consumption. The predecessors to today's cruise lines were the great ocean liners of the past. Las Vegas casinos had precursors such as the great casino at Monte Carlo. Shopping malls can be traced back to the markets of ancient Greece and Rome.

At the same time, these new means also exhibit a number of important and demonstrable differences.

The means of consumption are part of a broader set of phenomena related to goods and services: production, distribution advertising, marketing, sales, individual taste, style, fashion. Our concern is with the process leading up to, and perhaps including, an exchange of money (or equivalents such as checks, electronic debits to bank or credit card accounts, and so on) for goods or services between buyers and sellers. This is often dealt with under the rubric of shopping, but our interests are broader than that and include the consumer's relationship with not only shops and malls but also theme parks, casinos, and cruise lines, and other settings including athletic stadiums, universities, hospitals, and museums, which surprisingly are coming to resemble the more obvious new means of consumption. In most cases an exchange occurs; people do purchase and receive goods and services. This process may take place instantaneously or over a long period of time and may involve many steps—perception of a want, arousal of a desire by an advertisement, study of available literature (e.g., *Consumer Reports*), comparison of available options, and ultimately perhaps an actual purchase. Of course, the process need not stop there; it is not unusual for people to take things home, find them wanting, return them, and perhaps begin the process anew.

Many of the new settings have attracted a great deal of attention individually, but little has been said about them collectively. I undertake to analyze them not only because of their growing importance and inherent interest, but also because they have played a central role in greatly increasing, and dramatically altering the nature of, consumption. Americans, especially, consume very differently and we consume on a larger scale, at least in part because of the new means of consumption. Further, these settings are important beyond their role in the consumption process. Many of the settings considered here—for example, McDonald's®, Wal-Mart®, Disney—have become some of America's, and the world's, most powerful popular icons. My net is cast more widely than even Disney's reach to include discount malls, superstores, warehouse stores, Las Vegas casinos (which are increasingly Disneyesque), and so on.

. . . In addition to the inherent interest of a major, often spectacular, social change in the realm of consumption, the new means of consumption are involved in a variety to developments that are designed to get more of us to spend time and money in consumption. Admittedly, most of us are eager to spend money in these settings. Others may feel that they are devoting too much time and money to consuming in these settings. In either case, it makes sense to understand the ways in which those in charge of the new means of consumption are tempting us.

CATHEDRALS OF CONSUMPTION

The new means of consumption can be seen as *"cathedrals of consumption"*—that is, they have an enchanted, sometimes even sacred, religious character for many people. In order to attract ever-larger numbers of consumers, such cathedrals of consumption need to offer, or at least appear to offer, increasingly magical, fantastic, and enchanted settings in which to consume. Sometimes this magic is produced quite intentionally, whereas in other cases it is a result of a series of largely unforeseen developments. A worker involved in the opening of McDonald's in Moscow spoke of it "as if it were the Cathedral in Chartres . . . a place to experience 'celestial joy.'" A visit to Disney World has been depicted as "the middle-class hajj, the compulsory visit to the sunbaked city," and analogies have been drawn between a trip to Disney World and pilgrimages to religious sites such as Lourdes. Book superstores such as Barnes and Noble® and Borders® have been called "cathedrals for the printed word."

Shopping malls have been described as places where people go to practice their "consumer religion." It has been contended that shopping malls are more than commercial and financial enterprises; they have much in common with the religious centers of traditional civilizations. Like such religious centers, malls are seen as fulfilling peoples' need to connect with each other and with nature (trees, plants, flowers), as well as their need to participate in festivals. Malls provide the kind of centeredness traditionally provided by religious temples, and they are constructed to have similar balance, symmetry, and order. Their atriums usually offer connection to nature through water and vegetation. People gain a sense of community as well as more specific community services. Play is almost universally part of religious practice, and malls provide a place for people to frolic. Similarly, malls offer a setting in which people can partake in ceremonial meals. Malls clearly qualify for the label of cathedrals of consumption.

As is the case with religious cathedrals, the cathedrals of consumption are not only enchanted, they are also *highly rationalized*. As they attract more and more consumers, their enchantment must be reproduced over and over on demand. Furthermore, branches of the successful enchanted settings are opened across the nation and even the world with the result that essentially the same magic must be reproduced in a wide range of locations. To accomplish this, the magic has to be systematized so that it can be easily recreated from one time or place to another. However, it is difficult to reduce magic to corporate formulas that can be routinely employed at any time, in any place, and by anybody. Yet, if these corporations are to continue to attract increasing numbers of consumers who will spend more and more money on goods and services, that is just what they must be able to do. Although such rational, machine-like structures can have their enchanting qualities (food appears almost instantaneously, goods exist in unbelievable profusion), they are, in the main, disenchanting; they often end up *not* being very magical. There is a tendency for people to become bored and to be put off by too much machine-like efficiency in the settings in which they consume. The challenge for today's cathedrals of consumption (as for religious cathedrals), is *how to maintain enchantment in the face of increasing rationalization.*

Although the new means of consumption [are] described in terms of rationalization and enchantment (as well as disenchantment), it is important to recognize that they are not all equally rationalized or enchanting. Some are able to operate in a more machine-like manner than others. Similarly, some are able to take on a more enchanted quality than others. Disney World, a Las Vegas casino, or a huge cruise ship seem far more enchanted than the local McDonald's, Wal-Mart, or strip mall. In addition, specific settings may enchant some consumers and not others. For example, fast food restaurants and theme parks may enchant children far more than adults, although adults may be led by their children or grandchildren to participate—and to pay. Furthermore, enchantment tends to be something that declines over time as the novelty for consumers wears off. After nearly a half century of existence in the United States and proliferation into every nook and cranny of the nation, modern fast food restaurants offer very little enchantment to most adult American consumers. However, we should not forget that many adults found such restaurants quite enchanting when they first opened in the United States and they still do in other nations and cultures to which fast food outlets are relatively new arrivals. In sum, although we . . . describe the new means of consumption in terms of rationalization and enchantment, there is considerable variation among them, and over time, in their degree of rationalization and enchantment. . . .

DISCUSSION QUESTIONS

1. Give four specific examples of "cathedrals of consumption" that you have visited during the past year. What impact do they have on your life?
2. Ritzer argues that cathedrals of consumption are an increasingly common part of the American landscape. Identify the places in your community where you see these. How do they affect the overall character of the community?

INFOTRAC COLLEGE EDITION

You can use your access to InfoTrac College Edition to learn more about the subjects covered in this essay. Some suggested search terms include:

consumerism   McDonaldization
malling of America   shopping malls

23

The Gender of Sexuality

PEPPER SCHWARTZ AND VIRGINIA RUTTER

In this essay, the authors attempt to clarify the relationship of gender to sexuality. Specifically, they outline the social influences on sexual desire. They argue that sexuality, sexual orientation, and sexual identity are not simply about attraction to and desire for a biological man or woman. Instead, this essay points to the social construction of femininity and masculinity and the importance of social processes in determining sexual desire.

The gender of the person you desire is a serious matter seemingly fundamental to the whole business of romance. And it isn't simply a matter of whether someone is male or female; how well the person fulfills a lover's expectations of masculinity or femininity is of great consequence, as two examples from the movies illustrate.

In the movie *The Truth about Cats & Dogs* (1996), a man (Brian, played by Ben Chaplin) falls in love with a woman (Abby, played by Janeane Garofalo) over the phone, and she with him. They find each other warm, clever, charming, and intriguing. But she, thinking herself too plain asks her beautiful friend (Noelle, played by Uma Thurman) to impersonate her when the man and woman are scheduled to meet. The movie continues as a comedy of errors. Although the man becomes very confused about which woman he really desires, in the end the telephone lovers are united. The match depended on social matters far more than physical matters.

In the British drama *The Crying Game* (1994), Fergus, an Irish Republican Army underling, meets and falls in love with the lover (Dil) of Jody, a British soldier whom Fergus befriended prior to being ordered to execute him. The movie was about passionate love, war, betrayal, and, in the end, loyalty and commitment. Fergus seeks out Jody's girlfriend in London out of guilt and curiosity. But Fergus's guilt over Jody's death turns into love, and the pair become romantically and sexually involved. In the end, although Fergus is jailed for terrorist activities, Fergus and Dil have solidified their bond and are committed and, it seems, in love. The story of sexual conquest and love is familiar, but this particular story grabbed imaginations because of a single, crucial detail. Jody's girlfriend Dil, Fergus discovers, turns out to be (physically) a man. Although Fergus is horrified when he discovers his lover is biologically different from what he had expected, in the end their relationship survives.

These movies raise an interesting point about sexual desire. Although sex is experienced as one of the most basic and biological of activities, in human beings

From: Pepper Schwartz and Virginia Rutter. 1998. *The Gender of Sexuality.* Thousand Oaks, CA: Pine Forge Press, pp. 2–27. Reprinted with permission.

it is profoundly affected by things other than the body's urges. Who we're attracted to and what we find sexually satisfying is not just a matter of the genital equipment we're born with. . . .

On one level, sex can be regarded as having both a biological and a social context. The biological (and physiological) refers to how people use their genital equipment to reproduce. In addition, as simple as it seems, bodies make the experience of sexual pleasure available—whether the pleasure involves other bodies or just one's own body and mind. It should be obvious, however, that people engage in sex even when they do not intend to reproduce. They have sex for fun, as a way to communicate their feelings to each other, as a way to satisfy their ego, and for any number of other reasons relating to the way they see themselves and interact with others.

Another dimension of sex involves both what we do and how we think about it. **Sexual behavior** refers to the sexual acts that people engage in. These acts involve not only petting and intercourse but also seduction and courtship. Sexual behavior also involves the things people do alone for pleasure and stimulation and the things they do with other people. **Sexual desire,** on the other hand, is the motivation to engage in sexual acts. It relates to what turns people on. A person's **sexuality** consists of both behavior and desire.

The most significant dimension of sexuality is **gender.** Gender relates both to the biological and social contexts of sexual behavior and desire. People tend to believe they know whether someone is a man or a woman not because we do a physical examination and determine that the person is biologically male or biologically female. Instead, we notice whether a person is masculine or feminine. Gender is a social characteristic of individuals in our society that is only sometimes consistent with biological sex. Thus, animals, like people, tend to be identified as male and female in accordance with the reproductive function, but only people are described by their gender, as a man or a woman.

When we say something is **gendered** we mean that social processes have determined what is appropriately masculine and feminine and that gender has thereby become integral to the definition of the phenomenon. For example, marriage is a gendered institution: The definition of marriage involves a masculine part (husband) and a feminine part (wife). Gendered phenomena, like marriage, tend to appear "naturally" so. But, as recent debates about same-sex marriage underscore, the role of gender in marriage is the product of social processes and beliefs about men, women, and marriage. In examining how gender influences sexuality, moreover, you will see that gender rarely operates alone: Class, culture, race, and individual differences also combine to influence sexuality. . . .

DESIRE: ATTRACTION AND AROUSAL

The most salient fact about sex is that nearly everybody is interested in it. Most people like to have sex, and they talk about it, hear about it, and think about it. But some people are obsessed with sex and willing to have sex with anyone or anything. Others are aroused only by particular conditions and hold exacting criteria. For example, some people will have sex only if they are positive that they are in love, that their partner loves them, and that the act is sanctified by marriage. Others view sex as not much different from eating a sandwich. They neither love nor hate the sandwich; they are merely hungry, and they want something to satisfy that hunger. What we are talking about here are differences in desire. As you have undoubtedly noticed, people differ in what they find attractive, and they are also physically aroused by different things. . . .

Many observers argue that when it comes to sex, men and women have fundamentally different biological wiring. Others use the evidence to argue that culture has produced marked sexual differences among men and women. We believe, however, that it is hard to tease apart biological differences and social differences. As soon as a baby enters the world, it receives messages about gender and sexuality. In the United States, for example, disposable diapers come adorned in pink for girls and blue for boys. In case people aren't sure whether to treat the baby as masculine or feminine in its first years of life, the diaper signals them. The assumption is that girl babies really are different from boy babies and the difference ought to be displayed. This different treatment continues throughout life, and therefore a sex difference at birth becomes amplified into gender difference as people mature.

Gendered experiences have a great deal of influence on sexual desire. As a boy enters adolescence, he hears jokes about boys' uncontainable desire. Girls are told the same thing and told that their job is to resist. These gender messages have power not only over attitudes and behavior (such as whether a person grows up to prefer sex with a lover rather than a stranger) but also over physical and biological experience. For example, a girl may be discouraged from vigorous competitive activity, which will subsequently influence how she develops physically, how she feels about her body, and even how she relates to the adrenaline rush associated with physical competition. Hypothetically, a person who is accustomed to adrenaline responses experiences sexual attraction differently from one who is not. . . .

THE BIOLOGY OF DESIRE: NATURE'S EXPLANATION

Biology is admittedly a critical factor in sexuality. Few human beings fall in love with fish or sexualize trees. Humans are designed to respond to other humans. And human activity is, to some extent, organized by the physical equipment humans are born with. Imagine if people had fins instead of arms or laid eggs instead of fertilizing them during intercourse. Romance would look quite different.

Although biology seems to be a constant (i.e., a component of sex that is fixed and unchanging), the social world tends to mold biology as much as biology shapes humans' sexuality. Each society has its own rules for sex. Therefore, how people experience their biology varies widely. In some societies, women act intensely aroused and active during sex; in others, they have no concept of orgasm. In fact, women in some settings, when told about orgasm, do not even believe it exists, as anthropologists discovered in some parts of Nepal. Clearly, culture—not biology—is at work, because we know that orgasm is physically possible, barring damage to or destruction of the sex organs. Even ejaculation is culturally dictated. In some countries, it is considered healthy to ejaculate early and often; in others, men are told to conserve semen and ejaculate as rarely as possible. The biological capacity may not be so different, but the way bodies behave during sex varies according to social beliefs.

Sometimes the dictates of culture are so rigid and powerful that the so-called laws of nature can be overridden. Infertility treatment provides an example: For couples who cannot produce children "naturally," a several billion dollar industry has provided technology that can, in a small proportion of cases, overcome this biological problem.

Recently, in California, a child was born to a 63-year-old woman who had been implanted with fertilized eggs. The cultural emphasis on reproduction and parenthood, in this case, overrode the biological incapacity to produce children.

THE SOCIAL ORIGINS OF DESIRE

Your own experience should indicate that biology and genetics alone do not shape human sexuality. From the moment you entered the world, cues from the environment were telling you which desires and behaviors were "normal" and which were not. The result is that people who grow up in different circumstances tend to have different sexualities. Who has not had their sexual behavior influenced by their parents' or guardians' explicit or implicit rules? You may break the rules or follow them, but you can't forget them. On a societal level, in Sweden, for example, premarital sex is accepted, and people are expected to be sexually knowledgeable and experienced. Swedes are likely to associate sex with pleasure in this "sex positive" society. In Ireland, however, Catholics are supposed to heed the Church's strict prohibitions against sex outside of marriage, birth control, and the expression of lust. In Ireland the experience of sexuality is different from the experience of sexuality in Sweden because the rules are different. Certainly, biology in Sweden is no different from biology in Ireland, nor is the physical capacity to experience pleasure different. But in Ireland, nonmarital sex is clandestine and shameful. Perhaps the taboo adds excitement to the experience. In Sweden, nonmarital sex is acceptable. In the absence of social constraint, it may even feel a bit mundane. These culturally specific sexual rules and experiences arise from different norms, the well-known, unwritten rules of society.

Another sign that social influences play a bigger role in shaping sexuality than does biology is the changing notions historically of male and female differences in desire. Throughout history, varied explanations of male and female desire have been popular. At times, woman was portrayed as the stormy temptress and man the reluctant participant, as in the Bible story of Adam and Eve. At other times, women were seen as pure in thought and deed while men were voracious sexual beasts, as the Victorians would have it.

These shifting ideas about gender are the social "clothing" for sexuality. The concept of gender typically relies on a dichotomy of male versus female sexual categories, just as the tradition of women wearing dresses and men wearing pants has in the past made the shape of men and women appear quite different. Consider high heels, an on-again-off-again Western fashion. Shoes have no innate sexual function, but high heels have often been understood to be "sexy" for women, even though (or perhaps because) they render women less physically agile. (Of course, women cope. As Ginger Rogers, the 1940s movie star and dancing partner to Fred Astaire, is said to have quipped, "I did every thing Fred did, only backwards and in high heels.") Social norms of femininity have at times rendered high heels fashionable. So feminine are high heels understood to be that a man in high heels, in some sort of visual comedy gag, guarantees a laugh from the audience. Alternatively, high heels are a required emblem of femininity for cross-dressing men.

Such distinctions are an important tool of society; they provide guidance to human beings about how to be a "culturally correct" male or female. Theoretically, society could "clothe" its members with explicit norms of sexuality that de-emphasize difference and emphasize similarity or even multiplicity. Picture unisex hairstyles and men and women both free to wear skirts or pants, norms that prevail from time to time in some subcultures. What is remarkable about dichotomies is that even when distinctions, like male and female norms of fashion, are reduced, new ways to assert an ostensibly essential difference between men and women arise. Societies' rules, like clothes, are changeable. But societies' entrenched taste for constructing differences between men and women persists.

The Social Construction of Sexuality

Social constructionists believe that cues from the environment shape human beings from the moment they enter the world. The sexual customs, values, and expectations of a culture, passed on to the young through teaching and by example, exert a powerful influence over individuals . . .

In a heterogeneous and individualistic culture like North America, sexual socialization is complex. A society creates an "ideal" sexuality, but different families and subcultures have their own values. For example, even though contemporary society at large may now accept premarital sexuality, a given family may lay down the law: Sex before marriage is against the family's religion and an offense against God's teaching. A teenager who grows up in such a household may suppress feelings of sexual arousal or channel them into outlets that are more acceptable to the family. Or the teenager may react against her or his background, reject parental and community opinion, and search for what she or he perceives to be a more "authentic" self. Variables like birth order or observations of a sibling's social and sexual expression can also influence a person's development.

As important as family and social background are, so are individual differences in response to that background. In the abstract, people raised to celebrate their sexuality must surely have a different approach to enjoying their bodies than those who are taught that their bodies will betray them and are a venal part of human nature. Yet whether or not a person is raised to be at ease with physicality does not always help predict adult sexual behavior. Sexual sybarites and libertines may have grown up in sexually repressive environments, as did pop culture icon and Catholic-raised Madonna. Sometimes individuals whose families promoted sex education and free personal expression are content with minimal sexual expression.

Even with the nearly infinite variety of sexuality that individual experience produces, social circumstances shape sexual patterns. For example, research shows that people who have had more premarital sexual intercourse are likely to have more extramarital intercourse, or sex with someone other than their spouse (Blumstein and Schwartz 1983). Perhaps early experience creates a desire for sexual variety and makes it harder for a person to be monogamous. On the other hand, higher levels of sexual desire may generate both the premarital and extramarital propensities. Or perhaps nonmonogamous, sexually active individuals are "rule breakers" in other areas also, and resist not only the traditional rules of sex but also other social norms they encounter. Sexual history is useful for predicting sexual future, but it does not provide a complete explanation. . . .

Social Control of Sexuality

So powerful are norms as they are transmitted through both social structures and everyday life that it is impossible to imagine the absence of norms that control sexuality. In fact, most images of "liberated" sexuality involve breaking a social norm—say, having sex in public rather than in private. The social norm is always the reference point. Because people are influenced from birth by the social and physical contexts of sexuality, their desires are shaped by those norms. There is no such thing as a truly free sexuality. For the past two centuries in North America, people have sought "true love" through personal choice in dating and mating (D'Emilio and Freedman 1988). Although this form of sexual liberation has generated a small increase in the number of mixed pairs—interracial, interethnic, interfaith pairs—the rule of **homogamy,** or marrying within one's class, religion, and ethnicity, still constitutes one of the robust social facts of romantic life. Freedom to choose the person one loves turns out not to be as free as one might suppose.

Despite the norm of true love currently accepted in our culture, personal choice and indiscriminate sexuality have often been construed across cultures and across history as socially disruptive. Disruptions to the social order include liaisons between poor and rich; between people of different races, ethnicities, or faiths; and between members of the same sex. Traditional norms of marriage and sexuality have maintained social order by keeping people in familiar and "appropriate" categories. Offenders have been punished by ostracism, curtailed civil rights, or in some societies, death. Conformists are rewarded with social approval and material advantages. Although it hardly seems possible today, mixed-race marriage was against the law in the United States until 1967. Committed same-sex couples continue to be denied legal marriages, income tax breaks, and health insurance benefits; heterosexual couples take these social benefits for granted.

Some social theorists observe that societies control sexuality through construction of a dichotomized or gendered (male-female) sexuality (Foucault 1978). Society's rules about pleasure seeking and procreating are enforced by norms about appropriate male and female behavior. For example, saying that masculinity is enhanced by sexual experimentation while femininity is demeaned by it gives men sexual privilege (and pleasure) and denies it to women. Furthermore, according to Foucault, sexual desire is fueled by the experience of privilege and taboo regarding sexual pleasure. That is, the very rules that control sexual desire shape it and even enhance it. The social world could just as plausibly concentrate on how much alike are the ways that men and women experience sex and emphasize how broadly dispersed sexual conduct is across genders. However, social control turns pleasure into a scarce resource and endows leaders who regulate the pleasure of others with power. . . .

Society's interest in controlling sexuality is expressed in the debates regarding sex education. Debates about sex education in grade school and high school illustrate the importance to society of both the control of desire and its social construction. The debates raise the question, does formal learning about sex increase or deter early sexual experimentation? The point is, opponents and proponents of sex education all want to know how to control sexuality in young people. Those who favor sex education hold that children benefit from early, comprehensive information about sex, in the belief that people learn about sexuality from birth and are sexual at least from the time of puberty. Providing young people with an appropriate vocabulary and accurate information both discourages early sexual activity and encourages safe sexual practices for those teenagers who, according to the evidence, will not be deterred from sexual activity (Sexuality Information and Education Council of the United States [SIECUS] 1995). On the other hand, opponents of sex education are intensely committed to the belief that information about sex changes teenagers' reactions and values and leads to early, and what they believe are inappropriate, sexual behaviors (Whitehead 1994). Conservative groups hold that sex education, if it occurs at all, should emphasize abstinence as opposed to practical information.

These conflicting points of view about sex education are both concerned with managing adolescent sexual desire. Conservatives fear that education creates desire; liberals feel that information merely enables better decision making. So who is correct? In various studies, a majority of both conservative and liberal sex education programs have demonstrated little effect on behavior. Conservatives believe these results prove the programs' lack of worth. Liberals believe the studies prove that many programs are not good enough, usually because they do not include the most important content. Furthermore, liberals point out that sex education has increased contraceptive (including condom) use, which is crucial to public health goals of reducing sexually transmitted disease and unwanted pregnancy. Other research indicates that comprehensive sex education actually tends to delay the age of first intercourse and does not intensify desire or escalate sexual behavior. There is no evidence that comprehensive sex education promotes or precipitates early teen sexual activity (Kirby et al. 1994).

The passionate debate about sex education is played out with high emotions. Political ideology, parental fears, and the election strategies of politicians all influence this mode of social control. In the final analysis, however, teaching about sex clearly does not have an intense impact on the pupil. Students' response to sex education varies tremendously at the individual level. In terms of trends within groups, however, it appears that sex education tends to delay sexual activity and makes teenage sex safer when it happens.

To summarize, social constructionists believe that a society influences sexual behavior through its norms. Some norms are explicit, such as laws against adult sexual activity with minors. Others are implicit, such as norms of fidelity and parental responsibility. In a stable, homogeneous society, it is relatively easy to understand such rules. But in a changing, complex society like the United States, the rules may be in flux or indistinct. Perhaps this ambiguity is what makes some issues of sexuality so controversial today. . . .

Sexual Identity and Orientation

Sexual identity and **sexual orientation.** . . . are used to mean a variety of things. We use these terms to refer to how people tend to classify themselves sexually—either as **gay, lesbian, bisexual,** or **straight.** Sexual behavior and sexual desire may or may not be consistent with sexual identity. That is, people may identify themselves as heterosexual, but desire people of the same sex—or vice versa.

It is hard to argue with the observation that human desire is, after all, organized. Humans do not generally desire cows or horses (with, perhaps, the exception of Catherine the Great, the Russian czarina who purportedly came to her demise while copulating with a stallion). More to the point, humans are usually quite specific about which sex is desirable to them and even whether the object of their desire is short or tall, dark or light, hairy or sleek.

In the United States, people tend to be identified as either **homosexual** or **heterosexual.** Other cultures (and prior eras in the United States) have not distinguished between these two sexual orientations. However, our culture embraces the perspective that, whether gay or straight, one has an essential, inborn desire, and it cannot change. Many people seem convinced that homosexuality is an essence rather than a sexual act. . . .

In other words, even among gay men and lesbians, it is assumed that they will desire opposite-gendered people, even if they are of the same sex.

Historians have chronicled in Western culture the evolution of homosexuality from a behavior into an identity (e.g., D'Emilio and Freedman 1988). In the past, people might engage in same-gender sexuality, but only in the twentieth century has it become a well-defined (and diverse) lifestyle and self-definition. Nevertheless, other evidence shows that homosexual identity has existed for a long time. The distinguished historian John Boswell (1994) believes that homosexuals as a group and homosexuality as an identity have existed from the very earliest of recorded history. He used evidence of early Christian same sex "marriage" to support his thesis. Social scientist Fred Whitman (1983) has looked at homosexuality across cultures and declared that the evidence of a social type, including men who use certain effeminate gestures and have diverse sexual tastes, goes far beyond any one culture. Geneticist Dean Hamer provides evidence that sexual attraction may be genetically programmed, suggesting that it has persisted over time and been passed down through generations.

On the other side of the debate is the idea that sexuality has always been invented and that sexual orientations are socially created. A gay man's or lesbian's sexual orientation has been created by a social context. Although this creation takes place in a society that prefers dichotomous, polarized categories, the social constructionist vision of sexuality at least poses the possibility that sexuality could involve a continuum of behavior that is matched by a continuum of fantasy, ability to love, and sense of self. . . .

NOTES

Blumstein, P. and P. Schwartz. 1983. *American Couples: Money, Work and Sex*. New York: William Morrow.

Boswell, J. 1994. *Same-Sex Unions in Pre-Modern Europe*. New York: Villard.

D'Emilio, J. D. and H. Freedman. 1988. *Intimate Matters: A History of Sexual in America*. New York: Harper & Row.

Foucault, M. 1978. *A History of Sexuality: Vol. 1. An Introduction*. New York: Pantheon.

Kirby, D., L. Short, J. Collins, D. Rugg, L. Kolbe, M. Howard, B. Miller, F. Sonenstein, and L. S. Zabin. 1994. "School-Based Programs to Reduce Sexual Risk Behaviors: A Review of Effectiveness." *Public Health Reports* 109:339–60.

Sexuality Information and Education Council of the United States. 1995. *A Report on Adolescent Sexuality*. New York: SIECUS.

Whitehead, B. D. 1994. "The Failure of Sex Education." *Atlantic Monthly*, October, pp. 55–80.

Whitman, F. 1983. "Culturally Invariable Properties of Male Homosexualities: Tentative Conclusions from Cross-Cultural Research. *Archives of Sexual Behavior* 12:207–26.

DISCUSSION QUESTIONS

1. What do Schwartz and Rutter mean when saying that sexuality is socially constructed? What evidence do you see of this process in the media that you watch, hear, or read?

2. Consider the characteristics you are attracted to in a romantic partner. How many of the items on this list are biological? How many are associated with the social expectation of femininity or masculinity?

INFOTRAC COLLEGE EDITION

You can use your access to InfoTrac College Edition to learn more about the subjects covered in this essay. Some suggested search terms include:

gay men
homophobia
lesbian experience

monogamy
sexual orientation
social construction of sexuality

24

Masculinity as Homophobia

MICHAEL S. KIMMEL

Michael Kimmel argues that American men are socialized into a very rigid and limiting definition of masculinity. He states that men fear being ridiculed as too feminine by other men and this fear perpetuates homophobic and exclusionary masculinity. He calls for politics of inclusion or the broadening definition of manhood to end gender struggle.

The great secret of American manhood is: *We are afraid of other men.* Homophobia is a central organizing principle of our cultural definition of manhood. Homophobia is more than the irrational fear of gay men, more than the fear that we might be perceived as gay. "The word 'faggot' has nothing to do with homosexual experience or even with fears of homosexuals," writes David Leverenz (1986). "It comes out of the depths of manhood: a label of ultimate contempt for anyone who seems sissy, untough, uncool" (p. 455). Homophobia is the fear that other men will unmask us, emasculate us, reveal to us and the world that we do not measure up, that we are not real men. We are afraid to let other men see that fear. Fear makes us ashamed, because the recognition of fear in ourselves is proof to ourselves that we are not as manly as we pretend, that we are, like the young man in a poem by Yeats, "one that ruffles in a manly pose for all his timid heart." Our fear is the fear of humiliation. We are ashamed to be afraid . . .

The fear of being seen as a sissy dominates the cultural definitions of manhood. It starts so early. "Boys among boys are ashamed to be unmanly," wrote one educator in 1871 (cited in Rotundo, 1993, p. 264). I have a standing bet with a friend that I can walk onto any playground in America where 6-year-old boys are happily playing and by asking one question, I can provoke a fight. That question is simple: "Who's a sissy around here?" Once posed, the challenge is made. One of two things is likely to happen. One boy will accuse another of being a sissy, to which that boy will respond that he is not a sissy, that the first boy is. They may have to fight it out to see who's lying. Or a whole group of boys will surround one boy and all shout "He is! He is!" That boy will either burst into tears and run home crying, disgraced, or he will have to take on several boys at once, to prove that he's not a sissy. (And what will his father or older brothers tell him if he chooses to run home crying?) It will be some time before he regains any sense of self-respect.

Violence is often the single most evident marker of manhood. Rather it is the willingness to fight, the desire to fight. The origin of our expression that one has a chip on one's shoulder lies in the practice of an adolescent boy in the country or small town at the turn of the century, who would literally walk around with a chip of wood balanced on his shoulder—a signal of his readiness to fight with anyone who would take the initiative of knocking the chip off (see Gorer, 1964, p. 38; Mead, 1965).

As adolescents, we learn that our peers are a kind of gender police, constantly threatening to unmask us as feminine, as sissies. One of the favorite tricks when I was an adolescent was to ask a boy to look at his fingernails. If he held his palm toward his face and curled his fingers back to see them, he passed the test. He'd looked at his nails "like a man." But if he held the back of his hand away from his face, and looked at his fingernails with arm outstretched, he was immediately ridiculed as a sissy.

As young men we are constantly riding those gender boundaries, checking the fences we have constructed on the perimeter, making sure that nothing even remotely feminine might show through. The possibilities of being unmasked are everywhere. . . . Even the most seemingly insignificant thing can pose a threat or activate that haunting terror. On the day the students in my course "Sociology of Men and Masculinities" were scheduled to discuss homophobia and male-male friendships, one student provided a touching illustration. Noting that it was a beautiful day, the first day of spring after the brutal northeast winter, he decided to wear shorts to class. "I had this really nice pair of new Madras shorts," he commented. "But then I thought to myself, these shorts have lavender and pink in them. Today's class topic is homophobia. Maybe today is not the best day to wear these shorts."

Our efforts to maintain a manly front cover everything we do. What we wear. How we talk. How we walk. What we eat. Every mannerism, every movement contains a coded gender language. Think, for example, of how you would answer the question: How do you "know" if a man is homosexual? When I ask this question in classes or workshops, respondents invariably provide a pretty standard list of stereotypically effeminate behaviors. He walks a certain way, talks a certain way, acts a certain way. He's very emotional; he shows his feelings. One woman commented that she "knows" a man is gay if he really cares about her; another said she knows he's gay if he shows no interest in her, if he leaves her alone.

Now alter the question and imagine what heterosexual men do to make sure no one could possibly get the "wrong idea" about them. Responses typically refer to the original stereotypes, this time as a set of negative rules about behavior. Never dress that way. Never talk or walk that way. Never show your feelings or get emotional. Always be prepared to demonstrate sexual interest in women that you meet, so it is impossible for any woman to get the wrong idea about you. In this sense, homophobia, the fear of being perceived as gay, as not a real man, keeps men exaggerating all the traditional rules of masculinity, including sexual predation with women. Homophobia and sexism go hand in hand. . . .

POWER AND POWERLESSNESS IN THE LIVES OF MEN

. . . Manhood is equated with power—over women, over other men. Everywhere we look, we see the institutional expression of that power—in state and national legislatures, on the boards of directors of every major U.S. corporation or law firm, and in every school and hospital administration. Women have long understood this, and feminist women have spent the past three decades challenging both the public and the private expressions of men's power and acknowledging their fear of men. Feminism as a set of theories both explains women's fear of men and empowers women to confront it both publicly and privately. Feminist women have theorized that masculinity is about the drive for domination, the drive for power, for conquest.

This feminist definition of masculinity as the drive for power is theorized from women's point of view. It is how women experience masculinity. But it assumes a symmetry between the public and the private that does not conform to men's experiences. Feminists observe that women, as a group, do not hold power in our society. They also observe that individually, they, as women, do not feel powerful. They feel afraid, vulnerable. Their observation of the social reality and their individual experiences are therefore symmetrical. Feminism also observes that men, as a group, are in power. Thus, with the same symmetry, feminism has tended to assume that individually men must feel powerful.

This is why the feminist critique of masculinity often falls on deaf ears with men. When confronted with the analysis that men have all the power, many men react incredulously. "What do you mean, men have all the power?" they ask. "What are you talking about? My wife bosses me around. My kids boss me around. My boss bosses me around. I have no power at all! I'm completely powerless!"

Men's feelings are not the feelings of the powerful, but of those who see themselves as powerless. These are the feelings that come inevitably from the discontinuity between the social and the psychological, between the aggregate analysis that reveals how men are in power as a group and the psychological fact that they do not feel powerful as individuals. They are the feelings of men who were raised to believe themselves entitled to feel that power, but do not feel it. No wonder many men are frustrated and angry. . . .

Often the purveyors of the mythopoetic men's movement, that broad umbrella that encompasses all the groups helping men to retrieve this mythic deep manhood, use the image of the chauffeur to describe modern man's position. The chauffeur appears to have the power—he's wearing the uniform, he's in the driver's seat, and he knows where he's going. So, to the observer, the chauffeur looks as though he is in command. But to the chauffeur himself, they note, he is merely taking orders. He is not at all in charge.

Despite the reality that everyone knows chauffeurs do not have the power, this image remains appealing to the men who hear it at these weekend workshops. But there is a missing piece to the image, a piece concealed by the framing of the image in terms of the individual man's experience. That missing piece is that the person who is giving the orders is also a man. Now we have a relationship *between* men—between men giving orders and other men taking those orders. The man who identifies with the chauffeur is entitled to be the man giving the orders, but he is not. ("They," it turns out, are other men.)

The dimension of power is now reinserted into men's experience not only as the product of individual experience but also as the product of relations with other men. In this sense, men's experience of powerlessness is *real*—the men actually feel it and certainly act on it—but it is not *true,* that is, it does not accurately describe their condition. In contrast to women's lives, men's lives are structured around relationships of power and men's differential access to power, as well as the differential access to that power of men as a group. Our imperfect analysis of our own situation leads us to believe that we men need more power, rather than leading us to support feminists' efforts to rearrange power relationships along more equitable lines. . . .

Why, then, do American men feel so powerless? Part of the answer is because we've constructed the rules of manhood so that only the tiniest fraction of men come to believe that they are the biggest of wheels, the sturdiest of oaks, the most virulent repudiators of femininity, the most daring and aggressive. We've managed to disempower the overwhelming majority of American men by other means—such as discriminating on the basis of race, class, ethnicity, age, or sexual preference.

Masculinist retreats to retrieve deep, wounded masculinity are but one of the ways in which American men currently struggle with their fears and their shame. Unfortunately, at the very moment that they work to break down the isolation that governs men's lives, as they enable men to express those fears and that shame, they ignore the social power that men continue to exert over women and the privileges from which they (as the middle-aged, middle-class white men who largely make up these retreats) continue to benefit—regardless of their experiences as wounded victims of oppressive male socialization.

Others still rehearse the politics of exclusion, as if by clearing away the playing field of secure gender identity of any that we deem less than manly—women, gay men, nonnative-born men, men of color—middle-class, straight, white men can reground their sense of themselves without those haunting fears and that deep shame that they are unmanly and will be exposed by other men. This is the manhood of racism, of sexism, of homophobia. It is the manhood that is so chronically insecure that it trembles at the idea of lifting the ban on gays in the military, that is so threatened by women in the workplace that women become the targets of sexual harassment, that is so deeply frightened of equality that it must ensure that the playing field of male competition remains stacked against all newcomers to the game.

Exclusion and escape have been the dominant methods American men have used to keep their fears of humiliation at bay. The fear of emasculation by other men, of being humiliated, of being seen as a sissy, is the leitmotif in my reading of the history of American manhood. Masculinity has become a relentless test by which we prove to other men, to women, and ultimately to ourselves, that we have successfully mastered the part. The restlessness that men feel today is nothing new in American history; we have been anxious and restless for almost two centuries. Neither exclusion nor escape has ever brought us the relief we've sought, and there is no reason to think that either will solve our problems now. Peace of mind, relief from gender struggle, will come only from a politics of inclusion, not exclusion, from standing up for equality and justice, and not by running away.

NOTES

Gorer, G. (1964). *The American people: A study in national character.* New York: Norton.

Leverenz, D. (1986, Fall). Manhood, humiliation and public life: Some stories. *Southwest Review,* 71.

Mead, M. (1965). *And keep your powder dry.* New York: William Morrow.

Rotundo, E. A. (1993). *American manhood: Transformations in masculinity from the revolution to the modern era.* New York: Basic Books.

DISCUSSION QUESTIONS

1. Kimmel discusses men's fear of being called a "sissy." Can you think of other examples where men criticize each other's manhood? What are some other terms used to denote femininity as a negative attribute in men?

2. How is manhood defined in other cultures? Is the U.S. ideal of manhood more or less rigid than other examples you identify?

INFOTRAC COLLEGE EDITION

You can use your access to InfoTrac College Edition to learn more about the subjects covered in this essay. Some suggested search terms include:

homophobia

manhood and masculinity

men's movement

patriarchy

politics of exclusion

politics of inclusion

25

The Approach-Avoidance Dance:
Men, Women, and Intimacy

LILLIAN B. RUBIN

In this essay, Lillian Rubin distinguishes women's and men's views on intimacy. The cultural expectation that men think with their heads and women feel with their hearts leads to conflict between the sexes. The essay outlines the social and cultural basis for the gender differences in emotional expression.

Intimacy. We hunger for it, but we also fear it. We come close to a loved one, then we back off. A teacher I had once described this as the "go away a little closer" message, I call it the approach-avoidance dance.

The conventional wisdom says that women want intimacy, men resist it. And I have plenty of material that would *seem* to support that view. Whether in my research interviews, in my clinical hours, or in the ordinary course of my life, I hear the same story told repeatedly. "He doesn't talk to me," says a woman. "I don't know what she wants me to talk about," says a man. "I want to know what he's feeling," she tells me. "I'm not feeling anything," he insists. "Who can feel nothing?" she cries. "I can," he shouts. As the heat rises, so does the wall between them. Defensive and angry, they retreat—stalemated by their inability to understand each other.

Women complain to each other all the time about not being able to talk to their men about the things that matter most to them—about what they themselves are thinking and feeling, about what goes on in the hearts and minds of the men they're relating to. And men, less able to expose themselves and their conflicts—those within themselves or those with the women in their lives—either turn silent or take cover by holding women up to derision. It's one of the norms of male camaraderie to poke fun at women, to complain laughingly about the mystery of their minds, wonderingly about their ways. Even Freud did it when, in exasperation, he asked mockingly, "What do women want? Dear God, what do they want?". . .

The expression of such conflicts would seem to validate the common understandings that suggest that women want and need intimacy more than men do—that the issue belongs to women alone; that, if left to themselves, men would not suffer it. But things are not always what they seem. And I wonder: "If men would renounce intimacy, what is their stake in relationships with women?"

Some would say that men need women to tend to their daily needs—to prepare their meals, clean their houses, wash their clothes, rear their children—so that they can be free to attend to life's larger problems. And, given the traditional structure of roles in the family, it has certainly worked that way most of the time. But, if that were all men seek, why is it that, even when they're not relating to women, so much of their lives is spent in search of a relationship with another, so much agony experienced when it's not available?

These are difficult issues to talk about—even to think about—because the subject of intimacy isn't just complicated, it's slippery as well. Ask yourself: What is intimacy? What words come to mind, what thoughts?

It's an idea that excites our imagination, a word that seems larger than life to most of us. It lures us, beckoning us with a power we're unable to resist. And, just because it's so seductive, it frightens us as well—seeming sometimes to be some mysterious force from outside ourselves that, if we let it, could sweep us away.

But what is it we fear?

Asked what intimacy is, most of us—men and women—struggle to say something sensible, something that we can connect with the real experience of our lives. "Intimacy is knowing there's someone who cares about the children as much as you do." "Intimacy is a history of shared experience. It's sitting there having a cup of coffee together and watching the eleven o'clock news.". . . "It's talking when you're in the bathroom." "It's knowing we'll begin and end each day together."

These seem the obvious things—the things we expect when we commit our lives to one another in a marriage, when we decide to have children together. And they're not to be dismissed as inconsequential. They make up the daily experience of our lives together, setting the tone for a relationship in important and powerful ways. It's sharing such commonplace, everyday events that determines the temper and the texture of life, that keeps us living together even when other aspects of the relationship seem less than perfect. . . .

These ways in which a relationship feels intimate on a daily basis are only one part of what we mean by intimacy, however—the part that's most obvious, the part that doesn't awaken our fears. At a lecture where I spoke of these issues recently, one man commented also, "Intimacy is putting aside the masks we wear in the rest of our lives." A murmur of assent ran through the audience of a hundred or so. Intuitively we say, "yes." Yet this is the very issue that also complicates our intimate relationships.

On the one hand, it's reassuring to be able to put away the public persona—to believe we can be loved for who we *really* are, that we can show our shadow side without fear, that our vulnerabilities will not be counted against us. "The most important thing is to feel I'm accepted just the way I am," people will say.

But there's another side. For, when we show ourselves thus without the masks, we also become anxious and fearful. "Is it possible that someone could love the *real* me?" we're likely to ask. Not the most promising question for the further development of intimacy, since it suggests that, whatever else another might do or feel, it's we who have trouble loving ourselves. Unfortunately, such misgivings are not usually experienced consciously. We're aware only that our discomfort has risen, that we feel a need to get away. For the person who has seen the "real me" is also the one who reflects back to us an image that's usually not wholly to our liking. We get angry at that, first at ourselves for not living up to our own expectations, then at the other who becomes for us the mirror of our self-doubts—a displacement of hostility that serves intimacy poorly.

There's yet another level—one that's further below the surface of consciousness, therefore, one that's much more difficult for us to grasp, let alone to talk about. I'm referring to the differences in the ways in which women and men deal with their inner emotional lives—differences that create barriers between us that can be high indeed. It's here that we see how those early childhood experiences of separation and individuation—the psychological tasks that were required of us in order to separate from mother, to distinguish ourselves as autonomous persons, to internalize a firm sense of gender identity—take their toll on our intimate relationships.

Stop a woman in mid-sentence with the question, "What are you feeling right now?" and you might have to wait a bit while she reruns the mental tape to capture the moment just passed. But, more than likely, she'll be able to do it successfully. More than likely, she'll think for a while and come up with an answer.

The same is not true of a man. For him, a similar question usually will bring a sense of wonderment that one would even ask it, followed quickly by an uncomprehending and puzzled response, "What do you mean?" he'll ask. "I was just talking," he'll say.

I've seen it most clearly in the clinical setting where the task is to get to the feeling level—or, as one of my male patients said when he came into therapy, to "hook up the head and the gut." Repeatedly when therapy begins, I find myself having to teach a man how to monitor his internal states—how to attend to his thoughts and feelings, how to bring them into consciousness. In the early stages of our work, it's a common experience to say to a man, "How does that feel?," and to see a blank look come over his face. Over and over, I find myself listening as a man speaks with calm reason about a situation which I know must be fraught with pain. "How do you feel about that?" I'll ask. "I've just been telling you," he's likely to reply, "No," I'll say, "you've told me what happened, not how you *feel* about it." Frustrated, he might well respond. "You sound just like my wife."

It would be easy to write off such dialogues as the problems of men in therapy, of those who happen to be having some particular emotional difficulties. But it's not so, as any woman who has lived with a man will attest. Time and again women complain: "I can't get him to verbalize his feelings." "He talks, but it's always intellectualizing." "He's so closed off from what he's feeling, I don't know how he lives that way." . . .

To a woman, the world men live in seems a lonely one—a world in which their fears of exposing their sadness and pain, their anxiety about allowing their vulnerability to show, even to a woman they love, is so deeply rooted inside them that, most often, they can only allow it to happen "late at night in the dark."

Yet, if we listen to what men say, we will hear their insistence that they *do* speak of what's inside them, *do* share their thoughts and feelings with the women they love. "I tell her, but she's never satisfied," they complain. "No matter how much I say, it's never enough," they grumble.

From both sides, the complaints have merit. The problem lies not in what men don't say, however, but in what's not there—in what, quite simply, happens so far out of consciousness that it's not within their reach. For men have integrated all too well the lessons of their childhood—the experiences that taught them to repress and deny their inner thoughts, wishes, needs, and fears; indeed, not even to notice them. It's real, therefore, that the kind of inner thoughts and feelings that are readily accessible to a woman generally are unavailable to a man. When he says, "I don't know what I'm feeling," he isn't necessarily being intransigent and withholding. More than likely, he speaks the truth.

Partly that's a result of the ways in which boys are trained to camouflage their feelings under cover of an exterior of calm, strength, and rationality. Fears are not manly. Fantasies are not rational. Emotions, above all, are not for the strong, the sane, the adult. Women suffer them, not men—women, who are more like children with what seems like their never-ending preoccupation with their emotional life. But the training takes so well because of their early childhood experience when, as very young boys, they had to shift their identification from mother to father and sever themselves from their earliest emotional connection. Put the two together and it does seem like suffering to men to have to experience that emotional side of themselves, to have to give it voice. . . .

These differences in the psychology of women and men are born of a complex interaction between society and the individual. At the broadest social level is the rending of thought and feeling that is such a fundamental part of Western thought. Thought, defined as the ultimate good, has been assigned to men; feeling, considered at best a problem, has fallen to women.

So firmly fixed have these ideas been that, until recently, few thought to question them. For they were built into the structure of psychological thought as if they spoke to an eternal, natural, and scientific truth.

We equate the emotional with the nonrational.

This notion, shared by both women and men, is a product of the fact that they were born and reared in this culture. But there's also a difference between them in their capacity to apprehend the *logic* of emotions—a difference born in their early childhood experiences in the family, when boys had to repress so much of their emotional side and girls could permit theirs to flower, . . . It should be understood: Commitment itself is not a problem for a man; he's good at that. He can spend a lifetime living in the same family, working at the same job—even one he hates. And he's not without an inner emotional life. But when a relationship requires the sustained verbal expression of that inner life and the full range of feelings that accompany it, then it becomes burdensome for him. He can act out anger and frustration inside the family, it's true. But ask him to express his sadness, his fear, his dependency—all those feelings that would expose his vulnerability to himself or to another—and he's likely to close down as if under some compulsion to protect himself.

All requests for such intimacy are difficult for a man, but they become especially complex and troublesome in relations with women. It's another of those paradoxes. For, to the degree that it's possible for him to be emotionally open with anyone, it is with a woman—a tribute to the power of the childhood experience with mother. Yet it's that same early experience and his need to repress it that raises his ambivalence and generates his resistance.

He moves close, wanting to share some part of himself with her, trying to do so, perhaps even yearning to experience again the bliss of the infant's connection with a woman. She responds, woman style—wanting to touch him just a little more deeply, to know what he's thinking, feeling, fearing, wanting. And the fear closes in—the fear of finding himself again in the grip of a powerful woman, of allowing her admittance only to be betrayed and abandoned once again, of being overwhelmed by denied desires.

So he withdraws.

It's not in consciousness that all this goes on. He knows, of course, that he's distinctly uncomfortable when pressed by a woman for more intimacy in the relationship, but he doesn't know why. And, very often, his behavior doesn't please him any more than it pleases her. But he can't seem to help it.

DISCUSSION QUESTIONS

1. Interview some of your friends and family, asking them to define intimacy. Is there a difference in the definitions provided by men and by women? What do these differences (if any) represent?

2. There are more men's movements emerging today that claim to help men get in touch with their emotions. What is society's reaction to these movements? Do we see a change in the way men express their feelings? Are there still negative reactions to "emotional" men?

INFOTRAC COLLEGE EDITION

You can use your access to InfoTrac College Edition to learn more about the subjects covered in this essay. Some suggested search terms include:

dating relationships   intimacy
gender identity   Sigmund Freud

26

Pills and Power Tools

SUSAN BORDO

This essay examines the issue of male impotence and the use of Viagra to treat it. The author discusses how American culture has associated men's sexual ability with strength, machine-like reliability, and even violence. She points to illustrations in popular culture that equate men's sexual activity with performance rather than emotional attachment.

Viagra. "The Potency Pill," as *Time* magazine's cover describes it. Since it went on sale, it has had "the fastest takeoff of a new drug" that the RiteAid drugstore chain has ever seen. It is all over the media. Users are jubilant, claiming effects that last through the night, youth restored, better "quality" erections. "This little pill is like a package of dynamite," says one.

Some even see Viagra as a potential cure for social ills. Bob Guccione, publisher of *Penthouse,* hails the drug as "freeing the American male libido" from the emasculating clutches of feminism. This diagnosis does not sit very comfortably with current medical wisdom, which has declared impotence to be a physiological problem. I, like Guccione, am skeptical of that declaration—but would suggest a deeper meditation on what has put the squeeze on male libido.

Think, to begin with, of the term *impotence.* It rings with disgrace, humiliation—and it was not the feminists who invented it. Writer Philip Lopate, in an essay on his body, says that merely to say the word out loud makes him nervous. Yet remarkably, *impotence*—rather than the more forgiving, if medicalized, *erectile dysfunction*—is still a common nomenclature among medical researchers. *Frigidity,* with its suggestion that the woman is cold, like some barren tundra, went by the board a while ago. But *impotence,* no less loaded with ugly gender implications, remains. Lenore Tiefer, who researches medical terminology, suggests that we cannot let go of *impotence* because to do so would force us to also let go of *potency* and the cultural mythology that equates male sexuality with power. But to hold on to that mythology, men must pay a steep price.

Impotence. Unlike other disorders, impotence implicates the whole man, not merely the body part. He is impotent. Would we ever say about a man with a headache "He is a headache?" Yet this is just what we do with impotence, as Warren Farrell notes in *Why Men Are the Way They Are.* "We make no attempt to separate impotence from the total personality." Then, we expect the personality to perform like a machine.

From: *Men and Masculinities* 1: 87–90.

That expectation of men is embedded throughout our culture. Think of our slang terms, so many of which encase the penis, like a cyborg, in various sorts of metal or steel armor. Big rig. Blow torch. Bolt. Cockpit. Crank. Crowbar. Destroyer. Dipstick. Drill. Engine. Hammer. Hand tool. Hardware. Hose. Power tool. Rod. Torpedo. Rocket. Spear. Such slang—common among teenage boys—is violent in what it suggests the machine penis can do to another, "softer" body. But the terms are also metaphorical protection against the failure of potency. A human organ of flesh and blood is subject to anxiety, ambivalence, uncertainty. A torpedo or rocket, on the other hand, would never let one down.

Contemporary urologists have taken the metaphor of man the machine even further. Erectile functioning is "all hydraulics," says Irwin Goldstein of the Boston University Medical Center, scorning a previous generation of researchers who stressed psychological issues. But if it is all a matter of fluid dynamics, why keep the term *impotent* whose definitions (according to *Webster's Unabridged*) are "want or power," "weakness," "lack of effectiveness, helplessness" and (appearing last) "lack of ability to engage in sexual intercourse." In keeping the term *impotence,* the drug companies, it seems, get to have it both ways: reduce a complex human condition to a matter of chemistry while keeping the old shame machine working, helping to assure the flow of men to their doors.

We live in a culture that encourages men to think of their sexuality as residing in their penises and that gives men little encouragement to explore the rest of their bodies. The beauty of the male body has finally been brought out of the cultural closet by Calvin Klein, Versace, and other designers. But notice how many of those new underwear ads aggressively direct our attention to the (often extraordinary) endowments of the models. Many of the models stare coldly, challengingly at the viewer, defying the viewer's gaze to define them in any way other than how they have chosen to present themselves: powerful, armored, emotionally impenetrable. "I am a rock," their bodies seem to proclaim. Commercial advertisements depict women stroking their necks, their faces, their legs, lost in sensual reverie, taking pleasure in touching themselves—all over. Similar poses with men are very rare. Touching oneself languidly, lost in the sensual pleasure of the body, is too feminine, too "soft," for a real man. Crotch-grabbing, thrusting, putting it "in your face"—that is another matter.

There is a fascinating irony in the fact that although it is women whose bodies are most sexually objectified by this culture, women's bodies are permitted much greater sexual expression in their cultural representations than men's. In sex scenes, the moaning and writhing of the female partner have become the conventional cinematic mode for heterosexual ecstasy and climax. The male's participation largely gets represented via caressing hands, humping buttocks, and—on rare occasions—a facial expression of intense concentration. She is transported to another world: he is the pilot of the ship that takes her there. When men are shown being transported themselves, it is usually being played for comedy (as in Al Pacino's shrieks in *Frankie and Johnny,* Eddie Murphy's moaning in *Boomerang,* Kevin Kline's contortions in *A Fish Called Wanda*), or it is coded to suggest that something is not quite normal with the man—he is sexually enslaved, for example (as with Jeremy Irons in *Damage*). Men's bodies in the movies are action-hero toys, power tools—wind them up and watch them perform.

Thankfully, the equation between penis and power tool is now being questioned in other movies. Earlier this year, *The Full Monty* brought us a likable group of unemployed workers in Sheffield, England, who hatch the moneymaking scheme of displaying all in a male strip show and learn what it is like to be what feminist theorists call "the object of the gaze." Paul Thomas Anderson's *Boogie Nights* told the story of the rise and fall (so to speak) of a mythically endowed young porn star, Dirk Diggler, who does fine so long as he is the most celebrated stallion in the stable but loses his grip in the face of competition. On the surface, the film is about a world far removed from the lives of most men, a commercial underground where men pray for "wood" and lose their jobs unless they can achieve erection on command. On a deeper level, however, the world of the porn actor is simply the most literalized embodiment—and a perfect metaphor—for a masculinity that demands constant performance from men.

Even before he takes up a career that depends on it. Diggler's sense of self is constellated around his penis: he pumps up his ego by looking in the mirror and—like a coach mesmerizing the team before a game—intoning mantras about his superior gifts. That works well, so long as he believes it. But unlike a real power tool, the motor of male self-worth cannot simply be switched on and off. In the very final shot of the movie, we see Diggler's fabled organ itself. It is a prosthesis, actually (a fact that annoyed several men I know until I pointed out that it was no more a cheat than implanted breasts passing for the real thing). But prosthesis or not and despite its dimensions, it is no masterful tool. It points downward, weighted with expectation, with shame, looking tired and used.

Beginning with the French film *Ridicule* (in which an aristocrat, using his penis as an instrument of vengeance, urinates on the lap of another man), we have seen more unclothed penises in films this year than ever before. But what is groundbreaking about *Boogie Nights* is not that it displays a nude penis, but that it so unflinchingly exposes the job that the mythology of unwavering potency does on the male body. As long as the fortress holds, the sense of power may be intoxicating; but when it cracks—as it is bound to do at some point—the whole structure falls to pieces. Those of whom such constancy is expected (or who require it of themselves) are set up for defeat and humiliation.

Unless, of course, he pops his little pill whenever "failure" threatens. I have no desire to withhold Viagra from the many men who have been deprived of the ability to get an erection by accidents, diabetes, cancer, and other misfortunes to which the flesh—or psyche—is heir. I would just like CNN and *Time* to spend a fraction of the time they devote to describing "how Viagra cures" to looking at how our culture continues to administer the poison for whose effects we now claim a cure. Let us note, too, that the official medical definition of erectile dysfunction (like the definitions of depression and attention deficit disorder) has broadened coincident with the development of new drugs. Dysfunction is no longer defined as "inability to get an erection" but inability to get an erection that is adequate for "satisfactory sexual performance." Performance. Not pleasure. Not feeling. Performance.

Some of what we now call impotence may indeed be physiological in origin: some may be grounded in deep psychic fears and insecurities. But sometimes, too, a man's penis may simply be instructing him that his feelings are not in synch with the job he is supposed to do—or with the very fact that it's a "job." So, I like Philip Lopate's epistemological metaphor for the penis much better than the machine images. Over the years, he has come to appreciate, he writes, that his penis has its "own specialized form of intelligence." The penis knows that it is not a torpedo, no matter what a culture expects of it or what drugs are relayed to its blood vessels. If we accept that, the notion that a man requires understanding and tolerance when he does not "perform" would go by the wayside ("It's OK. It happens" still assumes that there is something to be excused.) So, too, would the idea that there ought to be one model for understanding nonarousal. Sometimes, the penis's "specialized intelligence" should be listened to rather than cured.

Viagra, unfortunately, seems to be marketed—and used—with the opposite message in mind. Now men can perform all night! Do their job no matter how they feel! (The drug does require some degree of arousal, but minimal.) The hype surrounding the drug encourages rather than deconstructs the expectation that men perform like power tools with only one switch—on or off. Until this expectation is replaced by a conception of manhood that permits men and their penises a full range of human feeling, we will not yet have the kind of "cure" we really need.

DISCUSSION QUESTIONS

1. Given all the publicity around Viagra, can we make the argument that impotence is now a less shameful disorder for men? What does the existence of such a drug say about American society? Is the culture around sex changing?

2. Compare the issue of infertility with the issue of impotence. How are these two issues similarly treated in society? Is infertility viewed differently in women than in men?

INFOTRAC COLLEGE EDITION

You can use your access to InfoTrac College Edition to learn more about the subjects covered in this essay. Some suggested search terms include:

impotence   sexual expression
masculine sexuality   Viagra®

27

The Functions of Crime

EMILE DURKHEIM

This classic essay, written in 1895 and translated many times since, points to crime as an inevitable part of society. Durkheim's main functionalist thesis that criminal behavior exists in all social settings is still the theoretical basis for many sociological inquiries into crime and deviance.

If there is a fact whose pathological nature appears indisputable, it is crime. All criminologists agree on this score. Although they explain this pathology differently, they none the less unanimously acknowledge it. However, the problem needs to be treated less summarily.

. . . Crime is not only observed in most societies of a particular species, but in all societies of all types. There is not one in which criminality does not exist, although it changes in form and the actions which are termed criminal are not everywhere the same. Yet everywhere and always there have been men who have conducted themselves in such a way as to bring down punishment upon their heads. If at least, as societies pass from lower to higher types, the crime rate (the relationship between the annual crime figures and population figures) tended to fall, we might believe that, although still remaining a normal phenomenon, crime tended to lose that character of normality. Yet there is no single ground for believing such a regression to be real. Many facts would rather seem to point to the existence of a movement in the opposite direction. From the beginning of the century statistics provide us with a means of following the progression of criminality. It has everywhere increased, and in France the increase is of the order of 300 percent. Thus there is no phenomenon which represents more incontrovertibly all the symptoms of normality, since it appears to be closely bound up with the conditions of all collective life. To make crime a social illness would be to concede that sickness is not something accidental, but on the contrary derives in certain cases from the fundamental constitution of the living creature. This would be to erase any distinction between the physiological and the pathological. It can certainly happen that crime itself has normal forms; this is what happens, for instance, when it reaches an excessively high level. There is no doubt that this excessiveness is pathological in nature. What is normal is simply that criminality exists, provided that for each social type it does not reach or go beyond a certain level which it is perhaps not impossible to fix in conformity with the previous rules.

From: Emile Durkheim. 1982. *The Rules of Sociological Method.* Steven Lukes (Ed.)
Translated by W. D. Halls, New York; The Free Press. A division of Macmillan, pp. 64–75.

We are faced with a conclusion which is apparently somewhat paradoxical. Let us make no mistake: to classify crime among the phenomena of normal sociology is not merely to declare that it is an inevitable though regrettable phenomenon arising from the incorrigible wickedness of men; it is to assert that it is a factor in public health, an integrative element in any healthy society. At first sight this result is so surprising that it disconcerted even ourselves for a long time. However, once that first impression of surprise has been overcome it is not difficult to discover reasons to explain this normality and at the same time to confirm it.

In the first place, crime is normal because it is completely impossible for any society entirely free of it to exist.

Crime, consists of an action which offends certain collective feelings which are especially strong and clear-cut. In any society, for actions regarded as criminal to cease, the feelings that they offend would need to be found in each individual consciousness without exception and in the degree of strength requisite to counteract the opposing feelings. Even supposing that this condition could effectively be fulfilled, crime would not thereby disappear; it would merely change in form, for the very cause which made the well-springs of criminality to dry up would immediately open up new ones.

Indeed, for the collective feelings, which the penal law of a people at a particular moment in its history protects, to penetrate individual consciousnesses that had hitherto remained closed to them, or to assume greater authority - whereas previously they had not possessed enough - they would have to acquire an intensity greater than they had had up to then. The community as a whole must feel them more keenly, for they cannot draw from any other source the additional force which enables them to bear down upon individuals who formerly were the most refractory. . . .

In order to exhaust all the logically possible hypotheses, it will perhaps be asked why this unanimity should not cover all collective sentiments without exception, and why even the weakest sentiments should not evoke sufficient power to forestall any dissentient voice. The moral conscience of society would be found in its entirety in every individual, endowed with sufficient force to prevent the commission of any act offending against it, whether purely conventional failings or crimes. But such universal and absolute uniformity is utterly impossible, for the immediate physical environment in which each one of us is placed, our hereditary antecedents, the social influences upon which we depend, vary from one individual to another and consequently cause a diversity of consciences. It is impossible for everyone to be alike in this matter, by virtue of the fact that we each have our own organic constitution and occupy different areas in space. This is why, even among lower peoples where individual originality is very little developed, such originality does however exist. Thus, since there cannot be a society in which individuals do not diverge to some extent from the collective type, it is also inevitable that among these deviations some assume a criminal character. What confers upon them this character is not the intrinsic importance of the acts but the importance which the common consciousness ascribes to them. Thus if the latter is stronger and possesses sufficient authority to make these divergences very weak in absolute terms, it will also be more sensitive and exacting. By reacting against the slightest deviations with an energy which it elsewhere employs against those what are more weighty, it endues them with the same gravity and will brand them as criminal.

Thus crime is necessary. It is linked to the basic conditions of social life, but on this very account is useful, for the conditions to which it is bound are themselves indispensable to the normal evolution of morality and law.

Indeed today we can no longer dispute the fact that not only do law and morality vary from one social type to another, but they even change within the same type if the conditions of collective existence are modified. Yet for these transformations to be made possible, the collective sentiments at the basis of morality should not prove unyielding to change, and consequently should be only moderately intense. If they were too strong, they would no longer be malleable. Any arrangement is indeed an obstacle to a new arrangement; this is even more the case the more deep-seated the original arrangement. The more strongly a structure is articulated, the more it resists modification; this is as true for functional as for anatomical patterns. If there were no crimes, this condition would not be fulfilled, for such a hypothesis presumes that collective sentiments would have attained a degree of intensity unparalleled in history. Nothing is good indefinitely and without limits. The authority which the moral consciousness enjoys must not be excessive, for otherwise no one would dare to attack it and it would petrify too easily into an immutable form. For it to evolve, individual originality must be allowed to manifest itself. But so that the originality of the idealist who dreams of transcending his era may display itself, that of the criminal, which falls short of the age, must also be possible. One does not go without the other.

Nor is this all. Beyond this indirect utility, crime itself may play a useful part in this evolution. Not only does it imply that the way to necessary changes remains open, but in certain cases it also directly prepares for these changes. Where crime exists, collective sentiments are not only in the state of plasticity necessary to assume a new form, but sometimes it even contributes to determining beforehand the shape they will take on. Indeed, how often is it only an anticipation of the morality to come, a progression towards what will be! . . . The freedom of thought that we at present enjoy could never have been asserted if the rules that forbade it had not been violated before they were solemnly abrogated. However, at the time the violation was a crime, since it was an offence against sentiments still keenly felt in the average consciousness. Yet this crime was useful since it was the prelude to changes which were daily becoming more necessary. . . .

From this viewpoint the fundamental facts of criminology appear to us in an entirely new light. Contrary to current ideas, the criminal no longer appears as an utterly unsociable creature, a sort of parasitic element, a foreign, unassimilable body introduced into the bosom of society. He plays a normal role in social life. For its part, crime must no longer be conceived of as an evil which cannot be circumscribed closely enough. Far from there being cause for congratulation when it drops too noticeably below the normal level, this apparent progress assuredly coincides with and is linked to some social disturbance. Thus the number of crimes of assault never falls so low as it does in times of scarcity. Consequently, at the same time, and as a reaction, the theory of punishment is revised, or rather should be revised. If in fact crime is a sickness, punishment is the cure for it and cannot be conceived of otherwise; thus all the discussion aroused revolves round knowing what punishment should be to fulfil its role as a remedy. But if crime is in no way pathological, the object of punishment cannot be to cure it and its true function must be sought elsewhere. . . .

DISCUSSION QUESTIONS

1. According to Durkheim's theory, criminal behavior exists in all societies. Consider the possibility of a society without the ability to punish criminal behavior (no prisons, no courts). How would individuals respond to crime? What informal social control mechanisms would help to maintain order?

2. How could you use Durkheim's theory as the basis for a research project on deviant behavior? What hypotheses could you test that would challenge or support the functionalist view of crime?

INFOTRAC COLLEGE EDITION

You can use your access to InfoTrac College Edition to learn more about the subjects covered in this essay. Some suggested search terms include:

criminality   normality
functionalism   pathology

28

The Medicalization of Deviance

PETER CONRAD AND JOSEPH W. SCHNEIDER

This essay outlines the social construction of social deviance. The authors specifically refer to the medical profession as redefining certain deviant behaviors as "illness," rather than as "badness." They argue that the "medicalization of deviance changes the social response to such behavior to one of treatment rather than punishment."

From: Peter Conrad and Joseph W. Schneider. 1992. *Deviance and Medicalization: From Badness to Sickness*. Philadelphia: Temple University Press, pp. 28–37.

Consider the following situations. A woman rides a horse naked through the streets of Denver claiming to be Lady Godiva and after being apprehended by authorities, is taken to a psychiatric hospital and declared to be suffering from a mental illness. A well-known surgeon in a Southwestern city performs a psychosurgical operation on a young man who is prone to violent outbursts. An Atlanta attorney, inclined to drinking sprees, is treated at a hospital clinic for his disease, alcoholism. A child in California brought to a pediatric clinic because of his disruptive behavior in school is labeled hyperactive and is prescribed methylphenidate (Ritalin) for his disorder. A chronically overweight Chicago housewife receives a surgical intestinal bypass operation for her problem of obesity. Scientists at a New England medical center work on a million-dollar federal research grant to discover a heroin-blocking agent as a "cure" for heroin addiction. What do these situations have in common? In all instances medical solutions are being sought for a variety of deviant behaviors or conditions. We call this "the medicalization of deviance" and suggest that these examples illustrate how medical definitions of deviant behavior are becoming more prevalent in modern industrial societies like our own. The historical sources of this medicalization, and the development of medical conceptions and controls for deviant behavior, are the central concerns of our analysis.

Medical practitioners and medical treatment in our society are usually viewed as dedicated to healing the sick and giving comfort to the afflicted. No doubt these are important aspects of medicine. In recent years the jurisdiction of the medical profession has expanded and encompasses many problems that formerly were not defined as medical entities. . . . There is much evidence for this general viewpoint—for example, the medicalization of pregnancy and childbirth, contraception, diet, exercise, child development norms—but our concern here is more limited and specific. Our interests focus on the medicalization of deviant behavior: the defining and labeling of deviant behavior as a medical problem, usually an illness and mandating the medical profession to provide some type of treatment for it. Concomitant with such medicalization is the growing use of medicine as an agent of social control, typically as medical intervention. Medical intervention as social control seeks to limit, modify, regulate, isolate, or eliminate deviant behavior with medical means and in the name of health. . . .

Conceptions of deviant behavior change, and agencies mandated to control deviance change also. Historically there have been great transformations in the definition of deviance—from religious to state-legal to medical-scientific. Emile Durkheim (1893/1933) noted in *The Division of Labor in Society* that as societies develop from simple to complex, sanctions for deviance change from repressive to restitutive or, put another way, from punishment to treatment or rehabilitation. Along with the change in sanctions and social control agent there is a corresponding change in definition or conceptualization of deviant behavior. For example, certain "extreme" forms of deviant drinking (what is now called alcoholism) have been defined as sin, moral weakness, crime, and most recently illness. . . . In modern industrial society there has been a substantial growth in the prestige, dominance, and jurisdiction of the medical profession (Freidson, 1970). It is only within the last century that physicians have become highly organized, consistently trained, highly paid, and sophisticated in their therapeutic techniques and abilities. . . . The medical profession dominates the organization of health care and has a virtual monopoly on anything that is defined as medical treatment, especially in terms of what constitutes "illness" and what is appropriate medical intervention. . . . Although Durkheim did not predict this medicalization, perhaps in part because medicine of his time was not the scientific, prestigious, and dominant profession of today, it is clear that medicine is the central restitutive agent in our society.

EXPANSION OF MEDICAL JURISDICTION
OVER DEVIANCE

When treatment rather than punishment becomes the preferred sanction for deviance, an increasing amount of behavior is conceptualized in a medical framework as illness. As noted earlier, this is not unexpected, since medicine has always functioned as an agent of social control, especially in attempting to "normalize" illness and return people to their functioning capacity in society. Public health and psychiatry have long been concerned with social behavior and have functioned traditionally as agents of social control (Foucault, 1965; Rosen, 1972). What is significant, however, is the expansion of this sphere where medicine functions in a social control capacity. In the wake of a general humanitarian trend, the success and prestige of modern biomedicine, the technological growth of the 20th century, and the diminution of religion as a viable agent of control, more and more deviant behavior has come into the province of medicine. In short, the particular, dominant designation of deviance has changed; much of what was badness (i.e., sinful or criminal) is now sickness. Although some forms of deviant behavior are more completely medicalized than others (e.g., mental illness), recent research has pointed to a considerable variety of deviance that has been treated within medical jurisdiction: alcoholism, drug addiction, hyperactive children, suicide, obesity, mental retardation, crime, violence, child abuse, and learning problems, as well as several other categories of social deviance. Concomitant with medicalization there has been a change in imputed responsibility for deviance: with badness the deviants were considered responsible for their behavior, with sickness they are not, or at least responsibility is diminished (see Stoll, 1968). The social response to deviance is "therapeutic" rather than punitive. Many have viewed this as "humanitarian and scientific" progress; indeed, it often leads to "humanitarian and scientific" treatment rather than punishment as a response to deviant behavior. . . .

A number of broad social factors underlie the medicalization of deviance. As psychiatric critic Thomas Szasz (1974) observes, there has been a major historical shift in the manner in which we view human conduct:

With the transformation of the religious perspective of man into the scientific, and in particular the psychiatric, which became fully articulated during the nineteenth century, there occurred a radical shift in emphasis away from viewing man as a *responsible agent acting in and on the world* and toward viewing him *as a responsive organism being acted upon* by biological and social "forces." (p. 149)

This is exemplified by the diffusion of Freudian thought, which since the 1920s has had a significant impact on the treatment of deviance, the distribution of stigma, and the incidence of penal sanctions.

Nicholas Kittrie (1971), focusing on decriminalization, contends that the foundation of the therapeutic state can be found in determinist criminology, that it stems from the *parens patriae* power of the state (the state's right to help those who are unable to help themselves), and that it dates its origin with the development of juvenile justice at the turn of the century. He further suggests that criminal law has failed to deal effectively (e.g., in deterrence) with criminals and deviants, encouraging a use of alternative methods of control. Others have pointed out that the strength of formal sanctions is declining because of the increase in geographical mobility and the decrease in strength of traditional status groups (e.g., the family) and that medicalization offers a substitute method for controlling deviance (Pitts, 1968). The success of medicine in areas like infectious disease has led to rising expectations of what medicine can accomplish. In modern technological societies, medicine has followed a technological imperative—that the physician is responsible for doing everything possible for the patient—while neglecting such significant issues as the patient's rights and wishes and the impact of biomedical advances on society (Mechanic, 1973). Increasingly sophisticated medical technology has extended the potential of medicine as social control, especially in terms of psychotechnology (Chorover, 1973). Psychotechnology includes a variety of medical and quasimedical treatments or procedures: psychosurgery, psychoactive medications, genetic engineering, disulfiram (Antabuse), and methadone. Medicine is frequently a pragmatic way of dealing with a problem (Gusfield, 1975). Undoubtedly the increasing acceptance and dominance of a scientific world view and the increase in status and power of the medical profession have contributed significantly to the adoption and public acceptance of medical approaches to handling deviant behavior.

THE MEDICAL MODEL AND "MORAL NEUTRALITY"

The first "victories" over disease by an emerging biomedicine were in the infectious diseases in which specific causal agents—germs—could be identified. An image was created of disease as caused by physiological difficulties located *within* the human body. This was the medical model. It emphasized the internal and biophysiological environment and deemphasized the external and social psychological environment.

There are numerous definitions of "the medical model." . . . We adopt a broad and pragmatic definition: the medical model of deviance locates the source of deviant behavior within the individual, postulating a physiological, constitutional, organic, or, occasionally, psychogenic agent or condition that is assumed to cause the behavioral deviance. The medical model of deviance usually, although not always, mandates intervention by medical personnel with medical means as treatment for the "illness." Alcoholics Anonymous, for example, adopts a rather idiosyncratic version of the medical model—that alcoholism is a chronic disease caused by an "allergy" to alcohol—but actively discourages professional medical intervention. But by and large, adoption of the medical model legitimates and even mandates medical intervention.

The medical model and the associated medical designations are assumed to have a scientific basis and thus are treated as if they were morally neutral (Zola, 1975). They are not considered moral judgments but rational, scientifically verifiable conditions. . . . Medical designations *are* social judgments, and the adoption of a medical model of behavior, a political decision. When such medical designations are applied to deviant behavior, they are related directly and intimately to the moral order of society. In 1851 Samuel Cartwright, a well-known Southern physician, published an article in a prestigious medical journal describing the disease "drapetomania," which only affected slaves and whose major symptom was running away from the plantations of their white masters (Cartwright, 1851). Medical texts during the Victorian era routinely described masturbation as a disease or addiction and prescribed mechanical and surgical treatments for its cure (Comfort, 1967; Englehardt, 1974). Recently many political dissidents in the Soviet Union have been designated mentally ill, with diagnoses such as "paranoia with counterrevolutionary delusions" and "manic reformism," and hospitalized for their opposition to the political order (Conrad, 1977). Although these illustrations may appear to be extreme examples, they highlight the fact that all medical designations of deviance are influenced significantly by the moral order of society and thus cannot be considered morally neutral. . . .

Even after a social definition of deviance becomes accepted or legitimated, it is not evident what particular type of problem it is. Frequently there are intellectual disputes over the causes of the deviant behavior and the appropriate methods of control. These battles about deviance designation (is it sin, crime, or sickness?) and control are battles over turf: Who is the appropriate definer and treater of the deviance? Decisions concerning what is the proper deviance designation and hence the appropriate agent of social control are settled by some type of political conflict.

How one designation rather than another becomes dominant is a central sociological question. In answering this question, sociologists must focus on claims-making activities of the various interest groups involved and examine how one or another attains ownership of a given type of deviance or social problem and thus generates legitimacy for a deviance designation. Seen from this perspective, public facts, even those which wear a "scientific" mantle are treated as products of the groups or organizations that produce or promote them rather than as accurate reflections of "reality." The adoption of one deviance designation or another has consequences beyond settling a dispute about social control turf.

. . . When a particular type of deviance designation is accepted and taken for granted, something akin to a paradigm exists. There have been three major deviance paradigms: deviance as sin, deviance as crime, and deviance as sickness. When one paradigm and its adherents become the ultimate arbiter of "reality" in society, we say a hegemony of definitions exists. In Western societies, and American society in particular, anything proposed in the name of science gains great authority. In modern industrial societies, deviance designations have become increasingly medicalized. We call the change in designations from badness to sickness the medicalization of deviance. . . .

NOTES

Cartwright, S. W. Report on the diseases and physical peculiarities of the negro race. *N.O. Med. Surg. J.,* 1851, 7, 691–715.

Chorover, S. "Big Brother and psychotechnology." *Psychol. Today,* 1973, 7, 43–54 (Oct.).

Comfort, A. *The anxiety makers.* London: Thomas Nelson & Sons, 1967.

Conrad, P. Soviet dissidents, ideological deviance, and mental hospitalization. Presented at Midwest Sociological Society Meetings, Minneapolis, 1977.

Durkheim, E. *The division of labor in society.* New York: The Free Press, 1933. (Originally published, 1893).

Englehardt, H. T., Jr. The disease of masturbation: Values and the concept of disease. *Bull. Hist. Med.,* 1974. 48, 234–248 (Summer).

Foucault, M. *Madness and civilization.* New York: Random House, Inc. 1965.

Freidson, E. *Profession of medicine.* New York: Harper & Row, Publishers Inc. 1970.

Gusfield, J.R. Categories of ownership and responsibility in social issues: Alcohol abuse and automobile use. *J. Drug Issues,* 1975. 5, 285–303 (Fall).

Kittrie, N. *The right to be different: deviance and enforced therapy.* Baltimore: Johns Hopkins University Press, 1971.

Mechanic, D. Health and illness in technological societies. *Hastings Center Stud.* 1973. *1*(3), 7–18.

Pitts, J. Social control: The concept. In D. Sills (Ed.), *International Encyclopedia of Social Sciences.* (Vol. 14). New York: Macmillan Publishing Co., Inc. 1968.

Rosen, G. The evolution of social medicine. In H. E. Freeman, S. Levine, & L. Reeder (Eds.), *Handbook of medical sociology* (2nd ed.). Englewood Cliffs, N.J.: Prentice-Hall, Inc. 1972.

Stoll, C.S. Images of man and social control. *Soc. Forces,* 1968. 47, 119–127 (Dec.).

Szasz, T. *Ceremonial chemistry.* New York: Anchor Books, 1974.

Zola, I. K. In the name of health and illness: On some socio-political consequences of medical influence. *Soc. Sci. Med.,* 1975. 9, 83–87.

DISCUSSION QUESTIONS

1. Alcoholism is an example of a deviant behavior being medicalized. How has this altered the understanding of and treatment of alcoholism? How does the involvement of health professionals in the treatment of alcoholism influence societal reaction to excessive drinking?

2. Consider the debate over the use of castrating rapists. How does this illustrate the transformation of understanding rape as a move "from badness to sickness"? What assumptions guide the suggestion that rapists should be castrated as a way of stopping rape?

INFOTRAC COLLEGE EDITION

You can use your access to InfoTrac College Edition to learn more about the subjects covered in this essay. Some suggested search terms include:

medicalization

social constructionism

political interest groups

social control

sanction

29

Construction of Deafness

HARLAN LANE

This essay outlines the different constructions of deafness as a social problem. On the one hand, deafness has been historically viewed as an individual disability. On the other hand, the author presents the argument that the deaf are a linguistic minority. This second construction of deafness as a minority recognizes deaf culture and promotes a different social response to deaf people.

Deafness, too, has had many constructions; they differ with time and place. Where there were many deaf people in small communities in the last century, on Martha's Vineyard, for example, as in Henniker, New Hampshire, deafness was apparently not seen as a problem requiring special intervention. Most Americans had quite a different construction of deafness at that time, however: it was an individual affliction that befell family members and had to be accommodated within the family. The great challenge facing Thomas Gallaudet and Laurent Clerc in their efforts to create the first American school for the deaf was to persuade state legislatures and wealthy Americans of quite a different construction which they had learned in Europe: Deafness was not an individual but a social problem, deaf people had to be brought together for their instruction, special "asylums" were needed. Nowadays, two constructions of deafness in particular are dominant and compete for shaping deaf peoples' destinies. The one construes deaf as a category of disability; the other construes deaf as designating a member of a linguistic minority. There is a growing practice of capitalizing Deaf when referring specifically to its second construction. . . .

From: Leanard J. Davis (ed.). 1997. *The Disability Studies Reader.* New York: Routledge, pp. 153–171. Reprinted with permission.

DISABILITY VS. LINGUISTIC MINORITY

Each construction has a core client group. No one disputes the claim of the hearing adult become deaf from illness or aging that he or she has a disability and is not a member of Deaf culture. Nor, on the other hand, has any one yet criticized Deaf parents for insisting that their Deaf child has a distinct linguistic and cultural heritage. The struggle between some of the groups adhering to the two constructions persists across the centuries (Lane, 1984) in part because there is no simple criterion for identifying most childhood candidates as clients of the one position or the other. More generally, we can observe that late deafening and moderate hearing loss tend to be associated with the disability construction of deafness while early and profound deafness involve an entire organization of the person's language, culture and thought around vision and tend to be associated with the linguistic minority construction.

In general, we identify children as members of a language minority when their native language is not the language of the majority. Ninety percent of Deaf children, however, have hearing parents who are unable to effectively model the spoken language for most of them. Advocates of the disability construction contend these are hearing-impaired children whose language and culture (though they may have acquired little of either) are in principle those of their parents; advocates of the linguistic minority construction contend that the children's native language, in the sense of primary language, must be manual language and that their life trajectory will bring them fully into the circle of Deaf culture. Two archetypes for these two constructions, disability and linguistic minority, were recently placed side by side before our eyes on the U.S. television program, "Sixty Minutes." On the one hand, seven-year-old Caitlin Parton, representing the unreconstructed disability-as-impairment: presented as a victim of a personal tragedy, utterly disabled in communication by her loss of hearing but enabled by technology, and dedicated professional efforts (yes, we meet the surgeon), to approach normal, for which she yearns, as she herself explains. On the other hand, Roslyn Rosen, then president of the National Association of the Deaf, from a large Deaf family, native speaker of ASL, proud of her status as a member of a linguistic minority, insistent that she experiences life and the world fully and has no desire to be any different (*Sixty Minutes,* 1992). . . .

The construction of the deaf child as disabled is legitimized early on by the medical profession and later by the special education and welfare bureaucracy. When the child is sent to a special educational program and obliged to wear cumbersome hearing aids, his or her socialization into the role of disabled person is promoted. In face-to-face encounters with therapists and teachers the child learns to cooperate in promoting a view of himself or herself as disabled. Teachers label large numbers of these deaf children emotionally disturbed or learning disabled (Lane, 1992). Once labeled as "multiply handicapped" in this way, deaf children are treated differently—for example, placed in a less demanding academic program where they learn less, so the label is self-validating. In the end, the troubled-persons industry creates the disabled deaf person.

DEAF AS LINGUISTIC MINORITY

From the vantage point of Deaf culture, deafness is not a disability (Jones & Pullen, 1989). British Deaf leader Paddy Ladd put it this way: "We wish for the recognition of our right to exist as a linguistic minority group . . . Labeling us as disabled demonstrates a failure to understand that we are not disabled in any way within our own community" (Dant & Gregory, 1991, p. 14). . . .

Nevertheless, many in the disability rights movement, and even some Deaf leaders, have joined professionals in promoting the disability construction of all deafness. To defend this construction, one leading disability advocate, Vic Finkelstein, has advanced the following argument based on the views of the people directly concerned: Minorities that have been discriminated against, like blacks, would refuse an operation to eliminate what sets them apart, but this is not true for disabled people: "every (!) disabled person would welcome such an operation" (*Finkelstein's exclamation point*). And, from this perspective, Deaf people, he maintains, "have more in common with other disability groups than they do with groups based upon race and gender" (Finkelstein, 1991, p. 265). However, in fact, American Deaf people are more like blacks in that most would refuse an operation to eliminate what sets them apart (as Dr. Rosen did on "Sixty Minutes"). One U.S. survey of Deaf adults asked if they would like an implant operation so they could hear; more than eight out of 10 declined (Evans, 1989). When the magazine *Deaf Life* queried its subscribers, 87 percent of respondents said that they did not consider themselves handicapped.

There are other indications that American Deaf culture simply does not have the ambivalence that, according to Abberley, is called for in disability: "Impairment must be identified as a bad thing, insofar as it is an undesirable consequence of a distorted social development, at the same time as it is held to be a positive attribute of the individual who is impaired" (Abberley, 1987, p. 9). American Deaf people (like their counterparts in many other nations) think cultural Deafness is a good thing and would like to see more of it. Expectant Deaf parents, like those in any other language minority, commonly hope to have Deaf children with whom they can share their language, culture and unique experiences. One Deaf mother from Los Angeles recounted to a researcher her reaction when she noticed that her baby did not react to Fourth of July fireworks: "I thought to myself, 'She must be deaf.' I wasn't disappointed; I thought, 'It will be all right. We are both deaf, so we will know what to do' (Becker, 1980, p. 55). . . .

Finkelstein acknowledges that many Deaf people reject the label "disabled" but he attributes it to the desire of Deaf people to distance themselves from social discrimination. What is missing from the construction of deafness is what lies at the heart of the linguistic minority construction: Deaf culture. Since people with disabilities are themselves engaged in a struggle to change the construction of disability, they surely recognize that disabilities are not "lying there in the road" but are indeed socially constructed. Why is this not applied to Deaf people? Not surprisingly, deafness is constructed differently in Deaf cultures than it is in hearing cultures.

Advocates of the disability construction for all deaf people, use the term "deaf community" to refer to all people with significant hearing impairment, on the model of "the disability community." So the term seems to legitimate the acultural perspective on Deaf people. When Ladd (*supra*) and other advocates of the linguistic minority construction speak of the Deaf community, however, the term refers to a much smaller group with a distinct manual language, culture, and social organization. . . .

Deaf cultures do not exist in a vacuum. Deaf Americans embrace many cultural values, attitudes, beliefs and behaviors that are part of the larger American culture and, in some instances, that are part of ethnic minority cultures such as African-American, Hispanic-American, etc. Because hearing people have obliged Deaf people to interact with the larger hearing society in terms of a disability model, that model has left its mark on Deaf culture. In particular, Deaf people frequently have found themselves recipients of unwanted special services provided by hearing people. "In terms of its economic, political and social relations to hearing society, the Deaf minority can be viewed as a colony" (Markowicz & Woodward, 1978, p. 33). As with colonized peoples, some Deaf people have internalized the "other's" (disability) construction of them alongside their own cultural construction (Lane, 1992). For example, they may be active in their Deaf club and yet denigrate skilled use of ASL as "low sign"; "high sign" is a contact variety of ASL that is closer to English-language word order. The Deaf person who uses a variety of ASL marked as English frequently has greater access to wider resources such as education and employment. Knowing when to use which variety is an important part of being Deaf (Johnson & Erting, 1989). Granted that culturally Deaf people must take account of the disability model of deafness, that they sometimes internalize it, and that it leaves its mark on their culture, all this does not legitimize that model—any more than granting that African-Americans had to take account of the construction of the slave as property, sometimes internalized that construction, and found their culture marked by it legitimizes that construction of their ethnic group. . . .

It is undeniable that culturally Deaf people have great common cause with people with disabilities. Both pay the price of social stigma. Both struggle with the troubled-persons industries for control of their destiny. Both endeavor to promote their construction of their identity in competition with the interested (and generally better funded) efforts of professionals to promote *their* constructions. And Deaf people have special reasons for solidarity with people with hearing impairments; their combined numbers have created services, commissions and laws that the DEAF-WORLD alone probably could not have achieved. Solidarity, yes, but when culturally Deaf people allow their special identity to be subsumed under the construct of disability they set themselves up for wrong solutions and bitter disappointments.

It is because disability advocates think of Deaf children as disabled that they want to close the special schools and absurdly plunge Deaf children into hearing classrooms in a totally exclusionary program called inclusion. It is because government is allowed to proceed with a disability construction of cultural Deafness that the U.S. Office of Bilingual Education and Minority Language Affairs has refused for decades to provide special resources for schools with large numbers of

ASL-using children although the law requires it to do so for children using any other non-English language. It is because of the disability construction that court rulings requiring that children who do not speak English receive instruction initially in their best language have not been applied to ASL-using children. It is because of the disability construction that the teachers most able to communicate with Britain's Deaf children are excluded from the profession on the pretext that they have a disqualifying disability. It is because lawmakers have been encouraged to believe by some disability advocates and prominent deaf figures that Deaf people are disabled that, in response to the Gallaudet Revolution, the U.S. Congress passed a law, not recognizing ASL or the DEAF-WORLD as a minority, but a law establishing another institute of *health,* The National Institute on Deafness and Other Communications Disorders [sic], operated by the deafness troubled persons industry, and sponsoring research to reduce hereditary deafness. It is because of the disability construction that organizations *for* the Deaf (e.g., the Royal National Institute for the Deaf) are vastly better funded by government that organizations *of* the Deaf (e.g., the British Deaf Association).

One would think that people with disabilities might be the first to grasp and sympathize with the claims of Deaf people that they are victims of a mistaken identity. People with disabilities should no more resist the self-construction of culturally Deaf people, than Deaf people should subscribe to a view of people with disabilities as tragic victims of an inherent flaw.

CHANGING TO THE LINGUISTIC MINORITY CONSTRUCTION

Suppose our society were generally to adopt a disability construction of deafness for most late-deafened children and adults and a linguistic minority construction of Deaf people for most others, how would things change? The admirable Open University course, *Issues in Deafness* (1991) prompted these speculations.

1. Changing the construction changes the legitimate authority concerning the social problem. In many areas, such as schooling, the authority would become Deaf adults, linguists and sociologists, among others. There would be many more service providers from the minority: Deaf teachers, foster and adoptive parents, information officers, social workers, advocates. Non-Deaf service providers would be expected to know the language, history, and culture of the Deaf linguistic minority.

2. Changing the construction changes how behavior is construed. Deaf people would be expected to use ASL (in the U.S.) and to have interpreters available; poor speech would be seen as inappropriate.

3. Changing the construction may change the legal status of the social problem group. Most Deaf people would no longer claim disability benefits or services under the present legislation for disabled people. The services to which the Deaf linguistic minority has a right in order to obtain equal treatment

under the law would be provided by other legislation and bureaucracies. Deaf people would receive greater protection against employment discrimination under civil rights laws and rulings. Where there are special provisions to assist the education of linguistic minority children, Deaf children would be eligible.

4. Changing the construction changes the arena where identification and labeling take place. In the disability construction, deafness is medicalized and labeled in the audiologist's clinic. In the construction as linguistic minority, deafness is viewed as a social variety and would be labeled in the peer group.

5. Changing the construction changes the kinds of intervention. The Deaf child would not be operated on for deafness but brought together with other Deaf children and adults. The disability construction orients hearing parents to the question, what can be done to mitigate my child's impairment? The linguistic minority construction presents them with the challenge of insuring that their child has language and role models from the minority (Hawcroft, 1991).

NOTES

Abberley, P. (1987) The concept of oppression and the development of a social theory of disability. *Disability, Handicap and Society,* 2, pp. 5–19.

Becker, G. (1980) *Growing Old in Silence* (Berkeley, University of California Press).

Cant, T. & Gregory, S. (1991) Unit 8. The social construction of deafness in Open University (eds.) *Issues in Deafness* (Milton Keynes, Open University).

Evans, J. W. (1989) Thoughts on the psychosocial implications of cochlear implantation in children, in E. Owens & D. Kessler, (eds.) *Cochlear Implants in Young Deaf Children.* (Boston, Little, Brown).

Finkelstein, V. (1991) 'We' are not disabled, 'you' are, in: S. Gregory & G. M. Hartley, (eds.) *Constructing Deafness.* London, (Eds.), Pinter.

Hawcroft, L. (1991) Block 2, unit 7, Whose welfare? in: Open University (eds.) *Issues in Deafness* (Milton Keynes, Open University).

Johnson, R. E. & Erting, C. (1989) Ethnicity and socialization in a classroom for deaf children, in C. Lucas (ed.) *The sociolinguistics of the Deaf Community,* pp. 41–84 (New York, Academic Press).

Jones, L. & Pullen, G. (1989) 'Inside we are all equal': a European social policy survey of people who are deaf, in L. Barton (ed.) *Disability and Dependency* (Bristol, PA, Taylor & Francis/Falmer Press).

Lane, H. (1984) *When the Mind Hears: a history of the deaf.* (New York, Random House).

Lane, H. (1992) *The Mask of Benevolence: disabling the deaf community* (New York, Alfred Knopf).

Markowicz, H. & Woodward, J. (1978) Language and the maintenance of ethnic boundaries in the deaf community, *Communication and Cognition,* 11, pp. 29–38.

Open University (1991) *Issues in deafness.* (Milton Keynes, Open University).

Sixty Minutes (1992) Caitlin's Story, 8 November.

DISCUSSION QUESTIONS

1. Compare and contrast the two constructions of deafness presented in this essay. How does the disability construction influence the "treatment" of deaf people? If we consider deafness as a cultural minority, will this change individual interaction with deaf people, and, if so, how?

2. Another linguistic minority issue in the news in recent years is the use of Ebonics. Can you draw any parallels between the use of American Sign Language (ASL) and the Ebonics debate?

INFOTRAC COLLEGE EDITION

You can use your access to InfoTrac College Edition to learn more about the subjects covered in this essay. Some suggested search terms include:

American sign language
disability

linguistic minority
social construction

30

Children, Prostitution, and Identity

A Case Study from a Tourist Resort in Thailand

HEATHER MONTGOMERY

In this essay, the author discusses child prostitution in a small village in Thailand. She challenges the conventional stigmatization of child prostitution as simply an evil exploitation of young girls. Although not ignoring the harm done to children, she shows how child prostitutes and their families reject the stigma of prostitution, seeing a lack of contribution to family's income as also deviant.

From: Kamala Kempadoo and Jo Doezema, eds. 1998. *Global Sex Workers: Rights, Resistance, and Redemption.* New York: Routledge, pp. 139–150. Reprinted with permission.

Child prostitution is viewed as an evil which must be eradicated by all means possible. This is a perfectly understandable response and one with which few people, except possibly pedophiles or others with obvious ulterior motives, would disagree. However with all the concern about child prostitution, individual children who sell sex have been largely overlooked. . . . In all the information received from the media about child prostitution, there is little about the children themselves. The unvoiced implications are always that only adults can represent children, that only adults can fully understand the situation, and that children may be prostitutes but they cannot understand it and certainly cannot analyze it. There is therefore little for them to say. As knowledge of the extent and nature of child abuse in the West grows, there is even greater confidence in speaking on behalf of child prostitutes: the effects of prostitution on them are stated with conviction.

My contention in this chapter is not that child prostitution is acceptable or that it is in any way beneficial to the children concerned in the long term. What was apparent, however, from my own fieldwork was how very differently campaigners against child prostitution and the children who actually sold sex conceptualized prostitution. The children in my study had very different means of seeing themselves and what they did and they constructed their social and personal identities in various ways. Prostitution was how they earned money but it said little about more pertinent issues such as loyalty, filial duty and private morality. It is these qualities that were important to the children and prostitution, far from negating them, actually reinforced them.

ATTITUDES TO PROSTITUTION

Attitudes to prostitution in general, and child prostitution in particular, are never neutral and inevitably the latter is viewed as a problem. Although the source of the problem is differently located for child prostitutes abroad and child prostitutes at home, the image of children selling their bodies for profit arouses indignation and outrage. It is a powerful metaphor for many things; societal decay, moral degeneracy, capitalism and patriarchy in its extreme form and the exploitation of Third World peoples by Westerners, in the form of both business and tourism. To study child prostitution in developing countries therefore, especially from the supposedly dispassionate and neutral standpoint of the observer anthropologist, is extremely difficult because there is a great deal of pressure to examine it as a prelude to stopping it. It is very hard to disassociate research from policy and therefore child prostitution must always be cast in terms of being a problem needing a solution, either the problem of how to stop Westerners going abroad and buying sex or how to stop children in overseas countries becoming prostitutes. There is a tendency to perceive prostitution as a moral issue, set apart from the wider political economy. It is seen as an aberration which demands immediate action and quick solutions. Prostitutes themselves, both adults and children, are viewed in a particular light which either regards them as depraved individuals in a functioning society or oppressed characters in a base world. Either way, the effects of economics on their lives and the amount of control over what they do are glossed over and ignored. . . . There is a moral bias inherent in most studies of prostitution so that even those who want to study prostitution sympathetically come to it with a set of expectations that are fulfilled because they are looking for them. . . .

It is extremely difficult to study the children's lives objectively and to say that children do know something of "hope or joy or uplift" without accusations of condoning child abuse which is automatically equated with child prostitution. . . . The desire for information is for stories which emphasize the degradation and abuse of children not the mundane aspects of their lives or even the areas of their lives away from prostitution. . . . Although sex with children is widely condemned, it is still an issue which causes great, if appalled, curiosity. There is a continuing fascination with it which is fed by the media and the NGOs and it is clearly an issue which both repels and allures. While newspaper articles often claim to be raising awareness, they can also titillate. This public interest in child sex undoubtedly exists but it is not the straightforward reflection of outrage that it claims to be. It rarely raises awareness and frequently has the affect of harnessing prurient horror for political ends which often substitute understanding for sensationalism and moral outrage. Child prostitution is cast as a clear cut case of good and evil while ignoring the wider political economy that allows child prostitution to flourish. . . .

A CASE STUDY OF CHILD PROSTITUTION

During 1993 and 1994, I was based in a small community in a tourist resort town in Thailand which earned its income through the prostitution of the children that lived there. It was a small community of less than one hundred and fifty people and it was situated on the edge of a tourist town. It was a poor community without running water and only intermittent electricity but the people there rented land and had built up their own houses. There is a stereotype of child prostitution that claims that they are tricked into leaving home, or sold by impoverished parents into a brothel where they are repeatedly raped and terrorized into servicing up to twenty clients a night. . . . It is undeniable that child prostitution is risky and dangerous in Thailand and many children are caught in situations which present a great threat. The children with whom I worked were not in any of these categories however, and they present a different, but rarely acknowledged model of child prostitution. They were technically "free" in that they were not debt-bonded or kept in brothels and they lived with their parents. They were perhaps not typical of all child prostitutes in Thailand but they are an important group whose lives and identities challenge many of the expected stereotypes of child prostitution.

There were sixty-five children under the age of fifteen in this slum and at least forty had worked as prostitutes at some point. These children worked because they felt a strong obligation towards their families and believed that it was their duty to support their parents financially. Income generating opportunities were extremely limited in the slum and outside it: there was little regulated, legal work available. The land on which they lived was poor and would not support raising crops and they did not have the education or training to take on even menial jobs in the resort. In these circumstances, prostitution, especially with foreign clients, was the only job which brought in enough money and many children turned to prostitution as a way of fulfilling their perceived obligations. Poverty is an often-cited cause of prostitution and in this community it certainly played a part. However, poverty and prostitution should not be linked too simplistically. Most children from poor families are not prostitutes and poverty is not necessarily the root cause of all child prostitution. In the case of the children in this community, it is far more pertinent to examine obligation and its effect on prostitution. The most powerful mitigating circumstance for many children was not that they were earning money because they were poor but that they were earning money to help their parents.

To outsiders, the lifestyle of the villagers looked extremely unpleasant and squalid and certainly the villagers themselves never romanticized it but there was an internal dynamic to the life of the community that enabled the people there to continue with their own logic and set of ethics. None of the children liked prostitution but they did have strategies for rationalizing it and coming to terms with it. They had found an ethical system whereby the public selling of their bodies did not affect their private sense of humanity and virtues. When I asked one thirteen-year-old about selling her body, she replied "it's only my body" but when I asked about the difference between adultery and prostitution, she would tell me that adultery was very wrong. In her eyes, adultery was a betrayal of a private relationship whereas prostitution was simply done for money. She could make a clear conceptual difference between her body and what happened to it and what she perceived to be her innermost "self." . . . These children could delineate clear boundaries between what happened to their bodies and what affected their personal sense of identity and morality. Selling sex was not immoral because it violated no ethical codes. Betraying family members, failing to provide for parents or cheating on spouses or boyfriends was roundly condemned but exchanging sex for money especially when that money was used for moral ends, carried no stigma. . . .

STRATAGEMS AND CONTROL

The conditions that the children lived in were extremely difficult. Their lack of education, their poverty and ill health all made them vulnerable to abuse and exploitation by those with more power. Yet, prostitution is always seen as the ultimate indignity and one which is far beyond all others. It is also always seen as one which they passively accept. In all the accounts of child prostitution that I have read in Thailand and overseas . . . the children's passivity is always emphasized. They are weak, helpless and have no control at all over their own lives. Yet, . . . resistance, even in the harshest circumstances, is possible and does occur. What is seen as passivity by outsiders may in fact be a form of protest. While direct confrontation is rarely in the interests of the weak and powerless, an unwilling compliance is not a sign that they have given up or that they passively accept what is happening. Rather it is an acceptance that their options are limited and their position weak, but even within those limits, they do not have to believe in what they have to do. They do not have to accept the dominant ideology that identifies them and stigmatizes them as prostitutes when, in their own terms, they are dutiful and much-loved family members.

The children that I knew did not passively accept this abuse and having a sense of control over some aspects of their lives was fundamental to them. For some, prostitution was a bad option with better pay than other bad options while others complied with their family's wishes that they become prostitutes as a way of showing their filial duty. Others responded more actively and were aware that there are levels of inequality within their own circles which they could exploit and use to form power bases. The older children formed entourages of younger children that they could control and grant favors to in the knowledge that these younger children were then indebted to them. In this way, prestige, status and power were built up as certain children could command the time and attention of others. It may not seem a very great power to outsiders but it indicated the skillful way in which the children sought to optimize their status and make use of the limited options. . . .

For people who are poor and powerless, prostitution does not seem a unique and ultimate horror (as many outsiders view it) or something they have to be forced into but one difficult choice among many. They can be forced into many forms of work that they dislike such as scavenging or collecting garbage, neither of which pay nearly as well as prostitution. It is unlikely that given a full range of choices and options, many of the children would choose prostitution, but these options have never been open to them. Their choices are limited but they do attempt to exercise these, despite the limitations. They choose not to represent themselves as victims and not to align themselves with negative connotations that others have placed on prostitution. Their power and ability to change their situation are extremely constrained but their constructions of their own identities and ways of viewing the world show a clear difference between them and the passive victims that child prostitutes are constantly assumed to be.

CONCLUSION

The children that I worked with were undoubtedly exploited and forced into lifestyles that exposed them to many forms of abuse and oppression. Whatever the children said about prostitution, it is difficult to examine other ways of seeing their situation without encountering suggestions that understanding is the same as condoning that which is indefensible. It is hard not to sound like a moral relativist and argue that if the children do not see abuses, then no abuse has occurred. Personally I felt that these children were exploited but that this exploitation came not through prostitution but through their general poverty and social exclusion. There has been no intention or attempt to justify that in writing this chapter but I do not wish to position it as the ultimate evil. The children did not have a choice as to whether they were exploited or not or between prostitution and work-free childhood. If they were not prostitutes, they still would have been impoverished and probably forced into the illegal labor market in a sweat shop or as a scavenger. These children were neglected by the state and given few viable options by their society. In this situation examining prostitution in isolation from other economic and social choices is pointless and leads only to narrow moralistic arguments about whether prostitution is "right" or whether any prostitute, either adult or child "really" chooses prostitution. Child prostitution has been a major cause of concern in recent years but there has been no widening of the debate and instead, campaigning groups have simply become increasingly shrill in denouncing it. Yet despite the passion that child prostitution arouses, the children themselves have been largely silent. Many people are speaking in their name but very few people have listened to them and know who they are or how they perceive what they do. . . . I have not given any definitive ways of viewing child prostitution or made claims that all child prostitutes think of themselves in these ways. Rather, by giving a context to the children's lives, it is possible to understand better what motivates and sustains them and their families. Through this contextualization, I have suggested other ways of understanding children who works as prostitutes and added complexities to what is becoming an increasingly simplistic debate.

DISCUSSION QUESTIONS

1. The author argues that prostitution is not considered morally deviant by the children in her study. Can we apply this argument to adult prostitution in this country?

2. What societal factors lead Americans to view child prostitution as much more deviant than adult prostitution? What about male prostitution?

INFOTRAC COLLEGE EDITION

You can use your access to InfoTrac College Edition to learn more about the subjects covered in this essay. Some suggested search terms include:

child prostitution   political economy
family loyalty   sex workers
global sex trade   stigma

31

The Rich Get Richer
and the Poor Get Prison

JEFFREY H. REIMAN

This essay challenges the reader to view the criminal justice system from a radically different angle. Specifically, Jeffrey Reiman argues that the corrections system and broader criminal justice policy in the United States simply provide the illusion of fighting crime. In reality, he argues that criminal justice policies reinforce public fears of crimes committed by the poor. These policies, in turn, help to maintain a "criminal class" of disadvantaged people.

A criminal justice system is a mirror in which a whole society can see the darker outlines of its face. Our ideas of justice and evil take on visible form in it, and thus we see ourselves in deep relief. Step through this looking glass to view the American criminal justice system—and ultimately the whole society it reflects—from a radically different angle of vision.

In particular, entertain the idea that the goal of our criminal justice system is not to eliminate crime or to achieve justice, *but to project to the American public a visible image of the threat of crime as a threat from the poor.* To do this, the justice system must maintain the existence of a sizable population of poor criminals. To do this, it must fail in the struggle to eliminate the crimes that poor people commit, or even to reduce their number dramatically. Crime may, of course, occasionally decline, as it has recently—*but not because of criminal justice policies.* . . .

In recent years, we have tripled our prison population and, in cities like New York, allowed the police new freedom to stop and search people they suspect. No one can deny that if you lock enough people up, and allow the police greater and greater power to interfere with the liberty and privacy of citizens, you will eventually prevent some crime that might otherwise have taken place. . . . When I say that criminal justice policy is failing to reduce crime, I mean that it is failing to reduce it in any substantial way, that it is failing to make more than a marginal difference. But it is failing nonetheless, because our rates of crime remain extremely high, our crime-reduction strategies do not touch on the social causes of crime, and our citizens remain extremely fearful about criminal victimization, even after the recent declines. . . .

From: Jeffrey H. Reiman. 1997. *The Rich Get Richer and The Poor Get Prison?* Needham Heights, MA: Allyn & Bacon, pp. 1–9. Reprinted with permission.

Some years ago, I taught a seminar for graduate students titled "The Philosophy of Punishment and Rehabilitation." Many of the students were already working in the field of corrections as probation officers or prison guards or halfway-house counselors. First we examined the various philosophical justifications for legal punishment, and then we directed our attention to the actual functioning of our correctional system. For much of the semester we talked about the myriad inconsistencies and cruelties and overall irrationality of the system. We discussed the arbitrariness with which offenders are sentenced to prison and the arbitrariness with which they are treated once there. We discussed the lack of privacy and the deprivation of sources of personal identity and dignity, the ever-present physical violence, as well as the lack of meaningful counseling or job training within prison walls. We discussed the harassment of parolees, the inescapability of the "ex-con" stigma, the refusal of society to let a person finish paying his or her "debt to society," and the nearly total absence of meaningful noncriminal opportunities for the ex-prisoner. We confronted time and again the bald irrationality of a society that builds prisons to prevent crime knowing full well that they do not and that does not even seriously try to rid its prisons and post release practices of those features that guarantee a high rate of *recidicism:* the return to crime by prison alumni. How could we fail so miserably? We are neither an evil nor a stupid nor an impoverished people. How could we continue to bend our energies and spend our hard-earned tax dollars on cures we know are not working?

Toward the end of the semester I asked the students to imagine that, instead of designing a correctional system to reduce and prevent crime, we had to design one that would maintain a stable and visible "class" of criminals. What would it look like? The response was electrifying. In briefer and somewhat more orderly form, here is a sample of the proposals that emerged in our discussion:

First. It would be helpful to have laws on the books against drug use or prostitution or gambling—laws that prohibit acts that have no unwilling victim. This would make many people "criminals" for what they regard as normal behavior and would increase their need to engage in *secondary* crime (the drug addict's need to steal to pay for drugs, the prostitute's need for a pimp because police protection is unavailable, and so on).

Second. It would be good to give police, prosecutors, and/or judges broad discretion to decide who got arrested, who got charged, and who got sentenced to prison. This would mean that almost anyone who got as far as prison would know of others who committed the same crime but who either were not arrested or were not charged or were not sentenced to prison. This would assure us that a good portion of the prison population would experience their confinement as arbitrary and unjust and thus respond with rage, which would make them more "antisocial," rather than respond with remorse, which would make them feel more bound by social norms.

Third. The prison experience should be not only painful but also demeaning. The pain of loss of liberty might deter future crime. But demeaning and emasculating prisoners by placing them in an enforced childhood characterized by no privacy and no control over their time and actions, as well as by the constant threat of rape or assault, is sure to overcome any deterrent effect by weakening whatever capacities a prisoner had for self-control. Indeed, by humiliating and brutalizing prisoners we can be sure to increase their potential for aggressive violence.

Fourth. Prisoners should neither be trained in a marketable skill nor provided with a job after release. Their prison records should stand as a perpetual stigma to discourage employers from hiring them. Otherwise, they might be tempted *not* to return to crime after release.

Fifth. Ex-offenders' sense that they will always be different from "decent citizens," that they can never finally settle their debt to society, should be reinforced by the following means. They should be deprived for the rest of their lives of rights, such as the right to vote. They should be harassed by police as "likely suspects" and be subject to the whims of parole officers who can at any time threaten to send them back to prison for things no ordinary citizens could be arrested for, such as going out of town or drinking or fraternizing with the "wrong people."

And so on.

In short, *asked to design a system that would maintain and encourage the existence of a stable and visible "class of criminals," we "constructed" the American criminal justice system.*

. . . The practices of the criminal justice system keep before the public the *real* threat of crime and the *distorted* image that crime is primarily the work of the poor. The value of this *to those in positions of power* is that it deflects the discontent and potential hostility of Middle America away from the classes above them and toward the classes below them. If this explanation is hard to swallow, it should be noted in its favor that it not only explains our dismal failure to make a significant dent in crime but also explains why the criminal justice system functions in a way that is biased against the poor at every stage from arrest to conviction. Indeed, even at the earlier stage, when crimes are defined in law, the system primarily concentrates on the predatory acts of the poor and tends to exclude or deemphasize the equally or more dangerous predatory acts of those who are well off. In sum, I will argue that *the criminal justice system fails to reduce crime substantially while making it look as if crime is the work of the poor.* It does this in a way that conveys the image that the real danger to decent, law-abiding Americans comes from below them, rather than from above them, on the economic ladder. This image sanctifies the status quo with its disparities of wealth, privilege, and opportunity and thus serves the interests of the rich and powerful in America—the very ones who could change criminal justice policy if they were really unhappy with it.

Therefore, it seems appropriate to ask you to look at criminal justice "through the looking glass." On the one hand, this suggests a reversal of common expectations. Reverse your expectations about criminal justice and entertain the notion that the system's real goal is the very reverse of its announced goal. On the other hand, the figure of the looking glass suggests the prevalence of image over reality. My argument is that the system functions the way it does *because it maintains a particular image of crime: the image that it is a threat from the poor.* Of course, for this image to be believable there must be a reality to back it up. The system must actually fight crime—or at least some crime—but only enough to keep it from getting out of hand and to keep the struggle against crime vividly and dramatically in the public's view—never enough to substantially reduce or eliminate crime.

I call this outrageous way of looking at criminal justice policy the *Pyrrhic defeat* theory. A "Pyrrhic victory" is a military victory purchased at such a cost in troops and treasure that it amounts to a defeat. The Pyrrhic defeat theory argues that the failure of the criminal justice system yields such benefits to those in positions of power that it amounts to success. . . .

The Pyrrhic defeat theory has several components. Above all, it must provide an explanation of *how* the failure to reduce crime substantially could benefit anyone—anyone other than criminals, that is. I argue that the failure to reduce crime substantially broadcasts a potent *ideological* message to the American people, a message that benefits and protects the powerful and privileged in our society by legitimating the present social order with its disparities of wealth and privilege and by diverting public discontent and opposition away from the rich and powerful and onto the poor and powerless.

To provide this benefit, however, not just any failure will do. It is necessary that the failure of the criminal justice system take a particular shape. *It must fail in the fight against crime while making it look as if serious crime and thus the real danger to society is the work of the poor.* The system accomplishes this both by what it does and by what it refuses to do. I argue that the criminal justice system refuses to label and treat as a crime a large number of acts that produces as much or more damage to life and limb as the crimes of the poor. Even among the acts treated as crimes, the criminal justice system is biased from start to finish in a way that guarantees that *for the same crimes* members of the lower classes are much more likely than members of the middle and upper classes to be arrested, convicted, and imprisoned—thus providing living "proof" that crime is a threat from the poor.

Our criminal justice system is characterized by beliefs about what is criminal, and beliefs about how to deal with crime, that predate industrial society. Rather than being anyone's conscious plan, the system reflects attitudes so deeply embedded in tradition as to appear natural. To understand why it persists even though it fails to protect us, all that is necessary is to recognize that, on the one hand, those who are the most victimized by crime are not those in positions to make and implement policy. Crime falls more frequently and more harshly on the poor than on the better off. On the other hand, there are enough benefits to the wealthy from the identification of crime with the poor and system's failure to reduce crime that those with the power to make profound changes in the system feel no compulsion nor see any incentive to make them. In short, the criminal justice system came into existence in an earlier epoch and persists in the present because, even though it is failing, indeed because of the way it fails, it generates no effective demand for change. When I speak of the criminal justice system as "designed to fail," I mean no more than this. I call this explanation of the existence and persistence of our failing criminal justice system the *historical inertia* explanation. . . .

DISCUSSION QUESTIONS

1. What does Reiman mean in arguing that the current criminal justice system works to maintain a class of criminals? Do you agree or disagree that our corrections system fails to rehabilitate and fails to deter crime?

2. If you had the power to change the corrections system in the United States, what changes would you make to help reduce and prevent crime?

INFOTRAC COLLEGE EDITION

You can use your access to InfoTrac College Edition to learn more about the subjects covered in this essay. Some suggested search terms include:

corrections
criminal justice policies
criminal justice system

recidivism
secondary crime

32

The Communist Manifesto

KARL MARX AND FRIEDRICH ENGELS

The analysis of the class system under capitalism, as developed by Marx and Engels, continues to influence sociological understanding of the development of capitalism and the structure of the class system. In this classic essay, first published in 1848, Marx and Engels define the class system in terms of the relationships between capitalism, the bourgeoisie, and the proletariat. Their analysis of the growth of capitalism and its influence on other institutions continue to provide a compelling portrait of an economic system based on the pursuit of profit.

1. BOURGEOIS AND PROLETARIANS

The history of all hitherto existing society is the history of class struggles. . . .

Modern industry has established the world market, for which the discovery of America paved the way. This market has given an immense development to commerce, to navigation, to communication by land. This development has, in its turn, reacted on the extension of industry; and in proportion as industry, commerce, navigation, railways extended, in the same proportion the bourgeoisie developed, increased its capital, and pushed into the background every class handed down from the Middle Ages.

We see, therefore, how the modern bourgeoisie is itself the product of a long course of development, of a series of revolutions in the modes of production and of exchange. . . .

The bourgeoisie has at last, since the establishment of modern industry and of the world market, conquered for itself, in the modern representative state, exclusive political sway. The executive of the modern state is but a committee for managing the common affairs of the whole bourgeoisie.

The bourgeoisie, historically, has played a most revolutionary part.

The bourgeoisie, wherever it has got the upper hand, has put an end to all feudal, patriarchal, idyllic relations. It has pitilessly torn asunder the motley feudal ties that bound man to his 'natural superiors,' and has left remaining no other nexus between man and man than naked self-interest, than callous 'cash payment.' It has

From: Karl Marx and Frederick Engels. 1998. *Manifesto of the Communist Party.* With introduction by Eric Hobsbawm. New York: Verso, pp. 33–51. Used with permission.

drowned the most heavenly ecstasies of religious fervour, of chivalrous enthusiasm, of philistine sentimentalism, in the icy water of egotistical calculation. It has resolved personal worth into exchange value, and in place of the numberless indefeasible chartered freedoms, has set up that single, unconscionable freedom - free trade. In one word, for exploitation, veiled by religious and political illusions, it has substituted naked, shameless, direct, brutal exploitation.

The bourgeoisie has stripped of its halo every occupation hitherto honoured and looked up to with reverent awe. It has converted the physician, the lawyer, the priest, the poet, the man of science, into its paid wage labourers.

The bourgeoisie has torn away from the family its sentimental veil, and has reduced the family relation to a mere money relation. . . .

The need of a constantly expanding market for its products chases the bourgeoisie over the whole surface of the globe. It must nestle everywhere, settle everywhere, establish connections everywhere.

The bourgeoisie has through its exploitation of the world market given a cosmopolitan character to production and consumption in every country. To the great chagrin of reactionists, it has drawn from under the feet of industry the national ground on which it stood. All old-established national industries have been destroyed or are daily being destroyed. They are dislodged by new industries, whose introduction becomes a life and death question for all civilized nations, by industries that no longer work up indigenous raw material, but raw material drawn from the remotest zones; industries whose products are consumed, not only at home, but in every quarter of the globe. In place of the old wants, satisfied by the productions of the country, we find new wants, requiring for their satisfaction the products of distant lands and climes. In place of the old local and national seclusion and self-sufficiency, we have intercourse in every direction, universal interdependence of nations. And as in material, so also in intellectual production. The intellectual creations of individual nations become common property. National one-sidedness and narrow-mindedness become more and more impossible, and from the numerous national and local literatures, there arises a world literature.

The bourgeoisie, by the rapid improvement of all instruments of production, by the immensely facilitated means of communication, draws all, even the most barbarian, nations into civilization. The cheap prices of its commodities are the heavy artillery with which it batters down all Chinese walls, with which it forces the barbarians' intensely obstinate hatred of foreigners to capitulate. It compels all nations, on pain of extinction, to adopt the bourgeois mode of production; it compels them to introduce what it calls civilization into their midst, i.e., to become bourgeois themselves. In one word, it creates a world after its own image.

The bourgeoisie has subjected the country to the rule of the towns. It has created enormous cities, has greatly increased the urban population as compared with the rural, and has thus rescued a considerable part of the population from the idiocy of rural life. Just as it has made the country dependent on the towns, so it has made barbarian and semi-barbarian countries dependent on the civilized ones, nations of peasants on nations of bourgeois, the East on the West.

The bourgeoisie keeps more and more doing away with the scattered state of the population, of the means of production, and of property. It has agglomerated population, centralized means of production, and has concentrated property in a few hands. The necessary consequence of this was political centralization. Independent, or but loosely connected provinces, with separate interests, laws, governments and systems of taxation, became lumped together into one nation, with one government, one code of laws, one national class interest, one frontier and one customs tariff. . . .

The weapons with which the bourgeoisie felled feudalism to the ground are now turned against the bourgeoisie itself.

But not only has the bourgeoisie forged the weapons that bring death to itself; it has also called into existence the men who are to wield those weapons – the modern working class – the proletarians.

In proportion as the bourgeoisie, i.e., capital, is developed, in the same proportion is the proletariat, the modern working class, developed – a class of labourers, who live only so long as they find work, and who find work only so long as their labour increases capital. These labourers, who must sell themselves piecemeal, are a commodity, like every other article of commerce, and are consequently exposed to all the vicissitudes of competition, to all the fluctuations of the market.

Owing to the extensive use of machinery and to division of labour, the work of the proletarians has lost all individual character, and, consequently, all charm for the workman. He becomes an appendage of the machine, and it is only the most simple, most monotonous, and most easily acquired knack, that is required of him. Hence, the cost of production of a workman is restricted, almost entirely, to the means of subsistence that he requires for his maintenance, and for the propagation of his race. But the price of a commodity, and therefore also of labour, is equal to its cost of production. In proportion, therefore, as the repulsiveness of the work increases, the wage decreases. Nay more, in proportion as the use of machinery and division of labour increases, in the same proportion the burden of toil also increases, whether by prolongation of the working hours, by increase of the work exacted in a given time or by increased speed of the machinery, etc.

Modern industry has converted the little workshop of the patriarchal master into the great factory of the industrial capitalist. Masses of labourers, crowded into the factory, are organized like soldiers. As privates of the industrial army they are placed under the command of a perfect hierarchy of officers and sergeants. Not only are they slaves of the bourgeois class, and of the bourgeois state; they are daily and hourly enslaved by the machine, by the overseer, and, above all, by the individual bourgeois manufacturer himself. The more openly this despotism proclaims gain to be its end and aim, the more petty, the more hateful and the more embittering it is.

The less the skill and exertion of strength implied in manual labour, in other words, the more modern industry becomes developed, the more is the labour of men superseded by that of women. Differences of age and sex have no longer any distinctive social validity for the working class. All are instruments of labour, more or less expensive to use, according to their age and sex. . . .

But with the development of industry the proletariat not only increases in number; it becomes concentrated in greater masses, its strength grows, and it feels that strength more. The various interests and conditions of life within the ranks of the proletariat are more and more equalized, in proportion as machinery obliterates all distinctions of labour, and nearly everywhere reduces wages to the same low level. The growing competition among the bourgeois, and the resulting commercial crises, make the wages of the workers ever more fluctuating. The unceasing improvement of machinery, ever more rapidly developing, makes their livelihood more and more precarious; the collisions between individual workmen and individual bourgeois take more and more the character of collisions between two classes. Thereupon the workers begin to form combinations (trade unions) against the bourgeois. . . .

This organization of the proletarians into a class, and consequently into a political party, is continually being upset again by the competition between the workers themselves. But it ever rises up again, stronger, firmer, mightier. It compels legislative recognition of particular interests of the workers, by taking advantage of the divisions among the bourgeoisie itself. . . .

Altogether, collisions between the classes of the old society further, in many ways, the course of development of the proletariat. The bourgeoisie finds itself involved in a constant battle: at first with the aristocracy; later on, with those portions of the bourgeoisie itself, whose interests have become antagonistic to the progress of industry; at all times, with the bourgeoisie of foreign countries. In all these battles it sees itself compelled to appeal to the proletariat, to ask for its help, and thus to drag it into the political arena. The bourgeoisie itself, therefore, supplies the proletariat with its own elements of political and general education, in other words, it furnishes the proletariat with weapons for fighting the bourgeoisie.

Further, as we have already seen, entire sections of the ruling classes are, by the advance of industry, precipitated into the proletariat, or are at least threatened in their conditions of existence. These also supply the proletariat with fresh elements of enlightenment and progress.

Finally, in times when the class struggle nears the decisive hour, the process of dissolution going on within the ruling class, in fact within the whole range of old society, assumes such a violent, glaring character, that a small section of the ruling class cuts itself adrift, and joins the revolutionary class, the class that holds the future in its hands. Just as, therefore, at an earlier period, a section of the nobility went over to the bourgeoisie, so now a portion of the bourgeoisie goes over to the proletariat, and in particular, a portion of the bourgeois ideologists, who have raised themselves to the level of comprehending theoretically the historical movement as a whole.

Of all the classes that stand face to face with the bourgeoisie today, the proletariat alone is a really revolutionary class. The other classes decay and finally disappear in the face of modern industry; the proletariat is its special and essential product. . . .

DISCUSSION QUESTIONS

1. What evidence do you see in contemporary society of Marx and Engels' claim that the need for a constantly expanding market means that capitalism "nestles everywhere"?
2. How do Marx and Engels depict the working class and what evidence do you see of their argument by looking at the contemporary labor market?

INFOTRAC COLLEGE EDITION

You can use your access to InfoTrac College Edition to learn more about the subjects covered in this essay. Some suggested search terms include:

class analysis
communism
global capitalism

Marxist theory
proletarianization
working class

33

Great Divides

THOMAS M. SHAPIRO

Thomas Shapiro provides an overview of the dimensions of social stratification commonly studied by sociologists. In contrast to the "American Dream" suggesting that everyone has an equal chance to succeed, class, race, and gender shape the opportunities that people have in the U.S. system of stratification. Shapiro also identifies some of the social changes that are currently influencing the structure of opportunity in the United States.

We all know that "the American Dream" means economic opportunity, social mobility, and material success. It means that all of us do better than our parents did; that our standard of living improves over the course of our lifetimes; and that we can own our homes. It also means that regardless of background or origin, all of us have a chance to succeed. As Americans, we do not favor the idea that some people or groups are privileged by their background, while others are systematically blocked from success. Economic opportunity and social equality are thus twin pillars of the American belief system.

From: Thomas M. Shapiro (ed.). 1998. *Great Divides: Readings in Social Inequality in the United States.* Mountain View, CA: Mayfield Publishing Co., pp. 1–6. Reprinted with permission.

In the decades following the end of World War II—from 1945 to the early 1970s—the American Dream became a reality for millions of families and individuals in our society. People were able to buy homes, purchase new automobiles, take vacations, and see their standard of living steadily rise. At the same time, our society made strides toward improved social conditions for minorities and greater equality among groups of people. For these changes, many observers credit pressure from the Civil Rights movement, the women's movement, labor unions, and other social organizations and movements, along with changes in public policy and a growing, prosperous economy. A better standard of living and a narrowing gap between rich and poor—together, these changes reinforced vital elements of the American credo. Many Americans assumed that material life would continue to improve indefinitely and that the economic and social gaps among groups would continue to narrow.

In the 1970s, however, the economy slowed down and began to stagnate. The standard of living began to fall, poverty began to increase, and membership in the middle class became more tenuous. Just to stay in the same place, many families had to adapt, as more women sought work outside the home, living spaces became smaller, and time for leisure activities and for families getting together was reduced as people worked harder for longer hours. The 1980s saw a return to some of the inequalities of the past, as measured by comparisons of the relative material positions of African and European Americans and by comparisons of women's income with men's. Many observers began to feel that the United States was losing ground in its struggle against social inequality.

The changes we are seeing in our society during the 1990s—stagnating living standards, increasing poverty, a precarious middle class, and a growing gap between rich and poor—are probably the result of the specific way that economic restructuring is taking place in the United States. Economic restructuring is one way in which corporations, businesses, bureaucracies, individuals, and various levels of government respond to the challenge of keeping the United States a preeminent nation in the emerging global economy. Restructuring includes a variety of strategies, policies, and practices—including corporate and government downsizing, deindustrialization, movement of production from the central city to the suburbs, and changes from a permanent to a contingent workforce, from local to global production, from Fordist (mass assembly line) production to more flexible and decentralized production, and from a manufacturing-based to a service-based economy. Together with other factors, including the increasing diversity of the American population, economic restructuring is giving a new shape to social and economic inequality in our society. . . .

SOCIAL STRATIFICATION

Strata means layers, or hierarchy; *social stratification* is a process or system by which groups of people are arranged into a hierarchical social structure. Dimensions of power and powerlessness undergird this hierarchy and influence subsequent opportunities for rewards. Consequently, people have differential access to—and control over—prospects, rewards, and whatever is of value in society at any given time, based on their hierarchical positions, primarily because of social factors. Social stratification, an expression of social inequality, is so pervasive in American society that an entire field of sociology is devoted to its study.

Social-stratification systems are based primarily on either ascribed status or achieved status. *Ascribed status* is a social position typically designated or given to each person at birth. In a society with ascribed status, differential opportunities, rewards, privileges, and power are provided to individuals according to criteria fixed at birth. *Achieved status* is a social position gained as a result of ability or effort. This type of stratification is evident in all industrial societies, including the United States.

Given that social stratification is an expression of social inequality, how does inequality result, in turn, from social stratification? One leading scholar has proposed that inequality is produced by two different kinds of matching processes: "The jobs, occupations, and social roles in society are first matched to 'reward packages' of unequal value"; then, individuals are sorted and matched to particular jobs, occupations, and social roles through training and other institutional processes (Grusky, 1994, p. 3). Both parts of this matching process have been the subject of much investigation in sociology, as many inquiries have probed these two questions (Fischer et al., 1996, p. 7): (1) What determines how much people get for performing various economic roles and tasks? (2) What social and institutional processes determine who gets ahead and who falls behind in the competition for positions of unequal value?

Other questions sociologists ask about stratification include the following: How is ascribed status constructed over time? What are the institutional processes and practices that shape ascriptive stratification? To what extent does an ascribed status circumscribe people's opportunities and rewards? To what extent is achieved status fixed or open? What determines how and whether individuals are able to move through the occupational and wage structure in a system characterized by achieved status? When people can move through such a structure, how do they move? . . .

Social Stratification in the United States

Ascriptive stratification based on gender is found in nearly every society, and racial and ethnic stratification is almost as widespread. Nevertheless, social stratification in the United States is based primarily on achieved status, at least in theory. When this nation was formed, the founders deliberately distinguished it from nations where life chances and social rank were determined by birth. A core element of the American credo is that life chances are determined largely by talent, skill, hard work, and achievement. We believe that everyone has a fair shot at whatever is valued or prized and that no individual or group is unfairly advantaged or disadvantaged.

This belief does not mean that we expect everyone to achieve equal results; rather, we expect that everyone is starting with the same opportunities for achieving these different outcomes. Indeed, we tend to see differences in material success as the legitimate result of playing by the agreed-on rules. Although our national history is ambiguous about our implementation of social inequality, we normally take great exception when systemic and systematic differences in achievement clearly and directly result from public policy, varying or hidden rules, discrimination, or differential rewards for similar accomplishments. These pernicious factors produce what we think of as inequality.

Despite our egalitarian values and beliefs, social inequality has been an enduring fact of life and politics in the United States. Some groups of people have sufficient power—through family, neighborhood, school, or community—to maintain higher economic class positions and higher social status in American society. People in these groups have the ability to get and stay ahead in the competition for success. Further, social inequality has always been integrally bound up with three dimensions of social stratification: socioeconomic class, race and ethnicity, and gender. Divisions based on these three dimensions are deeply embedded in the social structures and institutions that define our lives, so these three constructs must be at the center of any analysis of social inequality. The integration of these constructs is not simple, however, because we lack both a common understanding of them and an agreement as to their significance in the structure of social inequality.

An example illustrates this lack of common ground. Whenever I ask my students what they mean by *class*, they say they are sure that classes exist in the United States and that a lot of economic inequality, privilege, and disadvantage results from class structure. However, they become much less certain when I ask them what determines class status. My most recent group of students suggested a number of ways to determine class status, including income, wealth, education, job, and neighborhood, as well as how many members of a given family were working. Even when we focused on one criterion for class that is often used— income—they could not agree on how much income put people into which class. This example suggests the lack of common understanding about class. . . . The difficulties involved in analyzing the influence of ethnicity (and race) and gender on social inequality probably are even greater than those for analyzing class. This analysis is more difficult because there is little agreement regarding the existence or significance of ethnic (and racial) and gender inequality.

Dimensions of Inequality in the United States

Class, race and ethnicity, and gender shape the history, experiences, and opportunities of people in the United States. As a leading social theorist indicated, we should view class, race, and gender as different and interrelated, with interlocking levels of domination, not as discrete dimensions of stratification (Collins, 1990). Thus, even though the following discussion introduces class, race and ethnicity, and gender as separate concepts, many . . . examine these dimensions as simultaneous, interrelated, and interlocking means of configuring people's social relations and life opportunities.

Class   A *class* is a group of people who share the same economic or social status, life chances, and outlook on life. A *class system* is a system of social stratification in which social status is determined by the ownership and control of resources and by the kinds of work people do. The two major sociological explanations of class derive from its two most influential contributors, Karl Marx and

Max Weber. In Marx's theory, social classes are defined by their distinct relationship to the means of production—that is, by whether people own the means of production (the capitalists) or sell their labor to earn a living (the workers). People's role in social life and their place in society are fixed by their place in the system of production. In Marx's theoretical perspective, the classes that dominate production also dominate other institutions in society, from schools and the mass media to the institutions that make and enforce rules.

German sociologist Max Weber also believed that divisions between capitalists and workers and their assigned classes were the driving force of social organization. For Weber, however, Marx's theory of social stratification was too strongly driven by the single motor of economics and by where an individual was positioned in the production process. In addition to a person's economic position, Weber included social status and *party* (i.e., coordinated political action) as different bases of power, independent of (but closely related to) economics. Weber's multidimensional perspective examines wealth, prestige, and power.

In the social sciences, the debate over class has not been whether classes exist; rather, essential theoretical perspectives flow from these two different ways (Marx's or Weber's) of understanding and constructing class. These two theoretical perspectives on class are fundamental to an understanding of social stratification and social inequality. . . .

Race and Ethnicity   Like class, race and ethnicity are important dimensions of social stratification and social inequality. Although the terms *race* and *ethnicity* are often used interchangeably, they do not refer to the same thing. Both concepts are complex, and both defy easy definition. In the past, *race* was usually defined as a category of people sharing genetically transmitted traits deemed significant by society. However, this simple view does not hold up when we take into consideration the complex biology of genetic inheritance, migration, intermarriage, and the resulting wide variation within so-called racial groups. Today, most social scientists view *race* as a far more subjective (and shifting) social category than the fixed definitions of the past, wherein people are labeled by themselves or by others as belonging to a group based on some physical characteristic, such as skin color or facial features. Racial-formation theory, which emphasizes the shifting meanings and power relationships inherent in notions of race, defines *race* "as a concept that signifies and symbolizes sociopolitical conflicts and interests in reference to different types of human bodies" (Winant, 1994, p. 115). The concept of race, then, has both biological and social components. Examples of race that have been used in the past are Caucasian, Asian, and African.

The concept of ethnicity is closely related to that of race. An *ethnic group* can be defined as a category of people distinguished by their ancestry, nationality, traditions, or culture. Examples of ethnic groups in the United States are Puerto Ricans, Japanese Americans, Cuban Americans, Irish Americans, and Lebanese Americans. Ethnicity is a cultural and social construct, and people's ethnic categories may be either self-chosen or assigned by outsiders to the group. Because characteristics such as culture, traditions, religion, and language are less visible and more changeable than skin color or facial features, ethnicity is even more arbitrary and subjective than is race.

The distinction between race and ethnicity is important. The basis for the social construction of race is primarily (though not entirely) biological; for ethnicity, it is primarily (though not entirely) cultural and social. Race is usually visible to an observer; ethnicity is usually a guess. The historical discourse about race in the United States is charged with notions of difference, of superiority and inferiority, of domination and subordination. Ideas and practices surrounding race—especially the deeply embedded divisions between African Americans and Caucasian (European) Americans—go to the core of the American experience of social inequality. . . .

Gender   *Gender,* perhaps the oldest and deepest division in social life, may be defined as the set of social and cultural characteristics associated with biological sex—being female or male—in a particular society. Like race and ethnicity, gender is socially constructed (whereas biological sex is not). It is rooted in society's belief that females and males are naturally distinct and opposed social beings. These beliefs are translated into reality when people are assigned to different and often unequal political, social, and economic positions based on their sex. It is common for societies to separate adult work, family, and civic roles by gender and to prepare girls and boys differently for those roles. The result is socially discrete gender roles.

Our society provides a great deal of occupational segregation by gender. The division of labor by gender is hierarchical, such that males occupy positions accorded higher prestige and value than females do. So-called women's work is matched with inferior reward packages and low status.

Many feminists consider the sexual division of labor to be the primary source of gender stratification, along with the socialization and institutional processes that prepare females and males for different lives. These processes include the institutions of *patriarchy,* the institutional arrangements that bestow power and privilege on male roles and occupations, thereby allowing men to perpetuate their political, social, and economic advantages. Put another way, patriarchy is the ability of men to control the laws and institutions of society and to command superior status and reward packages. . . .

EXPLORING SOCIAL INEQUALITY
IN AN INSTITUTIONAL CONTEXT: EDUCATION

We have introduced some basic ideas about the three major dimensions of social inequality and social stratification in the United States—class, race and ethnicity, and gender. Discussing them one at a time is a simple and orderly way to begin, but it is critical to keep in mind that these dimensions do not function in isolation. Rather, their dynamic interaction is highly complex. Class, race and ethnicity, and gender are simultaneous, intersecting, and sometimes crosscutting systems of relationships and meanings.

For example, even though we might use the term *middle class* to refer to all people of a certain income level (or amount of wealth, or occupation, or educational level), African Americans of that income level—the black middle class—occupy a far more precarious position in our society than do members of the white middle class. In considering these two dimensions, we may ask, Do all middle-class people have the same interests? Is their social status a function more of class or of race? How do they identify themselves—as middle class or as African or European Americans?

Other examples of the intersection of the three dimensions abound. In a study conducted in several nations, employed married women identified their socio-economic class positions more on the basis of their spouses' jobs than on their own familial or educational backgrounds or their own occupations (Erickson and Goldthorpe, 1992). Another study showed that European American women in the United States tend to mention gender alone when asked how they identify themselves, whereas African American women tend to emphasize both race and gender (Rubin, 1994).

Evidence points to growing conflict among different ethnic minorities of the same social class or community in the United States, perhaps as a result of contracting opportunities and limited resources. For example, in Los Angeles and in New York City, many Korean immigrants run grocery, liquor, produce, and fish retail businesses that are heavily concentrated in ethnic-minority neighborhoods. These Korean American merchants often act as an intermediate ethnic minority interposed between low-income Latino and African American minority consumers and high-income European American owners of large companies. The Korean Americans encounter conflicts with both their ethnic-minority customers and their ethnic-majority suppliers. As a result, intergroup conflicts, tension, misunderstandings, and violence erupt all too often in Los Angeles, New York, and other ethnically diverse cities (Min, 1996).

. . . Education has traditionally been seen as a remedy for inequality in the United States. Americans believe that if society creates more educational opportunities for the disadvantaged, greater equality will eventually result. Thus, people who seek greater social equality often turn to education as an arena in which to institute change. Similarly, those who seek to maintain our society's system of social stratification may also try to manipulate the educational system to preserve the status quo. Hence, education becomes an arena for conflict between groups seeking change toward greater social equality and groups seeking to maintain the current social inequalities. . . .

NOTES

Collins, Patricia Hill. 1990. *Black Feminist Thought.* New York: Routledge.

Erickson, Robert, and John H. Goldthorpe. 1992. "Individual or Family? Results from Two Approaches to Class Assignment." *Acta Sociologica* 35: 95–105.

Fischer, Claude S., Michael Hout, Martin Sanchez Jankowski, Samuel R. Lucas, Ann Swidler, and Kim Voss. 1996. *Inequality by Design: Cracking the Bell Curve Myth.* Princeton, NJ: Princeton University Press.

Grusky, David B., Ed. 1994. *Social Stratification: Class, Race, and Gender in Sociological Perspective.* Boulder, CO: Westview Press.

Min, Pyong Gap. 1996. *Caught in the Middle: Korean Merchants in America's Multiethnic Cities.* Berkeley and Los Angeles: University of California Press.

Rubin, Lillian B. 1994. *Families on the Fault Line.* New York: HarperCollins.

Winant, Howard. 1994. *Racial Conditions.* Minneapolis: University of Minnesota Press.

DISCUSSION QUESTIONS

1. What specific changes in the late twentieth century does Shapiro cite as evidence of economic restructuring in the United States? How do you see these changes as affecting the opportunities of diverse groups in this country?

2. What does Shapiro mean when he says that class, race, ethnicity, and gender are crosscutting systems of relationships and meanings? How is this exemplified by the example of the middle class?

INFOTRAC COLLEGE EDITION

You can use your access to InfoTrac College Edition to learn more about the subjects covered in this essay. Some suggested search terms include:

Black middle class
class system
corporate downsizing
deindustrialization

living wage campaign
manufacturing-based economy
service-based economy

34

Wealth Matters

DALTON M. CONLEY

By contrasting the experiences of two different families, Dalton Conley illustrates the great difference in socioeconomic status that differences in wealth, even at a modest level, can produce. Wealth and the ability to pass it on also creates a cumulative difference in the status of different groups in society. Conley shows how understanding differences in wealth between racial groups is an important part of understanding the relationship between race and class inequality.

From: Dalton M. Conley. 1999. *Being Black, Living in the Red: Race, Wealth, and Social Policy in America.* Berkeley: University of California Press, pp. 1–7. Reprinted with permission.

I n 1994, the median white family held assets worth more than seven times those of the median nonwhite family. Even when we compare white and minority families at the same income level, whites enjoy a huge advantage in wealth. For instance, at the lower end of the income spectrum (less than $15,000 per year), the median African American family has no assets, while the equivalent white family holds $10,000 worth of equity. At upper income levels (greater than $75,000 per year), white families have a median net worth of $308,000, almost three times the figure for upper-income African American families ($114,600).

. . . First, why does this wealth gap exist and persist over and above income differences? Second, does this wealth gap explain racial differences in areas such as education, work, earnings, welfare, and family structure? In short, this [essay] examines where race *per se* really matters in the post-civil rights era and where race simply acts as a stand-in for that dirty word of American society: class. The answers to these questions have important implications for the debate over affirmative action and for social policy in general.

An alternative way to conceptualize what this . . . is about is to contrast the situations of two hypothetical families. Let's say that both households consist of married parents, in their thirties, with two young children. Both families are low-income—that is, the total household income of each family is approximately the amount that the federal government has "declared" to be the poverty line for a family of four (with two children). In 1996, this figure was $15,911.

Brett and Samantha Jones (family 1) earned about $12,000 that year. Brett earned this income from his job at a local fast-food franchise (approximately two thousand hours at a rate of $6 per hour). He found himself employed at this low-wage job after being laid off from his relatively well-paid position as a sheet metal worker at a local manufacturing plant, which closed because of fierce competition from companies in Asia and Latin America. After six months of unemployment, the only work Brett could find was flipping burgers alongside teenagers from the local high school.

Fortunately for the Jones family, however, they owned their own home. Fifteen years earlier, when Brett graduated from high school, married Samantha, and landed his original job as a sheet metal worker, his parents had lent the newlyweds money out of their retirement nest egg that enabled Brett and Samantha to make a 10 percent down payment on a house. With Samantha's parents cosigning—backed by the value of their own home—the newlyweds took out a fifteen-year mortgage for the balance of the cost of their $30,000 home. Although money was tight in the beginning, they were nonetheless thrilled to have a place of their own. During those initial, difficult years, an average of $209 of their $290.14 monthly mortgage payment was tax deductible as a home mortgage interest deduction. In addition, their annual property taxes of $800 were completely deductible, lowering their taxable income by a total of $3,308 per year. This more than offset the payments they were making to Brett's parents for the $3,000 they had borrowed for the down payment.

After four years, Brett and Samantha had paid back the $3,000 loan from his parents. At that point, the total of their combined mortgage payment ($290.14), monthly insurance premium ($50), and monthly property tax payment ($67), minus the tax savings from the deductions for mortgage interest and local property taxes, was less than the $350 that the Smiths (family 2) were paying to rent a unit the same size as the Joneses' house on the other side of town.

That other neighborhood, on the "bad" side of town, where David and Janet Smith lived, had worse schools and a higher crime rate and had just been chosen as a site for a waste disposal center. Most of the residents rented their housing units from absentee landlords who had no personal stake in the community other than profit. A few blocks from the Smiths' apartment was a row of public housing projects. Although they earned the same salaries and paid more or less the same monthly costs for housing as the Joneses did, the Smiths and their children experienced living conditions that were far inferior on every dimension, ranging from the aesthetic to the functional (buses ran less frequently, large supermarkets were nowhere to be found, and class size at the local school was well over thirty).

Like Brett Jones, David Smith had been employed as a sheet metal worker at the now-closed manufacturing plant. Unfortunately, the Smiths had not been able to buy a home when David was first hired at the plant. With little in the way of a down payment, they had looked for an affordable unit at the time, but the real estate agents they saw routinely claimed that there was just nothing available at the moment, although they promised to "be sure to call as soon as something comes up. . . ." The Smiths never heard back from the agents and eventually settled into a rental apartment.

David spent the first three months after the layoffs searching for work, drawing down the family's savings to supplement unemployment insurance—savings that were not significantly greater than those of the Joneses, since both families had more or less the same monthly expenses. After several months of searching, David managed to land a job. Unfortunately, it was of the same variety as the job Brett Jones found: working as a security guard at the local mall, for about $12,000 a year. Meanwhile, Janet Smith went to work part time, as a nurse's aide for a home health care agency, grossing about $4,000 annually.

After the layoffs, the Joneses experienced a couple of rough months, when they were forced to dip into their small cash savings. But they were able to pay off the last two installments of their mortgage, thus eliminating their single biggest living expense. So, although they had some trouble adjusting to their lower standard of living, they managed to get by, always hoping that another manufacturing job would become available or that another company would buy out the plant and re-open it. If worst came to worst, they felt that they could always sell their home and relocate in a less expensive locale or an area with a more promising labor market.

The Smiths were a different case entirely. As renters, they had no latitude in reducing their expenses to meet their new economic reality, and they could not afford their rent on David's reduced salary. The financial strain eventually proved too much for the Smiths, who fought over how to structure the family budget. After a particularly bad row when the last of their savings had been spent, they decided to take a break; both thought life would be easier and better for the children if Janet moved back in with her mother for a while, just until things turned around economically—that is, until David found a better-paying job. With no house to anchor them, this seemed to be the best course of action.

Several years later, David and Janet Smith divorced, and the children began to see less and less of their father, who stayed with a friend on a "temporary" basis. Even though together they had earned more than the Jones family (with total incomes of $16,000 and $12,000, respectively), the Smiths had a rougher financial, emotional, and family situation, which, we may infer, resulted from a lack of property ownership.

What this comparison of the two families illustrates is the inadequacy of relying on income alone to describe the economic and social circumstances of families at the lower end of the economic scale. With a $16,000 annual income, the Smiths were just above the poverty threshold. In other words, they were not defined as "poor," in contrast to the Joneses, who were. Yet the Smiths were worse off than the Joneses, despite the fact that the U.S. government and most researchers would have classified the Jones family as the one who met the threshold of neediness, based on that family's lower income.

These income-based poverty thresholds differ by family size and are adjusted annually for changes in the average cost of living in the United States. In 1998, more than two dozen government programs—including food stamps, Head Start, and Medicaid—based their eligibility standards on the official poverty threshold. Additionally, more than a dozen states currently link their needs standard in some way to this poverty threshold. The example of the Joneses and the Smiths should tell us that something is gravely wrong with the way we are measuring economic hardship—poverty—in the United States. By ignoring assets, we not only give a distorted picture of life at the bottom of the income distribution but may even create perverse incentives.

Of course, we must be cautious and remember that the Smiths and the Joneses are hypothetically embellished examples that may exaggerate differences. Perhaps the Smiths would have divorced regardless of their economic circumstances. The hard evidence linking modest financial differences to a propensity toward marital dissolution is thin; however, a substantial body of research shows that financial issues are a major source of marital discord and relationship strain. It is also possible that the Smiths, with nothing to lose in the form of assets, might have easily slid into the world of welfare dependency. A wide range of other factors, not included in our examples, affect a family's well-being and its trajectory. For example, the members of one family might have been healthier than those of the other, which would have had important economic consequences and could have affected family stability. Perhaps one family might have been especially savvy about using available resources and would have been able to take in boarders, do under-the-table work, or employ another strategy to better its standard of living. Nor do our examples address educational differences between the two households.

But I have chosen not to address all these confounding factors for the purpose of illustrating the importance of asset ownership *per se*. Of course, homeownership, savings behavior, and employment status all interact with a variety of other measurable and unmeasurable factors. This interaction, however, does not take away from the importance of property ownership itself.

. . . In order to understand a family's well-being and the life chances of its children—in short, to understand its class position—we not only must consider income, education, and occupation but also must take into account accumulated wealth (that is, property, assets, or net worth). . . . While the importance of wealth is the starting point, . . . [the] end point is the impact of the wealth distribution on racial inequality in America. As you might have guessed, an important detail is missing from the preceding description of the two families: the Smiths are black and have fewer assets than the Joneses, who are white.

At all income, occupational, and education levels, black families on average have drastically lower levels of wealth than similar white families. The situation of the Smiths may help us to understand the reason for this disparity of wealth between blacks and whites. For the Smiths, it was not discrimination in hiring or education that led to a family outcome vastly different from that of the Joneses; rather, it was a relative lack of assets from which they could draw. In contemporary America, race and property are intimately linked and form the nexus for the persistence of black-white inequality.

Let us look again at the Smith family, this time through the lens of race. Why did real estate agents tell the Smiths that nothing was available, thereby hindering their chances of finding a home to buy? This well-documented practice is called "steering," in which agents do not disclose properties on the market to qualified African American home seekers, in order to preserve the racial makeup of white communities—with an eye to maintaining the property values in those neighborhoods. Even if the Smiths had managed to locate a home in a predominantly African American neighborhood, they might well have encountered difficulty in obtaining a home mortgage because of "redlining," the procedure by which banks code such neighborhoods "red"—the lowest rating—on their loan evaluations, thereby making it next to impossible to get a mortgage for a home in these districts. Finally, and perhaps most important, the Smiths' parents were more likely to have been poor and without assets themselves (being black and having been born early in the century), meaning that it would have been harder for them to amass enough money to loan their children a down payment or to cosign a loan for them. The result is that while poor whites manage to have, on average, net worths of over $10,000, impoverished blacks have essentially no assets whatsoever.

Since wealth accumulation depends heavily on intergenerational support issues such as gifts, informal loans, and inheritances, net worth has the ability to pick up both the current dynamics of race and the legacy of past inequalities that may be obscured in simple measures of income, occupation, or education. This thesis has been suggested by the work of sociologists Melvin Oliver and Thomas Shapiro in their recent book *Black Wealth/White Wealth*. They claim that wealth is central to the nature of black-white inequality and that wealth—as opposed to income, occupation, or education—represents the "sedimentation" of both a legacy of racial inequality as well as contemporary, continuing inequities. . . .

Certain tenacious racial differences—such as deficits in education, employment, wages, and even wealth itself among African Americans—will turn out to be indirect effects, mediated by class differences. In other words, it is not race *per se* that matters directly; instead, what matters are the wealth levels and class positions that are associated with race in America. In this manner, racial differences in income and asset levels have come to play a prominent role in the perpetuation of black-white inequality in the United States.

This is not to say that race does not matter; rather, it maps very well onto class inequality, which in turn affects a whole host of other life outcomes. In fact, when class is taken into consideration, African Americans demonstrate significant net *advantages* over whites on a variety of indicators (such as rates of high school graduation, for instance). In this fact lies the paradox of race and class in contemporary America—and the reason that both sides of the affirmative action debate can point to evidence to support their positions. . . .

DISCUSSION QUESTIONS

1. What is the difference in *income* and *wealth* and why is this significant in understanding stratification by race?
2. Sociologists have long debated the relationship between race and class. What relationship between race and class is shown by the evidence in Dalton Conley's research?

INFOTRAC COLLEGE EDITION

You can use your access to InfoTrac College Edition to learn more about the subjects covered in this essay. Some suggested search terms include:

food stamps
home ownership (and race)
income gap

median family income
redlining
wealth

35

The Shredded Net

The End of Welfare as We Knew It

VALERIE POLAKOW

The repeal of Aid to Families with Dependent Children (AFDC, the long-standing welfare system in the United States) has left many without the public assistance that welfare was intended to provide. Polakow argues that the elimination of welfare leaves many vulnerable, particularly women and children, and also reflects the particular interests of a capitalist market economy based on the high value placed on privatization of many social services.

From: Valerie Polakow. 1997. *Sage Race Abstracts* 22, No. 3.

As a Democratic president and a Republican-controlled House and Senate joined forces to end welfare for the "good of the poor," one of the last remaining vestiges of the federal social safety net—a minimal guarantee of public assistance for poor children—was decisively shredded. In a society where child poverty and family homelessness have become endemic characteristics, the signing of the new welfare legislation and its passage into law on 22 August 1996 was the culmination of a continuing decades-long public policy assault on poor single mothers with dependent children. However, this legislation, which repeals Title IVa of the Social Security Act of 1935, is the most far-reaching and radical of any antiwelfare initiative, which Senator Daniel Patrick Moynihan characterized as "the first step in dismantling a social contract that has been in place in the United States since at least the 1930s. Do not doubt that social security itself. . . . will be next. . . .We would care for the elderly, the unemployed, the dependent children. Drop the latter; watch the others fall."[1]

In a "residual" welfare state such as the United States, market sovereignty is the foundation for the argument that the state should play a minimal role in the distribution of welfare and that individuals are essentially responsible for contracting their own welfare.[2] During the past decade, the boundaries of the public space have contracted, with narrowed public commitments and a discourse that has fostered "self-reliance." Limited public benefits have always been organized in means-tested assistance schemes that target particular segments of the population in stigmatizing ways. In such a residual welfare state, women alone—divorced, separated, or unmarried mothers—are a constituency at risk, disproportionately poor and facing an alarming array of obstacles and impediments to their autonomy.[3]

Furthermore, the impact of global capitalism on domestic infrastructures has been far-reaching; exporting jobs, downsizing, outsourcing, and increasing numbers of nonunionized contract workers all represent the face of the meaner, leaner, global economy. The North American Free Trade Agreement (NAFTA), which was implemented in 1994, caused seventy-five thousand U.S. workers to lose jobs and displaced more than a million and a half Mexican workers. In addition, as corporations sell the notion that international competition requires greater "flexibility," more and more part-time workers without benefits are hired, and corporate lobbying of lawmakers promotes government deregulation, the cutting of overtime pay, and the gutting of union shops. In all three NAFTA countries—the United States, Canada, and Mexico—real wages are actually far below increases in productivity. In Mexico real wages in May 1996 were 35 percent below pre-1994 levels, and in Canada and the United States real wages have stagnated and the proportion of full-time workers in poverty continues to grow.[4] While public infrastructures crumble and income inequality soars, powerful global corporations have expanded beyond the boundaries of the nation-state, and, increasingly, beyond the boundaries of governments. Korten points out that between 1980 and 1994 the *Fortune* 500 companies shed 4.4 million jobs, while sales increased 1.4 times and assets increased 2.3 times! Hence, he argues, in an unfettered market economy, "which responds to money, not needs, the rich win this competition every time. . . .The average CEO of a large corporation now receives a compensation package of more than 3.7 million per year. Those same corporations employ 1/20th of 1% of the world's population, but they control 25% of the world's output and 70% of world trade. Of the world's hundred largest economies, fifty are now corporations."[5]

Within the United States, the march toward privatization inexorably continues— schools, prisons, city services, transportation—and, since the passage of the new welfare law, privatized welfare, in which giant corporations vie for contracts to run welfare-to-work programs, is creating what *The New York Times* described as "the business opportunity of a lifetime."[6] The new law permits states to buy both welfare services and gatekeepers who screen eligibility and determine benefits. Lockheed Martin, the formidable $30-billion "giant of the weapons industry" and a major beneficiary of corporate welfare subsidies, plans to close one of its plants in New Jersey; as one congressman put it, "They would be getting a subsidy to lay off these folks and then could be getting additional money from the government to help these people get off welfare."[7]

As such "wealthfare" flourishes and public entitlements are reduced or eliminated, what happens to those who live in the broadening swath of the Other America? What is life like under a shredded net?

LIFE UNDER THE SHREDDED NET

I lived in public housing in Detroit, I stayed there for two years, and I know the debilitating factors that it takes just to survive. People have tried and tried and tried to find jobs; it's a vicious circle and after you get the door slammed in your face, you don't have the will to go out there and fight anymore. . . . I got a little slip in my son's pocket saying I owed $900 in child care and saying don't bring my children back until that is paid. . . . I couldn't do it so I tried to go back on welfare. It was either pay my rent or pay child care. . . so I called social services and asked them about getting my benefits again and my worker said I had to wait two months, so I stopped work in November and she said I wouldn't get any benefits until January. I asked my worker, "How'm I supposed to live?" and she said, "It's not for me to tell you!"
—Tanya, a single mother of two young children.[8]

At the present time in the United States, 30 percent of all families with children under eighteen are single-parent families, with single-mother families constituting the vast majority, or 82 percent. In addition, more than 40 percent of single-mother families are living in poverty.[9] However, the largest constituency of poor Americans is children: in 1995, 15.3 million children lived in poverty in the United States, and the younger in age, the higher the risk. While 21 percent of children under eighteen are classified as living in poverty, 25 percent of children under six and 27 percent under three live in poverty, the majority of whom reside in poor, single-mother households.[10]

The term *feminization of poverty* has been widely used to describe the particular plight of women in the United States who experience occupational segregation and workplace discrimination, disproportionately occupy part-time service-sector jobs with few or no benefits, perform unpaid domestic work at home and retain primary care of children. As single mothers, they are disproportionately poor and face an alarming array of obstacles that threaten their family stability.[11] By the late 1980s women and their children had become a significant majority of America's poor.[12] Unmet needs in housing, health care, and child care have coalesced to form a triple crisis where single mothers, as both providers for and nurturers of their children, cannot sustain family viability when they are low-wage earners, with no family support provisions in place to act as a buffer against the ravages of the market economy. In 1995, a full-time minimum wage salary was $8,800 a year— $3,000 less than under the federal poverty line for a single-mother family of three. And while the minimum wage was raised by 50 cents to $4.75 an hour in October 1996, full-time minimum wage work still yields a wage well below the poverty line. The child care crisis compounds the situation, with many low-wage mothers unable to afford the high costs of private child care.

Single mothers also make up 70–90 percent of homeless families nationwide,[13] and more than 50 percent of homeless families become homeless because the mother flees domestic violence.[14] One in four homeless persons is now a child younger than eighteen.[15] Furthermore, reports from the Food and Research Action Center indicate that four million children under twelve go hungry during part of each month, and there are now ten million children who have no health insurance coverage.

It is clear that the increasing economic vulnerability of the single-mother family represents a child welfare crisis of growing magnitude that has been further exacerbated by recent welfare repeal legislation. Mothers now struggling to survive on public assistance are being ordered to pay their dues and take up paid or unpaid work—part of the punitive rehabilitation of the "welfare culture" that defines the new "opportunity society," promoted by Newt Gingrich and his fellow Republican supporters, in which the poor are to assume personal responsibility for their family's poverty. Practically speaking, that means that mothers with dependent children are coerced into taking up low-wage jobs, or "voluntary" indentured servitude, as a condition for receiving any benefits, with no guarantee of adequate child care services or subsidies. The blatant failure of the new personal-responsibility legislation to address the daily survival needs of economically vulnerable single mothers clearly threatens their children's basic physical, social, and emotional development.

Unlike other industrialized democracies, the United States, as a residual welfare state, has no national health care system and no universal family support policies such as paid maternity and parental leave, national child/family allowances, or a national subsidized child care system for infants and preschool children.[16] Head Start, the national early childhood intervention program for poor children developed during the "War on Poverty" of the 1960s, was available to only 36 percent of income-eligible children during 1995.[17]

Despite the dismal record of the United States in developing public policies that provide for the basic health, shelter, and daily living needs of its most vulnerable citizens—poor women and their children—female and child poverty is still cast as a "moral" problem, tied to public rhetoric about "family values" and "family breakdown," which in turn is used to rationalize further cuts in public assistance. The structural evidence of poverty wages, contingent no-benefit jobs, corporate downsizing, racial and gender discrimination, and the growing number of job-poor isolated and destitute communities[18] is ignored in favor of "blaming the victim" discourses that reduce poor women to caricatured public parasites or promiscuous sluts, living it up at taxpayer expense. Rarely are their grim survival struggles chronicled, or their children's traumatic lives examined, as part of the consequences of a trickle-down market economy that offers few protections to those who live in the other America.[19]

While Social Security for elderly Americans is still considered an earned entitlement, a Social Security system for children (who in turn must depend on their parents' benefits) forms part of a completely different discourse in the United States, because the parents, specifically poor mothers of dependent children, are viewed as undeserving of government support. The discourse about poor mothers and their children is couched in "moral" and "personal responsibility" frames and, more recently, in terms of "productive work." Taking care of one's own children, however, is not viewed as productive work if one is poor, and having children when poor is perceived as both irresponsible and/or immoral.

It is significant to note that during the House debates in 1995 on the original welfare reform bill, ironically titled the Personal Responsibility Act, which was proposed as part of the Republican's Contract with America, poor mothers receiving public assistance were publicly vilified by Republican lawmakers as "breeding mules," as "alligators," and as "monkeys." Earlier, House Ways and Means chair Representative Clay Shaw Jr., who shepherded welfare repeal legislation through the House, stated, "It may be like hitting a mule with a two-by-four, but you've got to get their attention."[20] When the bill reached the Senate, Senator Phil Gramm demanded, "We've got to get a provision that denies more and more cash benefits to women who have more and more babies while on welfare."[21]

The antiwelfare and "underclass" punitive discourses that have targeted single mothers and pregnant teens have promoted a continuing public perception of poverty as a private and behavioral condition leading to proliferating racialized and sexualized fictions about *them;* the causes of family poverty are seen as rooted in failed and fallen women, failed children, and a failed work ethic—but not in the actual consequence of public policies that produce family poverty by omission and commission. The ravages of a market economy; the lack of affordable, quality child care; the lack of safe and affordable family housing; the absence of universal health care; and the traumas of domestic violence—all coalesce to create a tangled web of obstacles that entrap poor single mothers in poverty, destitution, and homelessness. Hence, a national family policy (or absence thereof) is a potent force in women's lives, either weaving a safety net to support family viability or shredding the net and destabilizing family existence.

In the Scandinavian countries, for example, where family policies are comparatively strong and where universal entitlements such as paid parental leave, health care, child care, absent-parent child support, and housing subsidies are in place, chronic single mother and child poverty has largely been eliminated.[22] Yet, despite the evidence of abysmal life conditions for the poorest Americans, the minimal safety net that formerly existed in the United States has now been dismantled.

The welfare repeal law that passed the House and Senate with bipartisan support in 1996 could hardly be more emblematic of a postmodern market individualism—eliminating "big government" and public responsibility in favor of the surreal privatized self—where to have is to be. . . .

NOTES

1. *New York Times,* 2 August 1996.

2. Gösta Esping-Anderson and Walter Korpi, "From Poor Relief to Institutional Welfare States: The Development of Scandinavian Social Policy," in *The Scandinavian Model: Welfare States and Welfare Research,* ed. R. Erikson, E. J. Hansen, S. Ringen, and H. Uusitalo (Armonk, N.Y.: M. E. Sharpe, 1987), 39–74.

3. Valerie Polakow, "Savage Distributions: Welfare Myths and Daily Lives," *Sage Race Relations Abstracts* 19, no. 4 (1994): 3–29.

4. Sarah Anderson, John Cavanagh, and David Ranney, "NAFTA: Trinational Fiasco," *Nation,* 15/22 July 1996, 26–28.

5. David C. Korten, "The Limits of the Earth," *Nation,* 15/22 July 1996, 16, 18.

6. *New York Times,* 15 September 1996.

7. Ibid., A14.

8. Excerpt from an interview conducted by the author and published in *Nation,* 1 May 1995.

9. U.S. Bureau of the Census, "Poverty in the United States," *Current Population Reports.* Series P60-194 (Washington, D.C.: Government Printing Office, 1995).

10. Children's Defense Fund, *The State of America's Children: Yearbook.* Washington, D.C.: Children's Defense Fund, 1996).

11. Barbara Ehrenreich and Frances Fox Piven, "The Feminization of Poverty: When the Family Wage System Breaks Down," *Dissent* 31 (Summer 1984): 162–68; Gertrude S. Goldberg and Eleanor Kremen, *The Feminization of Poverty: Only in America?* (New York: Greenwood Press, 1990); Linda Gordon, ed., *Women, the State and Welfare* (Madison: University of Wisconsin Press, 1990); Diana Pearce, "The Feminization of Poverty: Women, Work and Welfare," *Urban and Social Change Review* 11, nos. 1–2

(1978): 28–36; Valerie Polakow, *Lives on the Edge: Single Mothers and their Children in the Other America* (Chicago: University of Chicago Press, 1993); Valerie Polakow, "On a Tightrope without a Net," *Nation,* May 1995, 590–92.

12. U.S. Bureau of the Census, "Poverty in the United States," *Current Population Reports.* Series P60-163 (Washington, D.C.: Government Printing Office, 1989).

13. Ellen L. Bassuk, "Who Are the Homeless Families? Characteristics of Sheltered Mothers and Children," *Community Mental Health Journal* 26 (1990): 425–34; Marcia Steinbock, "Homeless Female-headed Families: Relationships at Risk," *Marriage and Family Review* 20 (1995): 143–59.

14. National Clearinghouse for the Defense of Battered Women, *Statistics Packet,* 3rd ed. (Philadelphia: National Clearinghouse for the Defense of Battered Women, 1994).

15. National Law Center on Homelessness and Poverty, *A Foot in the Schoolhouse Door.* (Washington, D.C.: National Law Center on Homelessness and Poverty, 1995).

16. Sheila B. Kamerman and Alfred J. Kahn, eds., *Child Care, Parental Leave, and the Under 3's: Policy Innovation in Europe* (New York: Auburn House, 1991); Sheila B. Kamerman and Alfred J. Kahn, *Starting Right: How America Neglects Its Youngest Children and What We Can Do about It* (New York: Oxford University Press, 1995); Polakow, *Lives on the Edge.*

17. Children's Defense Fund, *The State of America's Children: Yearbook* (1996).

18. James Jennings, "Persistent Poverty in the United States: Review of Theories and Explanations," *Sage Race Relations Abstracts* 19, no. 1 (1994): 5–34; William J. Wilson, "Work," *New York Times Magazine,* 18 August 1996.

19. Polakow, "Savage Distributions."

20. Quoted in Jason DeParle, "Momentum Builds for Cutting Back Welfare System," *New York Times,* 13 November 1994.

21. Quoted in Robin Toner, "Senate Passes Bill to Abolish Guarantees of Aid for the Poor," *New York Times,* 20 September 1995.

22. Sheila Kamerman and Alfred Kahn, *Child Care, Parental Leave, and the Under 3's;* Kamerman and Kahn, *Starting Right;* Valerie Polakow, "Family Policy in the United States and Denmark: A Cross-national Study of Discourse and Practice," *Early Education and Development* 8, no. 3 (1997): 242–60.

DISCUSSION QUESTIONS

1. What are the cultural beliefs and economic forces underlying the move to reduce so-called public entitlements? How does this result in the attitude about poverty that it is the fault of the victim?

2. What does *the feminization of poverty* mean and what evidence does Polakow provide of this phenomenon?

INFOTRAC COLLEGE EDITION

You can use your access to InfoTrac College Edition to learn more about the subjects covered in this essay. Some suggested search terms include:

feminization of poverty
Personal Responsibility and
    Work Reconciliation Act
poverty line
Temporary Assistance for Needy
    Families (TANF)

underclass
welfare reform
workfare

36

The Garment Industry in the Restructuring Global Economy

EDNA BONACICH, LUCIE CHENG, NORMA CHINCHILLA, NORA HAMILTON, AND PAUL ONG

These authors describe the consequences of the new processes in the global economy, particularly as they affect different classes of people in different international locales. The authors note the different interpretations that various scholars have given to the process of globalization—some seeing its positive effects, others being more critical of the impact of globalization on women workers, immigrants, and the working class.

Global integration, a long-standing feature of the world economy, is currently undergoing a restructuring. Generally, until after World War II, the advanced industrial countries of western Europe and the United States dominated the world economy and controlled most of its industrial production. The less-developed countries tended to concentrate in the production of raw materials. Since the late 1950s, and accelerating rapidly in the 1980s, however, industrial production has shifted out of the West, initially to Japan, then to the Asian NICs (newly industrializing countries—namely, Hong Kong, Taiwan, South Korea, and Singapore), and now to almost every country of the world. Less-developed countries are not manufacturing mainly for the domestic market or following a model of "import substitution"; rather, they are manufacturing for export, primarily to developed countries, and pursuing a development strategy of export-led industrialization. What we are witnessing has been termed by some a "new international division of labor" [Fröbel, Heinrichs, and Kreye 1980).

The developed countries are faced with the problem of "deindustrialization" in terms of traditional manufacturing, as their manufacturing base is shifted to other, less-developed countries (Bluestone and Harrison 1982). At the same time, they are faced with a massive rise in imports that compete with local industries' products, moving to displace them. This shift is accompanied by the rise of a new kind of transnational corporation (TNC). Of course, TNCs have existed since the beginning of the European expansion, but they concentrated mainly on the production of agricultural goods and raw materials and, in the postwar period, on manufacturing for the host country market. The new TNCs are global firms that are able to use advanced communications and transportation technology to coordinate manufacturing in multiple locations simultaneously. They engage in "off-

From: Edna Bonacich, Lucie Cheng, Norma Chinchilla, Nora Hamilton and Paul Ong, (eds.). 1994. *Global Production: The Apparel Industry in the Pacific Rim.* Philadelphia: Temple University Press, pp. 3–18. Reprinted with permission.

shore sourcing" to produce primarily for the home market (Grunwald and Flamm 1985; Sklair 1989).

TNCs sometimes engage in direct foreign investment, but globalized production does not depend on it. They can arrange for production in numerous locations through other, looser connections, such as subcontracting and licensing. In other words, TNCs can set up complex networks of global production without owning or directly controlling their various branches.

The nation-state has increasingly declined as an economic unit, with the result that states are often unable to control the actions of powerful TNCs. The TNCs are supragovernmental actors that make decisions on the basis of profit-making criteria without input from representative governments. Of course, strong states are still able to exercise considerable influence over trade policies and over the policies of the governments of developing countries.

Some scholars have used the concept of "commodity chains" to describe the new spatial arrangements of production (Gereffi and Korzeniewicz 1994). The concept shows how design, production, and distribution are broken down and geographically dispersed, with certain places serving as centers within the chain. Power is differentially allocated along the chain, and countries and firms vie to improve their position in the chain.

Focusing on the geographic aspects of global production also has led to the concept of "global cities" (Sassen 1991). These are coordination centers for the global economy, where planning takes place. They house the corporate headquarters of TNCs, as well as international financial services and a host of related business services. These cities have become the "capitals" of the new global economy.

Another way to view the restructuring is to see it as the proletarianization of most of the world. People who had been engaged primarily in peasant agriculture or in other forms of noncapitalist production are now being incorporated into the industrial labor force. Many of these people are first-generation wage-workers, and a disproportionate number of them are women. These "new" workers sometimes retain ties to noncapitalist sectors and migrate between them and capitalist employment, making their labor cheaper than that of fully proletarianized workers. But even if they are not attached to noncapitalist sectors, first-generation workers tend to be especially vulnerable to exploitative conditions. Thus, an important feature of the new globalization is that TNCs are searching the world for the cheapest available labor and are finding it in developing countries.

Countries pursuing export-led industrialization typically follow strategies that encourage the involvement of foreign capital. They offer incentives, including tax holidays and the setting up of export processing zones (EPZs), where the bureaucracy surrounding importing and exporting is curtailed; sometimes they also promise cheap and controllable labor. Countries using this development strategy do not plan to remain the providers of cheap labor for TNCs, however: they hope to move up the production ladder, gaining more economic power and control. They want to shift from labor-intensive manufactures to capital-intensive, high-technology goods. They hope to follow the path of Japan and the Asian NICs and become major economic players in the global economy.

Sometimes participation in global capitalist production is foisted on nations by advanced-industrial countries and/or suprastate organizations such as the World Bank and the International Monetary Fund (IMF), where advanced countries wield a great deal of influence. The United States, in particular, has backed regimes that support globalized production and has pushed for austerity programs that help to make labor cheap. At the same time, developed countries, including the United States, have been affected by the restructured global economy. Accompanying the rise in imports and deindustrialization has been a growth in unemployment and a polarization between the rich and the poor (Harrison and Bluestone 1988). This trend has coincided with increased racial polarization, as people of color have faced a disproportionate impact from these developments.

A rise in immigration from less-developed to more-developed countries has also accompanied globalization. The United States, for example, has experienced large-scale immigration from the Caribbean region and from Asia, two areas pursuing a manufacturing-for-export development strategy. At least part of this immigration is a product of globalization, as people are dislocated by the new economic order and are forced to emigrate for survival (Sassen 1988). Dislocations occur not only because global industries displace local ones (as in the case of agribusiness displacing peasants), but also because austerity programs exacerbate the wage gap between rich and poor countries (making the former ever more desirable). Political refugees, often from countries where the United States has supported repressive regimes, have added to the rise in immigration as well. Finally, some immigration results when people move to service global enterprise as managers, trade representatives, or technicians.

In the advanced countries, the immigration of workers has created a "Third World within." In this case, the newly created proletariat is shifting location. These immigrants play a part in the efforts of the advanced countries to hold on to their industries, by providing a local source of cheap labor to counter the low labor standards in competing countries.

In sum, we are seeing a shakeup of the old world economic order. Some countries have used manufacturing for export as a way to become major economic powers (Appelbaum and Henderson 1992; Gereffi and Wyman 1990). These countries now threaten U.S. dominance. Other countries are trying to pursue this same path, but it is not clear whether they will succeed. Meanwhile, despite the fact that the United States is suffering some negative consequences from the global restructuring, certain U.S.-based TNCs are deeply implicated in the process and benefit from it.

CONTRASTING VIEWS OF RESTRUCTURING

The new globalization receives different interpretations and different evaluations (Gondolf, Marcus, and Dougherty 1986). Some focus on the positive side; they see global production as increasing efficiency by allowing each country to specialize in its strengths. Less-developed countries are able to provide low-cost, unskilled labor while developed countries provide management, technical, and financial resources. Together they are able to maximize the efficient use of resources. The result is that more goods and services are produced more cheaply, to the benefit of all. Consumers, in particular, are seen as the great beneficiaries of globalized production, because of the abundance of low-cost, higher-quality goods from which to choose.

Globalization can be seen as part of the new system of flexible specialization (Piore and Sabel 1984). Consumer markets have become more differentiated, making the old, industrial system of mass production in huge factories obsolete. To be competitive today, a firm must be able to produce small batches of differentiated goods for diverse customers. Globalization contributes to this process by enabling firms to produce a vast range of products in multiple countries simultaneously.

Another aspect of the positive view is to see the entrance of less-developed countries into manufacturing for export as a step toward their industrialization and economic development. Although countries may enter the global economy at a tremendous disadvantage, by participating in exports they are able to accumulate capital and gradually increase their power and wealth. Japan and the Asian NICs have demonstrated the possibilities; now other countries can follow a similar path.

Although workers in the advanced countries may suffer some dislocation by the movement of industry abroad, in the long run they are seen to be beneficiaries of this process. While lower-skilled, more labor-intensive jobs will move to the developing countries, the advanced countries will gain higher-technology jobs, as well as jobs in coordinating and managing the global economy. Thus workers in the advanced countries will be "pushed up" to more middle-class positions, servicing and directing the workers in the rest of the world. Moreover, as other countries develop, their purchasing power will increase, leading to larger markets for the products of the developed countries. Growth in exports means growth in domestic production, and thus growth in domestic employment.

Those who favor globalization also note its inevitability. The economic logic that is propelling global production is immensely powerful. Technology allows globalization, and competition forges it; there is really no stopping the process, so the best one can do is adapt on the most favorable terms possible. Nations feel they must get into the game quickly so as not to be left behind.

A favorable standpoint on globalization is typically coupled with an optimistic view of the effects of immigration. Like new nations entering the global economy, immigrant workers are seen as having to suffer in the short run in order to make advances in the future. Instead of being viewed as exploited, the immigrants are seen as being granted an opportunity—one that they freely choose—to better their life circumstances. They may start off being paid low wages because they lack marketable skills, but with time, they or their children will acquire such skills and will experience upward mobility.

In general, a positive view of globalization is accompanied by a belief in the benefits of markets and free trade. The market, rather than political decision making, should, it is felt, be the arbiter of economic decision making. This favorable and inevitable view of globalization is by far the most predominant approach. It is promoted by the U.S. government, by the TNCs, by many governments in developing countries, and by various international agencies. This position receives considerable support from academics, especially economists, who provide governmental agencies with advice. It is the dominant world policy.

There is, however, a less sanguine interpretation of globalization voiced by U.S. trade unionists and many academics who study development, labor, women, inequality, and social class (Castells and Henderson 1987; Kamel 1990; Kolko 1988; Peet 1987; Ross and Trachte 1990; Sklair 1989). In general, their view is that globalization has a differential class impact: globalization is in the interests of capitalists, especially capitalists connected with TNCs, and of sectors of the capitalist class in developing nations. But the working class in both sets of countries is hurt, especially young women workers, who have become the chief employees of the TNCs (Fernandez-Kelly 1983; Fuentes and Ehrenreich 1983; Mies 1986; Nash and Fernandez-Kelly 1983).

Some argue that globalization is part of a response to a major crisis that has emerged in the advanced capitalist countries. In particular, after the post–World War II boom, the economies of these countries stagnated and profits declined; stagnation was blamed on the advances made by workers under the welfare state. Capital's movement abroad, which was preceded in the United States by regional relocation, is an effort to cut labor costs, weaken unions, and restore profitability. Put generally, globalization can be seen, in part, as an effort to discipline labor.

Globalization enables employers to pit workers from different countries against one another. Regions and nations must compete to attract investment and businesses. Competitors seek to undercut one another by offering the most favorable conditions to capital. Part of what they seek to offer is quality, efficiency, and timeliness, but they also compete in terms of providing the lowest possible labor standards: they promise a low-cost, disciplined, and unorganized work force. Governments pledge to ensure these conditions by engaging in the political repression of workers' movements (Deyo 1989).

The disciplining of the working class that accompanies globalization is not limited to conditions in the workplace. It also involves a cutback in state social programs. For example, in the United States, under the Reagan-Bush administrations, efforts were made to curtail multiple programs protecting workers' standard of living; these tax-based programs were seen as hindering capital accumulation. The argument was made that if these funds were invested by the private sector, everyone would benefit, including workers. This same logic has been imposed on developing countries; they have been granted aid and loans on the condition that they engage in austerity programs that cut back on social spending. The impact of such cutbacks is that workers are less protected from engaging in bargains of desperation when they enter the work force.

This view of globalization is accompanied by a pessimism about the policy of export-led development. Rather than believing that performing assembly for TNCs will lead to development, critics fear that it is another form of dependency, with the advanced capitalist countries and their corporations retaining economic (and political) control over the global economy (Bello and Rosenfeld 1990).

Critics also note a negative side to immigrants' experiences (Mitter 1986; Sassen 1988). They see the immigration of workers as, in part, a product of globalization and TNC activity, as workers in less-developed countries find their means of livelihood disrupted by capitalist penetration. Immigrants are thus not just people seeking a better life for themselves, but often those "forced" into moving because thet have lost the means to survive. On arrival in the more-advanced economies, they are faced with forms of coercion, including immigration regulations, racism, and sexism, that keep them an especially disadvantaged work force. Especially coercive is the condition of being an undocumented immigrant. Critics point out that those who favor globalization promote the free movement of commodities and capital, but not the free movement of labor, in the form of open borders. Political restrictions on workers add to the weakening of the working class.

In sum, the critical perspective sees globalization as an effort to strengthen the hand of capital and weaken that of labor. The favorable view argues that the interests of capital and labor are not antagonistic and that everyone benefits from capital accumulation, investment, economic growth, and the creation of jobs. Critics, on the other hand, contend that certain classes benefit at the expense of others, and that, even if workers in poor countries do get jobs, these jobs benefit the capitalists much more than they do the workers, and also hurt the workers in the advanced capitalist countries through deindustrialization.

Where does the truth lie? . . .

To a certain extent, one's point of view depends on geographic location. Generally, Asian countries, especially the NICs, appear to be transforming themselves from dependencies into major actors and competitors in the global economy, leading to an optimism about the effects of globalization. This optimism, however, blots out the suffering and labor repression that is still occurring for some workers in these countries, despite the rise in standard of living for the majority.

On the other hand, the Caribbean region generally faces a harsher reality, in part because the closeness and dominance of the United States pose special problems for these countries. They are more likely to get caught in simple assembly for the TNCs, raising questions about whether manufacturing for export will be transformable into broader economic development. Of course, some in these countries are firm believers in this policy and are pursuing it avidly, but there are clear signs that many workers are severely exploited in the process. . . .

Other confusing issues remain. For example, do women benefit from their movement into the wage sector (proletarianization) as a result of globalization? A case can be made that working outside the home and earning money gives women new-found power in their relations with men. It can also be argued, however, that these women remain under patriarchal control, but that now, in addition to their fathers and husbands, they are under the control of male bosses. They have double and even triple workloads, as they engage in wage labor, domestic labor, and often industrial homework and other forms of informalized labor (Ward 1990).

The two points of view lead to different politics. Those who hold the favorable outlook advocate working for the breakdown of all trade and investment barriers and to pushing rapidly ahead toward global integration. Critics are not trying to stem these forces completely, but rather, are attempting to set conditions on them. For example, globalization should be allowed only if labor and environmental standards are protected in the process. Similarly, the rights of workers to form unions should be safeguarded, so that business cannot wantonly pit groups of workers against one another. . . .

NOTES

Appelbaum, Richard P., and Jeffrey Henderson, eds. 1992. *States and Development in the Asian Pacific Rim.* Newbury Park, CA.: Sage.

Bello, Walden, and Stephanie Rosenfeld. 1990. *Dragons in Distress: Asia's Miracle Economies in Crisis.* San Francisco: Institute for Food and Development Policy.

Castells, Manuel, and Jeffrey Henderson. 1987. "Technoeconomic Restructuring, Sociopolitical Processes, and Spatial Transformation: A Global Perspective." In *Global Restructuring and Territorial Development,* edited by Jeffrey Henderson and Manuel Castells, 1–17. London: Sage.

Deyo, Frederic C. 1989. Beneath the Miracle: *Labor Subordination in the New Asian Industrialism.* Berkeley: University of California Press.

Fernandez-Kelly, M. Patricia. 1983. *For We Are Sold, I and My People: Women and Industry in Mexico's Frontier.* Albany: State University of New York Press.

Fröbel, Folker, Jürgen Heinrichs, and Otto Kreye. 1980. *The New International Division of Labour: Structural Unemployment in Industrialised Countries and Industrialisation in Developing Countries.* Cambridge: Cambridge University Press.

Fuentes, Annette, and Barbara Ehrenreich. 1983. *Women in the Global Factory.* Boston: South End Press.

Gereffi, Gary, and Miguel Korzeniewicz, eds. 1994. *Commodity Chains and Global Capitalism.* Westport, CT.: Greenwood Press.

Gereffi, Gary, and Donald L. Wyman, eds. 1990. *Manufacturing Miracles: Paths of Industrialization in Latin America and East Asia.* Princeton: Princeton University Press.

Gondolf, Edward W., Irwin M. Marcus, and James P. Daugherty. 1986. *The Global Economy: Divergent Perspectives on Economic Change.* Boulder, CO: Westview Press.

Grunwald, Joseph, and Kenneth Flamm. 1985. *The Global Factory: Foreign Assembly in International Trade.* Washington, D.C.: Brookings Institution.

Harrison, Bennett, and Barry Bluestone, 1988. *The Great U-Turn: Corporate Restructuring and the Polarizing of America.* New York: Basic Books.

Kamel, Rachael. 1990. *The Global Factory: Analysis and Action for a New Economic Era.* Philadelphia: American Friends Service Committee.

Kolko, Joyce. 1988. *Restructuring the World Economy.* New York: Pantheon.

Mies, Maria. 1986. *Patriarchy and Accumulation on a World Scale: Women in the International Division of Labor.* London: Zed Books.

Mitter, Swasti. 1986. *Common Fate, Common Bond: Women in the Global Economy.* London: Pluto Press.

Nash, June, and M. Patricia Fernandez-Kelly, eds. 1983. *Women, Men, and the International Division of Labor.* Albany: State University of New York Press.

Peet, Richard, ed. 1987. *International Capitalism and Industrial Restructuring.* Boston: Allen and Unwin.

Piore, Michael J., and Charles F. Sabel. 1984. *The Second Industrial Divide: Possibilities for Prosperity.* New York: Basic Books.

Ross, Robert J. S., and Kent C. Trachte. 1990. *Global Capitalism: The New Leviathan.* Albany: State University of New York Press.

Sassen, Saskia. 1988. *The Mobility of Labor and Capital: A Study in International*

Investment and Labor Flow. Cambridge: Cambridge University Press.

———. 1991. *The Global City: New York, London, Tokyo.* Princeton: Princeton University Press.

Sklair, Leslie. 1989. *Assembling for Development: The Maquila Industry in Mexico and the United States.* London: Unwin Hyman.

DISCUSSION QUESTIONS

1. What is a *commodity chain?* What evidence do you see of commodity chains in the wardrobe that you wear?

2. Compare and contrast the two perspectives on the new globalization that the authors describe. How do proponents of the positive and critical views of globalization view immigration?

INFOTRAC COLLEGE EDITION

You can use your access to InfoTrac College Edition to learn more about the subjects covered in this essay. Some suggested search term include:

commodity chains
guest workers
immigration

international (or global) division of labor
newly industrializing countries
transnational corporation

37

The American Belief System Concerning Poverty and Welfare

WILLIAM JULIUS WILSON

By analyzing social and economic inequality in different western European contexts, Wilson identifies the factors in global economic development that are producing joblessness and urban poverty. His analysis provides a comparative context for understanding how racial

From: William Julius Wilson. 1996. *When Work Disappears.* New York: Knopf, pp. 149–155.
Reprinted with permission.

and ethnic tensions within the United States can be attributed to global changes that are af-
fecting western European societies, as well. His article also shows how global stratification is
resulting from changes in technology and employment patterns.

Historically, Europe and the United States have contrasted sharply in terms of the nature of urban inequality. No European city has experienced the level of concentrated poverty and racial and ethnic segregation that is typical of American metropolises. Nor does any European city include areas that are as physically isolated, deteriorated, and prone to violence as the inner-city ghettos of urban America. In short, there is no real European equivalent to the plight of the American ghetto. Nonetheless, the situation is changing rapidly. European cities are beginning to experience many of the problems of urban social dislocations—including increasing unemployment, concentrated poverty, and racial and ethnic conflicts—that have traditionally plagued American cities.

European cites differ in the origins of their racial and ethnic problems. The rising ethnic conflicts in eastern Europe which have accompanied the breakdown of the Soviet empire stem from nationalist political struggles in the period of democratization, not from immigration. However, there are significant differences even among the countries of western Europe. Since the late 1960s, northern European economies have received workers from Turkey, the Maghreb countries of northwest Africa, North Africa, the Middle East, and former British, Dutch, and French territories. These influxes of immigrants have widened the cultural experiences of European nations, and differences between immigrants and indigenous populations are disturbing the traditionally homogeneous makeup of communities.

In discussions of urban poverty and racial tensions, the European countries that have most frequently been contrasted with the United States are Britain, France, Holland, and Belgium. All of these countries were once colonial powers, and racial groups from former colonies have been prominently represented among recent immigrants. A number of other countries, including Switzerland, Norway, Sweden, and Germany, have seen an influx of immigrants who have come mainly as laborers from southern Europe and as refugees from countries outside Europe. Finally, the southern part of Europe, which for most of the mid-twentieth century represented "sending countries" (countries whose own citizens often became immigrant workers), is now experiencing a fairly rapid growth of immigration from the Maghreb countries of northwest Africa.

The immigrants in all of these countries are disproportionately represented in areas that feature the highest levels of unemployment. Although they are far from matching the depth and severity of the social dislocations that plague the inner-city ghettos of the United States, many inner-city communities and outer-city public housing estates in countries like France and the Netherlands have been cut off from mainstream labor market institutions and informal job networks, creating the vicious cycle of weak labor force attachment, growing social exclusion, and rising tensions.

Although these European inner-city communities still feature many indigenous residents and are therefore more mixed than the nearly black-only ghettos of the United States, their population is invariably drawn disproportionately from the various first- and second-generation immigrant minorities. Trends in a number of European countries suggest the beginnings of the kind of urban social polarization typical of American metropolises.

More important, however, the economic and industrial restructuring of Europe, which includes the decline of traditional manufacturing areas, has decreased the need for unskilled immigrant labor. Thousands of the immigrants who had been recruited during periods of national labor shortages have now been laid off. A substantial number of the new jobs in the next several decades will require levels of training and education that are beyond the reach of most immigrant minorities. Being the last hired and the first fired, immigrant minorities were unemployed at rates that soared during the late 1970s through the 1980s and reached levels ranging from 25 to 50 percent in the cities with unprecedented high levels of unemployment overall. This is not unlike the situation facing residents of the inner-city ghettos of the United States. The economic origins of the problems are also similar, and they deserve a closer look.

It is important to recognize that much of the sharp rise in inner-city joblessness in the United States and the growth of unemployment in Europe and Canada stems from the swift technological changes in the global economy. First of all, there has been a decline in mass production. The effects of this change are perhaps most clearly seen in the U.S. economy, which has traditionally enjoyed rapid growth in productivity and living standards. . . . The mass production system benefited from large quantities of cheap natural resources, economies of scale, and processes that generated higher uses of productivity through shifts in market forces and that caused improvements in one industry to lead to advancements in others.

The skill requirements of the mass production system were reflected in the system of learning. Public schools in the United States were principally designed to provide low-income native and immigrant students with the basic literacy and numeracy skills required for routine work in mass production factories, service industries, or farms. On the other hand, families with professional, technical, and managerial backgrounds had access to elite learning processes in public and private schools and were able to utilize family connections and experiences to prepare their children for higher-paying occupations. "The school-to-work transition processes were informal and largely family related, but were perpetuated through the formal learning system."

Today's close interaction between technology and international competition has eroded the basic institutions of the mass production system. In the last several decades, almost all of the improvements in productivity have been associated with technology and human capital, thereby drastically reducing the importance of physical capital and natural resources. Moreover, in the traditional mass production system only a few highly educated professional, technical, and managerial workers were needed because most of the work "was routine and could be performed by workers who needed only basic literacy and numeracy." Accordingly, workers with limited education were able to take home wages that were comparatively high by international and historical standards. "This was especially true after the New Deal policies of the 1930s provided social safety nets for those who were not expected to work, collective bargaining to improve workers' share of system's gains, and monetary and fiscal policies to help the mass production system running at relatively low levels of unemployment."

At the same time that changes in technology are producing new jobs, however, they are making many others obsolete. The workplace has been revolutionized by technological changes that range from the development of robotics to the creation of information highways. A widening gap between the skilled and unskilled workers is developing because education and training are more important than ever. Indeed, because of low levels of education, more and more unskilled workers are out of work or poorly paid in both the United States and Europe. While educated workers are benefiting from the pace of technological change, less skilled workers, such as those found in many inner-city neighborhoods, face the growing threat of job displacement. For example, highly skilled designers, engineers, and operators are needed for the jobs created by the development of a new set of computer-operated machine tools, an advance that also *eliminates* jobs for those trained only for manual, assembly-line work. Also, advances in word processing have increased the demand for those who not only know how to type but can operate specialized software as well; the need for routine typists and secretaries is, accordingly, reduced.

The impact of technological change has been heightened by international competition. In order to adjust to changing markets and technology, competitive systems were forced to become more flexible. Companies can compete more effectively in the international market either by improving productivity and quality or by reducing workers' income. The easier approach is this latter low-wage strategy, which the United States has tended to follow. According to Ray Marshall, the former secretary of labor in the Carter administration, "Most other industrialized nations have rejected this strategy because it implies lower and more unequal wages, with serious political, social, and economic implications." To encourage western European and Japanese companies to follow high-wage strategies, several actions have been taken. First, governments have worked to create a national consensus concerning recovery strategies—perhaps the most important step toward effective changes. Companies were then pushed to pursue high-wage strategies in a political and regulatory environment that included adjustment processes to shift resources from low- to high-productivity activities, trade and industrial policies, wage regulations, relatively generous family support systems and unemployment compensation, and universal national health care. But with a recent sharp rise in unemployment there is growing pressure on European nations to adopt the low-wage strategy pursued by the United States.

A few years ago when unemployment rose in all the major industrial democracies during the worldwide recession, "the conventional wisdom was that the upturn in the business cycle would solve most of the problem." When the United States entered a period of economic recovery in the early 1990s and the worst of the recession was over in Europe, there remained a scarcity of new high-wage jobs on both sides of the Atlantic, a situation that created fears that the duration of the shortage would be lengthy.

In the highly integrated global marketplace of today, economies can grow, stock markets can rise, corporate profits can soar, and yet many workers may remain unemployed or underemployed. Why? Because "capital and technology are now so mobile that they do not always create good jobs in their own backyard." Corporate cutbacks, made in an effort to streamline operations for the global economy, have added to the jobless woes of many workers. In short, economic growth today does not necessarily produce good jobs.

In comparisons between the United States and other industrialized nations, the problem of jobs can be seen in different dimensions. Europe and Canada have continued to increase wages and benefits for their workers, but at the price of high levels of unemployment for the many who are kept afloat with generous unemployment insurance. In 1994, 11 percent of the European workforce, or 18 million people, were unemployed. By contrast, the United States, which moved out of the recession earlier, has created more jobs mainly by getting workers to accept low-paying employment. The result has been a widening gap in the United States between the highest- and lowest-paid workers.

Because of Europe's relatively munificent social safety nets, many eschew the kind of low-paying jobs taken by low-skilled workers in the United States. They rely instead on relatively generous unemployment compensation. "In most of the European Union, an unemployed worker can receive close to $1,000 a month indefinitely. In addition, Europe's highly regulated job markets, where it is very difficult to dismiss workers, have left employers wary of creating jobs." This has led not only to a steady growth in unemployment and increases in wages and social security contributions but to a less and less competitive European industry as well.

The U.S. economy created 35 million jobs between 1973 and 1991. Europe added just 8 million during the same period despite having about one-third more people. . . . Most of the new jobs in the United States were in the service sector—including those for workers with limited education and training disproportionately held by women. The growth of employment in the service sector reduced, but did not offset, the overall job declines for less educated workers in the U.S. central cities. Indeed, most of the job growth occurred outside central cities.

In the more flexible labor market in the United States, real wages grew each year by less than .5 percent, whereas annual real wage growth in Europe rose three times as fast. "Minimum wage levels tend to be higher in Europe, with cost-of-living increases and four-week annual vacations virtually guaranteed." Total compensation—wages, health benefits, vacations—for the typical U.S. worker in manufacturing has either remained flat or declined since the mid-1970s, while it has increased by 40 percent for a European worker in a comparable job. Average compensation for a manufacturing worker in the United States is $16 an hour; in Germany it is $26. During the 1980s, the wages for low-income U.S. workers who lacked any college education dropped 15 percent when adjusted for inflation, whereas those for comparable workers in Europe increased 15 to 20 percent.

There is mounting pressure on European countries to make some adjustments in order to become more globally competitive. Some European observers have argued that there will have to be a massive reduction in the continent's labor force. Others emphasize that production will have to be transferred to lower-wage parts of the world. Both strategies will lead to even greater unemployment. Still other strategists claim that one way to combat the growing problems of unemployment, including those stemming from industrial cost-cutting of the kind engaged in by American industries, is to develop the service sector, even if the jobs do carry lower wages. In so doing, restrictions on zoning laws and store-opening hours would have to be relaxed. Also, the welfare state would have to be altered by lowering taxes and nonwage contributions that significantly increase total labor costs. This strategy would create more low-wage jobs, but it would also, as in the United States, lead to greater inequality among workers.

A recent report based on interviews with leading government officials, corporate executives, economists, and union leaders in Europe reveals pessimism about the efficacy of future action to address the problem of global competition. These leaders feel that Europe lacks both the time and the political will to change social and political habits enough to enable the private sector to better compete in the global economy and thus create enough new jobs to contain the unemployment. "Most everyone agrees on the desirability of slashing regulatory red tape, taxes, and the nonwage labor costs that support Europe's encrusted social protection programs, but few believe it will happen any time soon." . . .

DISCUSSION QUESTIONS

1. How does Wilson link the growing problems of social and economic polarization within European countries to joblessness and urban poverty? How does this compare to the situation in the United States?

2. What specific developments in society does Wilson suggest are producing joblessness across different nations? How have these changes been propelled by international competition?

INFOTRAC COLLEGE EDITION

You can use your access to InfoTrac College Edition to learn more about the subjects covered in this essay. Some suggested search terms include:

economic restructuring   unemployment insurance
informal job networks   urban poverty
inner city joblessness

38

The Lexus and the Olive Tree

THOMAS L. FRIEDMAN

Thomas Friedman uses the metaphor of the Lexus and the olive tree to refer to the modern changes that are transforming the world economy, coupled with presence of traditional cultural values that are rapidly being transformed by globalization. Here he identifies how societies are changing under of globalization and how this process is transforming world affairs.

The globalization system, unlike the Cold War system, is not static, but a dynamic ongoing process: globalization involves the inexorable integration of markets, nation-states and technologies to a degree never witnessed before—in a way that is enabling individuals, corporations and nation-states to reach around the world farther, faster, deeper and cheaper than ever before, and in a way that is also producing a powerful backlash from those brutalized or left behind by this new system.

The driving idea behind globalization is free-market capitalism—the more you let market forces rule and the more you open your economy to free trade and competition, the more efficient and flourishing your economy will be. Globalization means the spread of free-market capitalism to virtually every country in the world. Globalization also has its own set of economic rules—rules that revolve around opening, deregulating and privatizing your economy.

Unlike the Cold War system, globalization has its own dominant culture, which is why it tends to be homogenizing. In previous eras this sort of cultural homogenization happened on a regional scale—the Hellenization of the Near East and the Mediterranean world under the Greeks, the Turkification of Central Asia, North Africa, Europe and the Middle East by the Ottomans, or the Russification of Eastern and Central Europe and parts of Eurasia under the Soviets. Culturally speaking, globalization is largely, though not entirely, the spread of Americanization—from Big Macs to iMacs to Mickey Mouse—on a global scale.

Globalization has its own defining technologies: computerization, miniaturization, digitization, satellite communications, fiber optics and the Internet. And these technologies helped to create the defining perspective of globalization. If the defining perspective of the Cold War world was "division," the defining perspective of globalization is "integration." The symbol of the Cold War system was a wall, which divided everyone. The symbol of the globalization system is a World Wide Web, which unites everyone. The defining document of the Cold War

From: Thomas L. Friedman. 1999. *The Lexus and the Olive Tree.* New York: Farrar Straus Giroux, pp. 7–12. Reprinted with permission.

system was "The Treaty." The defining document of the globalization system is "The Deal."

Once a country makes the leap into the system of globalization, its elites begin to internalize this perspective of integration, and always try to locate themselves in a global context. I was visiting Amman, Jordan, in the summer of 1998 and having coffee at the Inter-Continental Hotel with my friend Rami Khouri, the leading political columnist in Jordan. We sat down and I asked him what was new. The first thing he said to me was: "Jordan was just added to CNN's worldwide weather highlights." What Rami was saying was that it is important for Jordan to know that those institutions which think globally believe it is now worth knowing what the weather is like in Amman. It makes Jordanians feel more important and holds out the hope that they will be enriched by having more tourists or global investors visiting. . . .

While the defining measurement of the Cold War was weight—particularly the throw weight of missiles—the defining measurement of the globalization system is speed—speed of commerce, travel, communication and innovation. The Cold War was about Einstein's mass-energy equation, $e = mc^2$. Globalization is about Moore's law, which states that the computing power of silicon chips will double every eighteen to twenty-four months. In the Cold War, the most frequently asked question was: "How big is your missile?" In globalization, the most frequently asked question is: "How fast is your modem?" . . .

The speed by which your latest invention can be made obsolete or turned into a commodity is now lightning quick. Therefore, only the paranoid, only those who are constantly looking over their shoulders to see who is creating something new that will destroy them and then staying just one step ahead of them, will survive. Those countries that are most willing to let capitalism quickly destroy inefficient companies, so that money can be freed up and directed to more innovative ones, will thrive in an era of globilization. Those which rely on governments to protect them from such creative destruction will fall behind in the era.

If the defining anxiety of the Cold War was fear of annihilation from an enemy you knew all too well in a world struggle that was fixed and stable, the defining anxiety in globalization is fear of rapid change from an enemy you can't see, touch or feel—a sense that your job, community or workplace can be changed at any moment by anonymous economic and technological forces that are anything but stable.

In the Cold War we reached for the hot line between the White House and the Kremlin—a symbol that we were all divided but at least someone, the two superpowers, was in charge. In the era of globalization we reach for the Internet—a symbol that we are all connected but nobody is in charge. The defining defense system of the Cold War was radar—to expose the threats coming from the other side of the wall. The defining defense system of the globalization era is the X-ray machine—to expose the threats coming from within.

Globalization also has its own demographic pattern—a rapid acceleration of the movement of people from rural areas and agricultural lifestyles to urban areas and urban lifestyles more intimately linked with global fashion, food, markets and entertainment trends.

Last, and most important, globalization has its own defining structure of power, which is much more complex than the Cold War structure. The Cold War system was built exclusively around nation-states, and it was balanced at the center by two superpowers: the United States and the Soviet Union.

The globalization system, by contrast, is built around three balances, which overlap and affect one another. The first is the traditional balance between nation-states. In the globalization system, the United States is now the sole and dominant superpower and all other nations are subordinate to it to one degree or another. The balance of power between the United States and the other states still matters for the stability of this system. And it can still explain a lot of the news you read on the front page of the papers, whether it is the containment of Iraq in the Middle East or the expansion of NATO against Russia in Central Europe.

The second balance in the globalization system is between nation-states and global markets. These global markets are made up of millions of investors moving money around the world with the click of a mouse. I call them "the Electronic Herd," and this herd gathers in key global financial centers, such as Wall Street, Hong Kong, London and Frankfurt, which I call "the Supermarkets." The attitudes and actions of the Electronic Herd and the Supermarkets can have a huge impact on nation-states today, even to the point of triggering the downfall of governments. You will not understand the front page of newspapers today— whether it is the story of the toppling of Suharto in Indonesia, the internal collapse in Russia or the monetary policy of the United States—unless you bring the Supermarkets into your analysis.

The United States can destroy you by dropping bombs and the Supermarkets can destroy you by downgrading your bonds. The United States is the dominant player in maintaining the globalization gameboard, but it is not alone in influencing the moves on that gameboard. This globalization gameboard today is a lot like a Ouija board—sometimes pieces are moved around by the obvious hand of the superpower, and sometimes they are moved around by hidden hands of the Supermarkets.

The third balance that you have to pay attention to in the globalization system—the one that is really the newest of all—is the balance between individuals and nation-states. Because globalization has brought down many of the walls that limited the movement and reach of people, and because it has simultaneously wired the world into networks, it gives more power to individuals to influence both markets and nation-states than at any time in history. So you have today not only a superpower, not only Supermarkets, but, . . . you have Super-empowered individuals. Some of these Super-empowered individuals are quite angry, some of them quite wonderful—but all of them are now able to act directly on the world stage without the traditional mediation of governments, corporations or any other public or private institutions. . . .

Nation-states, and the American superpower in particular, are still hugely important today, but so too now are Supermarkets and Super-empowered individuals. You will never understand the globalization system, or the front page of the morning paper, unless you see it as a complex interaction between all three of these actors: states bumping up against states, states bumping up against Supermarkets, and Supermarkets and states bumping up against Super-empowered individuals. . . .

DISCUSSION QUESTIONS

1. Compare and contrast the globalization system that Friedman describes as marking the contemporary world with that which characterized the Cold War.
2. What does Friedman identify as the three balances in the globalization system and how are these transforming U.S. society?

INFOTRAC COLLEGE EDITION

You can use your access to InfoTrac College Edition to learn more about the subjects covered in this essay. Some suggested search terms include:

cultural homogeneity
multinational corporation
nation-states

superpower
technology and globalization

39

The Nanny Chain

ARLIE RUSSELL HOCHSCHILD

Arlie Hochschild identifies the "nanny chain" as a global system of work in which women workers from poor nations provide the "care work" for more privileged workers in other parts of the world. This pattern of labor is transforming social relations of care worldwide and, according to Hochschild, makes care and love a commodity that is transferred and exchanged in the world market.

Vicky Diaz, a 34-year-old mother of five, was a college-educated schoolteacher and travel agent in the Philippines before migrating to the United States to work as a housekeeper for a wealthy Beverly Hills family and as a nanny for their two-year-old son. Her children, Vicky explained to Rhacel Parrenas,

> were saddened by my departure. Even until now my children are trying to convince me to go home. The children were not angry when I left because they were still very young when I left them. My husband could not get angry either because he knew that was the only way I could seriously help him raise our children, so that our children could be sent to school. I send them money every month.

From: Arlie Russell Hochschild, "The Nanny Chain" (2000). The American Prospect, 3 January 2000, pp. 33–36.

In her forthcoming book *Servants of Globalization,* Parrenas, an affiliate of the Center for Working Families at the University of California, Berkeley, tells an important and disquieting story of what she calls the "globalization of mothering." The Beverly Hills family pays "Vicky" (which is the pseudonym Parrenas gave her) $400 a week, and Vicky, in turn, pays her own family's live-in domestic worker back in the Philippines $40 a week. Living like this is not easy on Vicky and her family. "Even though it's paid well, you are sinking in the amount of your work. Even while you are ironing the clothes, they can still call you to the kitchen to wash the plates. It . . . [is] also very depressing. The only thing you can do is give all your love to [the two-year-old American child]. In my absence from my children, the most I could do with my situation is give all my love to that child."

Vicky is part of what we could call a global care chain: a series of personal links between people across the globe based on the paid or unpaid work of caring. A typical global care chain might work something like this: An older daughter from a poor family in a third world country cares for her siblings (the first link in the chain) while her mother works as a nanny caring for the children of a nanny migrating to a first world country (the second link) who, in turn, cares for the child of a family in a rich country (the final link). Each kind of chain expresses an invisible human ecology of care, one care worker depending on another and so on. A global care chain might start in a poor country and end in a rich one, or it might link rural and urban areas within the same poor country. More complex versions start in one poor country and extend to another slightly less poor country and then link to a rich country.

Global care chains may be proliferating. According to 1994 estimates by the International Organization for Migration, 120 million people migrated—legally or illegally—from one country to another. That's 2 percent of the world's population. How many migrants leave loved ones behind to care for other people's children or elderly parents, we don't know. But we do know that more than half of legal migrants to the United States are women, mostly between ages 25 and 34. And migration experts tell us that the proportion of women among migrants is likely to rise. All of this suggests that the trend toward global care chains will continue.

How are we to understand the impact of globalization on care? If, as globalization continues, more global care chains form, will they be "good" care chains or "bad" ones? Given the entrenched problem of third world poverty—which is one of the starting points for care chains—this is by no means a simple question. But we have yet to fully address it, I believe, because the world is globalizing faster than our minds or hearts are. We live global but still think and feel local.

FREUD IN A GLOBAL ECONOMY

Most writing on globalization focuses on money, markets, and labor flows, while giving scant attention to women, children, and the care of one for the other. Most research on women and development, meanwhile, draws a connection between, say, World Bank loan conditions and the scarcity of food for women and children in the third world, without saying much about resources expended on caregiving. Much of the research on women in the United States and Europe focuses on a chainless, two-person picture of "work-family balance" without considering the child care worker and the emotional ecology of which he or she is a part. Fortunately, in recent years, scholars such as Ernestine Avila, Evelyn Nakano Glenn, Pierette Hondagneu-Sotelo, Mary Romero, and Rhacel Parrenas have produced some fascinating research on domestic workers. Building on this work, we can begin to focus on the first world end of the care chain and begin spelling out some of the implications of the globalization of love.

One difficulty in understanding these implications is that the language of economics does not translate easily into the language of psychology. How are we to understand a "transfer" of feeling from one link in a chain to another? Feeling is not a "resource" that can be crassly taken from one person and given to another. And surely one person can love quite a few people; love is not a resource limited the same way oil or currency supply is. Or is it?

Consider Sigmund Freud's theory of displacement, the idea that emotion can be redirected from one person or object to another. Freud believed that if, for example, Jane loves Dick but Dick is emotionally or literally unavailable, Jane will find a new object (say, John, Dick and Jane's son) onto which to project her original feeling for Dick. While Freud applied the idea of displacement mainly to relations within the nuclear family, the concept can also be applied to relations extending far outside it. For example, immigrant nannies and au pairs often divert feelings originally directed toward their own children toward their young charges in this country. As Sau-ling C. Wong, a researcher at the University of California, Berkeley, has put it, "Time and energy available for mothers are diverted from those who, by kinship or communal ties, are their more rightful recipients."

If it is true that attention, solicitude, and love itself can be "displaced" from one child (let's say Vicky Diaz's son Alfredo, back in the Philippines) onto another child (let's say Tommy, the son of her employers in Beverly Hills), then the important observation to make here is that this displacement is often upward in wealth and power. This, in turn, raises the question of the equitable distribution of care. It makes us wonder, is there—in the realm of love—an analogue to what Marx calls "surplus value," something skimmed off from the poor for the benefit of the rich?

Seen as a thing in itself, Vicky's love for the Beverly Hills toddler is unique, individual, private. But might there not be elements in this love that are borrowed, so to speak, from somewhere and someone else? Is time spent with the first world child in some sense "taken" from a child further down the care chain? Is the Beverly Hills child getting "surplus" love, the way immigrant farm workers give us surplus labor? Are first world countries such as the United States importing maternal love as they have imported copper, zinc, gold, and other ores from third world countries in the past?

This is a startling idea and an unwelcome one, both for Vicky Diaz, who needs the money from a first world job, and for her well-meaning employers, who want someone to give loving care to their child. Each link in the chain feels she is doing the right thing for good reasons—and who is to say she is not?

But there are clearly hidden costs here, costs that tend to get passed down along the chain. One nanny reported such a cost when she described (to Rhacel Parrenas) a return visit to the Philippines: "When I saw my children, I thought, 'Oh children do grow up even without their mother.' I left my youngest when she was only five years old. She was already nine when I saw her again but she still wanted for me to carry her [weeps]. That hurt me because it showed me that my children missed out on a lot."

Sometimes the toll it takes on the domestic worker is overwhelming and suggests that the nanny has not displaced her love onto an employer's child but rather has continued to long intensely for her own child. As one woman told Parrenas, "The first two years I felt like I was going crazy. . . . I would catch myself gazing at nothing, thinking about my child. Every moment, every second of the day, I felt like I was thinking about my baby. My youngest, you have to understand, I left when he was only two months old. . . . You know, whenever I receive a letter from my children, I cannot sleep. I cry. It's good that my job is more demanding at night."

Despite the anguish these separations clearly cause, Filipina women continue to leave for jobs abroad. Since the early 1990s, 55 percent of migrants out of the Philippines have been women; next to electronic manufacturing, their remittances make up the major source of foreign currency in the Philippines. The rate of female emigration has continued to increase and includes college-educated teachers, businesswomen, and secretaries. In Parrenas's study, more than half of the nannies she interviewed had college degrees and most were married mothers in their 30s.

Where are men in this picture? For the most part, men—especially men at the top of the class ladder—leave child-rearing to women. Many of the husbands and fathers of Parrenas's domestic workers had migrated to the Arabian peninsula and other places in search of better wages, relieving other men of "male work" as construction workers and tradesmen, while being replaced themselves at home. Others remained at home, responsible fathers caring or helping to care for their children. But some of the men tyrannized their wives. Indeed, many of the women migrants Parrenas interviewed didn't just leave; they fled. As one migrant maid explained:

> You have to understand that my problems were very heavy before I left the Philippines. My husband was abusive. I couldn't even think about my children, the only thing I could think about was the opportunity to escape my situation. If my husband was not going to kill me, I was probably going to kill him. . . . He always beat me up and my parents wanted me to leave him for a long time. I left my children with my sister. . . . In the plane. . . I felt like a bird whose cage had been locked for many years. . . . I felt free. . . . Deep inside, I felt homesick for my children but I also felt free for being able to escape the most dire problem that was slowly killing me.

Other men abandoned their wives. A former public school teacher back in the Philippines confided to Parrenas: "After three years of marriage, my husband left me for another woman. My husband supported us for just a little over a year. Then the support was stopped. . . . The letters stopped. I have not seen him since." In the absence of government aid, then, migration becomes a way of coping with abandonment.

Sometimes the husband of a female migrant worker is himself a migrant worker who takes turns with his wife migrating. One Filipino man worked in Saudi Arabia for 10 years, coming home for a month each year. When he finally returned home for good, his wife set off to work as a maid in America while he took care of the children. As she explained to Parrenas, "My children were very sad when I left them. My husband told me that when they came back home from the airport, my children could not touch their food and they wanted to cry. My son, whenever he writes me, always draws the head of Fido the dog with tears on the eyes. Whenever he goes to Mass on Sundays, he tells me that he misses me more because he sees his friends with their mothers. Then he comes home and cries."

THE END OF THE CHAIN

Just as global capitalism helps create a third world supply of mothering, it creates a first world demand for it. The past half-century has witnessed a huge rise in the number of women in paid work—from 15 percent of mothers of children aged 6 and under in 1950 to 65 percent today. Indeed, American women now make up 45 percent of the American labor force. Three-quarters of mothers of children 18 and under now work, as do 65 percent of mothers of children 6 and under. In addition, a recent report by the International Labor Organization reveals that the average number of hours of work per week has been rising in this country.

Earlier generations of American working women would rely on grandmothers and other female kin to help look after their children; now the grandmothers and aunts are themselves busy doing paid work outside the home. Statistics show that over the past 30 years a decreasing number of families have relied on relatives to care for their children—and hence are compelled to look for nonfamily care. At the first world end of care chains, working parents are grateful to find a good nanny or child care provider, and they are generally able to pay far more than the nanny could earn in her native country. This is not just a child care problem. Many American families are now relying on immigrant or out-of-home care for their *elderly* relatives. As a Los Angeles elder-care worker, an immigrant, told Parrenas, "Domestics here are able to make a living from the elderly that families abandon." But this often means that nannies cannot take care of their own ailing parents and therefore produce an elder-care version of a child care chain—caring for first world elderly persons while a paid worker cares for their aged mother back in the Philippines.

My own research for two books, *The Second Shift* and *The Time Bind,* sheds some light on the first world end of the chain. Many women have joined the law, academia, medicine, business—but such professions are still organized for men who are free of family responsibilities. The successful career, at least for those who are broadly middle class or above, is still largely built on some key traditional components: doing professional work, competing with fellow professionals, getting credit for work, building a reputation while you're young, hoarding scarce time, and minimizing family obligations by finding someone else to deal with domestic chores. In the past, the professional was a man and the "someone else to deal with [chores]" was a wife. The wife oversaw the family, which—in preindustrial times, anyway—was supposed to absorb the human vicissitudes of birth, sickness, and death that the workplace discarded. Today, men take on much more of the child care and housework at home, but they still base their identity on demanding careers in the context of which children are beloved impediments; hence, men resist sharing care equally at home. So when parents don't have enough "caring time" between them, they feel forced to look for that care further down the global chain.

The ultimate beneficiaries of these various care changes might actually be large multinational companies, usually based in the United States. In my research on a Fortune 500 manufacturing company I call Amerco, I discovered a disproportionate number of women employed in the human side of the company: public relations, marketing, human resources. In all sectors of the company, women often helped others sort out problems—both personal and professional—at work. It was often the welcoming voice and "soft touch" of women workers that made Amerco seem like a family to other workers. In other words, it appears that these working mothers displace some of their emotional labor from their children to their employer, which holds itself out to the worker as a "family." So, the care in the chain may begin with that which a rural third world mother gives (as a nanny) the urban child she cares for, and it may end with the care a working mother gives her employees as the vice president of publicity at your company.

HOW MUCH IS CARE WORTH?

How are we to respond to the growing number of global care chains? Through what perspective should we view them?

I can think of three vantage points from which to see care chains: that of the primordialist, the sunshine modernist, and (my own) the critical modernist. The primordialist believes that our primary responsibility is to our own family, our own community, our own country. According to this view, if we all tend our own primordial plots, everybody will be fine. There is some logic to this point of view. After all, Freud's concept of displacement rests on the premise that some original first object of love has a primary "right" to that love, and second and third comers don't fully share that right. (For the primordialist—as for most all of us—those first objects are members of one's most immediate family.) But the primordialist is an isolationist, an antiglobalist. To such a person, care chains seem wrong—not because they're unfair to the least-cared-for children at the bottom of the chain, but because they are global. Also, because family care has historically been provided by women, primordialists often believe that women should stay home to provide this care.

The sunshine modernist, on the other hand, believes care chains are just fine, an inevitable part of globalization, which is itself uncritically accepted as good. The idea of displacement is hard for the sunshine modernists to grasp because in their equation—seen mainly in economic terms—the global market will sort out who has proper claims on a nanny's love. As long as the global supply of labor meets the global demand for it, the sunshine modernist believes, everything will be okay. If the primordialist thinks care chains are bad because they're global, the sunshine modernist thinks they're good for the very same reason. In either case, the issue of inequality of access to care disappears.

The critical modernist embraces modernity but with a global sense of ethics. When the critical modernist goes out to buy a pair of Nike shoes, she is concerned to learn how low the wage was and how long the hours were for the third world factory worker making the shoes. The critical modernist applies the same moral concern to care chains: The welfare of the Filipino child back home must be seen as some part, however small, of the total picture. The critical modernist sees globalization as a very mixed blessing, bringing with it new opportunities—such as the nanny's access to good wages—but also new problems, including emotional and psychological costs we have hardly begun to understand.

From the critical modernist perspective, globalization may be increasing inequities not simply in access to money—and those inequities are important enough—but in access to care. The poor maid's child may be getting less motherly care than the first world child. (And for that matter, because of longer hours of work, the first world child may not be getting the ideal quantity of parenting attention for healthy development because too much of it is now displaced onto the employees of Fortune 500 companies.) We needn't lapse into primordialism to sense that something may be amiss in this.

I see no easy solutions to the human costs of global care chains. But here are some initial thoughts. We might, for example, reduce the incentive to migrate by addressing the causes of the migrant's economic desperation and fostering economic growth in the third world. Thus one obvious goal would be to develop the Filipino economy.

But it's not so simple. Immigration scholars have demonstrated that development itself can *encourage* migration because development gives rise to new economic uncertainties that families try to mitigate by seeking employment in the first world. If members of a family are laid off at home, a migrant's monthly remittance can see them through, often by making a capital outlay in a small business or paying for a child's education.

Other solutions might focus on individual links in the care chain. Because some women migrate to flee abusive husbands, a partial solution would be to create local refuges from such husbands. Another would be to alter immigration policy so as to encourage nannies to bring their children with them. Alternatively, employers or even government subsidies could help nannies make regular visits home.

The most fundamental approach to the problem is to raise the value of caring work and to ensure that whoever does it gets more credit and money for it. Otherwise, caring work will be what's left over, the work that's continually passed on down the chain. Sadly, the value ascribed to the labor of raising a child has always been low relative to the value of other kinds of labor, and under the impact of globalization, it has sunk lower still. The low value placed on caring work is due neither to an absence of demand for it (which is always high) nor to the simplicity of the work (successful caregiving is not easy) but rather to the cultural politics underlying this global exchange.

The declining value of child care anywhere in the world can be compared to the declining value of basic food crops relative to manufactured goods on the international market. Though clearly more essential to life, crops such as wheat, rice, or cocoa fetch low and declining prices while the prices of manufactured goods (relative to primary goods) continue to soar in the world market. And just as the low market price of primary produce keeps the third world low in the community of nations, the low market value of care keeps low the status of the women who do it.

One way to solve this problem is to get fathers to contribute more child care. If fathers worldwide shared child care labor more equitably, care would spread laterally instead of being passed down a social-class ladder, diminishing in value along the way. Culturally, Americans have begun to embrace this idea—but they've yet to put it into practice on a truly large scale [see Richard Weissbourd, "Redefining Dad," *TAP*, December 6, 1999]. This is where norms and policies established in the first world can have perhaps the greatest influence on reducing costs along global care chains.

According to the International Labor Organization, half of the world's women between ages 15 and 64 are working in paid jobs. Between 1960 and 1980, 69 out of 88 countries for which data are available showed a growing proportion of women in paid work (and the rate of increase has skyrocketed since the 1950s in the United States, Scandinavia, and the United Kingdom). If we want developed societies with women doctors, political leaders, teachers, bus drivers, and computer programmers, we will need qualified people to help care for children. And there is no reason why every society cannot enjoy such loving paid child care. It may even remain the case that Vicky Diaz is the best person to provide it. But we would be wise to adopt the perspective of the critical modernist and extend our concern to the potential hidden losers in the care chain. These days, the personal is global.

DISCUSSION QUESTIONS

1. What does Hochschild mean by the *globalization of love* and how is this phenomenon linked to the status of women in the United States? in other parts of the world?

2. What different perspectives on the care chain does Hochschild identify? What solutions to the problem does each perspective suggest? What would you recommend?

INFOTRAC COLLEGE EDITION

You can use your access to InfoTrac College Edition to learn more about the subjects covered in this essay. Some suggested search terms include:

commercialization of feeling

domestic workers

migrant workers

migration and domestic labor

third world poverty

transnational families/mothering

40

The Souls of Black Folk

W. E. B. DU BOIS

W. E. B. Du Bois, the first African American Ph. D. from Harvard University, is a classic sociological analyst. In this well-known essay, he develops the idea that African Americans have a "double consciousness"—one that they must develop as a protective strategy to understand how Whites see them. Originally written in 1903, Du Bois also reflects on the long struggle for Black freedom.

Between me and the other world there is ever an unasked question: unasked by some through feelings of delicacy; by others through the difficulty of rightly framing it. All, nevertheless, flutter round it. They approach me in a half-hesitant sort of way, eye me curiously or compassionately, and then, instead of saying directly. How does it feel to be a problem? they say, I know an excellent colored man in my town; or, I fought at Mechanicsville; or, Do not these Southern outrages make your blood boil? At these I smile, or am interested, or reduce the boiling to a simmer as the occasion may require. To the real question, How does it feel to be a problem? I answer seldom a word. . . .

After the Egyptian and Indian, the Greek and Roman, the Teuton and Mongolian, the Negro is a sort of seventh son, born with a veil, and gifted with second-sight in this American world,—a world which yields him no true self-consciousness, but only lets him see himself through the revelation of the other world. It is a peculiar sensation, this double-consciousness, this sense of always looking at one's self through the eyes of others, of measuring one's soul by the tape of a world that looks on in amused contempt and pity. One ever feels his twoness,—an American, a Negro; two souls, two thoughts, two unreconciled strivings: two warring ideals in one dark body, whose dogged strength alone keeps it from being torn asunder.

The history of the American Negro is the history of this strife—this longing to attain self-conscious manhood, to merge his double self into a better and truer self. In this merging he wishes neither of the older selves to be lost. He would not Africanize America, for America has too much to teach the world and Africa. He would not bleach his Negro soul in a flood of white Americanism, for he knows that Negro blood has a message for the world. He simply wishes to make it possible for a man to be both a Negro and an American, without being cursed and spit upon by his fellows, without having the doors of Opportunity closed roughly in his face.

This, then, is the end of his striving: to be a co-worker in the kingdom of culture, to escape both death and isolation, to husband and use his best powers

From: W. E. B. Du Bois. 1989. *The Souls of Black Folk,* edited and with an introduction by Donald B. Gibson. New York: Penguin, pp. 3–12. Reprinted with permission.

and his latent genius. These powers of body and mind have in the past been strangely wasted, dispersed, or forgotten. The shadow of a mighty Negro past flits through the tale of Ethiopia the Shadowy and of Egypt the Sphinx. Throughout history, the powers of single black men flash here and there like falling stars, and die sometimes before the world has rightly gauged their brightness. Here in America, in the few days since Emancipation, the black man's turning hither and thither in hesitant and doubtful striving has often made his very strength to lose effectiveness, to seem like absence of power, like weakness. And yet it is not weakness—it is the contradiction of double aims. The double-aimed struggle of the black artisan—on the one hand to escape white contempt for a nation of mere hewers of wood and drawers of water, and on the other hand to plough and nail and dig for a poverty-stricken horde—could only result in making him a poor craftsman, for he had but half a heart in either cause. By the poverty and ignorance of his people, the Negro minister or doctor was tempted toward quackery and demagogy; and by the criticism of the other world, toward ideals that made him ashamed of his lowly tasks. The would-be black *savant* was confronted by the paradox that the knowledge people needed was a twice-told tale to his white neighbors, while the knowledge which would teach the white world was Greek to his own flesh and blood. The innate love of harmony and beauty that set the ruder souls of his people a-dancing and a-singing raised but confusion and doubt in the soul of the black artist; for the beauty revealed to him was the soul-beauty of a race which his larger audience despised, and he could not articulate the message of another people. This waste of double aims, this seeking to satisfy two unreconciled ideals, has wrought sad havoc with the courage and faith and deeds of ten thousand thousand people,—has sent them often wooing false gods and invoking false means of salvation, and at times has even seemed about to make them ashamed of themselves. . . .

The Nation has not yet found peace from its sins; the freedman has not yet found in freedom his promised land. Whatever of good may have come in these years of change, the shadow of a deep disappointment rests upon the Negro people—a disappointment all the more bitter because the unattained ideal was unbounded save by the simple ignorance of a lowly people. . . .

. . .Merely a concrete test of the underlying principles of the great republic is the Negro Problem, and the spiritual striving of the freedmen's sons is the travail of souls whose burden is almost beyond the measure of their strength, but who bear it in the name of an historic race, in the name of this the land of their fathers' fathers and in the name of human opportunity.

DISCUSSION QUESTIONS

1. What does Du Bois mean by "double consciousness" and how does this affect how African American people see themselves and others?

2. In the contemporary world, what examples do you see that Black people are still defined as "a problem," as Du Bois notes? How does this affect the Black experience?

INFOTRAC COLLEGE EDITION

You can use your access to InfoTrac College Edition to learn more about the subjects covered in this essay. Some suggested search terms include:

double consciousness   *The Philadelphia Negro*
Emancipation   W.E.B. Du Bois
slavery

41

Being More Than Black & White:

Latinos, Racism, and the Cultural Divides

ELIZABETH MARTINEZ

Racism in the United States is more than simply a matter of Black and White. It involves intense racism about other groups as well, especially Lations. Elizabeth Martinez advocates a broader framework involving many groups for studying and combating racism.

When Kissinger said years ago "nothing important ever happens in the south," he articulated a contemptuous indifference toward Latin America, its people, and their culture which has long dominated U.S. institutions and attitudes. Mexico may be great for a vacation, and some people like burritos but the usual image of Latin America combines incompetence with absurdity in loud colors. My parents, both Spanish teachers, endured decades of being told kids were better off learning French.

U.S. political culture is not only Anglo-dominated but also embraces an exceptionally stubborn national self-centeredness, with no global vision other than relations of domination. The U.S. refuses to see itself as one nation sitting on a continent with 20 others all speaking languages other than English and having the right not to be dominated.

Such arrogant indifference extends to Latinos within the U.S. The mass media complain, "People can't relate to Hispanics"—or Asians, they say. Such arrogant indifference has played an important role in invisibilizing La Raza (except where we become a serious nuisance or a handy scapegoat). It is one reason the U.S. harbors an exclusively white-on-Black concept of racism. It is one barrier to new thinking about racism which is crucial today. There are others.

From: *Z Magazine* 7 (May 1994): 56–60.

GOOD-BYE WHITE MAJORITY

In a society as thoroughly and violently racialized as the United States, white-Black relations have defined racism for centuries. Today the composition and culture of the U.S. are changing rapidly. We need to consider seriously whether we can afford to maintain an exclusively white/Black model of racism when the population will be 32 percent Latin/Asian/Pacific American and Native American—in short, neither Black nor white—by the year 2050. We are challenged to recognize that multi-colored racism is mushrooming, and then strategize how to resist it. We are challenged to move beyond a dualism comprised of two white supremacist inventions: Blackness and Whiteness.

At stake in those challenges is building a united anti-racist force strong enough to resist contemporary racist strategies of divde-and-conquer. Strong enough in the long run, to help defeat racism itself. Doesn't an exclusively Black/white model of racism discourage the perception of common interests among people of color and thus impede a solidarity that can challenge white supremacy? Doesn't it encourage the isolation of African Americans from potential allies? Doesn't it advise all people of color to spend too much energy understanding our lives in relation to Whiteness, and thus freeze us in a defensive, often self-destructive mode?

NO "OPPRESSION OLYMPICS"

For a Latina to talk about recognizing the multi-colored varieties of racism is not, and should not be, yet another round in the Oppression Olympics. We don't need more competition among different social groupings for that "Most Oppressed" gold. We don't need more comparisons of suffering between women and Blacks, the disabled and the gay, Latino teenagers and white seniors, or whatever. We don't need more surveys like the recent much publicized Harris Poll showing that different peoples of color are prejudiced toward each other—a poll patently designed to demonstrate that us coloreds are no better than white folk. (The survey never asked people about positive attitudes.)

Rather, we need greater knowledge, understanding, and openness to learning about each other's histories and present needs as a basis for working together. Nothing could seem more urgent in an era when increasing impoverishment encourages a self-imposed separatism among people of color as a desperate attempt at community survival. Nothing could seem more important as we search for new social change strategies in a time of ideological confusion.

My call to rethink concepts of racism in the U.S. today is being sounded elsewhere. Among academics, liberal foundation administrators, and activist-intellectuals, you can hear talk of the need for a new "racial paradigm" or model. But new thinking seems to proceed in fits and starts, as if dogged by a fear of stepping on toes, of feeling threatened, or of losing one's base. With a few notable exceptions, even our progressive scholars of color do not make the leap from perfunctorily saluting a vague multi-culturalism to serious analysis. We seem to have made little progress, if any, since Bob Blauner's 1972 book *Racial Oppression in America*. Recognizing the limits of the white-Black axis, Blauner critiqued White America's ignorance of and indifference to the Chicano/a experience with racism.

Real opposition to new paradigms also exists. There are academics scrambling for one flavor of ethnic studies funds versus another. There are politicians who cultivate distrust of others to keep their own communities loyal. When we hear, for example, of Black/Latino friction, dismay should be quickly followed by investigation. In cities like Los Angeles and New York, it may turn out that political figures scrapping for patronage and payola have played a narrow nationalist game, whipping up economic anxiety and generating resentment that sets communities against each other.

So the goal here, in speaking about moving beyond a bipolar concept of racism is to build stronger unity against white supremacy. The goal is to see our similarities of experience and needs. If that goal sounds naive, think about the hundreds of organizations formed by grassroots women of different colors coming together in recent years. Their growth is one of today's most energetic motions and it spans all ages. Think about the multicultural environmental justice movement. Think about the coalitions to save schools. Small rainbows of our own making are there, to brighten a long road through hellish times.

It is in such practice, through daily struggle together, that we are most likely to find the road to greater solidarity against a common enemy. But we also need a will to find it and ideas about where, including some new theory.

THE WEST GOES EAST

Until very recently, Latino invisibility—like that of Native Americans and Asian/Pacific Americans—has been close to absolute in U.S. seats of power, major institutions, and the non-Latino public mind. Having lived on both the East and West Coasts for long periods, I feel qualified to pronounce: an especially myopic view of Latinos prevails in the East. This, despite such data as a 24.4 percent Latino population of New York City alone in 1991, or the fact that in 1990 more Puerto Ricans were killed by New York police under suspicious circumstances than any other ethnic group. Latino populations are growing rapidly in many eastern cities and the rural South, yet remain invisible or stigmatized—usually both.

Eastern blinders persist. I've even heard that the need for a new racial paradigm is dismissed in New York as a California hangup. A black Puerto Rican friend in New York, when we talked about experiences of racism common to Black and brown, said "People here don't see Border Patrol brutality against Mexicans as a form of police repression," despite the fact that the Border Patrol is the largest and most uncontrolled police force in the U.S. It would seem that an old ignorance has combined with new immigrant bashing to sustain divisions today.

While the East (and most of the Midwest) usually remains myopic, the West Coast has barely begun to move away from its own denial. Less than two years ago in San Francisco, a city almost half Latino or Asian/Pacific American, a leading daily newspaper could publish a major series on contemporary racial issues and follow the exclusively Black-white paradigm. Although millions of TV viewers saw massive Latino participation in the April 1992 Los Angeles uprising, which included 18 out of 50 deaths and the majority of arrests, the mass media and most people labeled that event "a Black riot."

If the West Coast has more recognition of those who are neither Black nor White, it is mostly out of fear about the proximate demise of its white majority. A second, closely related reason is the relentless campaign by California Gov. Pete Wilson to scapegoat immigrants for economic problems and pass racist, unconstitutional laws attacking their health, education, and children's future. Wilson has almost single-handedly made the word "immigrant" mean Mexican or other Latino (and sometimes Asian). Who thinks of all the people coming from the former Soviet Union and other countries? The absolute racism of this has too often been successfully masked by reactionary anti-immigrant groups like FAIR blaming immigrants for the staggering African-American unemployment rate.

Wilson's immigrant bashing is likely to provide a model for other parts of the country. The five states with the highest immigration rates—California, Florida, New York, Illinois and Texas—all have a governor up for re-election in 1994. Wilson's tactics won't appear in every campaign but some of the five states will surely see intensified awareness and stigmatization of Latinos as well as Asian/Pacific Islanders.

As this suggests, what has been a regional issue mostly limited to western states is becoming a national issue. If you thought Latinos were just "Messicans down at the border," wake up—they are all over North Carolina, Pennsylvania and 8th Avenue Manhattan now. A qualitative change is taking place. With the broader geographic spread of Latinos and Asian/Pacific Islanders has come a nationalization of racist practices and attitudes that were once regional. The west goes east, we could say.

Like the monster Hydra, racism is growing some ugly new heads. We will have to look at them closely.

ROOTS OF RACISM AND LATINOS

A bipolar model of racism—racism as white on Black—has never really been accurate. Looking for the roots of racism in the U.S. we can begin with the genocide against American Indians which made possible the U.S. land base, crucial to white settlement and early capitalist growth. Soon came the massive enslavement of African people which facilitated that growth. As slave labor became economically critical, "blackness" became ideologically critical; it provided the very source of "whiteness" and the heart of racism. Franz Fanon would write, "colour is the most outward manifestation of race."

Had Native Americans had been a crucial labor force during those same centuries, living and working in the white man's sphere, our racist ideology might have evolved differently. "The tawny," as Ben Franklin dubbed them, might have defined the opposite of what he called "the lovely white." But with Indians decimated and survivors moved to distant concentration camps, they became unlikely candidates for this function. Similarly, Mexicans were concentrated in the distant West; elsewhere Anglo fear of them or need to control was rare. They also did not provide the foundation for a definition of whiteness.

Some anti-racist left activists have put forth the idea that only African Americans experience racism as such and that the suffering of other people of color results from national minority rather than racial oppression. From this viewpoint, the exclusively white/Black model for racism is correct. Latinos, then, experience exploitation and repression for reasons of culture and nationality—not for their "race." (It should go without saying . . . that while racism is an all-too-real social fact, race has no scientific basis.)

Does the distinction hold? This and other theoretical questions call for more analysis and more expertise than one article can offer. In the meantime, let's try on the idea that Latinos do suffer for their nationality and culture, especially language. They became part of the U.S. through the 1846–48 war on Mexico and thus a foreign population to be colonized. But as they were reduced to cheap or semi-slave labor, they quickly came to suffer for their "race"—meaning, as non-whites. In the Southwest of a super-racialized nation the broad parallelism of race and class embraces Mexicans ferociously.

The bridge here might be a definition of racism as "the reduction of the cultural to the biological," in the words of French scholar Christian Delacampagne now working in Egypt. Or: "racism exists wherever it is claimed that a given social status is explained by a given natural characteristic." We know that line: Mexicans are just naturally lazy and have too many children, so they're poor and exploited.

The discrimination, oppression and hatred experienced by Native Americans, Mexicans, Asian/Pacific Islanders, and Arab Americans are forms of racism. Speaking only of Latinos, we have seen in California and the Southwest, especially along the border, almost 150 years of relentless repression which today includes Central Americans among its targets. That history reveals hundreds of lynchings between 1847 and 1935, the use of counter-insurgency armed forces beginning with the Texas Rangers, random torture and murder by Anglo ranchers, forced labor, rape by border lawmen, and the prevailing Anglo belief that a Mexican life doesn't equal a dog's in value.

But wait. If color is so key to racial definition, as Fanon and others say, perhaps people of Mexican background experience racism less than national minority oppression because they are not dark enough as a group. For White America, shades of skin color are crucial to defining worth. The influence of those shades has also been internalized by communities of color. Many Latinos can and often want to pass for whites; therefore, White America may see them as less threatening than darker sisters and brothers.

Here we confront more of the complexity around us today, with questions like: What about the usually poor, very dark Mexican or Central American of strong Indian or African heritage? (Yes, folks, 200,000–300,000 Africans were brought to Mexico as slaves, which is far, far more than the Spaniards who came.) And what about the effects of accented speech or foreign name, characteristics that may instantly subvert "passing"?

What about those cases where a Mexican-American is never accepted, no matter how light-skinned, well-dressed or well-spoken? A Chicano lawyer friend coming home from a professional conference in suit, tie and briefcase found himself on a bus near San Diego that was suddenly stopped by the Border Patrol. An agent came on board and made a beeline through the all-white rows of passengers direct to my friend. "Your papers." The agent didn't believe Jose was coming from a U.S. conference and took him off the bus to await proof. Jose was lucky; too many Chicanos and Mexicans end up killed.

In a land where the national identity is white, having the "wrong" nationality becomes grounds for racist abuse. Who would draw a sharp line between today's national minority oppression in the form of immigrant-bashing, and racism?

None of this aims to equate the African American and Latino experiences; that isn't necessary even if it were accurate. Many reasons exist for the persistence of the white/Black paradigm of racism; they include numbers, history, and the psychology of whiteness. In particular they include centuries of slave revolts, a Civil War, and an ongoing resistance to racism that cracked this society wide open while the world watched. Nor has the misery imposed on Black people lessened in recent years. New thinking about racism can and should keep this experience at the center.

DISCUSSION QUESTIONS

1. Think of four examples of subtle White-to-Black racism. Now think of four examples of subtle White-to-Latino/a racism. What are the similarities and what are the differences?

2. The author advocates going beyond a White-Black paradigm of racism. Describe this paradigm briefly, and suggest names for a new racial paradigm.

INFOTRAC COLLEGE EDITION

You can use your access to InfoTrac College Edition to learn more about the subjects covered in this essay. Some suggested search terms include:

bipolar concept of racism   new racial paradigm
(Franz) Fanon   white/Black paradigm

42

Racial Formation

MICHAEL OMI AND HOWARD WINANT

Here Michael Omi and Howard Winant elaborate a new theory of race and racism, that of the racial formation *process. They argue that "race" in the United States is less a matter of physical traits, such as skin color, and more of a sociohistorical process by which racial categories are created. Categorization of individuals according to "race" is seen as a phenomenon of social structure.*

In 1982–83, Susie Guillory Phipps unsuccessfully sued the Louisiana Bureau of Vital Records to change her racial classification from black to white. The descendant of an 18th-century white planter and a black slave, Phipps was designated "black" in her birth certificate in accordance with a 1970 state law which declared anyone with at least 1/32nd "Negro blood" to be black.

The Phipps case raised intiguing questions about the concept of race, its meaning in contemporary society, and its use (and abuse) in public policy. Assistant Attorney General Ron Davis defended the law by pointing out that some type of racial classification was necessary to comply with federal record-keeping requirements and to facilitate programs for the prevention of genetic diseases. Phipps's attorney, Brian Begue, argued that the assignment of racial categories on birth certificates was unconstitutional and that the 1/32nd designation was inaccurate. He called on a retired Tulane University professor who cited research indicating that most Louisiana whites have at least 1/20th "Negro" ancestry.

In the end, Phipps lost. The court upheld the state's right to classify and quantify racial identity. . . .

Phipps's problematic racial identity, and her effort to resolve it through state action, is in many ways a parable of America's unsolved racial dilemma. It illustrates the difficulties of defining race and assigning individuals or groups to racial categories. It shows how the racial legacies of the past—slavery and bigotry—continue to shape the present. It reveals both the deep involvement of the state in the organization and interpretation of race, and the inadequacy of state institutions to carry out these functions. It demonstrates how deeply Americans both as individuals and as a civilization are shaped, and indeed haunted, by race.

Having lived her whole life thinking that she was white, Phipps suddenly discovers that by legal definition she is not. In U.S. society, such an event is indeed catastrophic. But if she is not white, of what race is she? The *state* claims that she is black, based on its rules of classification. . . and another state agency, the court,

From: Michael Omi and Howard Winant. 1994. *Racial Formation in the United States,* revised ed. New York: Routledge, pp. 53–61. Reprinted with permission.

upholds this judgment. But despite these classificatory standards which have imposed an either-or logic on racial identity, Phipps will not in fact "change color." Unlike what would have happened during slavery times if one's claim to whiteness was successfully challenged, we can assume that despite the outcome of her legal challenge, Phipps will remain in most of the social relationships she had occupied before the trial. Her socialization, her familial and friendship networks, her cultural orientation, will not change. She will simply have to wrestle with her newly acquired "hybridized" condition. She will have to confront the "Other" within.

The designation of racial categories and the determination of racial identity is no simple task. For centuries, this question has precipitated intense debates and conflicts, particularly in the U.S.—disputes over natural and legal rights, over the distribution of resources, and indeed, over who shall live and who shall die.

A crucial dimension of the Phipps case is that it illustrates the inadequacy of claims that race is a mere matter of variations in human physiognomy, that it is simply a matter of skin color. But if race cannot be understood in this manner, how *can* it be understood? We cannot fully hope to address this topic—no less than the meaning of race, its role in society, and the forces which shape it—in one chapter, nor indeed in one book. Our goal in this chapter, however, is far from modest: we wish to offer at least the outlines of a theory of race and racism.

WHAT IS RACE?

There is a continuous temptation to think of race as an *essence*, as something fixed, concrete, and objective. And there is also an opposite temptation: to imagine race as a mere *illusion*, a purely ideological construct which some ideal nonracist social order would eliminate. It is necessary to challenge both these positions, to disrupt and reframe the rigid and bipolar manner in which they are posed and debated, and to transcend the presumably irreconcilable relationship between them.

The effort must be made to understand race as an unstable and "decentered" complex of social meanings constantly being transformed by political struggle. With this in mind, let us propose a definition: *race is a concept which signifies and symbolizes social conflicts and interests by referring to different types of human bodies.* Although the concept of race invokes biologically based human characteristics (so-called "phenotypes"), selection of these particular human features for purposes of racial signification is always and necessarily a social and historical process. In contrast to the other major distinction of this type, that of gender, there is no biological basis for distinguishing among human groups along the lines of race. . . Indeed, the categories employed to differentiate among human groups along racial lines reveal themselves, upon serious examination, to be at best imprecise, and at worst completely arbitrary.

If the concept of race is so nebulous, can we not dispense with it? Can we not "do without" race, at least in the "enlightened" present? This question has been posed often, and with greater frequency in recent years. . . .An affirmative answer would of course present obvious practical difficulties: it is rather difficult to jettison widely held beliefs, beliefs which moreover are central to everyone's identity and understanding of the social world. So the attempt to banish the concept as an archaism is at best counterintuitive. But a deeper difficulty, we believe, is inherent in the very formulation of this schema, in its way of posing race as a *problem,* a misconception left over from the past, and suitable now only for the dustbin of history.

A more effective starting point is the recognition that despite its uncertainties and contradictions, the concept of race continues to play a fundamental role in structuring and representing the social world. The task for theory is to explain this situation. It is to avoid both the utopian framework which sees race as an illusion we can somehow "get beyond," and also the essentialist formulation which sees race as something objective and fixed, a biological datum. Thus we should think of race as an element of social structure rather than as an irregularity within it; we should see race as a dimension of human representation rather than an illusion. These perspectives inform the theoretical approach we call racial formation.

Racial Formation

We define *racial formation* as the sociohistorical process by which racial categories are created, inhabited, transformed, and destroyed. Our attempt to elaborate a theory of racial formation will proceed in two steps. First, we argue that racial formation is a process of historically situated *projects* in which human bodies and social structures are represented and organized. Next we link racial formation to the evolution of hegemony, the way in which society is organized and ruled. Such an approach, we believe, can facilitate understanding of a whole range of contemporary controversies and dilemmas involving race, including the nature of racism, the relationship of race to other forms of differences, inequalities, and oppression such as sexism and nationalism, and the dilemmas of racial identity today.

From a racial formation perspective, race is a matter of both social structure and cultural representation. Too often, the attempt is made to understand race simply or primarily in terms of only one of these two analytical dimensions. . . . For example, efforts to explain racial inequality as a purely social structural phenomenon are unable to account for the origins, patterning, and transformation of racial difference.

Conversely, many examinations of racial difference—understood as a matter of cultural attributes *à la* ethnicity theory, or as a society-wide signification system, *à la* some poststructuralist accounts—cannot comprehend such structural phenomena as racial stratification in the labor market or patterns of residential segregation.

An alternative approach is to think of racial formation processes as occurring through a linkage between structure and representation. Racial *projects* do the ideological "work" of making these links. *A racial project is simultaneously an interpretation, representation, or explanation of racial dynamics, and an effort to reorganize and redistribute resources along particular racial lines.* Racial projects connect what race *means* in a particular discursive practice and the ways in which both social structures and everyday experiences are racially *organized*, based upon that meaning. Let us consider this proposition, first in terms of large-scale or macro-level social processes, and then in terms of other dimensions of the racial formation process.

Racial Formation as a Macro-Level Social Process   *To interpret the meaning of race is to frame it social structurally.* Consider for example, this statement by Charles Murray on welfare reform:

> My proposal for dealing with the racial issue in social welfare is to repeal every bit of legislation and reverse every court decision that in any way requires, recommends, or awards differential treatment according to race, and thereby put us back onto the track that we left in 1965. We may argue about the appropriate limits of government intervention in trying to enforce the ideal, but at least it should be possible to identify the ideal: Race is not a morally admissible reason for treating one person differently from another. Period. . . .

Here there is a partial but significant analysis of the meaning of race: it is not a morally valid basis upon which to treat people "differently from one another." We may notice someone's race, but we cannot act upon that awareness. We must act in a "color-blind" fashion. This analysis of the meaning of race is immediately linked to a specific conception of the role of race in the social structure: it can play no part in government action, save in "the enforcement of the ideal." No state policy can legitimately require, recommend, or award different status according to race. This example can be classified as a particular type of racial project in the present-day U.S.—a "neoconservative" one.

Conversely, *to recognize the racial dimension in social structure is to interpret the meaning of race.* Consider the following statement by the late Supreme Court Justice Thurgood Marshall on minority "set-aside" programs:

> A profound difference separates governmental actions that themselves are racist, and governmental actions that seek to remedy the effects of prior racism or to prevent neutral government activity from perpetuating the effects of such racism. . . .

Here the focus is on the racial dimensions of *social structure*—in this case of state activity and policy. The argument is that state actions in the past and present have treated people in very different ways according to their race, and thus the government cannot retreat from its policy responsibilities in this area. It cannot suddenly declare itself "color-blind" without in fact perpetuating the same type of differential, racist treatment. . . . Thus, race continues to signify difference and structure inequality. Here, racialized social structure is immediately linked to an interpretation of the meaning of race. This example too can be classified as a particular type of racial project in the present-day U.S.—a "liberal" one.

To be sure, such political labels as "neoconservative" or "liberal" cannot fully capture the complexity of racial projects, for these are always multiply determined, politically contested, and deeply shaped by their historical context. Thus, encapsulated within the neoconservative example cited here are certain egalitarian commitments which derive from a previous historical context in which they played a very different role, and which are rearticulated in neoconservative racial discourse precisely to oppose a more open-ended, more capacious conception of the meaning of equality. Similarly, in the liberal example, Justice Marshall recognizes that the contemporary state, which was formerly the architect of segregation and the chief enforcer of racial difference, has a tendency to reproduce those patterns of inequality in a new guise. Thus he admonishes it (in dissent, significantly) to fulfill its responsibilities to uphold a robust conception of equality. These particular instances, then, demonstrate how racial projects are always concretely framed, and thus are always contested and unstable. The social structures they uphold or attack, and the representations of race they articulate, are never invented out of the air, but exist in a definite historical context, having descended from previous conflicts. This contestation appears to be permanent in respect to race.

These two examples of contemporary racial projects are drawn from mainstream political debate; they may be characterized as center-right and center-left expressions of contemporary racial politics. . . We can, however, expand the discussion of racial formation processes far beyond these familiar examples. In fact, we can identify racial projects in at least three other analytical dimensions: first, the political spectrum can be broadened to include radical projects, on both the left and right, as well as along other political axes. Second, analysis of racial projects can take place not only at the macro-level of racial policy-making, state activity, and collective action, but also at the micro-level of everyday experience. Third, the concept of racial projects can be applied across historical time, to identify racial formation dynamics in the past. We shall now offer examples of each of these types of racial projects.

The Political Spectrum of Racial Formation   We have encountered examples of a neoconservative racial project, in which the significance of race is denied, leading to a "color-blind" racial politics and "hands off" policy orientation; and of a "liberal" racial project, in which the significance of race is affirmed, leading to an egalitarian and "activist" state policy. But these by no means exhaust the political possibilities. Other racial projects can be readily identified on the contemporary U.S. scene. For example, "far right" projects, which uphold biologistic and racist views of difference, explicitly argue for white supremacist policies. "New right" projects overtly claim to hold "color-blind" views, but covertly manipulate racial fears in order to achieve political gains. . . . On the left, "radical democratic" projects invoke notions of racial "difference" in combination with egalitarian politics and policy.

Further variations can also be noted. For example, "nationalist" projects, both conservative and radical, stress the incompatibility of racially defined group identity with the legacy of white supremacy, and therefore advocate a social structural solution of separation, either complete or partial. . . . Nationalist currents represent a profound legacy of the centuries of racial absolutism that initially defined the meaning of race in the U.S. Nationalist concerns continue to influence racial debate in the form of Afrocentrism and other expressions of identity politics.

Taking the range of politically organized racial projects as a whole, we can "map" the current pattern of racial formation at the level of the public sphere, the "macro-level" in which public debate and mobilization takes place. . . . But important as this is, the terrain on which racial formation occurs is broader yet.

Racial Formation as Everyday Experience   At the micro-social level, racial projects also link signification and structure, not so much as efforts to shape policy or define large-scale meaning, but as the applications of "common sense." To see racial projects operating at the level of everyday life, we have only to examine the many ways in which, often unconsciously, we "notice" race.

One of the first things we notice about people when we meet them (along with their sex) is their race. We utilize race to provide clues about *who* a person is. This fact is made painfully obvious when we encounter someone whom we cannot conveniently racially categorize—someone who is, for example, racially "mixed" or of an ethnic/racial group we are not familiar with. Such an encouter becomes a source of discomfort and momentarily a crisis of racial meaning.

Our ability to interpret racial meanings depends on preconceived notions of a racialized social structure. Comments such as, "Funny, you don't look black," betray an underlying image of what black should be. We expect people to act out their apparent racial identities; indeed we become disoriented when they do not. The black banker harassed by police while walking in casual clothes through his own well-off neighborhood, the Latino or white kid rapping in perfect Afro patois, the unending *faux pas* committed by whites who assume that the non-whites they encounter are servants or tradespeople, the belief that non-white colleagues are less qualified persons hired to fulfill affirmative action guidelines, indeed the whole gamut of racial stereotypes—that "white men can't jump," that Asians can't dance, etc., etc.—all testify to the way a racialized social structure shapes racial experience and conditions meaning. Analysis of such stereotypes reveals the always present, already active link between our view of the social structure—its demography, its laws, its customs, its threats—and our conception of what race means.

Conversely, our ongoing interpretation of our experience in racial terms shapes our relations to the institutions and organizations through which we are imbedded in social structure. Thus we expect differences in skin color, or other racially coded characteristics, to explain social differences. Temperament, sexuality, intelligence, athletic ability, aesthetic preferences, and so on are presumed to be fixed and discernible from the palpable mark of race. Such diverse questions as our confidence and trust in others (for example, clerks or salespeople, media figures, neighbors), our sexual preferences and romantic images, our tastes in music, films, dance, or sports, and our very ways of talking, walking, eating, and dreaming become racially coded simply because we live in a society where racial awareness is so pervasive. Thus in ways too comprehensive even to monitor consciously, and despite periodic calls—neoconservative and otherwise—for us to ignore race and adopt "color-blind" racial attitudes, skin color "differences" continue to rationalize distinct treatment of racially identified individuals and groups.

To summarize the argument so far: the theory of racial formation suggests that society is suffused with racial projects, large and small, to which all are subjected. This racial "subjection" is quintessentially ideological. Everybody learns some combination, some version, of the rules of racial classification, and of her own racial identity, often without obvious teaching or conscious inculcation. Thus are we inserted in a comprehensively racialized social structure. Race becomes "common sense"—a way of comprehending, explaining, and acting in the world. A vast web of racial projects mediates between the discursive or representational means in which race is identified and signified on the one hand, and the institutional and organizational forms in which it is routinized and standardized on the other. These projects are the heart of the racial formation process.

Under such circumstances, it is not possible to represent race discursively without simultaneously locating it, explicitly or implicitly, in a social structural (and historical) context. Nor is it possible to organize, maintain, or transform social structures without simultaneously engaging, once more either explicitly or implicitly, in racial signification. Racial formation, therefore, is a kind of synthesis, an outcome, of the interaction of racial projects on a society-wide level. These projects are, of course, vastly different in scope and effect. They include large-scale public action, state activities, and interpretations of racial conditions in artistic, journalistic, or academic fora, as well as the seemingly infinite number of racial judgments and practices we carry out at the level of individual experience.

Since racial formation is always historically situated, our understanding of the significance of race, and of the way race structures society, has changed enormously over time. The processes of racial formation we encounter today, the racial projects large and small which structure U.S. society in so many ways, are merely the present-day outcomes of a complex historical evolution. The contemporary racial order remains transient. By knowing something of how it evolved, we can perhaps better discern where it is heading.

DISCUSSION QUESTIONS

1. What is the "racial formation" process? How does it change the more traditional definition of "race" on the basis of skin color, hair texture, and other physical characteristics?

2. Discuss how and in what variety of ways the definition and categorization of persons by "race" is a social structural process.

INFOTRAC COLLEGE EDITION

You can use your access to InfoTrac College Edition to learn more about the subjects covered in this essay. Some suggested search terms include:

racial formation theory   racialization
racial project   The Phipps (legal) case
racial stereotypes

43

Immigrant America:
Who They Are and Why They Come

ALEJANDRO PORTES AND RUBÉN RUMBAUT

In this essay, Alejandro Portes and Rubén Rumbaut convey the varied yet similar experiences of some of the "new" immigrants of color (Mexican, South Vietnamese, Cuban, and East Indian) to the United States. Their experiences after arrival are seen as dependent upon their own socioeconomic successes, occupations, and intense personal ambition. They are ethnically a diverse population, and these new immigrants face discrimination and hardship upon their arrival.

In Guadalajara, Juan Manuel Fernández worked as a mechanic in his uncle's repair shop making the equivalent of $150 per month. At thirty-two and after ten years on the job, he decided it was time to go into business on his own. The family, his uncle included, was willing to help, but capital for the new venture was scarce. Luisa, Juan's wife, owned a small corner grocery shop; when money ran out at the end of the month, she often fed the family off the store's shelves. The store was enough to sustain her and her children but not to capitalize her husband's project. For a while, it looked as if Juan would remain a worker for life.

Today Juan owns his own auto repair shop, where he employs three other mechanics, two Mexicans and a Salvadoran. The shop is not in Guadalajara, however, but in Gary, Indiana. The entire family—Luisa, the two children, and a brother—have resettled there. Luisa does not work any longer because she does not speak English and because income from her husband's business is enough to support the family. The children attend school and already speak better English than their parents. They resist the idea of going back to Mexico.

From: Alejandro Portes and Rubén Rumbaut. 1996. *Immigrant America*, 2nd ed. Berkeley: University of California Press, pp. 1–9. Reprinted with permission.

Juan crossed the border on his own near El Paso in 1979. No one stopped him, and he was able to head north toward a few distant cousins and the prospect of a factory job. To his surprise, he found one easily and at the end of four months was getting double the minimum wage in steady employment. Almost every worker in the plant was Mexican, his foreman was Puerto Rican, and the language of work was uniformly Spanish. Three trips from Gary to Guadalajara during the next two years persuaded him that it made much better sense to move his business project north of the border. Guadalajara was teeming with repair shops of all sorts, and competition was fierce. "In Gary," he said, "many Mexicans would not get their cars fixed because they did not know how to bargain with an American mechanic." Sensing the opportunity, he cut remittances to Mexico and opened a local savings account instead.

During his last trip, the "migra" (border patrol) stopped him shortly after crossing; that required a costly second attempt two days later with a hired "coyote" (smuggler). The incident put a stop to the commuting. Juan started fixing cars out of a shed in front of his barrio home. Word got around that there was a reliable Spanish-speaking mechanic in the neighborhood. In a few months, he was able to rent an abandoned garage, buy some equipment, and eventually hire others. To stay in business, Juan has had to obtain a municipal permit and pay a fee. He pays his workers in cash, however, and neither deducts taxes from their wages nor contributes to Social Security for them. All transactions are informal and, for the most part, in cash.

Juan and Luisa feel a great deal of nostalgia for Mexico, and both firmly intend to return. "In this country, we've been able to move ahead economically, but it is not our own," she says. "The gringos will always consider us inferior." Their savings are not in the bank, as before the shop was rented, but in land in Guadalajara, a small house for his parents, and the goodwill of many relatives who receive periodic remittances. They figure that in ten years they will be able to return, although they worry about their children, who may be thoroughly Americanized by then. A more pressing problem is their lack of "papers" and the constant threat of deportation. Juan has devised ingenious ways to run the business, despite his illegal status, but it is a constant problem. A good part of his recent earnings is in the hands of an immigration lawyer downtown, who has promised to obtain papers for a resident's visa, so far without results.

At age twenty-six, Nguyen Van Tran was a young lieutenant in the army of the Republic of South Vietnam when a strategic retreat order from the ARVN high command quickly turned into the final root. Nguyen spent three years in communist reeducation camps, all the while attempting to conceal his past as a skilled electronics technician. He finally got aboard a boat bound for Malaysia and after two more years in a refugee camp arrived in Los Angeles in 1980. He had neither family nor friends in the city, but the government provided some resettlement aid and the opportunity to improve his English. At the end of a year, he had secured a job in a local electronics assembly plant, which brought in enough to support himself and his wife and child.

Seeing this plant double in a single year, Nguyen realized the opportunities opening up in electronics. He enrolled in the local community college at night and graduated with an associate degree in computer science. He pooled his savings with another Vietnamese technician and a Chinese engineer and in 1983 launched his own firm. Two years later Integrated Circuits, Inc. employed approximately three hundred workers; most were not Asians, but undocumented Mexican women. By 1985, the company sold about $20 million worth of semiconductors and other equipment to the local IBM plant and other large firms. ICI has even started its own line of IBM-compatible personal computers, the Trantex, which has sold well so far in the local market.

Nguyen, who is chairman of the company, sports a mustache, a sleek Mercedes, and a brand new name, George Best. Perhaps for fear of the "protection gangs" re-created by former Vietnamese policemen in Los Angeles, he has kept a low profile within the Vietnamese community. The name change is part of this approach. "Mr. Best" is not particularly nationalistic, nor does he dream of returning to Vietnam. He attributes his remarkable five-year ascent to hard work and a willingness to take risks. To underline the point, he has hung a large portrait of himself in his community college graduation gown behind his oversized desk. He and his wife are already U.S. citizens. They vote Republican, and he has recently joined the local chamber of commerce.

Lilia González-Fleites left Cuba at fifteen, sent alone by her formerly wealthy parents, who remained behind. The Catholic Welfare Agency received her in Miami, and she went to live with other refugee children in an orpharange in Kendall, Florida, until released to an aunt. She finished high school promptly and married, without her parents' consent, her boyfriend from Cuba, Tomás. There was little work in Miami, and the young couple accepted an offer from the Cuban Refugee Center to resettle them, along with the rest of Tomás's family, in North Carolina. Everyone found work in the tobacco and clothing factories except Lilia, whom Tomás kept at home. At eighteen, the formerly pampered girl found herself a cook and maid for Tomás's entire family.

By sheer luck, the same order of nuns who ran her private school in Havana had a college nearby. Lilia used her school connections to gain admittance with a small scholarship and found herself a part-time job. Those were hard years, working in one city and attending school in another. Tomás and Lilia rarely saw each other because he also decided to return to school while still working.

At age thirty-nine, Lilia is today a successful Miami architect. Divorced from Tomás, she has not remarried, instead pursuing her professional career with single-minded determination. When Cuban refugees finally abandoned their dreams of return, Lilia entered local politics, affiliating with the Republican party. She ran for state office in 1986 but was defeated. Undaunted, she remained active in the party and became increasingly prominent in south Florida political circles. More than an immigrant success story, she sees herself at the beginning of a public career that will bridge the gap between the Anglo and Cuban communities in south Florida. Her unaccented English, fierce loyalty to her adopted country, and ability to shift easily between languages and cultures bodes well for her political future.

After finishing medical school, Amitar Ray confronted the prospect of working *ad honorem* in one of the few well-equipped hospitals in Bombay or moving to a job in the countryside and to quick obsolescence in his career. He opted instead for preparing and taking the Educational Council for Foreign Medical Graduates (ECFMG) examination, administered at the local branch of the Indo-American Cultural Institute. He passed it on his second attempt. In 1972, there was a shortage of doctors in the United States, and U.S. consulates were directed to facilitate the emigration of qualified physicians from abroad.

Amitar and his wife, also a doctor, had little difficulty obtaining permanent residents' visas under the third preference of the U.S. immigration law, reserved for professionals of exceptional ability. He went on to specialize in anesthesiology and completed his residence at a public hospital in Brooklyn. After four years, nostalgia and the hope that things had improved at home moved the Rays to go back to India with their young daughter, Rita. The trip strengthened their professional and family ties, but it also dispelled any doubts as to where their future was. Medical vacancies were rare and paid a fraction of what, he earned as a resident in Brooklyn. More important, there were few opportunities to grew professionally because he would have had to combine several part-time jobs to earn a livelihood, leaving little time for study.

At fifty-one, Amitar is now associate professor of anesthesiology at a midwestern medical school; his wife has a local practice as an internist. Their combined income is in the six figures, affording them a very comfortable life-style. Their daughter is a senior at Bryn Mawr, and she plans to pursue a graduate degree in international relations. There are few Indian immigrants in the mid-sized city where the Rays live: thus, they have had to learn local ways in order to gain entry into American social circles. Their color is sometimes a barrier to close contact with white middle-class families, but they have cultivated many friendships among the local faculty and medical community.

Ties to India persist and are strengthened through periodic trips and through the professional help the Rays are able to provide to colleagues back home. They have already sponsored the immigration of two bright young physicians from their native city. More important, they make sure that information on new medical developments is relayed to a few selected specialists back home. However, there is little chance that they will return, even after retirement. Work and new local ties play a role in this, but the decisive factor is a thoroughly Americanized daughter whose present life and future have very little to do with India. Rita does not plan to marry soon; she is interested in Latin American politics, and her current goal is a career in the foreign service.

After a lapse of half a century, the United States has again become a country of immigration. In 1990, the foreign-born population reached 19.8 million or 7.9 percent of the total. Although a far cry from the situation eighty years earlier, when immigrants accounted for 14.7 percent of the American population, the impact of contemporary immigration is both significant and growing. Numerous books and articles have called attention to this revival and sought its causes—first in a booming American economy and second in the liberalized provisions of the 1965 immigration act. A common exercise is to compare this "new" immigration with the "old" inflow at the turn of the century. Similarities include the predominantly urban destination of most newcomers, their concentration in a few port cities, and their willingness to accept the lowest paid jobs. Differences are more frequently stressed, however, for the "old" immigration was overwhelmingly European and white; but the present inflow is, to a large extent, nonwhite and comes from countries of the Third World.

The public image of contemporary immigration has been colored to a large extent by the Third World origins of most recent arrivals. Because the sending countries are generally poor, many Americans believe that the immigrants themselves are uniformly poor and uneducated. Their move is commonly portrayed as a one-way escape from hunger, want, and persecution and their arrival on U.S. shores as not too different from that of the tired, "huddled masses" that Emma Lazarus immortalized at the base of the Statue of Liberty. The "quality" of the newcomers and their chances for assimilation are sometimes portrayed as worse because of their non-European past and the precarious legal status of many.

The reality is very different. The four previous cases, each a composite of real life experiences, are certainly not representative of all recent immigrants. Clearly, not all newcomers are doctors or skilled mechanics, and fewer still become politicians or millionaires. Still, these are not isolated instances. Underneath its apparent uniformity, contemporary immigration features a bewildering variety of origins, return patterns, and modes of adaptation to American society. Never before has the United States received immigrants from so many countries, from such different social and economic backgrounds, and for so many reasons. Although pre–World War I European immigration was by no means homogeneous, the differences between successive waves of Irish, Italians, Jews, Greeks, and Poles often pale by comparison with the current diversity. For the same reason, theories coined in the wake of the European's arrival at the turn of the century have been made obsolete by events during the last decades.

Increasingly implausible, for example, is the view of a uniform assimilation process that different groups undergo in the course of several generations as a precondition for their social and economic advancement. There are today first-generation millionaires who speak broken English, foreign-born mayors of large cities, and top-flight immigrant engineers and scientists in the nation's research centers; there are also those, at the other extreme, who cannot even take the first step toward assimilation because of the insecurity linked to an uncertain legal status.

. . . Many of the countries from which today's immigrants come have one of their largest cities in the United States. Los Angeles' Mexican population is next in size to those of Mexico City, Monterrey, and Guadalajara. Havana is not much larger than Cuban Miami, and Santo Domingo holds a precarious advantage over Dominician New York. This is not the case for all groups; others, such as Asian Indians, Laotians, Argentines, and Brazilians, are more dispersed throughout the country. Reasons for both these differences and other characteristics of contemporary immigrant groups are not well known—in part because of the recency of their arrival and in part because of the common expectation that their assimilation process would conform to the well-known European pattern. But immigrant America is a different place today from the America that emerged out of Ellis Island and grew up in the tenements of New York and Boston. . . .

DISCUSSION QUESTIONS

1. How do the "new" immigrants (primarily immigrants of color during the second half of the twentieth century) differ from the "old" immigrants of the first half of the twentieth century (who were primarily European Whites)? List some differences.

2. How does the social structure of the United States affect new immigrants and their chances for success?

INFOTRAC COLLEGE EDITION

You can use your access to InfoTrac College Edition to learn more about the subjects covered in this essay. Some suggested search terms include:

assimilation   new immigrant
"coyote"   undocumented workers
"migra"

44

What White Supremacists Taught a Jewish Scholar about Identity

ABBY L. FERBER

Here, Abby Ferber relates how White supremacist literature caused her to think about her own Jewish ethnic identity. The point is made that a significant determinant of one's own personal ethnic identity is how others *view you. Ferber also shows the connections between anti-Semitism, racism, and white supremacy.*

A few years ago, my work on white supremacy led me to the neo-Nazi tract *The New Order,* which proclaims: "The single serious enemy facing the white man is the Jew." I must have read that statement a dozen times. Until then, I hadn't thought of myself as the enemy.

From: *The Chronicle of Higher Education.* (May 7, 1999): B6–B7.

When I began my research for a book on race, gender, and white supremacy, I could not understand why white supremacists so feared and hated Jews. But after being immersed in newsletters and periodicals for months, I learned that white supremacists imagine Jews as the masterminds behind a great plot to mix races and, thereby, to wipe the white race out of existence.

The identity of white supremacists, and the white racial purity they espouse, requires the maintenance of secure boundaries. For that reason, the literature I read described interracial sex as "the ultimate abomination." White supremacists see Jews as threats to racial purity, the villains responsible for desegregation, integration, the civil-rights movement, the women's movement, and affirmative action—each depicted as eventually leading white women into the beds of black men. Jews are believed to be in control everywhere, staging a multipronged attack against the white race. For *WAR,* the newsletter of White Aryan Resistance, the Jew "promotes a thousand social ills. . .[f]or which you'll have to foot the bills."

Reading white-supremacist literature is a profoundly disturbing experience, and even more difficult if you are one of those targeted for elimination. Yet, as a Jewish woman, I found my research to be unsettling in unexpected ways. I had not imagined that it would involve so much self-reflection. I knew white supremacists were vehemently anti-Semitic, but I was ambivalent about my Jewish identity and did not see it as essential to who I was. Having grown up in a large Jewish community, and then having attended a college with a large Jewish enrollment, my Jewishness was invisible to me—something I mostly ignored. As I soon learned, to white supremacists, that is irrelevant.

Contemporary white supremacists define Jews as non-white: "not a religion, they are an Asiatic *race*, locked in a mortal conflict with Aryan man," according to *The New Order*. In fact, throughout white-supremacist tracts, Jews are described not merely as a separate race but as an impure race, the product of mongrelization. Jews, who pose the ultimate threat to racial boundaries, are themselves imagined as the product of mixed-race unions.

Although self-examination was not my goal when I began, my research pushed me to explore the contradictions in my own racial identity. Intellectually, I knew that the meaning of race was not rooted in biology or genetics, but it was only through researching the white-supremacist movement that I gained a more personal understanding of the social construction of race. Reading white supremacist literature, I moved between two worlds: one where I was white, another where I was the non-white seed of Satan; one where I was privileged, another where I was despised; one where I was safe and secure, the other where I was feared and thus marked for death.

According to white-supremacist ideology, I am so dangerous that I must be eliminated. Yet, when I put down the racist, anti-Semitic newsletters, leave my office, and walk outdoors, I am white.

Growing up white has meant growing up privileged. Sure, I learned about the historical persecution of Jews, overheard the hushed references to distant relatives lost in the Holocaust. I knew of my grandmother's experiences with anti-Semitism as a child of the only Jewish family in a Catholic neighborhood. But those were just stories to me. Reading white supremacists finally made the history real.

While conducting my research, I was reminded of the first time I felt like an "other." Arriving in the late 1980s for the first day of graduate school in the Pacific Northwest, I was greeted by a senior graduate student with the welcome: "Oh, you're the Jewish one." It was a jarring remark, for it immediately set me apart. This must have been how my mother felt, I thought, when, a generation earlier, a college classmate had asked to see her horns. Having lived in predominantly Jewish communities, I had never experienced my Jewishness as "otherness." In fact, I did not even *feel* Jewish. Since moving out of my parents' home, I had not celebrated a Jewish holiday or set foot in a synagogue. So it felt particularly odd to be identified by this stranger as a Jew. At the time, I did not feel that the designation described who I was in any meaningful sense.

But whether or not I define myself as Jewish, I am constantly defined by others that way. Jewishness is not simply a religious designation that one may choose, as I once naively assumed. Whether or not I see myself as Jewish does not matter to white supremacists.

I've come to realize that my own experience with race reflects the larger historical picture for Jews. As whites, Jews today are certainly a privileged group in the United States. Yet the history of the Jewish experience demonstrates precisely what scholars mean when they say that race is a social construction.

At certain points in time, Jews have been defined as a non-white minority. Around the turn of the last century, they were considered a separate, inferior race, with a distinguishable biological identity justifying discrimination and even genocide. Today, Jews are generally considered white, and Jewishness is largely considered merely a religious or ethnic designation. Jews, along with other European ethnic groups, were welcomed into the category of "white" as beneficiaries of one of the largest affirmative action programs in history—the 1944 GI Bill of Rights. Yet, when I read white-supremacist discourse, I am reminded that my ancestors were excluded from the dominant race, persecuted, and even killed.

Since conducting my research, having read dozens of descriptions of the murders and mutilations of "race traitors" by white supremacists, I now carry with me the knowledge that there are many people out there who would still wish to see me dead. For a brief moment, I think that I can imagine what it must feel like to be a person of color in our society. . . but then I realize that, as a white person, I cannot begin to imagine that.

Jewishness has become both clearer and more ambiguous for me. And the questions I have encountered in thinking about Jewish identity highlight the central issues involved in studying race today. I teach a class on race and ethnicity, and usually about midway through the course, students complain of confusion. They enter my course seeking answers to the most troubling and divisive questions of our time, and are disappointed when they discover only more questions. If race is not biological or genetic, what is it? Why, in some states, does it take just one black ancestor out of 32 to make a person legally black, yet those 31 white ancestors are not enough to make that person white? And, always, are Jews a race?

I have no simple answers. As Jewish history demonstrates, what is and is not a racial designation, and who is included within it, is unstable and changes over time—and that designation is always tied to power. We do not have to look far to find other examples: The Irish were also once considered non-white in the United States, and U.S. racial categories change with almost every census.

My prolonged encounter with the white-supremacist movement forced me to question not only my own assumptions about Jewish identity, but also my assumptions about whiteness. Growing up "white," I felt raceless. As it is for most white people, my race was invisible to me. Reflecting the assumption of most research on race at the time, I saw race as something that shaped the lives of people of color—the victims of racism. We are not used to thinking about whiteness when we think about race. Consequently, white people like myself have failed to recognize the ways in which our own lives are shaped by race. It was not until others began identifying me as the Jew, the "other," that I began to explore race in my own life.

Ironically, that is the same phenomenon shaping the consciousness of white supremacists: They embrace their racial identity at the precise moment when they feel their privilege and power under attack. Whiteness historically has equaled power, and when that equation is threatened, their own whiteness becomes visible to many white people for the first time. Hence, white supremacists seek to make racial identity, racial hierarchies, and white power part of the natural order again. The notion that race is a social construct threatens that order. While it has become an academic commonplace to assert that race is socially constructed, the revelation is profoundly unsettling to many, especially those who benefit most from the constructs.

My research on hate groups not only opened the way for me to explore my own racial identity, but also provided insight into the question with which I began this essay: Why do white supremacists express such hatred and fear of Jews? The ambiguity in Jewish racial identity is precisely what white supremacists find so threatening. Jewish history reveals race as a social designation, rather than a God-given or genetic endowment. Jews blur the boundaries between whites and people of color, failing to fall securely on either side of the divide. And it is ambiguity that white supremacists fear most of all.

I find it especially ironic that, today, some strict Orthodox Jewish leaders also find that ambiguity threatening. Speaking out against the high rates of intermarriage among Jews and non-Jews, they issue dire warnings. Like white supremacists, they fear assaults on the integrity of the community and fight to secure its racial boundaries, defining Jewishness as biological and restricting it only to those with Jewish mothers. For both white supremacists and such Orthodox Jews, intermarriage is tantamount to genocide.

For me, the task is no longer to resolve the ambiguity, but to embrace it. My exploration of white-supremacist ideology has revealed just how subversive doing so can be: Reading white-supremacist discourse through the lens of Jewish experience has helped me toward new interpretations. White supremacy is not a movement just about hatred, but even more about fear; fear of the vulnerability and instability of white identity and privilege. For white supremacists, the central goal is to naturalize racial identity and hierarchy, to establish boundaries.

Both my own experience and Jewish history reveal that to be an impossible task. Embracing Jewish identity and history, with all their contradiction, has given me an empowering alternative to white-supremacist conceptions of race. I have found that eliminating ambivalence does not require eliminating ambiguity.

DISCUSSION QUESTIONS

1. What is your own ethnic identity? What would you say determines it? In general, what factors or forces determine one's own personal ethnic identity in the United States?
2. Do you agree with Ferber's argument that the "other" is a significant determinant of one's personal ethnic identity?

INFOTRAC COLLEGE EDITION

You can use your access to InfoTrac College Edition to learn more about the subjects covered in this essay. Some suggested search terms include:

anti-Semitism   holocaust
ethnic identity   white supremacy

45

The Social Construction
of Gender

MARGARET L. ANDERSEN

In this essay, Margaret Andersen outlines the meaning of the "social construction of gender." She discusses the difference between the terms "sex" and "gender" and defines sexuality as it relates to both. After a brief discussion of the cultural basis of gender, the essay outlines the difference between a gender roles conceptualization of gender and the gendered institutions approach.

To understand what sociologists mean by the phrase *the social construction of gender,* watch people when they are with young children. "Oh, he's such a boy!" someone might say as he or she watches a 2-year-old child run around a room or shoot various kinds of play guns. "She's so sweet," someone might say while watching a little girl play with her toys. You can also see the social construction of gender by listening to children themselves or watching them play with each other. Boys are more likely to brag and insult other boys (often in joking ways) than are girls; when conflicts arise during children's play, girls are more likely than boys to take action to diffuse the conflict (McCloskey and Coleman, 1992; Miller, Danaber, and Forbes, 1986).

To see the social construction of gender, try to buy a gender-neutral present for a child—that is, one not specifically designed with either boys or girls in mind. You may be surprised how hard this is, since the aisles in toy stores are highly stereotyped by concepts of what boys and girls do and like. Even products such as diapers, kids' shampoos, and bicycles are gender stereotyped. Diapers for boys are packaged in blue boxes; girls' diapers are packaged in pink. Boys wear diapers with blue borders and little animals on them; girls wear diapers with pink borders with flowers. You can continue your observations by thinking about how we describe children's toys. Girls are said to play with dolls; boys play with action figures!

When sociologists refer to the **social construction of gender,** they are referring to the many different processes by which the expectations associated with being a boy (and later a man) or being a girl (later a woman) are passed on through society. This process pervades society, and it begins the minute a child is born. The exclamation "It's a boy!" or "It's a girl!" in the delivery room sets a

From: Margaret L. Andersen. 2000. *Thinking about Women: Sociological Perspectives on Sex and Gender.* Needham Heights, MA: Allyn and Bacon, pp. 19–24. Reprinted with permission.

course that from that moment on influences multiple facets of a person's life. Indeed, with the modern technologies now used during pregnancy, the social construction of gender can begin even before one is born. Parents or grandparents may buy expected children gifts that reflect different images, depending on whether the child will be a boy or a girl. They may choose names that embed gendered meanings or talk about the expected child in ways that are based on different social stereotypes about how boys and girls behave and what they will become. All of these expectations—communicated through parents, peers, the media, schools, religious organizations, and numerous other facets of society—create a concept of what it means to be a "woman" or be a "man." They deeply influence who we become, what others think of us, and the opportunities and choices available to us. The idea of the social construction of gender sees society, not biological sex differences, as the basis for gender identity. To understand this fully, we first need to understand some of the basic concepts associated with the social construction of gender and review some information about biological sex differences.

SEX, GENDER, AND SEXUALITY

The terms *sex, gender,* and *sexuality* have related, but distinct, meanings within the scholarship on women. **Sex** refers to the biological identity and is meant to signify the fact that one is either male or female. One's biological sex usually establishes a pattern of gendered expectations, although, . . . biological sex identity is not always the same as gender identity; nor is biological identity always as clear as this definition implies.

   Gender is a social, not biological, concept, referring to the entire array of social patterns that we associate with women and men in society. Being "female" and "male" are biological facts; being a woman or a man is a social and cultural process—one that is constructed through the whole array of social, political, economic, and cultural experiences in a given society. Like race and class, gender is a social construct that establishes, in large measure, one's life chances and directs social relations with others. Sociologists typically distinguish sex and gender to emphasize the social and cultural basis of gender, although this distinction is not always so clear as one might imagine, since gender can even construct our concepts of biological sex identity.

   Making this picture even more complex, **sexuality** refers to whole constellation of sexual behaviors, identities, meaning systems, and institutional practices that constitute sexual experience within society. This is not so simple a concept as it might appear, since sexuality is neither fixed nor unidimensional in the social experience of diverse groups. Furthermore, sexuality is deeply linked to gender relations in society. Here, it is important to understand that sexuality, sex, and gender are intricately linked social and cultural processes that overlap in establishing women's and men's experiences in society.

   Fundamental to each of these concepts is understanding the significance of culture. Sociologists and anthropologists define **culture** as "the set of definitions of reality held in common by people who share a distinctive way of life" (Kluck-hohn, 1962:52). Culture is, in essence, a pattern of expectations about what are appropriate behaviors and beliefs for the members of the society; thus, culture provides prescriptions for social behavior. Culture tells us what we ought to do, what we ought to think, who we ought to be, and what we ought to expect of others. . . .

The cultural basis of gender is apparent especially when we look at different cultural contexts. In most Western cultures, people think of *man* and *woman* as dichotomous categories—that is, separate and opposite, with no overlap between the two. Looking at gender from different cultural viewpoints challenges this assumption, however. Many cultures consider there to be three genders, or even more. Consider the Navaho Indians. In traditional Navaho society, the *berdaches* were those who were anatomically normal men but who were defined as a third gender and were considered to be intersexed. Berdaches married other men. The men they married were not themselves considered to be berdaches; they were defined as ordinary men. Nor were the berdaches or the men they married considered to be homosexuals, as they would be judged by contemporary Western culture. . . .

Another good example for understanding the cultural basis of gender is the *hijras* of India (Nanda, 1998). Hijras are a religious community of men in India who are born as males, but they come to think of themselves as neither men nor women. Like berdaches, they are considered a third gender. Hijras dress as women and may marry other men; typically, they live within a communal subculture. An important thing to note is that hijras are not born so; they choose this way of life. As male adolescents, they have their penises and testicles cut off in an elaborate and prolonged cultural ritual—a rite of passage marking the transition to becoming a hijra. . . .

These examples are good illustrations of the cultural basis of gender. Even within contemporary U.S. society, so-called "gender bending" shows how the dichotomous thinking that defines men and women as "either/or" can be transformed. Cross-dressers, transvestites, and transsexuals illustrate how fluid gender can be and, if one is willing to challenge social convention, how easily gender can be altered. The cultural expectations associated with gender, however, are strong, as one may witness by people's reactions to those who deviate from presumed gender roles. . . .

In different ways and for a variety of reasons, all cultures use gender as a primary category of social relations. The differences we observe between men and women can be attributed largely to these cultural patterns.

THE INSTITUTIONAL BASIS OF GENDER

Understanding the cultural basis for gender requires putting gender into a sociological context. From a sociological perspective, gender is systematically structured in social institutions, meaning that it is deeply embedded in the social structure of society. Gender is created, not just within family or interpersonal relationships (although these are important sources of gender relations), but also within the structure of all major social institutions, including schools, religion, the economy, and the state (i.e., government and other organized systems of authority such as the police and the military). These institutions shape and mold the experiences of us all.

Sociologists define **institutions** as established patterns of behavior with a particular and recognized purpose; institutions include specific participants who share expectations and act in specific roles, with rights and duties attached to them. Institutions define reality for us insofar as they exist as objective entities in our experience. . . .

Understanding gender in an institutional context means that gender is not just an attribute of individuals; instead, institutions themselves are *gendered*. To say that an institution is gendered means that the whole institution is patterned on specific gendered relationships. That is, gender is "present in the processes, practices, images and ideologies, and distribution of power in the various sectors of social life" (Acker, 1992:567). The concept of a gendered institution was introduced by Joan Acker, a feminist sociologist. Acker uses this concept to explain not just that gender expectations are passed to men and women within institutions, but that the institutions themselves are structured along gendered lines. **Gendered institutions** are the total pattern of gender relations—stereotypical expectations, interpersonal relationships, and men's and women's different placements in social, economic, and political hierarchies. This is what interests sociologists, and it is what they mean by the social structure of gender relations in society.

Conceptualizing gender in this way is somewhat different from the related concept of gender roles. Sociologists use the concept of social roles to refer to culturally prescribed expectations, duties, and rights that define the relationship between a person in a particular position and the other people with whom she or he interacts. For example, to be a mother is a specific social role with a definable set of expectations, rights, and duties. Persons occupy multiple roles in society; we can think of social roles as linking individuals to social structures. It is through social roles that cultural norms are patterned and learned. **Gender roles** are the expectations for behavior and attitudes that the culture defines as appropriate for women and men.

The concept of gender is broader than the concept of gender roles. *Gender* refers to the complex social, political, economic, and psychological relations between women and men in society. Gender is part of the social structure—in other words, it is institutionalized in society. *Gender roles* are the patterns through which gender relations are expressed, but our understanding of gender in society cannot be reduced to roles and learned expectations.

The distinction between gender as institutionalized and gender roles is perhaps most clear in thinking about analogous cases—specifically, race and class. Race relations in society are seldom, if ever, thought of in terms of "race roles." Likewise, class inequality is not discussed in terms of "class roles." Doing so would make race and class inequality seem like matters of interpersonal interaction. Although race, class, and gender inequalities are experienced within interpersonal interactions, limiting the analysis of race, class, or gender relations to this level of social interaction individualizes more complex systems of inequality; moreover, restricting the analysis of race, class, or gender to social roles hides the power relations that are embedded in race, class, and gender inequality (Lopata and Thorne, 1978).

Understanding the institutional basis of gender also underscores the interrelationships of gender, race, and class, since all three are part of the institutional framework of society. As a social category, gender intersects with class and race; thus, gender is manifested in different ways, depending on one's location in the race and class system. For example, African American women are more likely than White women to reject gender stereotypes for women, although they are more accepting than White women of stereotypical gender roles for children. Although this seems contradictory, it can be explained by understanding that African American women may reject the dominant culture's view while also hoping their children can attain some of the privileges of the dominant group (Dugger, 1988).

Institutional analyses of gender emphasize that gender, like race and class, is a part of the social experience of us all—not just of women. Gender is just as important in the formation of men's experiences as it is in women's (Messner, 1998). From a sociological perspective, class, race, and gender relations are systemically structured in social institutions, meaning that class, race, and gender relations shape the experiences of all. Sociologists do not see gender simply as a psychological attribute, although that is one dimension of gender relations in society. In addition to the psychological significance of gender, gender relations are part of the institutionalized patterns in society. Understanding gender, as well as class and race, is central to the study of any social institution or situation. Understanding gender in terms of social structure indicates that social change is not just a matter of individual will—that if we changed our minds, gender would disappear. Transformation of gender inequality requires change both in consciousness and in social institutions. . . .

NOTES

Acker, Joan. 1992. "Gendered Institutions: From Sex Roles to Gendered Institutions." *Contemporary Sociology* 21 (September): 565–569.

Dugger, Karen. 1988. "The Social Location of Black and White Women's Attitudes." *Gender & Society* 2 (December): 425–448.

Kluckhohn, C. 1962. *Culture and Behavior.* New York: Free Press.

Lopata, Helene Z., and Barrie Thorne. 1978. "On the Term 'Sex Roles.'" *Signs* 3 (Spring): 718–721.

McCloskey, Laura A., and Lerita M. Coleman. 1992. "Difference Without Dominance: Children's Talk in Mixed- and Same-Sex Dyads." *Sex Roles* 27 (September): 241–258.

Messner, Michael A. 1998. "The Limits of 'The Male Sex Role': An Analysis of the Men's Liberation and Men's Rights Movements' Discourse." *Gender & Society* 12 (June): 255–276.

Miller, D., D. Danaber, and D. Forbes. 1986. "Sex-related Strategies for Coping with Interpersonal Conflict in Children Five and Seven." *Development Psychology* 22: 543–548.

Nanda, Serena. 1998. *Neither Man Nor Woman: The Hijras of India.* Belmont, CA: Wadsworth.

DISCUSSION QUESTIONS

1. Walk through a baby store. Can you easily identify products for girls and for boys? Could you easily purchase clothing appropriate for either a boy or a girl?

2. Consider an occupation that is traditionally men's work or traditionally women's work. What happens when a member of the opposite sex works in that field? What stereotypes and derogatory assumptions do we make about a woman working in a man's occupation or a man working in a woman's occupation?

INFOTRACT COLLEGE EDITION

You can use your access to InfoTrac College Edition to learn more about the subjects covered in this essay. Some suggested search terms include:

culture

gender roles

gendered institutions

sexuality

social construction of gender

46

The Politics of Masculinities

MICHAEL A. MESSNER

Michael Messner uses personal accounts to illustrate three important considerations in the study of masculinity. Specifically, he discusses the institutionalized nature of male privilege that gives men as a group power over women as a group. Next, he argues that the limiting definition of masculinity in U.S. culture comes at some cost for men. Finally, he addresses differences and inequalities among men that give White, heterosexual men more power than others.

Not long ago, I was standing in line behind a woman and a man at the local car wash, waiting to pay the cashier. As the man, a thirtyish white guy wearing a tight tank top, paid his money, the female cashier asked him about the prominent Asian characters that were tattooed on his heavily muscled arm. "Is that Chinese?" she asked him. "No," he replied tersely, "it's Korean." She persisted in her curiosity: "What's it say?" He lifted the arm a bit and flexed: "It says, 'Fear No Man, Trust No Woman.' " An uncomfortable moment of silence passed as he got his change and left. Then, the next woman in line stepped

From: Michael A. Messner. 1997. *The Politics of Masculinities.* Thousand Oaks: Sage, pp. xiii–xv, 1–10. Reprinted with permission.

forward to pay her money, and the cashier said to her, "That's pretty scary!" "I don't know," the other woman replied. "I think they *all* should come with warning labels."

At first, I thought about this scene only in terms of how women today are often so adept at poking fun not only at some men's hyper-masculine posturing but also at the very real danger of violence that all men potentially represent. But as I thought more about it, I began to think and wonder about what might have compelled this man to inscribe—apparently permanently—this depressing message on his arm. "Fear No Man, Trust No Woman." Imagine the isolation, loneliness, and alienation that must underlie such a slogan.

And I began to ponder the "No Fear" slogan that has appeared lately on the baseball caps, T-shirts, and bumper stickers of boys and young men, seemingly everywhere. It seems to me that you don't see hundreds of thousands of people massed in the streets chanting for peace unless you already have war. And you don't have a whole generation of young males publicly proclaiming that they have "No Fear" unless there's something actually scaring the crap out of them.

What are men so afraid of today? Most obviously, they are afraid of other men's violence. Young African American males in particular are falling prey to each other's violence in epidemic proportions, but young males from all social groups feel increasingly vulnerable today. Less obvious, but just as ominous, are young men's worries and fears of an uncertain future. As deindustrialization has eliminated tens of thousands of inner-city jobs, as structural unemployment has risen, and as government has become increasingly unable and unwilling to provide hope, a higher and higher proportion of young males today see that the image of the male family breadwinner is increasingly unattainable for them.

It's actually getting harder and harder for a young male to figure out how to *be* a man. But this is not necessarily a bad thing. Young men's current fears of other men and the continued erosion of the male breadwinner role might offer a historic opportunity for men—individually and collectively—to reject narrow, limiting, and destructive definitions of masculinity and, instead, to create a more humane, peaceful, and egalitarian definition of manhood. . : .

INSTITUTIONALIZED PRIVILEGE

In the early 1970s, when I was an undergraduate, I took a course on social inequality in which I was confronted with research that showed that women in the paid labor force were earning about $.59 to the male's dollar. Even women who were working in the same occupations as men, I learned, were earning substantially less than their male counterparts. These facts radically contradicted what I had been taught about the United States; that this is a country of equal opportunity in which merit is rewarded independent of race, religion, or sex. In a term paper for this course, I explored the reasons for gender inequities in the workforce and stated passionately in my conclusion that it was only fair for women to have equal rights with men. My professor liked the paper, and I felt proud that I had taken a "profeminist" position.

The following summer, I was back in my hometown, working at my regular summer job as a recreation worker in city parks. With the exception of two full-time supervisors, the summer staff consisted of about 15 temporary workers who were, like me, college students. Perhaps a dozen or so of these workers were women, and three of us were men, giving the appearance, perhaps, that this was a female-dominated job. But what had not occurred to me at the time, or struck me as unusual, was that all of the women had been given 20- or 30-hour per week assignments at smaller city parks, whereas each of the men had been assigned 40-hour weeks at the larger parks. What's more, when opportunities for overtime work arose, the supervisors invariably invited the men to do the work. So I regularly chalked up 42, perhaps even 46, hours of work each week. One week, at a staff meeting, a supervisor routinely invited me and another man to come to the recreation center to do some overtime work. Before we had a chance to say yes, we were interrupted by one of the women workers, who firmly stated, "I don't know why the guys always get the extra hours; we women can do that work as well as them. It doesn't really seem fair." I immediately felt threatened and defensive and broke the uncomfortable moment of silence in the room by whispering—far too loudly, as it turned out—to my male coworker, "Who the hell does she think she is, Gloria Steinem?" In response, the woman worker glared and pointed her finger at me: "Don't talk about something you don't know anything about, Mike!"

Immediately, it ran through my mind that I *did,* in fact, know *a lot* about this topic. Why, I had just written this wonderful paper about how women workers are paid less than men and had taken the position that this should change. Why, then, when faced with a concrete situation where I could put that knowledge and those principles to work had I taken a defensive, reactionary position? In retrospect, I can see that I had not yet learned the difference between taking an intellectual position on an issue and actually integrating principles into my life ("the personal is political," my feminist friends would later teach me). But more important, I had not yet come to grips with the reality that men—especially white, heterosexual, middle-class men like myself—tend to take for granted certain *institutional privileges.* In this case, because I was a man, I was "just naturally" afforded greater opportunities than my female coworkers. Yes, I had to work hard, but I worked no harder than did the women. Yet, to receive equal treatment, the women had to stand up for themselves and make public claims based on values of justice and equal opportunity. I just had to show up. What strikes me about this in retrospect is how easy it is for members of a privileged group to remain ignorant of the ways that the social structures of which we are a part grant us privileges, often at the expense of others. . . .

. . . .Gender is a system of unequal—but shifting and at times contested—power relations between women and men (Connell, 1987). In the current historical moment, men's institutional privileges still persist, by and large, but they no longer can be entirely taken for granted. For the past three decades, women have organized to actively challenge unequal and unfair gender arrangements. This reality was brought home to me over 20 years ago quite dramatically by my female workmate.

THE COSTS OF MASCULINITY

In 1977, a major support undergirding my world suddenly gave way when my father died. It just didn't seem fair: A nonsmoker, moderate social drinker, and former athlete only a few pounds overweight, he appeared to have lived a fairly healthy life. Seemingly vibrant at the age of 56, he was just too young to die. But within a few short months, cancer quickly dropped the man who, to our family, had always seemed like the Rock of Gibraltar. And maybe, I can now see, that was part of the problem. As a high school and college football player in the 1930s and 1940s, then in the navy in World War II, he had been taught that a real man ignores his own pain and pays whatever price is necessary for the good of the team or country. Throughout his adulthood, this lesson was buttressed by his conservative Lutheranism, which taught him that a man's first responsibility was as a family breadwinner, which meant to work hard and sacrifice himself, day in and day out, for the good of his family.

Indeed, as my mother cared for me and my two sisters, my father worked very hard during the school year as a high school teacher and a coach and on summer "tours" in the navy reserve. He prided himself that he had never let a little cold or flu or a sore back keep him from work. He'd been taught to "play through the pain," to keep his complaints to himself, never to show his own hurt, pain, or fears. . . .

He died with nearly a year's worth of accumulated "sick leave" at the high school.

I've come to see my father's story as paradigmatic of the story of men in general. The promise of public status and masculine privilege comes with a price tag: Often, men pay with poor health, shorter lives, emotionally shallow relationships, and less time spent with loved ones. Indeed, the current gap between women's and men's life expectancy is about 7 years; men tend to consume tobacco and alcohol at higher rates than do women, resulting in higher rates of heart disease, cirrhosis of the liver, and lung cancer; men tend to be slower than women to ask for professional medical help; men tend to engage in violence and high-risk behavior at much higher rates than do women; and men are taught to downplay or ignore their own pain (Harrison, Chin, & Ficarratto, 1995; Sabo, 1994b; Sabo & Gordon, 1995; Stillion, 1995; Waldron, 1995). In short, conformity with narrow definitions of masculinity can be lethal for men. . . .

DIFFERENCES AND INEQUALITIES AMONG MEN

In the early 1980s, at one of the first National Conferences on Men and Masculinity, I sat with several hundred men and listened to a radical feminist male exhort all of us to "renounce masculinity" and "give up all of our male privileges" as we unite with women to work for a just and egalitarian world. Shortly after this moving speech, a black man stood up and angrily shouted, "When you ask me to give up my privileges as a man, you are asking me to give up something that white America has never allowed me in the first place! I've never been allowed to *be* a man in this racist society." After a smattering of applause and confused chatter, another man stood and said, "Yeah—I feel the same way as a gay man. My struggle is not to learn how to cry and hug other men. That's what you straight guys are all hung up on. I am oppressed in this homophobic society and need to empower myself to fight that oppression. I can't relate to your guilt-tripping us all into *giving up* our power. What power?"

This meeting illustrated one of the major issues faced by feminists, especially beginning in the 1980s: Women and men of color, gay men and lesbians, and differently abled people have all challenged the simplistic assumption that we can neatly discuss "women" and "men" as discrete categories within which members are assumed to share certain life experiences, life chances, and worldviews (Baca Zinn, Cannon, Higginbotham, & Dill, 1986; Collins, 1990; Wittig, 1992). In fact, although it may be true that "men, as a group, enjoy institutional privileges at the expense of women, as a group," men share very unequally in the fruits of these privileges. Indeed, one can make a good case that the economic, political, and legal constraints facing poor African American, Latino, or Native American men, institutionally disenfranchised disabled men, illegal immigrant men, and some gay men more than overshadow whatever privileges these people might have as men in this society (Anderson, 1990; Gerschick & Miller, 1995; Hondagneu-Sotelo & Messner, 1994; Nonn, 1995; Staples, 1995a). And the "costs of masculinity," such as poor health and shorter life expectancy are paid out disproportionately by socially and economically marginalized men (Sabo, 1995; Staples, 1995a).

When we examine gender relations along with race and ethnicity, social class, sexuality, and age as crosscutting, interrelated systems of power and inequality, it becomes clear that studying men and women is far more complicated than it might first have seemed. In fact, as R. W. Connell (1987) has argued, it makes little sense to talk of a singular masculinity (or femininity, for that matter) as did much of the 1970s "sex role" literature. Instead, Connell observes that at any given historical moment there are various and competing masculinities. Hegemonic masculinity, the form of masculinity that is dominant, expresses (for the moment) a successful strategy for the domination of women, and it is also constructed in relation to various marginalized and subordinated masculinities (e.g., gay, black, and working-class masculinities). . . .

NOTES

Anderson, E. (1990). *Streetwise: Race, class, and change in an urban community*. Chicago: University of Chicago Press.

Baca Zinn, M., Cannon, L. W., Higginbotham, E., & Dill, B. T. (1986). The costs of exclusionary practices in women's studies. *Signs: Journal of Women in Culture and Society, 11,* 290–303.

Collins, P. H. (1990). *Black feminist thought: Knowledge, consciousness, and the politics of empowerment.* Boston: Unwin Hyman.

Connell, R. W. (1987). *Gender and power.* Stanford, CA: Stanford University Press.

Gerschick, T. J., & Miller, A. S. (1995). Coming to terms: Masculinity and physical disability. In M. S. Kimmel & M. A. Messner (Eds.), *Men's lives* (3rd ed., pp. 262–276). Boston: Allyn & Bacon.

Harrison, J., Chin, J., & Ficarratto, T. (1995). Warning: Masculinity may be dangerous to your health. In M. S. Kimmel & M. A. Messner (Eds.), *Men's Lives* (3rd ed., pp. 237–249). Boston: Allyn & Bacon.

Hondagneu-Sotelo, P., & Messner, M. A. (1994). Gender displays and men's power: The "new man" and the Mexican immigrant man." In H. Brod & M. Kaufman (Eds.), *Theorizing masculinities* (pp. 200–218). Thousand Oaks, CA: Sage.

Nonn, T. (1995). Hitting bottom: Homelessness, poverty, and masculinity. In M. S. Kimmel & M. A. Messner (Eds.), *Men's lives* (3rd ed., pp. 225–234). Boston: Allyn & Bacon.

Sabo, D. F. (1994b). Pigskin, patriarchy, and pain. In M. A. Messner & D. F. Sabo (Ed.), *Sex, violence, and power in sports: Rethinking masculinity* (pp. 82–88). Freedom, CA: Crossing Press.

Sabo, D. F. (1995). Caring for men. In J. M. Cookfair (Ed.), *Nursing care in the community* (2nd ed., pp. 346–365). St. Louis: C.V. Mosby.

Sabo, D., & Gordon, D. F. (Eds.). (1995). *Men's health and illness: Gender, power, and the body.* Thousand Oaks, CA: Sage.

Staples, R. (1995a). Health among Afro-American males. In D. Sabo & D. F. Gordon (Eds.), *Men's health and illness: Gender, power, and the body* (pp. 212–138). Thousand Oaks, CA: Sage.

Stillion, J. M. (1995). Premature death among males: Extending the bottom line of men's health. In D. Sabo & D. F. Gordon (Eds.), *Men's health and illness: Gender, power, and the body* (pp. 46–67). Thousand Oaks, CA: Sage.

Waldron, I. (1995). Contributions of changing gender differences in behavior and social roles to changing gender differences in mortality. In D. Sabo & D. F. Gordon (Eds.), *Men's health and illness: Gender, power, and the body* (pp. 22–45). Thousand Oaks, CA: Sage.

Wittig, M. (1992). *The straight mind and other essays.* Boston: Beacon.

DISCUSSION QUESTIONS

1. What was your reaction to hearing about the "Million Man March" or another recent men's movement gathering? How do race and sexuality fit into men's movements? Are their any issues addressed by these groups that are universally a concern for all men?

2. In what ways can men's lives improve if we alter our conceptualization of masculinity? What can men learn from women's roles? What privilege or power would men need to give up for this to work?

INFOTRAC COLLEGE EDITION

You can use your access to InfoTrac College Edition to learn more about the subjects covered in this essay. Some suggested search terms include:

hegemonic masculinity   men's movement
institutionalized privilege

47

Lipstick Politics in Iran

FARZANEH MILANI

In this short essay, Farzaneh Milani portrays the gender inequality and power differential in Islamic Iran. The illustration points to seemingly inconsequential symbols of femininity (i.e., makeup) as powerful political statements. Milani's essay provides a clear example of the political and cultural context for gender identity.

In Iran, nothing is what it seems to be. There are layers upon layers of meaning attached to every word, to every gesture, to every action.

Take makeup. It is as fraught with political meanings and intentions as it has ever been. Women use it to signal their political ideology or to defy authority. I learned this lesson in July on a visit to Iran, when I found myself caught in the midst of riots in Teheran.

Accompanied by my friend Mariam, I had gone to the main bazaar to purchase a rug. After we finished our shopping, we decided to have a kebab at an old and established restaurant in the heart of the bazaar. We had not even touched our food when the restaurant's owner suddenly snapped off the lights and locked the door. A sense of horror filled the air. The walls of the restaurant were shaking as if there were an earthquake.

"The vigilantes have come to the bazaar; they're here," screamed one woman. Immediately I knew that the self-appointed morals police, ever so obsessed with the dress code for women, had attacked the bazaar.

While I sat paralyzed with fear, Mariam was deftly wiping off her lipstick with a paper napkin. One woman was covering her painted nails with thick, dark gloves. Another was covering her colorful head scarf with a black one she pulled out of her handbag.

A young woman next to me was putting on knee-length socks to hide her impeccably colored toenails, which showed through her sandals. Another middle-aged woman, with highlighted hair showing through her scarf, yelled: "I am sick and tired of all this. We have to free ourselves or die."

At the same time, a fight between supporters of the hard-liners and supporters of the reformers broke out in the men's section of the segregated restaurant. I felt trapped and terrified. Leaving the rug behind, we rushed to the door and persuaded the owner to let us out.

All the shops had closed, turning the beautiful bazaar into a wicked maze. After what seemed like an eternity, we reached a major street, hailed a cab and offered the driver an exorbitant fee.

From: *The New York Times* (August 19, 1999).

In the heat of that summer day, covered head to toe in my Islamic garb and drenched in sweat and panic, I found the locked, unair-conditioned cab a safe haven. Once we broke through the traffic gridlock and the bazaar district receded into the background, I sighed with relief and looked over my shoulder at Mariam.

I could not believe my eyes. She was reapplying her lipstick. Only half an hour ago she had frantically wiped off all traces of it. The skill and speed with which she had removed her lipstick and her haste and zeal now in reapplying it were astounding.

"Lipstick is not just lipstick in Iran," Mariam explained. "It transmits political messages. It is a weapon."

My friend was right. In the political history of modern Iran, doubts about modernity, about change, about relations with the West have always been projected upon a woman's body. In 1936, the Shah forced women to unveil themselves, and this was considered a mark of progress. In 1983, the Islamic Republic veiled women, and this signaled the reconstruction of an Islamic-Iranian identity.

Today, women still have to cover themselves, but they have become a vibrant political force. More and more of them are behind steering wheels, on motorcycles, in universities, in mosques, ascending the rungs of government. Their pictures are in newspapers and on television. Their participation in the artistic and literary arena is unprecedented.

Iranian women have successfully invaded male territories, although a dab of lipstick can still land them in jail. Perhaps the next victory will be ownership of their own bodies.

DISCUSSION QUESTIONS

1. What are some powerful political symbols of femininity in the United States society (both historically and contemporarily)? How have these symbols been used?

2. What should be the political position of the United States in response to gender oppression in Iran? How do we reconcile cultural differences and human rights?

INFOTRAC COLLEGE EDITION

You can use your access to InfoTrac College Edition to learn more about the subjects covered in this essay. Some suggested search terms include:

Islamic feminism   political ideology
Islamic-Iranian identity   women and Islam
Middle Eastern women

48

Ideological Racism
and Cultural Resistance:
Constructing Our Own Images

YEN LE ESPIRITU

This essay outlines the intersection between race and gender in the portrayal of Asian American women and men in American culture. Yen Le Espiritu discusses the competing images of Asian men as both emasculated and dangerous. Additionally, Asian women are typically portrayed as either over-sexualized exotics or servile and demure sex objects. Racialized stereotypes are used to create images of gender and sexuality among Asian people. This process maintains racial and cultural domination over Asian Americans.

Besides structural discrimination, Asian American men and women have been subject to ideological assaults. Focusing on the ideological dimension of Asian American oppression, this [essay] examines the cultural symbols—or what Patricia Hill Collins (1990) called "controlling images" (pp. 67–68)—generated by the dominant group to help justify the economic exploitation and social oppression of Asian American men and women over time. . . . The objectification of Asian Americans as the exotic and inferior "other" has never been absolute. Asian Americans have always, but particularly since the 1960s, resisted race, class, and gender exploitation not only through political and economic struggles but also through cultural activism. . . .

YELLOW PERIL, CHARLIE CHAN, AND SUZIE WONG

A central aspect of racial exploitation centers on defining people of color as "the other" (Said, 1979). The social construction of Asian American "otherness"— through such controlling images as the Yellow Peril, the model minority, the Dragon Lady, and the China Doll—is "the precondition for their cultural marginalization, political impotence, and psychic alienation from mainstream American society" (Hamamoto, 1994, p. 5). As indicated by these stereotypes, representations of gender and sexuality figure strongly in the articulation of

From: Yen Le Espiritu. 1997. *Asian American Women and Men.* Thousand Oaks, CA: Sage, pp. 86–107. Reprinted with permission racism. These racist stereotypes collapse gender and sexuality: Asian men have been constructed as hypermasculine, in the image of the "Yellow Peril," but also as effeminate, in the image of the "model minority," and Asian women have been depicted as superfeminine, in the images of the "China Doll," but also as castrating, in the image of the "Dragon Lady" (Mullings, 1994, pp. 279–280; Okihiro, 1995). . . . The gendering of ethnicity—the process whereby white ideology assigns selected gender characteristics to various ethnic "others"—casts Asian American men and women as simultaneously masculine and feminine but also as neither masculine nor feminine. On the one hand, as part of the Yellow Peril, Asian American men and women have been depicted as a *masculine* threat that needs to be contained. On the other hand, both sexes have been skewed toward the female side: an indication of the group's marginalization in U.S. society and its role as the complaint "model minority" in contemporary U.S. cultural ideology. Although an apparent disjunction, both the feminization and masculinization of Asian men and women exist to define and confirm the white man's superiority (Kim, 1990). . . .

The Racial Construction of Asian American Manhood

Like other men of color, Asian American men have been excluded from white-based cultural notions of the masculine. Whereas white men are depicted both as virile and as protectors of women, Asian men have been characterized both as asexual *and* as threats to white women. It is important to note the historical contexts of these seemingly divergent representations of Asian American manhood. The racist depictions of Asian men as "lascivious and predatory" were especially pronounced during the nativist movement against Asians at the turn of the century (Frankenberg, 1993, pp. 75–76). The exclusion of Asian women from the United States and the subsequent establishment of bachelor societies eventually reversed the construction of Asian masculinity from "hypersexual" to "asexual" and even "homosexual." The contemporary model-minority stereotype further emasculates Asian American men as passive and malleable. Disseminated and perpetuated through the popular media, these stereotypes of the emasculated Asian male construct a reality in which social and economic discrimination against these men appears defensible. As an example, the desexualization of Asian men naturalized their inability to establish conjugal families in pre-World War II United States. Gliding over race-based exclusion laws that banned the immigration of most Asian women and antimiscegenation laws that prohibited men of color from marrying white women, these dual images of the eunuch and the rapist attributed the "womanless households" characteristic of pre-war Asian America to Asian men's lack of sexual prowess and desirability. . . .

The motion picture industry has been key in the construction of Asian men as sexual deviants. In a study of Asians in the U.S. motion pictures, Eugene Franklin Wong (1978) maintained that the movie industry filmically castrates Asian males to magnify the superior sexual status of white males (p. 27). As on-screen sexual rivals of whites, Asian males are neutralized, unable to sexually engage Asian women and prohibited from sexually engaging white women. By saving the white woman from sexual contact with the racial "other," the motion picture industry protects the Anglo-American, bourgeois male establishment from any challenges to its hegemony (Marchetti, 1993, p. 218). At the other extreme, the industry has exploited one of the most potent aspects of the Yellow Peril discourses—the sexual danger of contact between the races—by concocting a sexually threatening portrayal of the licentious and aggressive Yellow Man lusting after the White Woman (Marchetti, 1993, p. 3). Heedful of the larger society's taboos against Asian male-white female sexual union, white male actors donning "yellowface"—instead of Asian male actors—are used in these "love scenes." Nevertheless, the message of the perverse and animalistic Asian male attacking helpless white women is clear (Wong, 1978). Though depicting sexual aggression, this image of the rapist, like that of the eunuch, casts Asian men as sexually undesirable. As Wong (1978) succinctly stated, in Asian male-white female relations, "There can be rape, but there cannot be romance" (p. 25). Thus, Asian males yield to the sexual superiority of the white males who are permitted filmically to maintain their sexual dominance over both white women and women of color.

White cultural and institutional racism against Asian males is also reflected in the motion picture industry's preoccupation with the death of Asians—a filmic solution to the threats of the Yellow Peril. In a perceptive analysis of Hollywood's view of Asians in films made from the 1930s to the 1960s, Tom Engelhardt (1976) described how Asians, like Native Americans are seen by the movie industry as inhuman invaders, ripe for extermination. He argued that the theme of the non-humanness of Asians prepares the audience to accept, without flinching, "the levelling and near-obliteration of three Asian areas in the course of three decades" (Engelhardt, 1976, p. 273). The industry's death theme, though applying to all Asians, is mainly focused on Asian males, with Asian females reserved for sexual purposes (Wong, 1978, p. 35). Especially in war films, Asian males, however advantageous their initial position, inevitably perish at the hands of the superior white males (Wong, 1978, p. 34).

The Racial Construction of Asian American Womanhood

Like Asian men, Asian women have been reduced to one-dimensional caricatures in Western representation. The condensation of Asian women's multiple differences into gross character types—mysterious, feminine, and non-white—obscures the social injustice of racial, class, and gender oppression (Marchetti, 1993, p. 71). Both Western film and literature promote dichotomous stereotypes of the Asian woman: Either she is the cunning Dragon Lady or the servile Lotus Blossom Baby (Tong, 1994, p. 197). Though connoting two extremes, these stereotypes are interrelated: Both eroticize Asian women as exotic "others"—sensuous, promiscuous, but untrustworthy. Whereas American popular culture denies "manhood" to Asian men, it endows Asian women with an excess of "womanhood," sexualizing them but also impugning their sexuality. In this process, both sexism and racism have been blended together to produce the sexualization of white racism (Wong, 1978, p. 260). Linking the controlling images of Asian men and women, Elaine Kim (1990) suggested that Asian women are portrayed as sexual for the same reason that men are asexual: "Both exist to define the white man's virility and the white man's superiority" (p. 70).

As the racialized exotic "others," Asian American women do not fit the white-constructed notions of the feminine. Whereas white women have been depicted as chaste and dependable, Asian women have been represented as promiscuous and untrustworthy. . . . The Asian woman was portrayed as the castrating Dragon Lady who, while puffing on her foot-long cigarette holder, . . . could poison a man as easily as she could seduce him.

At the opposite end of the spectrum is the Lotus Blossom stereotype, reincarnated throughout the years as the China Doll, the Geisha Girl, the War Bride, or the Vietnamese prostitute—many of whom are the spoils of the last three wars fought in Asia (Tajima, 1989, p. 309). Demure, diminutive, and deferential, the Lotus Blossom Baby is "modest, tittering behind her delicate ivory hand, eyes downcast, always walking ten steps behind her man, and, best of all, devot[ing] body and soul to serving him" (Ling, 1990, p. 11). Interchangeable in appearance and name, these women have no voice; their "nonlanguage" includes uninterpretable chattering, pidgin English, giggling, or silence (Tajima, 1989). These stereotypes of Asian women as submissive and dainty sex objects not only have impeded women's economic mobility but also have fostered an enormous demand for X-rated films and pornographic materials featuring Asian women in bondage, for "Oriental" bathhouse workers in U.S. cities, and for Asian mail-order brides (Kim, 1984, p. 64).

Sexism, Racism, and Love

The racialization of Asian manhood and womanhood upholds white masculine hegemony. Cast as sexually available, Asian women become yet another possession of the white man. In motion pictures and network television programs, interracial sexuality, though rare, occurs principally between a white male and an Asian female. A combination of sexism and racism makes this form of miscegenation more acceptable: Race mixing between an Asian male and a white female would upset not only racial taboos but those that attend patriarchal authority as well (Hamamoto, 1994, p. 39). Whereas Asian men are depicted as either the threatening rapist or the impotent eunuch, white men are endowed with the masculine attributes with which to sexually attract the Asian woman. Such popular television shows as *Gunsmoke* (1955–1975) and *How the West Was Won* (1978–1979) clearly articulate the theme of Asian female sexual possession by the white male. In these shows, only white males have the prerogative to cross racial boundaries and to choose freely from among women of color as sex partners. Within a system of racial and gender oppression, the sexual possession of women and men of color by white men becomes yet another means of enforcing unequal power relations (Hamamoto, 1994, p. 46).

The preference for white male–Asian female is also prevalent in contemporary television news broadcasting, most recently in the 1993–1995 pairing of Dan Rather and Connie Chung as coanchors of the *CBS Evening News*. Today, virtually every major metropolitan market across the United States has at least one Asian American female newscaster (Hamamoto, 1994, p. 245). While female Asian American anchorpersons—Connie Chung, Tritia Toyota, Wendy Tokuda, and Emerald Yeh—are popular television news figures, there is a nearly total absence of Asian American men. Critics argue that this is so because the white male hiring establishment, and presumably the larger American public, feels more comfortable (i.e., less threatened) seeing a white male sitting next to a minority female at the anchor desk than the reverse. Stephen Tschida of WDBJ-TV (Roanoke, Virginia), one of only a handful of male Asian American television news anchors, was informed early in his career that he did not have the proper "look" to qualify for the anchorperson position. Other male broadcast news veterans have reported being passed over for younger, more beauteous, female Asian Americans (Hamamoto, 1994, p. 245). This gender imbalance sustains the construction of Asian American women as more successful, assimilated, attractive, and desirable than their male counterparts.

The controlling images of Asian men and Asian women, exaggerated out of all proportion in Western representation, have created resentment and tension between Asian American men and women. Given this cultural milieu, many American-born Asians do not think of other Asians in sexual terms (Fung, 1994, p. 163). In particular, due to the persistent desexualization of the Asian male, many Asian females do not perceive their ethnic counterparts as desirable marriage partners (Hamamoto, 1992, p. 42). In so doing, these women unwittingly enforce the Eurocentric gender ideology that objectifies both sexes and racializes all Asians (see Collins, 1990, pp. 185–186). In a column to *Asian Week*, a weekly Asian American newspaper, Daniel Yoon (1993) reported that at a recent dinner discussion hosted by the Asian American Students Association at his college, the Asian American women in the room proceeded, one after another, to describe how "Asian American men were too passive, too weak, too boring, too traditional, too abusive, too domineering, too ugly, too greasy, too short, too . . . Asian. Several described how they preferred white men, and how they never had and never would date an Asian man" (p. 16). Partly as a result of the racist constructions of Asian American womanhood and manhood and their acceptance by Asian Americans, intermarriage patterns are high, with Asian American women intermarrying at a much higher rate than Asian American men. Moreover, Asian women involved in intermarriage have usually married white partners (Agbayani-Siewert & Revilla, 1995, p. 156; Min, 1995, p. 22; Nishi, 1995, p. 128). In part, these intermarriage patterns reflect the sexualization of white racism that constructs white men as the most desirable sexual partners, frowns on Asian male–white women relations, and fetishizes Asian women as the embodiment of perfect womanhood. Viewed in this light, the high rate of outmarriage for Asian American women is the "material outcome of an interlockng system of sexism and racism" (Hamamoto, 1992, p. 42).

CULTURAL RESISTANCE:
RECONSTRUCTING OUR OWN IMAGES

"One day/I going to write/about you," wrote Lois-Ann Yamanaka (1993) in "Empty Heart" (p. 548). And Asian Americans did write—"to inscribe our faces on the blank pages and screens of America's hegemonic culture" (Kim, 1993, p. xii). As a result, Asian Americans' objectification as the exotic aliens who are different from, and other than, Euro-Americans has never been absolute. Within the confines of race, class, and gender oppression, Asian Americans have maintained independent self-definitions, challenging controlling images and replacing them with Asian American standpoints. The civil rights and ethnic studies movements of the late 1960s were training grounds for Asian American cultural workers and the development of oppositional projects. Grounded in the U.S. black power movement and in anticolonial struggles of Third World countries, Asian American anti-hegemonic projects have been unified by a common goal of articulating cultural resistance. Given the historical distortions and misrepresentations of Asian Americans in mainstream media, most cultural projects produced by Asian American men and women perform the important tasks of correcting histories, shaping legacies, creating new cultures, constructing a politics of resistance, and opening spaces for the forcibly excluded (Kim, 1993, p. xiii; Fung, 1994, p. 165).

Fighting the exoticization of Asian Americans has been central in the ongoing work of cultural resistance. As discussed above, Asian Americans, however rooted in this country, are represented as recent transplants from Asia or as bearers of an exotic culture. . . . Asian American cultural workers simply do not accept the exotic, one-dimensional caricatures of themselves in U.S. mass media. . . .

Portraying Asian Americans in all our contradictions and complexities—as exiled, assimilated, rebellious, noble—Asian American cultural projects reveal heterogeneity rather than "producing regulating ideas of cultural unity or integration" (Lowe, 1994, p. 53). In so doing, these projects destabilize the dominant racist discourse that constructs Asians as a homogeneous group who are "all alike" and readily conform to "types" such as the Yellow Peril, the Oriental mastermind, and the sexy Suzie Wong (Lowe, 1991).

Asian American cultural projects also deconstruct the myth of the benevolent United States promised to women and men from Asia. Carlos Bulosan's *America Is in the Heart* (1943/1973), one of the core works of Asian American literature, challenges the narrative of the United States as the land of opportunity. Seduced by the promise of individual freedom through education, the protagonist Carlos discovers that as a Filipino immigrant in the United States, he is denied access to formal schooling. This disjunction between the promise of education and the unequal access of different racial and economic groups to that education—reinforced by Carlos's observations of the exploitation, marginality, and violence suffered by his compatriots in the United States—challenges his faith in the promise of U.S. democracy and abundance (Lowe, 1994, p. 56). . . .

To reject the myth of a benevolent United States is also to refute ideological racism: the justification of inequalities through a set of controlling images that attribute physical and intellectual traits to racially defined groups (Hamamoto, 1994, p. 3). In the 1980 autobiographical fiction *China Men*, Maxine Hong Kingston (1977) smashed the controlling image of the emasculated Asian man by foregrounding the legalized racism that turned immigrant Chinese "men" into "women" at the turn of the century. In his search for the Gold Mountain, the novel's male protagonist Tang Ao finds instead the Land of Women, where he is caught and transformed into an Oriental courtesan. Because Kingston reveals at the end of the legend that the Land of Women was in North America, readers familiar with Chinese American history will readily see that "the ignominy suffered by Tang Ao in a foreign land symbolizes the emasculation of Chinamen by the dominant culture"(Cheung, 1990, p. 240). . . .

More recently, Steven Okazaki's film *American Sons* (1995) tells the stories of four Asian American men who reveal how incidents of prejudice and bigotry shaped their identity and affected the way they perceived themselves and society. . . . Asian American men's increasing involvement in hip-hop—a highly masculinized cultural form and a distinctly American phenomenon—is yet another contemporary denouncement of the stereotype of themselves as "effeminate, nerdy, asocial foreigners" (Choe, 1996). . . .

Finally, Asian American cultural workers reject the narrative of salvation: the myth that Asian women (and a feminized Asia) are saved, through sexual relations with white men (and a masculinized United States), from the excesses of their own culture. Instead, they underscore the considerable potential for abuse in these inherently unequal relationships. Writing in Vietnamese, transplanted Vietnamese writer Tran Dieu Hang described the gloomy existence of Vietnamese women in sexist and racist U.S. society—an accursed land that singles out women, especially immigrant women, for oppression and violence. Her short story "Roi Ngay Van Moi" ("There Will Come New Days"; 1986) depicts the brutal rape of a young refugee woman by her American sponsor despite her tearful pleas in limited English (Tran, 1993, pp. 72–73). . . .

CONTROLLING IMAGES, GENDER, AND CULTURAL NATIONALISM

Cultural nationalism has been crucial in Asian Americans' struggles for self-determination. Emerging in the early 1970s, this unitary Asian American identity was primarily racial, male, and heterosexual. Asian American literature produced in those years highlighted Chinese and Japanese American male perspectives, obscuring gender and other intercommunity differences (Kim, 1993). Asian American male writers, concerned with recuperating their identities as men and as Americans, objectified both white and Asian women in their writings (Kim, 1990, p. 70). . . .

Because the racial oppression of Asian Americans involves the "feminization" of Asian men (Said, 1979), Asian American women are caught between the need to expose the problems of male privilege and the desire to unite with men to contest the overarching racial ideology that confines them both. As Cheung (1990) suggested, Asian American women may be simultaneously sympathetic and angry toward the men in their ethnic community: sensitive to the men's marginality but resentful of their sexism (p. 239). . . .

CONCLUSION

Ideological representations of gender and sexuality are central in the exercise and maintenance of racial, patriarchal, and class domination. In the Asian American case, this ideological racism has taken seemingly contrasting forms: Asian men have been cast as both hypersexual and asexual, and Asian women have been rendered both superfeminine and masculine. Although in apparent disjunction, both forms exist to define, maintain, and justify white male supremacy. The racialization of Asian American manhood and womanhood underscores the interconnections of race, gender, and class. As categories of difference, race and gender relations do not parallel but intersect and confirm each other, and it is the complicity among these categories of difference that enables U.S. elites to justify and maintain their cultural, social, and economic power. Responding to the ideological assaults on their gender identities, Asian American cultural workers have engaged in a wide range of oppositional projects to defend Asian American manhood and womanhood. In the process, some have embraced a masculinist cultural nationalism, a stance that marginalizes Asian American women and their needs. Though sensitive to the emasculation of Asian American men, Asian American feminists have pointed out that Asian American nationalism insists on a fixed masculinist identity, thus obscuring gender differences. Though divergent, both the nationalist and feminist positions advance the dichotomous stance of man or woman, gender or race or class, without recognizing the complex relationality of these categories of oppression. It is only when Asian Americans recognize the intersections of race, gender, and class that we can transform the existing hierarchical structure. . . .

NOTES

Agbayani-Siewert, P., & Revilla, L. (1995). Filipino Americans. In P. G. Min (Ed.), *Asian Americans: Contemporary trends and issues* (pp. 134–168). Thousand Oaks, CA: Sage.

Cheung, K.-K. (1990). The woman warrior versus the Chinaman pacific: Must a Chinese American critic choose between feminism and heroism? In M. Hirsch & E. F. Keller (Eds.), *Conflicts in feminism* (pp. 234–251). New York: Routledge.

Choe, Laura. 1996, February 10. "Versions." Asian Americans in Hip Hop. Paper presented at the California Studies Conference, Long Beach, CA.

Collins, P. H. (1990). *Black feminist thought: Knowledge, consciousness, and the politics of empowerment.* New York: Routledge.

Engelhardt, T. (1976). Ambush at Kamikaze Pass. In E. gee (Ed.), *Counterpoint: Perspectives on Asian America* (pp. 270–279). Los Angeles: University of California at Los Angeles, Asian American Studies Center.

Frankenberg, R. (1993). *White women, race matters: The social construction of whiteness.* Minneapolis: University of Minnesota Press.

Fung, R. (1994). Seeing yellow: Asian identities in film and video. In K. Aguilar-San Juan (Ed.), *The state of Asian America* (pp. 161–171). Boston: South End.

Hamamoto, D.Y. (1994). *Monitored peril: Asian Americans and the politics of representation.* Minneapolis: University of Minnesota Press.

Kim, E. (1984). Asian American writers: A bibliographical review." *American Studies International, 22,* 2.

Kim, E. (1990). "Such opposite creatures": Men and women in Asian American literature. *Michigan Quarterly Review, 29,* 68–93.

Kim, E. (1993). Preface in J. Hagedorn (Ed.), *Charlie Chan is dead: An anthology of contemporary Asian American fiction* (pp. vii–xiv). New York: Penguin.

Kingston, M. H. (1977). *The woman warrior.* New York: Vintage.

Ling, A. (1990). *Between worlds: Women writers of Chinese ancestry.* New York: Pergamon.

Lowe, L. (1991). Heterogeneity, hybridity, multiplicity: Marking Asian American difference. *Diaspora, 1,* 24–44.

Lowe, L. (1994). Canon, institutionalization, identity: contradictions for Asian American Studies. In D. Palumbo-Liu (Ed.), *The ethnic canon: Histories, institutions, and interventions* (pp. 48–68). Minneapolis: University of Minneapolis Press.

Marchetti, G. (1993). *Romance and the "Yellow Peril": Race, sex, and discursive strategies in Hollywood fiction.* Berkeley: University of California Press.

Min, P. G. (1995). Korean Americans. In P. G. Min (Ed.), *Asian Americans: Contemporary trends and issues* (pp. 199–231). Thousand Oaks, CA: Sage.

Mullings, L. (1994). Images, ideology, and women of color. In M. Baca Zinn & B. T. Dill (Eds.), *Women of color in U.S.* society (pp. 265–289). Philadelphia: Temple University Press.

Nishi, S. M. (1995). Japanese Americans. In P. G. Min (Ed.), *Asian Americans: Contemporary trends and issues* (pp. 95–133). Thousand Oaks, CA: Sage.

Okihiro, G.Y. (1995, November). *Reading Asian bodies, reading anxieties.* Paper presented at the University of California, San Diego Ethnic Studies Colloquium, La Jolla.

Said, E. (1979). *Orientalism.* New York: Random House.

Tajima, R. (1989). Lotus blossoms don't bleed: Images of Asian women. In Asian Women United of California (Ed.), *Making waves: An anthology of writings by and about Asian American women* (pp. 308–317). Boston: Beacon.

Tong, B. (1994). *Unsubmissive women: Chinese prostitutes in nineteenth-century San Francisco.* Norman: University of Oklahoma Press.

Tran, Q. P. (1993). Exile and home in contemporary Vietnamese American feminine writing. *Amerasia Journal, 19,* 71–83.

Wong, E. F. (1978). *On visual media racism: Asians in the American motion pictures.* New York: Arno.

Yamanaka, L. A. 1993. Empty heart. In J. Hagedorn (ed.), *Charlie Chan is dead: An anthology of contemporary Asian American fiction* (pp. 544–550). New York: Penguin.

Yoon, D. D. (1993, November 26). Asian American male: Wimp or what? *Asian Week,* p. 16.

DISCUSSION QUESTIONS

1. What are some examples of the sexualized portrayal of Asian American men and women in contemporary popular culture? What movies or television programs present Asian Americans and how are they portrayed?

2. Consider popular culture images of other racial-ethnic groups. How do race and gender intersect to create limiting portrayals of people?

INFOTRAC COLLEGE EDITION

You can use your access to InfoTrac College Edition to learn more about the subjects covered in this essay. Some suggested search terms include:

cultural resistance

model minority

racial construction of manhood/
　womanhood

structural discrimination

"Yellow Peril"

49

Cultural Images of Old Age

ELEANOR PALO STOLLER AND ROSE C. GIBSON

In this essay, Eleanor Palo Stoller and Rose Campbell Gibson discuss the negative images of older people in contemporary U.S. culture and summarize how older Americans respond to these images. They outline the images of older people related to gender, race, and social class and explain how older people reject negative stereotypes.

Old age is accompanied by physical changes. Our immune systems are less able to fight off disease, and our capacity to mobilize physical energy declines. Our hearing and vision become less acute. Changes in appearance, such as gray hair and wrinkled skin, become more prevalent. In the absence of disease, these changes seldom interfere with the ability to pursue desired activities or fulfill social obligations (Atchley, 1997). Age-related physical changes, however, occur within a cultural context that assigns meanings to physical characteristics, and often the symbolic rather than the functional consequences of age-related changes have the most negative impact on older people. Here, we will explore the impact of cultural images about old age and the ways older people respond to them. Our discussion will emphasize the following points:

1. The images of old age reflected in our culture and the ways in which these images vary along hierarchies based on gender, race, and class

2. The consequences of these images for older people and the strategies older people use to maintain positive self-concepts in the face of negative images or of positive images they cannot attain

3. The importance of considering variation within groups in interpreting comparisons between groups

Contemporary U.S. culture reflects mixed images of older people. On the positive side, older people have been viewed as wise, kind, understanding, family oriented, generous, happy, knowledgeable, and patriotic. On the negative side, they have been depicted as sad, forgetful, lonely, dependent, demanding, ill-tempered, nosy, complaining, senile, selfish, bored, and inflexible (Hummert, Garstka, Shaner, & Strahm, 1994; Schmidt & Boland, 1986). Negative images of aging are more common among people younger than 65 than among elders themselves. For example, in a recent national survey, a majority of people younger than 65 believed that older people were lonely (64%), felt they were not needed (57%),

From: Eleanor Palo Stoller and Rose C. Gibson, eds. 2000. *Worlds of Difference: Inequality in the Aging Experience*, 3rd ed. Thousand Oaks, CA: Pine Forge, pp. 75–86. Reprinted with permission.

and suffered from poor health (57%). In reality, only 6% of people over 65 said they were lonely, only 8% did not feel needed, and only 15% reported poor health (Speas & Obenshain, 1995).

Interest in ways to avoid the most negative aspects of aging has grown rapidly over the past several decades. Bookstores stock an expanding collection of titles advising older readers about fitness, nutrition, finances, and sexuality. Elderly Americans are advised to "keep their emotional balance," "stay active and involved," "pursue new interests," and "lead productive lives." Gerontologists focus our attention on the concept of *successful aging*, which the MacArthur Foundation Study of Successful Aging defines as a combination of physical health, high cognitive functioning, continued productivity, and active involvement in social relationships (Rowe & Kahn, 1997). Indeed, the book jacket of the MacArthur Foundation studies promises to "show you how the lifestyle choices you make now—more than heredity—determine your health and vitality." What was once described as a "roleless role" has been transformed into a prescription for activity and personal growth, for an awareness of expanded options and possibilities. . . .

Positive though these images of aging may be, they often depict a lifestyle predicated on class privilege. Recommendations to join an exercise class, learn ballroom dancing, or take up lap swimming imply sufficient discretionary income to purchase lessons or gain access to appropriate facilities. Retirement financial planning assumes assets to invest, and preventive health care recommendations require insurance or other financial resources to pay for physician visits and laboratory tests. Older people with meager incomes and poor health, as well as older people isolated in rural settings by lack of transportation or in inner cities by fear of crime, are hard pressed to fulfill these popular recommendations for so-called successful aging.

The images of successful aging reflected in popular guides are also more applicable to people in their 60s and 70s than to people in their 80s and 90s. Extreme old age is more often characterized by losses—loss of a future, of health and mobility, of cherished roles, of people one loves. Shortly before she died, Caroline Preston (as quoted in Thone, 1992) commented on the downside of these prescriptions for successful aging: "Such optimistic views of aging are as hard on us as our previous invisibility. We find ourselves yearning to be like people in these pictures and belabor ourselves for failing these role models" (p. 15). Norms for appropriate behavior in extreme old age are more nebulous. Ruth Raymond Thone (1992) suggests that emergent norms, at least for old women, can be captured in the phrase "aging gracefully," which she defines as being

> flexible, loving, a lady, positive, appreciative, accepting, tidy, active, open-minded, optimistic, proud of one's achievements, to have a zest for life, inner beauty, a sense of humor, to still learn and take risks, not to complain of one's aches and pains or other's faults, and not to worry about old age. (p. 13)

Not all elderly Americans define successful aging by positive adaptation and optimism. For many elders of color, aging is perceived "not as a series of adjustments, but rather as a process of survival" (Burton, Dilworth-Anderson, & Bengtson, 1991). . . .

As our development of the life course perspective reminded us, elderly Blacks have lived through a period of upheaval and change in the status of persons of color in the United States. Jim Crow patterns of segregation in public establishments have disappeared. Like Miss Jane Pittman, the title character in Ernest J. Gaines's novel, many can remember the first time they drank out of a public water fountain that had previously been restricted to Whites or sat at a previously segregated lunch counter. They remember "help wanted" advertisements in newspapers that searched for "a young White woman" or a "strong colored boy." For many elders of color, dreams that their children would achieve professional occupational status became realities. Yet public pronouncements regarding the declining significance of race clash with their everyday experiences and with official statistics regarding race and ethnic differences in levels of poverty, morbidity and mortality, and victimization by crime. . . .

IMAGES RELATED TO GENDER

Satchell Paige, who would never reveal his age but was probably the oldest man ever to play professional baseball, recognized old age as a social construct when he asked, "How old would you be if you didn't know how old you was?" For women in our society, the answer might well be "older than a man my age." Women are viewed as old at least a decade sooner than are men, and old age brings greater loss of status for women than for men. These gender differences in the social construction of age can be attributed to the emphasis on youth within our culture and to the association between youth and sexuality, especially for women (Andersen, 1997):

> A man's wrinkles will not define him as sexually undesirable until he reaches his late fifties. For him, sexual value is defined much more in terms of personality, intelligence, and earning power than physical appearance. Women, however, must rest their case largely on their bodies. (Bell, 1989, p. 236)

Given this double standard of aging, it is no wonder that many women are reluctant to reveal their ages, try to "pass" as younger than they are, and are complimented when told they don't "look their age" (Bell, 1989).

Although men also face negative stereotypes and loss of status as they get older, these experiences occur at a more advanced age. Compare, for example, the complexion of the female model who plans to fight aging "every step of the way" by using a popular moisturizer to that of the male model who touches up his graying hair. Negative consequences of aging for men focus more on occupational success than physical attractiveness (Andersen, 1997).

Negative though cultural images appear, older people do not internalize them uncritically. Older women retain an image of themselves grounded in the past, a sense of self that stays the same despite physical changes (Kaufman, 1986). . . .

Dean Rodeheaver and Joanne Stohs (1991) argue that such misperceptions are adaptive strategies that help older women overcome the effects of negative images of aging. Indeed, a survey by Ronald Goldsmith and Richard Heims (1992) found that 73% of people in their 60s, 83% of people in their 70s, and all of the respondents in their 80s thought they looked younger than their chronological age; no one in their 20s thought they looked younger than their chronological age.

Older people can diminish the impact of negative images by reinterpreting them from the standpoint of their own experiences. . . . For today's older women, who were told that happiness came from fulfilling the needs of people they love, old age can be a time of autonomy, a time for "rediscovering and recreating themselves" (Thone, 1992, p. 63).

IMAGES RELATED TO RACE/ETHNICITY

Images of older people also vary by race and ethnicity. In some cases, these images reflect stereotypes based on the group in general, whereas in other cases, they apply primarily to old people.

Classic descriptions of old age in American Indian cultures emphasized the respect accorded elderly people. Councils of elders were the centers of political decision making, and elders were responsible for transmitting culture across generations. The anthropologist Joan Weibel-Orlando (1989) studied contemporary characteristics of successful aging by interviewing American Indian elders. She identified six criteria, all of which reflected an active engagement in ethnic life. American Indian elders who ranked high on successful aging were involved in charitable activities with a wide range of kin, friends, and other coethnics. They evidenced a high level of involvement in community roles (e.g., head dancers at powwows, song and prayer leaders at sun dances, medicine men and women, tribal council members, family heads, and caretaking grandparents), roles for which elderly American Indians are revered and that provide service to the ethnic community. Good health in late life was also associated with successful aging. . . .

Older Blacks have confronted multiple negative images throughout their lives. Black women have been objectified as mammies, matriarchs, welfare recipients, and hot mommas (Collins, 1990), whereas Black men have been stereotyped as lazy workers, unstable husbands and fathers, and dangerous criminals. These controlling images are not the province of isolated individuals. Racist images have permeated political discourse. Public debates on Black families, for example, often use value-laden terms, such as "disorganization, maladjustment, and deterioration" (Andersen, 1997). . . .

Stereotypical images also emerge from the popularization of research results. Studies of older minority populations often involve statistical comparisons of people of color with Whites. Readers are shown, for example, that the average income of older Blacks is lower than the average income of older Whites, that the percentage of elderly Hispanic Americans living below the poverty level is higher than the percentage of elderly Whites, or that the probability of living in substandard housing is higher for American Indians than for older Whites. These statistical comparisons are accurate, and analyses of this type have been used successfully to justify allocating government dollars for public programs addressing the plight of low-income minority elders. Comparisons of group means, however, tell only part of the story. To fully comprehend the situation of any group of older people, we also need to consider differences within that group. . . .

Although some older Black women are surrounded by supportive children and extended kin, others are isolated or are themselves caring for grandchildren whose parents are struggling with drug addiction or economic survival. Many older Black women are destitute, but others enjoy considerable affluence. Although their risk of poverty is considerably higher than it is among either White women or Black men, many older Black women live above the poverty line. And although some older Black women lack the educational resources to negotiate medical and social welfare organizations, others pursue careers as nurses, social workers, lawyers, or physicians. . . .

Stereotypes need not always be negative in content to be detrimental in their consequences. The "model minority" myth applied to Asian Americans implies that unlike other people of color, Asians have "made it," that their diligence and hard work have been rewarded by economic success. This myth has several outcomes. First, like the composite of the "average Black woman," the image of the model minority ignores variation both among Asian American nationality groups in the United States and between generations within groups, thus obscuring the economic hardship and cultural isolation that many older Asian Americans experience. Studies reporting higher incomes among Asian American households often fail to adjust for the greater prevalence of multiple wage earners in Asian households and for the higher cost of living in geographic regions with high concentrations of Asian Americans (Woo, 1989). Second, they fail to report the situation of elderly households with low incomes, an omission that highlights again the importance of dispersion. The myth also overlooks the isolation experienced by recent elderly immigrants from Southeast Asia, who feel isolated not only by a different language and way of life but by the acculturation of their children and grandchildren. . . . Last, the model minority myth reinforces the ideology of a meritocracy and creates divisions among racial and ethnic groups. Claims that Asian Americans have succeeded because of values stressing hard work in school and career imply that the stratification system in the United States is open and fair and that other groups could be equally successful if only they had appropriate values.

As we saw with negative images based on gender, older people do not incorporate ethnic stereotypes uncritically. The combination of controlling images and a system of racial oppression taught minority elders the necessity of reflecting dominant images in many contacts outside their own communities. Although this behavior may have reinforced dominant group stereotypes, it did not necessarily mean that people of color had internalized negative stereotypes. Public displays of compliance with expectations of the dominant group often coexist with alternative conceptions of self. . . .

Resistance to controlling images and assertions of alternative definitions of self have been important themes in writings by Blacks. Patricia Hill Collins (1990) explains that these alternative definitions have "allowed Black women to cope with and, in most cases transcend the confines of race, class and gender oppression. . . . Most African American women simply don't define ourselves as Mammies, Matriarchs, welfare mothers, mules or sexually denigrated women" (p. 93). This process of reclaiming the power of self-definition is a major component in the struggle to maintain personal integrity in a climate of racism. . . .

IMAGES RELATED TO SOCIAL CLASS

Images of older people have always been tied to social class, but our cultural images of poverty and affluence in old age have changed in recent years. Along with poor health, poverty is one of the most dreaded aspects of old age. Images of destitute old men subsisting in skid row hotels or of "shopping bag ladies" carrying their lifetime of accumulated possessions in grocery carts have replaced images of the county poor farm in popular representations of poverty in late life. People today are less likely to blame older people for low economic status than they were earlier in this century. Pre-Depression beliefs that poverty in old age resulted from laziness and failure to save for the future have been displaced by images of older people as the "deserving poor," images that provide political justifications for income transfers, such as Social Security and medicare. These changes are recent, however, and today's older people grew up with the pre-Depression stereotypes of the elderly poor, stereotypes that blamed the victim. Values emphasizing hard work, thrift, and self-reliance can still undermine self-esteem and discourage acceptance of programs viewed as "welfare" or "poverty" among some old people, particularly those in the oldest cohorts.

Older people today are less likely to be poor than were older people in the past, and the economic situation of the average older American has improved dramatically over the past two decades, particularly in comparison to younger people (Quinn & Smeeding, 1993). . . . Consistent with the growing disparity between the rich and the poor in the general population, the most affluent 5% of Americans over age 70 own 27% of all the aggregate wealth of that age cohort.

The relative improvement in the economic status of the older population has given rise to a new class-related stereotype: the affluent older person who collects unearned and unneeded benefits at the expense of less affluent workers. Older people have also become scapegoats for the burgeoning budget deficit. According to this "intergenerational inequity" argument, elderly people receive too large a portion of public social expenditures, primarily through social security and medical care benefits. Advocates of this approach posited a causal connection between improvements in the economic status of elderly people and increasing poverty rates among children (Cornman & Kingson, 1996). . . .

This new stereotype of the affluent elderly population has also been reflected in advertising. Twenty years ago, old people were absent from advertisements, except those selling products such as laxatives and denture adhesives. Today, advertisements geared specifically to older consumers feature tennis rackets, luxury automobiles, and Caribbean cruises, as well as vitamins, health insurance, and (almost invisible) hearing aids. The models in these ads, although visibly gray (or, more likely, silver haired), are active, physically fit adults enthusiastically pursuing or sagely planning their retirement leisure.

These new marketing strategies provide a welcome contrast to the negative images of old age in past advertising, but they can also deflect attention from the needs of low-income elders and from the precarious economic positions of elders just below the poverty level. Although fewer than 9% of older Americans have incomes below the poverty level, the risk of poverty increases to 25% for elderly Blacks and to 23.5% for elderly Hispanics. Among elderly women living alone, these risks increase to 48% for Blacks and 49% of Hispanics, in comparison to 21% among Europoean Americans (U.S. Bureau of the Census). . . . Social Security has been a policy success story, lifting 15 million elderly Americans out of poverty and relieving adult children of the burden of supporting impoverished elderly parents. The Social Security Administration estimates that without social security, half of American retirees—and 60% of all women—would fall below the poverty threshold. For two thirds of elderly Americans, Social Security is their primary source of income. For 18%, it is their only source. . . .

NOTES

Andersen, M. L. 1997. *Thinking about women: Sociological perspectives on sex and gender* (4th ed.). Boston: Allyn & Bacon.

Atchley, R. C. 1997. *The social forces in later life: An introduction to social gerontology* (7th ed.). Belmont, CA: Wadsworth.

Bell, L. P. (1989). The double standard: Age. In J. Freeman (Ed.), *Women: A feminist perspective* (4th ed., pp. 236–244). Mountain View, CA: Northfield.

Burton, L., Dilworth-Anderson, P., & Bengtson, V. (1991, Fall-Winter). Creating culturally relevant ways of thinking about diversity and aging. *Generations, 15,* 67–72.

Collins, P. H. (1990). *Black feminist thought: Knowledge, consciousness, and the politics of empowerment.* Boston: Unwin Hyman.

Cornman, J., & Kingson, E. 1996. Trends, issues, perspectives and values for the aging of the baby boom cohorts." *The Gerontologist, 36*(1), 15–26.

Goldsmith, R., & Heims, R. (1992). Subjective age: A test of five hypotheses. *The Gerontologist, 32*(3), 312–317.

Hummert, M. L., Garstka, T., Shaner, J., & Strahm, S. 1994. "Stereotypes of the elderly held by young, middle-aged and elderly adults." *Journal of Gerontology, 49,* P240–P249.

Kaufman, S. R. 1986. *The ageless self.* Madison: University of Wisconsin Press.

Quinn, J., & Smeeding, T. (1993). The present and future economic well-being of the aged. In R. Burkhauser & D. Salisbury (Eds.), *Pensions in a changing economy* (pp. 5–18). Washington, DC: Employee Benefit Research Institute.

Rodeheaver, D., & Stohs, J. (1991). The adaptive misperception of age in older women: Sociocultural images and psychological mechanisms of control. *Educational Gerontology, 17,* 141–156.

Rowe, J., & Kahn, K. (1997). *Successful aging.* New York: Pantheon.

Schmidt, D., & Boland, S. (1986). Structure of perceptions of older adults: Evidence for multiple stereotypes. *Psychology and Aging, 1,* 255–260.

Speas, K., & Obenshain, B. 1995. *AARP Images: Aging in America: Final Report.* Washington, DC: American Association of Retired Persons.

Thone, R. R. (1992). *Women and aging: Celebrating ourselves.* New York: Haworth.

Weibel-Orlando, J. (1989). Elders and elderlies: Well-being in Indian Old Age. *American Indian Culture and Research Journal, 13*(3–4), 149–170.

Woo, D. (1989). The gap between striving and achieving: The case of Asian American women. In Asian Women United of California (Eds.), *Making waves: An anthology of writings by and about Asian American women* (pp. 185–194). Boston: Beacon.

DISCUSSION QUESTIONS

1. What are some typical portrayals of older Americans in the media? How are these images presented with regard to gender, race, and social class?

2. With a growing number of older Americans, how can the society develop an improved social system that cares for the elderly?

INFOTRAC COLLEGE EDITION

You can use your access to InfoTrac College Edition to learn more about the subjects covered in this essay. Some suggested search terms include:

ageism

life course perspective

model minority

retirement

50

From Stepping Stones
to Building Blocks

KATHLEEN SLEVIN AND C. ROY WINGROVE

In this essay, Kathleen Slevin and C. Roy Wingrove provide interview data to illustrate the success of older professional African American women. Their study of fifty retired professional African American women reveals a very different picture of older Black women than common stereotypes hold. The respondents in this study talk also about how they handled race and sex discrimination during their working years to achieve life satisfaction in their retirement years.

We met and interviewed [fifty retired professional African American] women in the summer and fall of 1993 and the spring of 1994. Inclusion in the study required that a woman be African American and a retired professional career person. We began the process with the help of two associates, both prominent African American women who live in different cities in Virginia. They understood the nature of our study and identified women who met our criteria. . . .

Our initial telephone contacts were met with enthusiasm. (Only one woman refused an interview.) All interviews were conducted in the homes of the women. The interviews averaged three hours, but some lasted five hours. With the exception of three women who were uncomfortable being recorded, all interviews were taped. At the close of each interview, we asked for additional names of retired professional or business women who might agree to be interviewed. Using this snowball approach, we continued interviewing until our sample size reached fifty. . . .

Although at the time of interview the women ranged in age from fifty-three to eighty-seven, only three were in their fifties. Thirty-six women were aged sixty-five or over, and the average age was sixty-nine. Only one self-employed businesswoman in our group lacks any college experience. Twenty-four women hold master's degrees, and nine have earned doctorates. Their preretirement profile shows that eight retired as college professors, and twenty-eight had held a variety of teaching/supervisory positions in elementary or secondary education. The remaining fourteen represented careers in state or federal government, medicine, nursing, public service, and

From: Kathleen Slevin, and C. Roy Wingrove. 1998. *From Stepping Stones to Building Blocks: The Life Experiences of Fifty African American Women.* New York: New York University Press, pp. 1–9, 123–125, 151–159. Reprinted with permission.

self-employment. Most of them had married, and twenty-five still live with their spouses. A majority of the seventeen widows and the six who are divorced or separated live alone, but seven of them share homes with a relative or companion. . . .

The stories these fifty women told transported us to the earlier decades of this century as they related how they came of age during the era of legal segregation. They told us what it was like to be educated and to work in both segregated and integrated worlds. They shared with us what it was like to retire from the world of paid employment, and they spoke of their joys and concerns in dealing with advancing years and what they liked and disliked about being older women. It became apparent that they are role models and sources of inspiration for women of *all* races. The ways in which they navigated challenges and continue to live active and productive lives can be instructive for *all* people. In fact, these Black women personified independence, strength, and a "can do" approach to life long before these traits were legitimized for White women by the women's movement. . . .

As well as providing a legacy of survival and resistance, these women point the way to creating a positive sense of self as older women in a society that devalues such women. Whatever their hurdles in younger life, whatever the challenges of growing old, these women lead lives that are graphic testimonies to successful aging. Thus, these women provide a rich legacy to future generations of women—Black and White—who will leave the workforce after a lifetime of continuous employment and who look for guidance in how to live productive and happy lives as they grow old.

The lives of these Black women also dispel the monolithic image of older African American women as being poor, uneducated, and sick inhabitants of blighted inner-city areas. Their individual lives taken collectively demand—in a way that statistics cannot—that we acknowledge the diversity that exists among older Black women and that we recognize the richness, complexity, and depth of their experiences. They are living proof that at least some elderly Black American women are doing quite well financially and socially and that not all of them experience old age ravished by illness and disability. . . .

Despite their privilege, these retired professional African American women lead lives still tainted by oppression. It is precisely their status as privileged African American women that provides us the opportunity to illustrate—in a manner that could not be achieved by looking at the lives of poor women—the tensions between privilege and power, on the one hand, and the oppression caused by the interacting forces of racism and sexism, on the other. . . .

After they had put in so many decades of hard work and struggle, we were gratified to hear the women say that life in retirement is good. They told us what retirement means to them and about the sense of freedom it brings. They told us about the financial and social preparations they made and how those preparations and plans are paying off. They spoke of their hopes and dreams for the future and how they are tempered by their concerns and anxieties about the present. We learned how these lives, so rich in experiences, are now being shared with others, especially younger generations, as these women quietly go about leaving their legacies of survival and resistance. Their work in the church, schools, and other community organizations leaves them little time to complain about growing old or to bemoan their personal losses. As they shared their accounts, we heard echoes of recurrent themes that have become a part of the fabric of their lives, and we noted again how their personal biographies are intimately linked to a particular social and historical context. And, of course, race, class, and gender remain today, as in the past, significant forces as their life stories continue to unfold. . . .

Although many of the women might characterize sexism as less virulent in its consequences than racism, patriarchy has also been a force to be reckoned with in their lives. In truth, the intersection of race and gendered systems of inequality created, and continues to create, a location for these African American women that places them at a disadvantage compared to similarly educated White men or, indeed, White women. Thus, their successes are particularly noteworthy because they were so hard earned and because they must be viewed through the prism of multiple jeopardies. . . .

When we first met the women of our study, they had already reached maturity, and they had come to appreciate their successes and to feel good about themselves and their lives. The path to such a positive sense of self has been long and often difficult. The confluence of many factors helped to establish their patterns of achievement. For example, historical circumstances created a different form of gender asymmetry for African Americans than for Whites. Because African American men faced barriers that denied them access to the same employment and wage opportunities that White men enjoyed, African American women did not have the luxury of relying on men as their sole source of livelihood. Even if African Americans subscribed to the White society's "cult of domesticity" that relegated women to the home, a racialized economic reality denied them that choice. Instead, African American women learned very early in life the essential lessons of self-sufficiency and economic independence. Thus, lack of choice born from racist structures created a major strength in these women. They experienced "forced liberation" from a least some of the restrictions of conventional gendered norms. . . .

By most measures, the women we interviewed are living proof that it is possible for women who attain relative privilege—in terms of their prior location in the labor market—to retire from the world of paid labor with resources adequate to ensure a satisfactory quality of life. Their quality of life is also intimately connected to their good health and to the fact that they are active in retirement.

We believe that the aging process has been largely positive for these women "because of," rather than "in spite of," their race and gender. Their professional pursuits were shaped by their being African American and female. Almost all were involved in education or in serving the needs of others. The race uplift work that was embedded in the Black communities of segregation became second nature to them. Thus, much of their paid labor of the past is connected to their retirement pursuits of today. The unpaid work through which they make major contributions to young and old provides them a major outlet for their talents and interests. It also allows them to feel positive about their activities. . . .

There are two final elements to these women's lives in retirement that deserve mention. First, a lifetime of living and of meeting multiple challenges has given them the wisdom, "the wealth of experiences," as one described it, that accompanies aging. This wisdom was a central theme in our discussions of being old, and specifically of being old woman. "I have sound advice and experiences that can be shared," was the way one retiree described her wisdom. Another woman used words that made clear that she had legacies to share with young people: "I have a lot of experiences to look back on and a lot to offer to young people." And then some women thought specifically in terms of what their earned wisdom meant for young African American women: "[I like] the experiences that I've had that I can share with younger women in order to help them cope because the hassles of life are behind me."

Finally, their life stories would be incomplete if we did not acknowledge the joie de vivre that characterizes their approach to living. Enjoying life was a recurrent theme in their discussions of aging and retirement. They engage in numerous activities that allow them to have fun, to feel vital, and to remain engaged. . . .

Being independent, positive, and forward looking, active in service to others—all are key elements in making the aging process a positive one for the women in this study. As African American women, they have lived long, rich lives that are testimony to their indominable spirit and to their "can do" attitude toward life as a double minority in American society. They live life to the fullest, and their regrets are few and far between. "Life has been good," as many of them testified. . . .

DISCUSSION QUESTIONS

1. Slevin and Wingrove see African American professional women as providing a model for others in retirement. What can White women and men learn from the experience of these women?

2. How does the image of elderly Black women presented in this essay contradict your image of elderly Black women? How were your preconceived notions of older African American women influenced by socioeconomic class assumptions?

INFOTRAC COLLEGE EDITION

You can use your access to InfoTrac College Edition to learn more about the subjects covered in this essay. Some suggested search terms include:

"cult of domesticity"   patriarchy
labor force participation   retirement
life course perspective   snowball sampling

51

Social Security and the Myth of the Entitlement "Crisis"

JILL QUADAGNO

Jill Quadagno presents the financial and political picture of Social Security and the "crisis" over entitlement. She explains public disillusionment with government and the inaccurate arguments against Social Security, Medicare, and Medicaid. She challenges the picture of "greedy entitlements" and an "unsustainable future" and briefly outlines proposals for improving the system.

Although public opinion surveys indicate high support for Social Security, confidence in the viability of the program has declined continuously. A 1986 survey by Cook and Barrett found that 96.7% of respondents favored maintaining (40%) or increasing (57.3%) Social Security benefits (Cook & Barrett, 1992). Similarly, 84% of the respondents to a 1994 poll by the American Association of Retired Persons (AARP) said Social Security benefits were "very important," and 88% opposed cutting benefits to reduce the federal deficit (AARP, 1994). By 1993, however, only 30% of the public felt confident that Social Security benefits would be paid throughout their retirement. Lack of confidence is especially low among young people (Friedland, 1994; Marmor, Mashaw, & Harvey, 1990).

The disparity between support and confidence partly reflects a pervasive distrust of government. . . . Trust in government has steadily eroded from 75% in 1958 to only 19% by 1994 (Skocpol, 1995). According to a 1993 Gallup survey, 88% of the public believe that the federal government routinely mismanages money, 80% that many elected officials are dishonest, and 70% that government employees are dishonest. If people believe that government is incompetent and government officials corrupt, it's not surprising that 81% believe that fraud and waste in the Social Security system will reduce (their) retirement benefits" (Friedland, 1994).

Public apprehension about Social Security's long-range viability reflects more than general distrust in government. It also reflects confusion generated by a public dialogue about such technocratic issues as the integrity of the trust fund as well as broader ideological questions about equity between generations. This dialogue has undermined public faith in social insurance and made tenable discussion of such radical options as means-testing and privatization.

From: *The Gerontologist* 36 (1996): 391–399. Reprinted with permission.

The most recent debate concerns an entitlement "crisis." The entitlement crisis combines the theme of generational equity with dire predictions about the deficit, the erosion of family income, and the future of the economy. This article first identifies two core themes of the entitlement crisis, that entitlement spending is crowding out discretionary spending and that current trends cannot be sustained. It questions the substantive basis of these themes and then critically evaluates two proposals for the restructuring of Social Security: means-testing and privatization.

SOCIAL SECURITY AND THE ENTITLEMENT "CRISIS"

In the summer of 1994, the American public was bombarded with news of an entitlement crisis. It began in the House of Representatives with an "A to Z" spending cut plan for across-the-board cuts in all entitlement programs. It gathered momentum with media reports of progress by the Bipartisan Commission on Entitlement and Tax Reform. Yet, until 1994, few knew what an entitlement was.

Part of the confusion arises from the range of definitions associated with the term *entitlement*. It has a legal meaning, a theoretical meaning, and a budgetary meaning. The concept of an entitlement grew out of the "new property" movement in legal thought in the 1960s when the courts ruled in regard to Aid to Families with Dependent Children that social welfare benefits were not gratuities that could be denied at will. Rather, according to the court decision, "beneficiaries have something akin to property rights in them and therefore have a right to due process in their distribution" (Weaver, 1985, p. 308). In the legal sense, then, the term *entitlement* confers a right to benefits.

The theoretical definition of entitlements emphasizes their distinction from means-tested programs in regard to how benefits are distributed and in terms of what their objectives are. Social Security is an entitlement because people obtain eligibility based on prior work history and because it is designed to maintain pre-retirement living standards. This differentiates it from means-tested programs where eligibility is determined by income and where the objective is to provide a minimal income floor (Marmor et al., 1990). As Esping-Anderson (1990, p. 85) notes in his discussion of the origins of social insurance programs, "The formula was to combine universal entitlements with high earnings-graduated benefits, thus matching welfare-state benefits and services to middle-class expectations."

In the purely budgetary sense, however, what distinguishes entitlements from other programs is that they are governed by formulas set in law and not subject to annual appropriations by Congress (Congressional Budget Office [CBO], 1994). This latter meaning has become the sole definition in the construction of the entitlement "crisis." Entitlements stand in distinction to two other federal budget categories, discretionary spending, which includes domestic and defense spending, and net interest on the debt. In the federal budget there are more than 100 programs defined as entitlements, the three largest being Social Security, Medicare, and Medicaid. . . .

Greedy Entitlements

The share of federal spending devoted to entitlements increased from 22.7% in 1963 to 47.3% by 1993. This message—that entitlements are crowding out spending for domestic programs—derives from the rising deficit, which doubled from 1981 to 1985 from $784 billion to $1,499 billion (U.S. House of Representatives, 1994). . . .

The implication is twofold. First, both entitlements and discretionary spending cannot continue to increase because they will drive up the deficit. Second, even if discretionary spending is drastically cut, no deficit reduction will occur because of wasteful entitlement spending.

There are two problems . . . The first is that entitlement spending has not experienced explosive growth, but rather, has been stable for more than a decade. The second is that shrinking discretionary spending is not the result of entitlement growth, but rather of tactics by conservatives to reduce the welfare state. . . .

The more valid measure of expenditure growth, that used by most economists and in all government documents, is the percent of gross domestic product (GDP). By this measure, entitlement spending has shown almost *no* growth; it was 11.3% of GDP in 1976 and 11.9% in 1994 (CBO, 1995). Social Security, the real target of this charge of rampant growth, has also remained steady at just over 4% of GDP since 1975. It will remain at this level until 2010, when it will rise by 2% of GDP as the baby boom generation retires. If there is an entitlement crisis due to rapid growth, it ended in 1975. . . .

Proponents of Social Security have sought to protect the program by weakening possibilities for program cuts. Programmatically, this tactic has succeeded but ideologically, it has provided critics with a new critique. The ensuing debate has contributed to public confusion over the status of the trust fund and further undermined confidence in the program.

The Unsustainable Future

With no immediate crisis in sight, conservative critics of Social Security have used the salable message of protecting the American dream for the next generation to describe a future crisis. This crisis' is the "finding" that "current trends are not sustainable." The "finding" is based on 30-year projections for Social Security, Medicare, Medicaid, federal employee's retirement benefits, and more than 100 other entitlements. The accuracy of the findings depends on the validity of the projections.

The model for such long-term projections is Social Security, which itself is subject to a variety of inevitable inaccuracies as well as political manipulation. The Social Security actuaries yearly make projections about the long-range solvency of the Social Security trust fund (Board of Trustees, 1994). These projections are based on assumptions about future economic and demographic trends. Their purpose is to estimate whether the system's resources and expenditures are somewhat aligned (Koitz, 1986).

According to present projections, the Social Security trust fund will be insolvent by 2031 (Koitz & Kollmann, 1995). These long-term projections have now become the grist for a new crisis, which combines fears about the economy with concern for future generations. As the National Taxpayers Union warns, we face "a huge financing gap that must be closed if tomorrow's promised benefits are to be paid at all" (Howe & Jackson, 1994)...The problem with this message is that a crisis 30 years in the future may undermine confidence in the program, but it is unlikely to create sentiment for change, especially when experts explain that modest changes would restore the Social Security system to long-range actuarial balance. With a Social Security crisis as a dubious political weapon, critics have instead used the estimates as a base for a larger entitlement crisis, an unsustainable future. The severity of the crisis depends on the accuracy of the projections. . . .

Although the entitlement crisis lacks a factual basis, the charts and measures used to define it have not only become embedded in public debates, they have also become accepted estimates in official government documents. . . .

Through most of the twentieth century, the image of a trust fund ensured beneficiaries that their payroll tax contributions were held in a separate account autonomous from other less trustworthy government activities. Over the past 20 years, this image has withered. Long-term projections of trust fund insolvency have become a potent symbol in the entitlement crisis.

PROPOSALS FOR RETRENCHMENT

Although the entitlement "crisis" has thus far had no programmatic impact, it has undermined further confidence in Social Security. As a result, critics have been able to make proposals for radical cuts that would have been unheard of even five years ago. Proposals now receiving serious attention include means-testing benefits and privatization.

Means-Testing

Recently, means-testing Social Security benefits has been proposed as a way to reduce the deficit and relieve younger workers of an unfair tax burden. Advocates of means-testing argue that paying benefits to wealthy older people can no longer be justified when the tax burden is borne by low income young workers. Means-testing is fair, they contend, because it would target benefits to those most in need. . . .

Privatization

Another option is the privatization of a portion of payroll taxes. In 1995, two prominent senators, Robert Kerrey (D-NE) and Alan Simpson (R–WY), introduced the Personal Investment Plan Act of 1995. As a revised version of a proposal in the *Final Report* of the Bipartisan Commission Entitlement and Tax Reform, the Act would allow workers the option of diverting 2% of their payroll taxes to their own personal investment plans. Employees would be allowed to invest their contributions either into an investment fund or into an Individual Retirement Account (IRA). The objective is to increase the national savings rate. . . .

Both means-testing and privatization would fundamentally alter the core feature that makes Social Security a program of social insurance—the redistribution of income—by providing higher income workers an alternative set of incentives. In the case of means-testing, higher income workers would no longer view their payroll taxes as contributions because they would receive no benefits. In the case of privatization, higher income workers, especially younger workers, would be most likely to opt out, and it is their taxes that subsidize the higher replacement rates of lower income workers. Both options would further undermine the moral framework that has sustained public support for Social Security by fracturing solidarity along lines of class and of generations.

CONCLUSION

In an era of mass-mediated political change, power struggles are not merely a matter of who gets what, but also of who defines what. The United States is presently engaged in a power struggle over the parameters of welfare state restructuring. Thus far, the battle has been waged at the ideological level. Although public support for Social Security remains high, confidence has declined, because discussion of options for reform have been narrowly circumscribed around budgetary issues. Other definitions of Social Security, as a program that provides an earned benefit, as a program that maintains preretirement living standards, and as a program that protects families over the life course have become extraneous to ongoing debates.

This is not to imply that the budgetary issues have no relevance. The debate over entitlements does have a material base, which is grounded in declining economic growth and declining family incomes. For 100 years, from 1870 to 1972, the American economy grew at an annual rate of 3.4% after inflation. Then between 1973 and 1994, it grew only 2.3% a year. Recent estimates of the cost of slow growth is a $12 trillion dollar loss of goods and services produced by the economy (Maddison, 1991).

The causes of declining economic growth are complex and subject to much debate among economists. Among the explanations proposed are that foreign competition has reduced the domestic market for mass-produced goods, that the offshore growth of U.S.-owned multinational corporations has helped to erode the domestic wage base and reduce employment, and that rising deficits have increased interest rates and reduced investment capital. Regardless of the cause, the consequences have been damaging to workers and families. In the past 20 years, average wages have fallen for most categories of workers; the poverty rate, even for those who work full time, has increased; and incomes for even the best educated have grown more slowly than in the past (Maddison, 1991, pp. 50–53).

These trends are not distinct to the United States but are occurring in most Western, capitalist democracies. They raise legitimate issues about how to protect vulnerable families and how to enhance economic growth. Yet, when discussion of these issues becomes absorbed into an entitlement "crisis," possibilities for rational problem solving dissipate. As *Washington Post* columnist Robert Samuelson (1996) recognizes, "(I)n politics words do matter. They help create a climate of opinion. The abuse of language subverts reasoned debate. Exaggeration, simplification and distortion are normal parts of political debate. But the more these excesses are compounded, the harder discussion becomes" (p. 5). Social constructions like an entitlement crisis thwart reasoned public discussion of social needs and legitimate options for responding to those needs.

NOTES

American Association of Retired Persons. (1994). *Public opinion on entitlement programs* (Research Report from AARP Research Division). Unpublished manuscript.

Board of Trustees of the Federal Old Age and Survivors Insurance and Disability Insurance Trust Fund. (1994). *Annual report.* Washington, DC: U.S. Government Printing Office.

Cook, F. L., & Barrett, E. J. (1992). *Support for the American welfare state: The views of Congress and the public.* New York: Columbia University Press.

Congressional Budget Office. (1994). *The economic and budget outlook: Fiscal years 1995–1999.* Washington, DC: U.S. Government Printing Office.

Congressional Budget Office. (1995). *The economic and budget outlook, fiscal years 1996–2000.* Washington, DC: U.S. Government Printing Office.

Esping-Anderson, G. (1990). *The three worlds of welfare capitalism.* Princeton, NJ: Princeton University Press.

Friedland, R. (1994). *When support and confidence are at odds: The public's understanding of the Social Security program.* Washington, DC: National Academy of Social Insurance.

Howe, N., & Jackson, R.(1994). *Entitlements and the aging of America.* Washington, DC: National Taxpayers Union Foundation.

Koitz, D. (1986). *Social Security: Its funding outlook and significance for government finance.* Washington, DC: Library of Congress.

Koitz, D., & Kollmann, G. (1995). *The financial outlook for Social Security and Medicare.* Washington, DC: Congressional Research Service.

Maddison, A. (1991). *Dynamic forces in capitalist development.* New York: Oxford University Press.

Marmor, T., Mashaw, J. L., & Harvey, P. (1990). *America's misunderstood welfare state.* New York: Basic Books.

Samuelson, R. J. (1996, January 8). You call this a revolution? *Washington Post National Weekly Edition*, p. 5.

Skocpol, T. (1995, January 24). *Why it happened: The rise and resounding demise of the Clinton Health Security Plan.* Paper presented at the Brookings Institution, conference on The Past and Future of Health Reform, Washington, DC.

U.S. House of Representatives. (1994). *Overview of entitlement programs, 1994 green book.* Committee on Ways and Means, 103d Congress, 2d Sess.

Weaver, K. (1985). Controlling entitlements. In J. Chubb and P. E. Peterson (Eds.), *The new direction in American politics* (pp. 307–341). Washington, DC: Brookings Institution.

DISCUSSION QUESTIONS

1. What is your understanding of Social Security benefits? How confident are you in the U.S. government's ability to provide for you in your retirement?

2. What is your opinion about entitlement? Do you believe your Social Security taxes taken from each paycheck should assure you benefits in retirement? Should Social Security be means-tested and/or privatized?

INFOTRAC COLLEGE EDITION

You can use your access to InfoTrac College Edition to learn more about the subjects covered in this essay. Some suggested search terms include:

generational equity
gross domestic product
means-testing

Medicaid/Medicare
privatization
Social Security

52

In the Name of the Family

JUDITH STACEY

The nuclear family evolved as the result of specific historical transformations, referred to here as modernization. Family patterns of male breadwinners and dependent women emerged in a particular context and are now being transformed by demographic and postindustrial trends. Yet, as Judith Stacey argues, nostalgia for families of the past often shapes family policies, leaving the society with inadequate supports for the diverse family forms that actually exist.

In most of Europe and North America the family has become nearly synonymous with the nuclear household unit made up of a married, heterosexual couple and their biological or adopted children. Although popular usage more fluidly adapts the concept to refer to all people related through blood marriage, or adoption, most Westerners do erroneously associate the family with nature and project it backward into a timeless past.

It is important to recognize . . . that the family is a product . . . of long historical transformations, generally referred to as modernization. Indeed, many historians employ the concept of the modern family, to describe the particular domestic arrangements which the family has come to designate. The modern family in the West developed historically out of a patriarchal, premodern family economy in which work and family life were thoroughly integrated. In the United States, the modern family system arose in the nineteenth century when industrialization turned men into breadwinners and women into homemakers by separating paid work from households. Beginning first among white middle-class people, this family pattern came to represent modernity and success. Indeed the American way of life came to be so identified with this family form that the trade union movement struggled for nearly a century to secure for male workers the material condition upon which it was based—the male breadwinner wage. However, not until the mid-twentieth century did significant percentages of industrial workers achieve this access to the male breadwinner nuclear family, and it has always exceeded the reach of the vast majority of African-Americans. Slaves were not allowed to marry and had no parental rights at all, and few African-American households have ever been able to afford a full-time homemaker. In fact, many African-American mothers have worked as domestic workers in the modern-family homes of relatively privileged whites.

The rise of the modern family system spelled the demise of the premodern, family economy which was explicitly patriarchal. Thus, it represented a shift in

From: Judith Stacey. 1997. *In the Name of the Family: Rethinking Values in the Postmodern Age.* Boston: Beacon Press, pp. 38–50. Reprinted with permission.

what sociologist Deniz Kanidyoti has called "patriarchal bargains." In the classical patriarchal bargain, women accept overt subordination in exchange for protection and secure social status. The modern patriarchal bargain sugarcoats this exchange by wrapping it in an ideology of separate spheres and romantic love. In place of premodern marriages, which were arranged, in whole or in part, by parents and kin for economic, political, and social purposes, modern men and women, seeking love and companionship, voluntarily bind themselves for life to the complementary object of their individual desires. Under the guise of a separate but equal division of labor between male breadwinners and female homemakers, women and children became increasingly dependent upon the earnings of men. The nineteenth century gave rise to cults of "true womanhood," celebrating domesticity and maternalism. This generated conceptions of femininity that continue to infuse Western family ideology. The development of analogous doctrines about the "tender years" of young children who need a specifically maternal form of love and care began to undermine earlier legal doctrines, which had treated children as patriarchal property.

U.S. family patterns became more predictable and homogeneous as the modern family system evolved in the nineteenth and twentieth centuries. High mortality and remarriage rates had kept premodern family patterns diverse and complex, but declines in mortality enabled increasing numbers of people to anticipate a normal family life course. By the mid-twentieth century, modern family life patterns, from birth through courtship, marriage, work, childrearing, and death had become so homogeneous, normative, and predictable that the family began to appear natural, universal and self-evident.

Social scientists are rarely impervious to the tacit cultural understandings of their times. During the post–World War II period, family sociologists in the United States developed a theory of family modernization that was rooted in the conviction that U.S. family history would prove to be a global model. Arguing that the modern nuclear family was ideally suited to support the functioning of industrial society, and that it was both a product of and handmaiden to Enlightenment progress and democracy, social scientists predicted that it would spread throughout the modernizing world. A product of Western cultural imperialism, the family modernization thesis presumed that the superiority of Western cultural forms would insure their eventual triumph over the "backward" nations and peoples of the globe. Indeed some family scholars came to argue that the early development of the modern nuclear family in the West facilitated the Western supremacy in developing capitalism.

So convinced have Western governments been of the superiority of their family patterns that they have often imposed their gender and family patterns on conquered peoples. The United States, for example, disrupted matrilineal and extended kin systems among several indigenous New World cultures by awarding land titles exclusively to male-headed, nuclear household units. In a similar fashion, Europeans have destructively imposed nuclear family principles on very different African kinship systems. In the Zambian copperbelt, for example, mineowners ignored and disrupted the actual extended kinship patterns of their workers by distributing benefits only to a worker's wife and children. More often, however, Westerners presumed that the global diffusion of the modern nuclear family system would come about automatically. These rather contradictory ideas about the family—that it is natural and universal, on the one hand, and that it is a sign and agent of Western superiority, on the other—continue to collide in popular and scholarly discourse.

CONTRADICTIONS OF THE FAMILY

We can gain some perspective on contemporary family turmoil by recognizing contradictions inherent in the ideology, principles, and practices of the modern family system, the most glaring of which is the tension between volition and coercion. The ideology of the modern family construes marital commitment as a product of the free will and passions of two equal individuals who are drawn to each other by romantic attraction and complementary emotional needs. However, the domestic division of labor of the modern family system, which made women economically dependent upon male earners, and the subordination of women, both de jure and de facto, provided potent incentives for women to choose to enter and remain in marriages, quite apart from their individual desires. And while men certainly have always enjoyed greater opportunities to pursue their emotional and sexual interests inside and outside of marriage, until quite recently cultural codes and material sanctions led most men to depend upon the personal, emotional, and social services of a full-time homemaker. Political satirist Barbara Ehrenreich has observed that the white middle classes in the United States are likely the only bourgeoisie in history to employ members of their own class as personal servants.

The relative acceptability of the contradiction between egalitarian principles of free love and companionship and inegalitarian forms of material and cultural coercion depended upon the availability and accessibility of a male breadwinner wage. Feminist historians have debated the degree to which working-class wives supported, resisted, or benefitted from the trade-union struggle that men conducted to earn wages sufficient to support fulltime homemakers and mothers. However, no matter who achieved this arrangement, which Heidi Hartmann has called a patriarchal-capitalist bargain negotiated between male factory owners and laborers, it has proven to be quite ephemeral. The majority of industrial workers did not earn enough to support a full-time housewife until the 1950s or 1960s, and soon after they did so, deindustrialization and post-industrialization conspired to eliminate their jobs and erode their earnings.

Thus, instability was written into the genetic code of the modern family system (on the "Y" chromosome), because its sustenance depended upon the wide availability of stable, liveable-wage jobs for men. As that strand of the bargain began to unravel during the 1970s and 1980s, the fragility of the entire gender and family order moved into full view, provoking widespread consternation over "family crisis" throughout advanced industrial societies.

During the past few decades, every developed industrial nation has experienced soaring divorce rates, falling birth rates, and rising rates of unmarried domestic partners, of step- and blended families, and of nonfamily households.

Alarmists who decry family decline in the United States often overlook the transnational character of these demographic trends. A 1977 Viennese study warned that if the rate of increase in European divorce rates during the 1970s were to continue until the year 2000, at that point 85 percent of all European marriages would end in divorce.

During this same period, the employment rates of women and men, formerly quite distinct, began to converge worldwide. Women, especially mothers of young children, now find it necessary to work for pay to support or contribute to the support of families that have been undermined by the loss of jobs and real earnings by men. The loss of steady work, or any work, for men at lower educational levels has been quite dramatic. While more than two-thirds of men with less than a high school education worked full time, year round during the 1970s, a decade later only half could find such steady work. A significant wage gap between men and women persists, but the normalization of female employment and the decline in jobs for men has reduced some of women's economic dependency on men, and thus, has weakened one coercive buttress of marriage.

That is one major reason why single motherhood is rising around the globe, and why increasing percentages of single mothers have never been married. Sitcom heroine Murphy Brown has become a controversial symbol of the family circumstances of a small, but rising number of affluent, professional women in the U.S. who are choosing to become single mothers rather than to forego motherhood entirely. In reality, the vast majority of single-mother families confront dire economic circumstances. At the same time that many women began choosing to become mothers alone, and for related reasons, birth rates were falling below replacement levels throughout the postindustrial world. It is particularly striking that women in Italy, an overwhelmingly Catholic country, now give birth to the smallest national average number of children in the advanced industrial world. On the other hand, birth rates have begun to rise in Sweden, despite its reputation as the leading country for family decline. The comparative level of security and confidence that prospective Swedish parents, particularly would-be mothers, derive from their nation's exceptionally progressive tax structure and social welfare provisions is the most likely explanation for this paradox. Meanwhile, *The New York Times* reports that "Eastern Germany's adults appear to have come as close to a temporary suspension of childbearing as any large population in the human experience," a response to the region's dire economic conditions since reunification. The state of Brandenburg has voted to offer parents a cash incentive of $650 per new child born.

Because global capitalism is governed by the endless search for profits through increased productivity and technological development, we can be certain that our only social constant is change. Social change is a permanent and endless feature of our world, and all we can know about the future of family life is that it too will continue to change. Recent developments in reproductive technology and genetic engineering offer glimpses of some of the most dramatic and radical implications of future family scenarios. *Junior*, a 1994 Christmas season family movie starring Arnold Schwarzenegger as a pregnant experimental scientist, (a movie which proved to be more popular with women than men), presages some of the redefinitions of family life in store as science completes its Faustian gift of separating sexuality, conception, gestation, procreation, marriage, childrearing, and parenting. Pregnant men and test-tube babies, once the standard fare of science fiction, now appear inevitable. We have already reached the point at which a man's sperm can fertilize one woman's ovum, which gestates in the uterus of a second woman, who, in turn, serves as a "surrogate" for yet a third woman, who plans to adopt and rear the offspring, with or without a second man or a fourth woman as co-parent. What and who is the mother, the father, or the family in such a world?

THE POSTMODERN FAMILY CONDITION

The astonishing transformations sketched above indicate that the particular patriarchal bargain of the modern family system has collapsed. Instead, we now forge our intimate lives within the terms of the postmodern family condition described earlier. At the current moment in Western family history, no single family pattern is statistically dominant, and our domestic arrangements have become increasingly diverse. Only a minority of U.S. households still contain married couples with children; and many of these include divorced and remarried adults. More children live with single mothers than in modern families containing a breadwinner dad and a full-time homemaker mom. Most features of the postmodern family condition are most prominent in the United States and Scandinavia. But demographic trends are similar throughout the highly industrialized world, with variations only in the degree, timing, and pace of the changes, but not in their direction. Once the family modernization thesis predicted that all the societies of the globe would converge toward a singular family system—the modern Western family system. Ironically, instead we are converging internationally toward the postmodern family condition of diversity, flux, and instability.

Under postmodern conditions, the social character of practices of gender, sexuality, parenting, and family life, which once appeared to be natural and immutable, become visible and politically charged. While similar demographic trends are dissolving the modern family system throughout the capitalist, industrialized world, national responses to the modern family crisis differ widely. Some societies have adapted to the decline of the male breadwinner family by devising generous social welfare policies that attempt to mitigate some of the destructive impact that marital fragility too often inflicts on children and the unequal burden it places on women. Again the Scandinavian countries, with Sweden and Norway in the lead, set the standards for innovative family support policies of this sort. In both nations, parents of either gender are entitled to apportion a full year's leave with 90 percent pay to take care of a newborn. Because so few fathers availed themselves of this benefit, both Sweden and Norway recently offered them added incentive to do so. Both countries now allow men, and only men, to receive an additional month of paid parental leave beyond the original twelve months, which men and women can allot as they choose. Moreover, Scandinavian workers enjoy

paid leave to care for sick children and relatives, as well as universal family allowances, health care, including sex education, contraception, and abortion services, and subsidized high-quality daycare. There are few deadbeat dads in these
Nordic nations, because the state assumes responsibility for collecting and distributing child care payments. As a result, while more than half of single-parent families in the United States live below the official poverty line, in Sweden only 2
percent do so. Most likely this is why Swedish women have been willing to bear
more children in recent years. Likewise, Sweden and Norway also followed Denmark's lead in legalizing a form of marriage for same-sex couples before this became a visible political issue in the United States.

Other affluent societies, however, have proven far more hostile to postmodern demographic and cultural changes. They are far less willing to assume
public responsibility for addressing the unjust and disruptive effects caused by
these changes. The United States is far and away the most extreme in this regard. Reflecting an exceptionally privatized economy, an individualistic culture, and racial antagonisms, social welfare for the poor in the United States
has always been comparatively stingy, punitive, and unpopular. Yet even this
meager system is currently being dismantled. The United States alone, among
18 advanced industrial nations, does not provide its citizens with universal
health coverage, family allowances, or paid parental leaves. In fact, it was not
until the Family Leave Act of 1993 that the right to take an unpaid three-
month maternity leave, which few families can afford to use, was mandated
for workers in firms with at least 50 employees. Welfare provisions in the
United States have always been means-tested, stigmatized, and niggardly. As a
result, a higher percentage of single-mother families in the United States as
well as a higher percentage of children in general, live in poverty than in any
advanced industrial nation. Conservative estimates of the numbers that current welfare reform legislation will add to this disturbing record have even
frightened Senator Moynihan, one of the original advocates of revising the
welfare system.

While family support policies in the United States are the weakest in the industrial world, no society has yet to come close to our expenditure of politicized
rhetoric over family crisis. The politics of gender, sexuality, reproduction, and
family here are the most polarized, militant, and socially divisive in the world,
precisely because social structural responses to the decline of the modern family
system have been so weak. This is an important reason why feminism, gay liberation, and backlash "profamily" movements are so vocal and influential across the
political spectrum.

Rampant nostalgia for the modern family system, or more precisely, for an
idealized version of a 1950s Ozzie and Harriet image of the family, has become
an increasingly potent ideological force in the United States, with milder versions
evident in Canada and England. Fundamentalist Christians and right-wing Republicans spearheaded the profamily movement that abetted the Reagan "revolution" of the 1980s. By the 1994 electoral season, however, even President Clinton
had embraced the ideology of an explicitly centrist campaign for family values
led by a small group of social scientists. This ongoing campaign portrays family

breakdown as the primary source of social malaise in the United States, blaming the decline of the married-couple family for everything from crime, violence, and declining educational standards to poverty, drug abuse, and sexually transmitted disease.

There seems to be nearly an inverse relationship between a nation's rhetorical concern over the plight of children in declining families and its willingness to implement policies to ease their suffering. This may appear paradoxical, if not hypocritical, but family support policies are consistent with the historical development of public responsibility for social welfare in each nation. They are strongest in parliamentary governments in which labor movements have achieved a significant voice. Lip service to the family, on the other hand, serves as a proxy for the private sphere and as a rationale for abdicating public responsibility for social welfare. Unfortunately, the more individualistic and market-oriented a society becomes, the more difficult it becomes to sustain family bonds. . . .

LET'S BURY "THE FAMILY"

The family indeed is dead, if what we mean by it is the modern family system in which units comprised of male breadwinner and female homemaker, married couples, and their offspring dominate the land. But its ghost, the ideology of the family, survives to haunt the consciousness of all those who refuse to confront it. It is time to perform a social autopsy on the corpse of the modern family system so that we may try to lay its troublesome sprit to rest. Perhaps, a proper memorial service for the family system we have lost can free us to address the diverse needs of people struggling to sustain intimate relationships under very difficult postmodern family conditions.

Adopting the pathologist's stance of hard-hearted, clinical detachment in this case can lead to an uncomfortable conclusion. Historically, all stable systems of marriage and family life have rested upon diverse measures of coercion and inequality. Family systems appear to have been most stable when women and men have been economically interdependent, when households served as units of production with sufficient resources to reproduce themselves, and when individuals lacked alternative means of economic, sexual, and social life. Family units of this sort have always been embedded in, supported, and sanctioned by wider sets of kinship, community, and religious ties. Disturbingly, all such family systems have been patriarchal. The stability of the modern family system, which represented a significant departure from several of these principles, depended upon the adequacy and reliability of the male family wage. However, the ceaseless development of capitalist industrialization, which disrupted the premodern patriarchal bargain, has now disrupted the modern one as well, and it will continue to disrupt postmodern familial regimes of any sort.

DISCUSSION QUESTIONS

1. What historical conditions shaped the development of the nuclear family and how have these changed in contemporary society?

2. Why does Stacey argue that there are contradictions in the ideology, principles, and practices of the modern family and what impact does this have on social policies on behalf of families?

INFOTRAC COLLEGE EDITION

You can use your access to InfoTrac College Edition to learn more about the subjects covered in this essay. Some suggested search terms include:

family diversity
male breadwinner
modern nuclear family

monogamy
postmodern families
single mothers

53

Divorce and Remarriage

TERRY ARENDELL

Increasingly common in family experience, divorce and remarriage are producing new family patterns and family experiences. Terry Arendell reviews patterns in divorce and remarriage, including the impact on children, custody arrangements, and economic consequences of divorce. In addition, she reviews some of the social dynamics found in stepfamilies.

DEMOGRAPHIC PATTERNS IN MARITAL DISSOLUTION AND REMARRIAGE

The divorce rate more than doubled between the early 1960s and mid-1970s. Despite some fluctuations in the annual divorce rates, more than 1 million marriages still are dissolved each year. If trends continue as anticipated, as many as three in five first marriages will end in legal dissolution, as they have since 1980.

From: Terry Arendell, ed. 1997. *Contemporary Parenting: Challenges and Issues:* Thousand Oaks, CA: Sage, pp. 154–195. Reprinted with permission.

Second marriages have a somewhat higher termination rate (Gottman, 1994; Kitson & Holmes, 1992; Martin & Bumpass, 1989).

Most likely to divorce are younger adults in shorter term marriages with dependent children. Indeed, children are involved in approximately two thirds of all divorces (U.S. Bureau of the Census, 1995a), and more than half of all children experience their parents' divorces before they reach 18 years of age (Cherlin & Furstenberg, 1994; Furstenberg & Cherlin, 1991; Martin & Bumpass, 1989). Nearly twice as many black children as white children born to married parents will experience parental divorce if trends persist as expected (Amato & Keith, 1991). Marital separation and dissolution rates among parents in other racial and ethnic groups, which generally have been lower than those among whites and blacks, also are increasing (U.S. Bureau of the Census, 1995b). Children who experience divorce spend an average of 5 years in single-parent homes (Glick & Lin, 1986); even among those whose custodial mothers remarry, about half spend 5 years with their mothers alone (Furstenberg, 1990).

Separation and divorce are not the only transitions in parents' marital status and household arrangements experienced by children. Even though remarriage rates are declining, with only about two-thirds of separated or divorced women and about three-fourths of men likely to remarry compared to three-fourths and four fifths, respectively, in the 1960s, more than one third of adults currently in first marriages will divorce and remarry before their youngest children reach age 18. Thus a high proportion of children will experience the remarriage of the parent, if not both, and the formation of a stepfamily or stepfamilies. Moreover, many children will experience the dissolution of a stepfamily when a parent and stepparent divorce. About one in six children will experience two divorces of the custodial parent before the child reaches age 18 (Furstenberg & Cherlin, 1991). Additionally, increasing numbers of adults, including those who are custodial parents of minor children, are cohabiting. Whether they eventually will marry remains to be seen (see Cherlin & Furstenberg, 1994).

Approximately 1 in 10 children in 1992 lived with a biological parent and a stepparent, and this proportion is expected to increase. About 15% of all children lived in blended families—homes in which children lived with at least one stepparent, stepsibling, or half-sibling. More children lived with at least one half-brother or half-sister than with a stepparent or with at least one stepsibling (Furukawa, 1994). Because the practice of cohabitation, or sharing domestic life and intimacy without legal marriage, is increasing steadily, the number of children who reside with a custodial parent and her or his adult partner, who presumably functions, at least to some extent, as a stepparent—*a quasi-stepparent*—probably is much higher than the official numbers indicate (Cherlin & Furstenberg, 1994, pp. 363–365). Because the large majority of children whose parents divorce live with their mothers, most residential stepparents and quasi-stepparents are men. Census data for 1991 show that among children in single-mother families (which includes never-married as well as divorced), 20% also lived with an adult male (related or unrelated) present in the household. About 37% of children living with a single father also lived with an adult female (related or unrelated) (Furukawa 1994, pp. 1–2). . . .

DIVORCE

Postdivorce Parenting

With respect to family functioning, marital dissolution can be a lengthy process, often underway years before the actual spousal separation occurs. Children show the effects of marital dissension and discontent long before divorce (e.g., Amato & Booth, 1996; Block, Block, & Gjerde, 1986, 1988; Shaw, Emery, & Tuer, 1993). Adjustment to the changes wrought by divorce itself can be a gradual and lengthy process, and many parents and children enter a "crisis" period after the marital separation that can last for several years (e.g., Chase-Lansdale & Hetherington, 1990; Hetherington, 1987, 1988; Morrison & Cherlin, 1995). Maccoby and Mnookin (1992), in the Stanford Custody Project, concluded that

> divorcing parents find it difficult to take the time and trouble required to negotiate with children over task assignments and joint plans. Under these conditions of diminished parenting, children tend to become bored, moody, and restless and to feel misunderstood; these reactions lead to an increase in behaviors that irritate their parents, and mutually coercive cycles ensue. (pp. 204–205)

A related phenomenon is single parents' lesser ability to make control demands on their children. Examining data from the National Survey of Families and Households, Thomson, McLanahan, and Curtin (1992) found that single parents of both sexes seem to be "structurally limited" in their ability to control and make demands on a child without the presence of another adult.

The extent and duration of uneven parenting, however, varies by families, with some family units adapting fairly rapidly to their altered circumstances and arrangements, achieving stable and healthy family functioning rather soon after divorce. Some units take much longer to find an equilibrium. Others have a delayed reaction, functioning well initially and then encountering adjustment difficulties (e.g., Kitson & Holmes, 1992). In addition, "some show intense and enduring deleterious outcomes" (Hetherington, 1993, p. 40). Whatever the pattern, parental functioning usually recovers over time, returning nearly to the level found in intact families (Hetherington, 1988; Hetherington, Cox, & Cox, 1982). That is, most family units formed by divorce establish workable and functional interactional processes (Maccoby & Mnookin, 1992; Wallerstein & Blakeslee, 1989).

One of the first major tasks facing parents in divorce is that of determining children's living arrangements as family members separate into two households. Most custody decisions occur with little discussion between the parents, and relatively few custody allocations are actually litigated. Yet the working out of parenting and parental relationships after divorce, including children's access to and involvement with the nonresidential parent if parenting is not shared, can be complicated and difficult, involving various changes and intraparental conflicts. Of the four relationships between married persons that must be altered in divorce—parental, economic, spousal, and legal (Maccoby & Mnookin, 1992)—the parental divorce is perhaps the most difficult to achieve (Ahrons & Wallisch, 1987, p. 228; see also Bohannon, 1970).

Custody Arrangements for Minor Children

Three residential patterns are available for children in divorce: maternal, paternal, and dual. Primary physical custody, maternal or paternal, is the situation in which children spend more than 10 overnights in a 2-week period with a particular parent (Mnookin, Maccoby, Albiston, & Depner, 1990, pp. 40–41). Dual or shared custody is defined as the situation in which "the children spend at least a third of their time in each household" (Maccoby & Mnookin, 1992, p. 203). Shared custody is unusual. Even in California, where dual custody probably is more common than anywhere else, only about one in six children actually lives in a shared custody situation. And in these circumstances, "more often than not" mothers handle the bulk of the managerial aspects of child rearing (Maccoby & Mnookin, 1992, p. 269).

As has been the case for most of this century, maternal custody is predominant; more than 85% of children whose parents are divorced are in the custody of their mothers (U.S. Bureau of the Census, 1995a). A somewhat higher proportion of offspring actually reside with their mothers because, in legally mandated dual-custody situations, children often spend relatively little time with their fathers (e.g., Maccoby & Mnookin, 1992; Seltzer, 1991; Seltzer & Bianchi, 1988). Overwhelmingly, then, it is mothers who become the primary parents in divorce. . . .

Economic Support of Children

Although the preseparation parenting division of labor persists after divorce with mothers doing most of the parenting, what does change is the economic providing for minor children. Whereas men's earned incomes provide the larger share of the economic resources available to intact married families, divorced custodial women assume most financial responsibilities for their offspring. The overwhelming body of scholarly research and governmental and other policy studies shows that fathers' contributions to the economic support of their children are much reduced after marital dissolution (e.g., Kellan, 1995; Maccoby & Mnookin, 1992) despite many men's claims to the contrary (e.g., Arendell, 1995). Approximately three fourths of divorced mothers have child support agreements, but only about half of those women receive the full amounts ordered in the agreements (Holden & Smock, 1991; Scoon-Rogers & Lester, 1995). One fourth receive no payment whatsoever, and the other one fourth receive irregular payments in amounts less than those ordered. According to the Congressional Research Service, only about $13 billion of the $34 billion in outstanding support orders was collected in 1993 (Kellan, 1995, p. 27). Moreover, child support payments amounted to only about 16% of the incomes of divorced mothers and their children in 1991. The average monthly child support paid by divorced fathers contributing economic support in 1991 was $302, amounting to $3,623 for the year (Scoon-Rogers & Lester, 1995). Fathers' limited or lack of financial contributions to the support of their children not residing with them is not offset by other kinds of assistance (Teachman, 1991, p. 360).

As a group, women's incomes drop more than 30% following divorce. About 40% of divorcing women lose more than half of their family incomes, whereas fewer than 17% of men experience this large a drop (Hoffman & Duncan, 1988). Men, in general, experience an increase in their incomes—an average of 15%—partially because they share less of their incomes with their children (Furstenberg & Cherlin, 1991; Kitson & Holmes, 1992; Maccoby & Mnookin, 1992). For many women, the financial hardships accompanying divorce become the overriding experience, affecting psychological well-being and parenting as well as dictating decisions such as where to live, what type of child care to use, and whether or not to obtain health care (Arendell, 1986; Kurz, 1995).

Children's economic well-being after divorce is directly related to their mothers' economic situations. Those living with single mothers are far more likely to be poor than are children in other living arrangements; families headed by single mothers are nearly six times as likely to be impoverished as are families having both parents present (U.S. Bureau of the Census, 1995a). This is not the experience of children being raised by single fathers because men's wages are higher than women's (Holden & Smock, 1991; Scoon-Rogers & Lester, 1995; Seltzer & Garfinkel, 1990); about one eighth of custodial fathers, compared to nearly two fifths of divorced custodial mothers, are poor (Scoon-Rogers & Lester, 1995). Divorced women and their children do not regain their predivorce standards of living until 5 years after the marital breakups. Women's decisions to remarry often involve economic considerations; the surest route to financial well-being for many women is remarriage, not their employment, even when it is full-time (Furstenberg & Cherlin, 1991; Kitson & Morgan, 1990). . . .

CHILD OUTCOMES IN DIVORCE

How children fare with divorce is a crucial question, one intimately related to issues of parenting. The arguments vary, with assertions ranging from children being irreparably damaged to children adapting successfully to divorce. Most research evidence suggests that a large majority of children adjust reasonably well to their parents' marital dissolutions. . . .

Some argue that the research findings on the effects of divorce on children are not so clear-cut (see, for review, Bolgar, Sweig-Frank, & Paris, 1995). But even those arguments are tempered when large data sets are the bases of analysis, especially those involving longitudinal studies and not just small, nonrepresentative samples (Amato & Booth, 1996). That a majority of children seem to cope with and adapt well to the change in their parents' marital status is particularly salient because many children enter the divorce phase already disadvantaged by exposure, often of long duration, to parental strife and conflict (Block et al., 1986, 1988; Chase-Lansdale & Hetherington, 1990). Furthermore, as numerous scholars point out, children who experience the dissolution of their parents' marriages may well have to cope with multiple adverse circumstances including family events prior to divorce (e.g., Furstenberg & Teitler, 1994). Allen (1993, p. 47), for example, argued that when scholars (and others) uncritically compare divorced families to nondivorced ones, they imply that two-parent intact families inevitably result in positive parenting outcomes. Other events that might be more detrimental than divorce itself, as she notes, are father abandonment; failure to pay child support; neglect; intersection of class, race, and gender with poverty; and women's inequality in traditional families.

Some earlier findings suggested that a child's sex and age mattered in postdivorce adjustment. Hetherington (1993, pp. 48–49), drawing from recent work, concluded that these variables—sex and age—are not pivotal factors in children's divorce responses and adjustments (see also Furstenberg & Teitler, 1994; Garasky, 1995). Sex differences in adverse responses, previously attributed to boys, disappeared in Hetherington's (1993) longitudinal study as children moved into adolescence. Where age mattered, it was for adolescents, all of whom showed somewhat increased problem behaviors. Children with divorced and remarried parents did show more such problem behaviors than did those whose parents remained married. "Adolescence often triggered problems in children from divorced and remarried families who had previously seemed to be coping well" (p. 49). Furstenberg and Teitler (1994) summarized findings pertaining to adolescents:

> The findings indicate that certain effects of divorce are quite persistent even when we consider a wide range of predivorce conditions. Early timing of sexual activity, nonmarital cohabitation, and high school dropout do appear to be more frequent for children from divorced families. (p. 188)

The researchers note that these outcomes may be a result of growing up in single-parent homes or witnessing parents' marital transitions, among other things, not just divorce itself (p. 188).

Also, in contrast to earlier arguments, being reared by same-sex parents appears not to be inherently beneficial to children (Powell & Downey, 1995). And, although it may seem counterintuitive, children's overall well-being in divorce does not seem related to the extent of involvement or quality of parent-child relationship with the noncustodial father (e.g., Amato & Keith, 1991; Bolgar et al., 1995; Furstenberg & Cherlin, 1991). . . .

REMARRIAGE

Research attention to stepparenting has increased, dramatically in the past 15 years as the divorce and remarriage rates have escalated and remain high. For instance, Coleman and Ganong (1990, p. 925) noted that there were only a handful of studies published prior to 1980 but more than 200 during the decade of the 1980s. The increased attention has continued into the 1990s (e.g., see Booth & Dunn, 1994).

The circumstances leading to the formation of stepfamilies vary. They especially include the marriages of formerly unmarried teen mothers, widowed parents, and divorced ones. Prior to the early 1970s, the death of a spouse was the principal prior circumstance leading a parent to remarry, not divorce as is now the case. Even just among those formed by divorced parents, stepparent families are diverse in composition. For instance, Dunn and Booth (1994, p. 220) noted that two scholars, Burgoyne and Clark (1982), had identified 26 different types. Children may reside with either a stepmother or a stepfather, although the latter is far more common given the preponderance of mother custody. Or, children may have a nonresidential stepparent. Additionally, children may have stepsiblings and half-siblings with whom they may or may not share residences. More specifically,

> somewhere between two-fifths and half of these children [whose parents remarry] will have a stepsibling, although most will not typically live with him or her. And for more than a quarter, a half-sibling will be born within four years. Thus, about two-thirds of children living in stepfamilies will have either half-siblings or stepsiblings. (Furstenberg, 1990, p. 154)

Depending on their cognitive developmental stage, children construct family relatedness with stepsiblings and half-siblings in various ways, adding to the complexity in understanding family relationships (Bernstein, 1988). . . .

The amount of domestic life and parenting shared with nonresidential family members can range greatly between family units and across time for particular children. Variations among stepfamilies occur, moreover, not only in their configurations but also in their functioning.

The remarriage of a divorced parent and creation of a stepfamily entail numerous disruptions and transitions. Altered by the entry of another adult into the family is the family system established by the custodial parent and children following divorce.

. . .Children, and sometimes the custodial parent, often resist a newcomer's efforts to exert authority and alter the existing family dynamics (Hetherington, 1993; Hetherington et al., 1992). Disruption is not limited to the relationship between the stepparent and stepchildren; it can involve the relationship between the custodial parent and children as well. Conflict within the original unit often increases (Brooks-Gunn, 1994, p. 179). Other problems may include a decline in parental supervision and responsibility as the parent divides her time between a new spouse and her children, shifting alliances between family members, and open tension and disputes between children and stepparent and between children of the original unit (e.g., Brooks-Gunn, 1994, p. 170; Hetherington, 1993; Hetherington et al., 1992). Nor are interpersonal tensions and difficulties limited to the residential unit. They may involve the noncustodial parent, his spouse, or other relatives, such as grandparents, aunts, or uncles. Dealing with the larger family context is an ongoing, lengthy, and demanding process (Mills, 1988; more generally, see Beer, 1988).

Some stepparents respond to children's resistance by becoming more authoritarian and dogmatic. Others, on the other hand, withdraw emotionally and cease their attempts to forge intimate relationships. They move to "exhibiting little warmth, control, or monitoring. [These] stepparents are not necessarily negative, they are just distant" (Brooks-Gunn, 1994, p. 179; Hetherington, 1993). Whatever the strategy assumed by stepparents, it has direct impacts on the home ambiance and parent-child relationships. In turn, these all affect the interactional dynamics between spouses; the effects become circular and interactive.

As with other kinds of family transitions, restabilization often follows the initial disequilibrium experienced by the newly formed stepfamily (Ahrons & Wallisch, 1987; Hetherington, 1993; Hetherington et al., 1992). The successful integration of a stepparent into a family is a gradual process, sometimes taking years (e.g., Papernow, 1988, p. 60). Not all families reach such a level; indeed, a large number of stepfamilies dissolve through divorce long before they ever approach the place of becoming smoothly functioning households. . . .

CONCLUSION

In conclusion, a sizable proportion of American children will experience their parents' divorce. The majority of these children will be parented predominantly by one parent, not by both parents. Many of these children also will experience the formation of a stepparent family when a parent remarries. In many families, both parents will remarry, resulting in situations where children have both a live-in and a live-out stepparent. And numerous children will experience another parental divorce. Children, then, are experiencing multiple transitions in the composition and arrangements of their families. Current evidence indicates that the vast majority of children adjust to these changes successfully. What is most crucial in children's well-being and positive outcomes, according to a growing body of research, is the quality and constancy of the parenting by the primary parent. Experiencing relatively low intraparental and other family conflict is crucial for children's adjustment to changing circumstances and positive development.

REFERENCES

Ahrons, C. R., & Wallisch, L. (1987). Parenting in the binuclear family relationships between biological and stepparents. In K. Pasley & M. Ihinger-Tallman (Eds.), *Remarriage and stepparenting: Current research and theory* (pp. 225–256). New York: Guilford.

Allen, K. R. (1993). The dispassionate discourse of children's adjustment to divorce. *Journal of Marriage and the Family, 55,* 46–50.

Amato, P. R., & Booth, A. (1996). A prospective study of divorce and parent-child relationships. *Journal of Marriage and the Family, 58,* 356–365.

Amato, P. R., & Keith, B. 1991. Parental divorce and the well-being of children: A meta-analysis. *Psychological Bulletin, 11,* 26–46.

Arendell, T. (1986). Mothers and divorce: Legal, economic, and social dilemmas. Berkley: University of California Press.

Arendell, T. (1995). *Fathers and divorce.* Thousand Oaks, CA: Sage.

Beer, W. R. (Ed.). (1988). *Relative strangers: Studies of stepfamilies processes.* Totowa, NJ: Rowman & Littlefield.

Bernstein, A. C. (1988). Unraveling the tangles: Children's understanding of stepfamily kinship. In W. Beer (Ed.), *Relative Strangers* (pp. 83–111). Totowa, NJ: Rowman & Littlefield.

Block, J. H., Block, J., & Gjerde, P. F. (1986). Personality of children prior to divorce: A prospective study. *Child Development, 57,* 827–840.

Block, J. H., Block, J., & Gjerde, P. F. (1988). Parental functioning and the home environment in families of divorce: Prospective and concurrent analyses. *Journal of American Academy of Child and Adolescent Psychiatry, 27,* 207–213.

Bohannon, P. (1970). Divorce and after. Garden City, NY: Doubleday.

Bolgar, R., Sweig-Frank, H., & Paris, J. (1995). Childhood antecedents of interpersonal problems in young adult children of divorce. *Journal of the American Academy of Child and Adolescent Psychiatry, 34*(2), 143–150.

Booth, A., & Dunn, J. (Eds.). (1994). *Stepfamilies: Who benefits? Who does not?* Mahwah, NJ: Lawrence Erlbaum.

Brooks-Gunn, J. (1994). Research on stepparenting families: Integrating disciplinary approaches and informing policy. In A. Booth & J. Dunn (Eds.), *Stepfamilies: Who benefits? Who does not?* (pp. 167–204). Mahwah, NJ: Lawrence Erlbaum.

Burgoyne, J., & Clark, D. (1982). Parenting in stepfamilies. In R. Chester, P. Diggory, & M. Sutherland (Eds.), *Changing patterns of child-bearing and child-rearing* (pp. 133–147). London: Academic Press.

Chase-Lansdale, P. L., & Hetherington, E. M. (1990). The impact of divorce on life-span development: Short and long term effects. In D. Featherman & R. Lerner (Eds.), *Life span development and behavior* (Vol. 10, pp. 105–150). Hillsdale, NJ: Lawrence Erlbaum.

Cherlin, A. J., & Furstenberg, F. F., Jr. (1994). Stepfamilies in the United States: A reconsideration. *Annual Review of Sociology, 20,* 359–381.

Coleman, M. & Ganong, L. H. (1990). Remarriage and stepfamily research in the 1980s: Increased interest in an old family form. Journal of Marriage and the Family, 52, 925–940.

Dunn, J., & Booth, A. (1994). Stepfamilies: An overview. In A. Booth & J. Dunn (Eds.), *Stepfamilies: Who benefits? Who does not?* (pp. 217–224). Mahwah, NJ: Lawrence Erlbaum.

Furstenberg, F. F. (1990). Coming of age in a changing family system. In S. Feldman & G. Elliot (Eds.), *At the threshold: The developing adolescent* (pp. 147–170). Cambridge, MA: Harvard University Press.

Furstenberg, F. F., & Cherlin, A. (1991). *Divided families: What happens to children when parents part.* Cambridge, MA: Harvard University Press.

Furstenberg, F., & Teitler, J. O. (1994). Reconsidering the effects of marital disruption: What happens to the children of divorce in early adulthood. *Journal of Family Issues, 15,* 173–190.

Furukawa, S. (1994). The diverse living arrangements of children: Summer 1991. In U.S. Bureau of the Census, *Current Population Reports* (Series P70–38). Washington, DC: Government Printing Office.

Garasky, S. (1995). The effects of family structure on educational attainment: Do the effects vary by the age of the child? *American Journal of Economics and Sociology, 54*(1), 89–106.

Glick, P. C., & Lin, S.-L. (1986). Recent changes in divorce and remarriage. *Journal of Marriage and the Family, 48,* 737–747.

Gottman, J. M. (1994). *What predicts divorce? The relationship between marital processes and marital outcomes.* Hillsdale, NJ: Lawrence Erlbaum.

Hetherington, E. M. (1987). Family relations six years after divorce. In K. Pasley & M. Ihinger-Tallman (Eds.), *Remarriage and stepparenting: Current research and theory* (pp. 185–205). New York: Guilford.

Hetherington, E. M. (1988). Parents, children, and siblings six years after divorce. In R. Hinde & J. Stevenson-Hinde (Eds.), *Relationships within families* (pp. 311–331). Cambridge, UK: Clarendon.

Hetherington, E. M. (1993). An overview of the Virginia Longitudinal Study of

Divorce and Remarriage with a focus on early adolescence. *Journal of Family Psychology, 7,* 39–56.

Hetherington, E. M. & Clingempeel, W. G., with Anderson, E., Deal, J., Hagan, M. S., Hollier, A., & Lindner, M. (1992). Coping with marital transitions: A family systems perspective. *Monographs of the Society for Research in Child Development, 57*(2-3, Serial No. 227), 1–14.

Hetherington, E., Cox, M., & Cox, R. (1982). Effects of parents and children. In M. Lamb (Ed.), *Nontraditional families: Parenting and child development* (pp. 233–288). Hillsdale, NJ: Lawrence Erlbaum.

Hoffman, S. D., & Duncan, G. D. (1988). What are the consequences of divorce? *Demography, 23,* 641–645.

Holden, K., & Smock, P. J. (1991). The economic costs of marital dissolution: Why do women bear a disproportionate cost? *Annual Review of Sociology, 17,* 51–78.

Kellan, S. (1995). Child custody and support. *Congressional Quarterly Researcher, 5*(2), 25–48.

Kitson, G. C., with Holmes, W. M. (1992). *Portrait of divorce: Adjustment to marital breakdown.* New York: Guilford.

Kitson, G., & Morgan, L. (1990), The multiple consequences of divorce: A decade review. *Journal of Marriage and the Family, 52,* 913–924.

Kurz, D. (1995). *For better or for worse: Mothers confront divorce.* New York: Routledge.

Maccoby, E. E., & Mnookin, R. H. (1992). *Dividing the child: Social and legal dilemnas of custody.* Cambridge, MA: Harvard University Press.

Martin, T. C., & Bumpass, L. L. (1989). Recent trends in marital disruption. *Demography, 26*(1), 37–51.

Mills, D. M. (1988). Stepfamilies in context. In W. Beer (Ed.), *Relative strangers* (pp. 1–29). Totowa, NJ: Rowman & Littlefield.

Mnookin, R., Maccoby, E. E., Albiston, C. R., & Depner, C. E. (1990). Private ordering revisited: What custodial arrangements are parents negotiating? In S. Sugarman & H. H. Kay (Eds.), *Divorce reform at the crossroads* (pp. 37–74). New Haven, CT: Yale University Press.

Morrison, D. R. & Cherlin, F. J. (1995). The divorce process and young children's well-being: A perspective analysis. Journal of Marriage and the Famly, 57, 800–812.

Papernow, P. L. (1988). Stepparent role development: From outsider to intimate. In W. Beer (Ed.), Relative strangers (pp. 54–82). Totowa, NJ: Rowman & Littlefield.

Powell, B., & Downey, D. B. (1995, August). *Well-being of adolescents in single-parent households: The case of the same-sex hypothesis.* Paper presented at the annual meeting of the American Sociological Association, Washington, DC.

Scoon-Rogers, L., & Lester, G. H. (1995). Child support for custodial mothers and fathers: 1991. In U.S. Bureau of the Census, *Current population reports* (Series P60–187). Washington, DC: Government Printing Office.

Seltzer, J. A. (1991). Relationships between fathers and children who live apart: The father's role after separation. *Journal of Marriage and the Family, 53,* 79–101.

Seltzer, J. A., & Bianchi, S. M. (1988). Children's contact with absent parents. *Journal of Marriage and the Family, 50,* 663–677.

Seltzer, J. A. & Garfinkel, I. (1990). Inequality in divorce settlements: An investigation of property settlements and child support awards. Social Science Research, 19, 82–111.

Shaw, D. S., Emery, R. E., & Tuer, M. D. (1993). Parental functioning and children's adjustment in families of divorce: A prospective study. *Journal of Abnormal Child Psychology, 21,* 119–134.

Teachman, J. (1991). Contributions to children by divorced fathers. Social Problems, 38, 358–371.

Thomson, E., McLanahan, S. S., & Curtin, R. B. (1992). Family structure, gender, and parental socialization. *Journal of Marriage and the Family, 54,* 368–378.

U.S. Bureau of the Census. (1995a). Child support for custodial mothers and fathers: 1991. *In Current population reports* (Series P60–187). Washington, DC: Government Printing Office.

U.S. Bureau of the Census. (1995b). *Statistical abstract of the United States,*

1994. Washington, DC: Government Printing Office.

Wallerstein, J., & Blakeslee, S. (1989). *Second chances: Men, women, and children a decade after divorce.* New York: Ticknor & Fields.

DISCUSSION QUESTIONS

1. What are the factors that research has identified as affecting children's well-being after divorce? Are there others that you would add?

2. Remarriage produces disruptions and transitions in family life that affect all family members. What are some of the processes that emerge in the creation of stepfamilies and how are different family members affected by these changes?

INFOTRAC COLLEGE EDITION

You can use your access to InfoTrac College Edition to learn more about the subjects covered in this essay. Some suggested search terms include:

child custody
child support
custodial parents
divorce

no-fault divorce
remarriage
stepfamilies

54

Doing Parenting:

Mothers, Care Work, and Policy

DEMIE KURZ

Women's increased labor force participation has created dual roles for most women—both at work and at home. Demi Kurz reviews the research on women's and men's responsibilities for "care work" in the context of the "second shift" that most women do. She also points to the effects of social speedup on men's and women's roles in families.

THE FAILURE TO SUPPORT MOTHERS AND CARETAKING

The Burdens of Care Work

. . . Currently, mothers face stress due to their dual roles of caring for children and working in the paid labor force. Although 70% of mothers with children under 18 years of age have now entered the labor force (Herz & Wootton, 1996), the previous domain of men, the majority of men have not substantially increased their contribution to household work. Mothers continue to do most of the work of raising children and the bulk of the household labor. Research indicates that wives, including ethnic minority women (Wyche, 1993), still are responsible for two thirds of household work or between 13 and 17 hours more each week of child care and housework than are husbands (Arendell, 1996; Blair & Johnson, 1992). Hochschild and Machung (1989) called this situation a "stalled revolution."

Whereas mothers now are doing a second shift of domestic work (Hochschild & Machung, 1989), it also is the case that there are men who are participating more in families, with some of them sharing equally in child care with their wives. In their samples, Gerson (1993) and Hochschild and Machung (1989) found some men who shared fully in child rearing and believed that fairness demanded that they share housework with their partners. Some researchers argue that since women have entered the workplace in large numbers, fathers' participation in domestic work has increased (Barnett & Rivers, 1996; Coltrane, 1996; Pleck, 1996). Coltrane (1996) claimed that, as women's participation in the labor

From: Terry Arendell, ed. 1997. *Contemporary Parenting: Challenges and Issues.* Thousand Oaks: Sage, pp. 92–118. Reprinted with permission.

market continues to increase, mothers will demand more of fathers, who will take on more household tasks, particularly child rearing, and will then come to experience more rewards in it. He also cited survey data showing that men in the United States and other industrialized countries now rank fatherhood as more important to them than paid work. Most researchers agree, however, that to date only a relatively small group of men have taken on serious fatherhood and household roles (see Arendell, 1996). Furthermore, Barnett and Rivers (1996) pointed out that women still are responsible for what they call "low-control" jobs, the household jobs that are most stressful because of the constant urgent deadlines, including meal preparation, grocery shopping, cleaning up after meals, and doing laundry. Barnett and Rivers contrasted these with the jobs men usually do, "higher control" jobs such as yard work, household repairs, and looking after the car, which are not as stressful because they do not have to be attended to every day and can be done at the discretion of those who do them.

The lack of affordable, quality child care services compounds the problems women face in providing care. Some parents send their children to day care centers, organized and run by individuals and groups in the private sector; some turn to neighborhood day care run out of homes. Programs such as HeadStart provide some day care for poor children. Unfortunately, however, the cost of most child care is high. In 1991, families living below the poverty level spent 27% of their incomes on child care, whereas those with average incomes spent 7% (U.S. Bureau of the Census, 1994, p. 27). The inability to pay for child care is one reason many single mothers must go on welfare (Kurz, 1995). In addition, some mothers, particularly those who are poorer, worry that the facilities of the day care centers their children attend are inadequate.

Several problems result from this lack of father participation and lack of support for doing care work. First, as Hochschild and Machung (1989) and others (Mirowsky & Ross, 1989; Steil, 1994) have suggested, negotiating work and family roles creates significant amounts of stress for mothers. Hochschild and Machung speak of a "speed-up" in family life that has occurred because working mothers now have to accomplish household tasks in much less time than they did when they were housewives. In her survey, Thoits (1986) found that working mothers experienced more anxiety than did any other group surveyed. When fathers do participate in household work, researchers have found that this help is the single most important factor in decreasing stress for working mothers (Hoffman, 1989). Because they rarely have such help, single mothers face even more stress than do married mothers (McLanahan & Adams, 1987).

Second, lack of participation by fathers in family work also creates friction between husbands and wives. Bergmann (1986) and Hochschild and Machung (1989) believe that although women have tried to negotiate for more help from their husbands, there is a limit to how far women will pressure their husbands to take on household work because of their fear that conflict over housework and children could result in divorce and the reduced standard of living it brings for mothers and children. Single mothers receive even less help from fathers with caretaking. According to recent figures from the National Survey of Families and Households, roughly 30% of divorced fathers did not see their children at all in the previous year, 60% saw their children several times or less during the year, and only 25% saw their children weekly (Seltzer, 1991). The more time passes after a divorce, the less fathers see their children (Furstenberg & Cherlin, 1991).

The third problem resulting from the delegation of caretaking work to women is that because they have so many family responsibilities in addition to their job responsibilities, many women find it difficult to advance in the workplace. Promotions require increased commitments of time, a scarce commodity for mothers who return home after work to a second shift of housework and child care. Some mothers remain at lower level jobs where employers are willing to accommodate their need for flexible hours. Others work part-time jobs to have more time to care for their children. Fewer than half of all employed women are full-time, year-round workers (Population Reference Bureau, 1993, p. 85). Part-time work can be particularly important for mothers with very young children or children with special needs. Unfortunately, however, part-time workers typically receive less than proportionate earnings and fringe benefits and also have more difficulty gaining promotions and higher paying jobs (Blau & Ferber, 1992, p. 184). . . .

SINGLE-MOTHER FAMILIES AND THE HIGH POVERTY RATE

A second very serious problem for family life is the poverty that single-mother families face. Two-parent families also can fall into poverty; however, far greater percentages of single-parent families live below the poverty level. As I illustrate in this section, their difficult economic situation highlights the costs of our system of distributing family resources. Although mothers' incomes now are essential to the financial well-being of most families, women still earn substantially less money than do men, both because women's salaries and wages are lower and because many more women than men work part-time. Women of color earn even less money and have fewer family resources to fall back on than do white women, making it that much more difficult for them to live outside of marriage on their own incomes. In the following paragraphs, I describe the situation of divorced and never-married single mothers.

As divorce has become widespread, researchers have documented the difficult economic situation that divorced women face. Nevertheless, many stereotypes of divorced women remain, such as the view that they gain large alimony settlements. In fact, only about one sixth of divorced women receive alimny or spousal maintenance awards, and many mothers face great economic difficulty after divorce (Weitzman, 1985), As noted previously, an astonishing 39% of divorced mothers with children age 18 or under live in poverty (U.S. Bureau of the Census, 1993, p.79).

The law puts divorced women in a vulnerable and dependent position. Laws regulating the distribution of assets at divorce do not address the fact that women are economically disadvantaged by their participation in marriages that fail. The amounts of money that women receive from marital assets and for child support do not reflect their contributions to their marriages or their needs for themselves and their children after their divorces (Babcock et al., 1996; Kurz, 1995). Only a minority of women receive child support, and amounts are generally low (Garfinkel, 1992; Roberts, 1994).

England and Kilbourne (1990) argued that the reason divorce is so economically disastrous for women is that the tasks in which they have invested so much time, child rearing and housekeeping, are not transferable when a marriage dissolves, unlike the job skills in which men have invested. At divorce, women receive no benefits or compensation for all of the caretaking work they have done. Furthermore, the fact that career assets are not recognized as marital property means that the primary wage earner, generally the husband, is permitted to keep most of the assets accumulated during marriage (Arendell, 1986; Babcock et al., 1996; Weitzman, 1985). Thus the husband does not suffer financially at divorce, whereas the wife, who has invested in her family and in her husband's career, is deprived of a return on her marital investment and must support herself and her children by working at wages that are not sufficient to support them at an adequate standard of living. The decline in the standard of living that divorced mothers face takes a severe toll on their lives (Arendell, 1986; Kurz, 1995), pushing some of them toward permanent downward mobility or even into poverty. Their children, whose well-being is tied to that of their mothers, suffer a similar fate. . . .

Because of the failure to support mothers with children outside of marriage, increasing numbers of the poor are women and children. This phenomenon has been referred to as the "feminization of poverty" (Pearce, 1993). Mothers of all backgrounds can be at risk of poverty. It is very important to underscore, however, that because they have lower salaries and fewer job prospects than do white women (Malveaux, 1985; U.S. Department of Labor, 1989), disproportionate numbers of women of color are poor, particularly minority single mothers. Therefore, the feminization of poverty cannot be viewed apart from the "racialization of poverty" (Wilkerson & Gresham, 1993).

CAUSES OF THE FAILURE TO PROVIDE CARE

It is critical to examine the causes of these two serious problems of family life and to understand them within a framework that addresses issues of gender, race, and class. Too frequently, family problems have been analyzed within a framework that looks to the loss of values as the cause of family problems or, as noted earlier, vague concepts of "family decline." Such frameworks lead to policies, based on the model of the traditional family, that fail to reflect the new realities of the entry of women into the workplace and the rise of single-mother families and that do not address the crises of care work or the rise in female poverty. The United States has no national child care policy, and there is little attempt by employers to give supports and benefits to part-time work. Those mothers who live outside of marriages or partnerships with men who contribute family income are not able to support themselves and their children at an adequate standard of living. As noted earlier, we provide these women with only meager help through a punitive welfare system.

Powerful material and ideological factors underlie the failure to provide assistance for families. Current economic interests contribute to the support of the traditional family as the preferred family form. As Marxist-, socialist-, and other types of feminists have noted, those owning and making profits in business and industry benefit incalculably from the unpaid domestic labor of mothers and other care-takers, particularly women of color (Glazer, 1987; Hartmann, 1981; Luxton, 1980; Smith, 1987); indeed, their profits depend on it. Business and industry make few attempts to support domestic life; rather, they contribute to the stress of family life by failing to make any significant accommodations to it. Many business groups lobbied against passage of the Family and Medical Leave Act, a bill that ultimately did pass but that provides less leave time to care for children and other family members than do bills in other industrialized countries (Reskin & Padavic, 1994) and that, unlike bills in other countries, provides no stipend for caretakers taking a family leave (Kittay, 1995).

In popular and policy circles, the ideal of the traditional family, based on deeply held ideologies about gender roles and the nature of male and female identities, also continues to be the model for family life. Traditional family ideol-ogy promotes the traditional role of women as the nurturers of children and other family members who need care. Men, still designated as breadwinners, are to be "independent" and free of caretaking responsibilities. Government policies always have assumed that mothers will be available at any time to take care of children (Abramowitz, 1988; Gordon, 1994). Although only a minority of families now take the form of the traditional family, it continues to be the predominant model of the ideal family, to the serious disadvantage of women. Some even want to make this family form a reality again. Politicians on the right, as well as some conservative and neoconservative thinkers (Blankenhorn, 1995; Murray, 1993; Popenoe, 1988), urge a return to "the true American family" to reinstate tradi-tional marriage as the "cure" for family problems and "family decline." In addi-tion, norms of male dominance dictate that women should live in relationships with men. Punitive policies toward single-mother families, and recent proposals by conservative policymakers and politicians to make it more difficult for couples to divorce, reflect the belief that women should not live independently of men. Such views deny the hardships of marriage and the roles of poverty, domestic vi-olence, and other structural factors in causing divorce or the breakup of male-fe-male unions (Kurz, 1995).

The ideology of the traditional family also promotes the stigmatization of sin-gle mothers. Although single mothers always have been stigmatized, they have been increasingly blamed for a whole host of social problems. The animosity to-ward single mothers is particularly evident in welfare debates and policies. Some commentators (Roberts, 1995; Sidel, 1996) believe that the demonization of sin-gle mothers is especially serious because we associate single motherhood with being black, another stigmatized status. Roberts (1995) believes that the myth of the black matriarch, the domineering female head of the black family, fuels the stigmatization of black single women. Although there is a stereotype that single mothers are black, in fact the rate of increase in single motherhood is greater among white women than among black women (U.S. Bureau of the Census, 1995). Thus, although racist beliefs and stereotypes are particularly harmful to women of color, they are in fact harmful to most other women as well because they obscure the need of many mothers for help with caretaking work.

STRATEGIES FOR CHANGE: COPARENTING
AND THE ROLE OF FATHERS

The lack of support for the care work that mothers do and the high poverty rate of single mothers clearly demonstrate that the current system of addressing dependency needs and providing care, based on the policy of relying on a male wage earner to support a traditional family structure, is deeply flawed. Given that the preferred traditional family form is inadequate to the task of providing care in an equitable manner, what alternative possibilities exist for the provision of the physical, financial, and social care of children? The strategy most often suggested is to increase fathers' involvement in family life. Support for this strategy is widespread and has come from many quarters: Feminists have long proposed that fathers share care work (Fineman, 1995b; Silverstein, 1996); a new neoconservative movement, led by David Blankenhorn, has undertaken a public campaign to encourage fathers to spend more time with families (Blankenhorn, 1995); and the right has increasingly promoted father involvement as the solution to family problems (Fineman, 1995b).

For several decades, the women's movement has promoted shared housework and coparenting as major strategies for achieving equality in marriage, one of the most important goals of the movement (Fineman, 1995b; Silverstein, 1996). In addition to promoting the ideal of equality, the women's movement views the goal of equality as having practical benefits. Many in the feminist movement hoped and still hope for an increase in shared parenting so that the burden of managing the household can be shared and mothers can experience less stress while taking care of their children, as well as more flexibility to negotiate for rewards in their work situations. Shared parenting has the potential to allow mothers to make more serious investments in careers and offers the promise of increased well-being for fathers and children.

Although in theory shared parenting would seem to be the ideal means of providing family care, as already discussed, equal participation by fathers has proven difficult to achieve, both within and outside of marriage. A variety of researchers have tried to determine what keeps fathers from becoming seriously involved in the lives of their children. A study of Swedish fathers indicates some of the obstacles to father participation in the care of children (Sandquist, 1987). Despite generous national legislation in Sweden promoting paternity leaves, fathers generally have not taken advantage of these leaves. Fathers in the Swedish study gave two major reasons for their reluctance to take paternity leaves. First, they felt no support from their workplaces for taking leaves to care for children. Second, when they did take on increased child care responsibilities at home, they felt isolated. . . .

Addressing the problem of lack of father involvement, Czapanskiy (1989) claimed that new custody laws do not treat mothers fairly. She argued that although joint custody statutes give fathers new rights and powers, they do not require fathers to participate in the activities that are the basis for joint decision making. They require no new duties or responsibilities. Thus, according to Czapanskiy, these statutes enable fathers to realize the promise of joint custody while mothers do not. Women must share control of decision making about their children with fathers while not necessarily receiving any help from fathers in the raising of the children.

There are other problems with policies that try to increase father visitation. In my study, a second group of mothers reported that they wanted less, not more, visitation. Of those mothers who reported that visitation took place, 29% reported a great deal of conflict with fathers. Many of these women believed that fathers were using visitation as a way in which to check up on and control them, and some were afraid of their ex-husbands. A number of them reported that their husbands had been violent during their marriages, and some after the separations, and many of these women were afraid of their exhusbands. Fully 50% of the women in the sample had experienced violence during their marriages, and some of these women also experienced violence during the separations. Women's fears and their experiences of violence point to the fact that mandatory joint custody can be harmful in some situations by making these mothers vulnerable to violence again. Arendell's (1995) study confirmed that some divorced fathers use visitation as a way in which to try to gain control over their former wives.

In addition to trying to increase fathers' participation in the social lives of their children, policymakers also have tried to make fathers more financially responsible for their children after divorce. They have done this by developing much strict child support systems. Interest in child support enforcement has grown, in part because it is seen as a way in which to keep women off of welfare (Fineman, 1995b; Kurz, 1995). As a result of the increased attention to child support, there is widespread resentment of fathers who do not support their children—"deadbeat dads" who father children and disappear. The most effective system of child support that has been developed is wage withholding, in which the wages of the noncustodial parent, usually the father, are automatically deducted from his or her paycheck and sent to the custodial parent, usually the matter (U.S. Department of Health and Human Services, 1994, p. 20). Automatic wage withholding, mandated by the Family Support Act to be implemented by 1994, will substantially increase the amount of child support that mothers receive (Garfinkel, 1992).

Unfortunately, however, implementation of wage withholding will take some time. Furthermore, even when wage withholding is implemented, 23% to 35% of mothers never will receive child support (Garfinkel, 1992). Based on reports from the women in my study, there are several factors that will prevent mothers from getting the child support they are due. First, some fathers still will be able to evade payment, particularly by working "under the table." Second, some mothers, particularly those who experience domestic violence during their marriages, will not apply for child support because they fear their ex-husbands will be violent toward them if they do. Finally, as others have argued (Roberts, 1994), some fathers cannot afford to pay child support. These circumstances demonstrate that a number of women will not be able to count on child support from fathers to provide them with an adequate standard of living.

To increase fathers' financial participation, states have begun to require mothers to locate and identify the fathers of their children to obtain child support, a procedure called *paternity establishment*. Requiring mothers to identify fathers to get child support is part of a general trend to try to make fathers take more financial responsibility for their children. Some states also now require mothers to find and identify fathers as a condition of receiving welfare benefits. Wisconsin, for example, makes full cooperation in paternity determinations of "nonmarital" children a condition of eligibility for receiving assistance (Fineman, 1996). Failure to cooperate disqualifies the caretaker for assistance, and "protective payments" for the child will be paid to "a person other than the person charged with the care of the dependent child" (Fineman, 1996).

Laws such as these can put an undue burden on mothers who have serious conflicts with their former partners, especially those who have suffered physical abuse. At this time, federal law requiring welfare mothers to identify fathers provides for a "good cause" exemption from naming the fathers. However, a determination of good cause is based on the needs of the child, rather than the mother, and is said to exist if the child is at risk of physical or emotional harm from the father or if the mother will suffer such harm from the father that she will be unable to adequately care for the child. These policies will put some mothers at risk. In a high percentage of cases, women seeking welfare have been physically abused and may be at further risk of abuse at the hands of the fathers (Raphael, 1995). Recent research has demonstrated that many women remain at risk of experiencing violence after separations (Kurz, 1996).

Although there are serious obstacles to shared parenting, there also are, of course, many benefits to shared parenting for mothers, fathers, and children, and it remains an important goal for many feminists and women's rights activists (Silverstein, 1996). When fathers coparent, mothers can spend more time in their workplaces, the lives of fathers are enriched, and children benefit not only from increased time with their fathers but also from seeing new models of fatherhood. Encouraging fathers to become involved in parenting should be a high societal priority. . . .

The Role of the State
and the American Response to Care Work

A variety of measures are required to support mothers and care work. First, the government must provide better family leave policies. Fortunately, in 1993 the United States passed its first family leave bill guaranteeing mothers and fathers 12 weeks of leave to care for newborns or sick family members. The family leave policy in the United States, however, has many shortcomings such as the fact, noted earlier, that it provides less time for family leave than similar bills in other industrialized countries and no pay for workers during their leaves. Kittay (1995) noted three additional limitations of the family leave bill that are particularly serious: the fact that leaves are unpaid; that they are available only to traditional husband-wife families, not nonmarried cohabiting adults, gay or lesbian families, or extended families; and that employers with fewer than 50 employees are not obligated to provide family leaves. A more extended, comprehensive family leave bill is critical to the future of families.

Second, families need subsidized day care to reduce family stress and so that women can find and keep suitable jobs. Based on a comparative study of the United States, with its limited day care policies, and other countries that have more comprehensive ones, Bergmann (1994) argued that a large government role in the provision of child care and after-school care would go a long way toward eliminating child poverty. Third, to enable all families, especially single-mother families, to maintain an adequate standard of living, women need higher wages. Women's low pay fails to keep large numbers of women above the poverty level and keeps others far below the median income level. Government policies also must provide better benefits for those mothers who work part-time to care for their children. Other economic reforms also are necessary including increasing the minimum wage, regulating wages and benefits in part time and contingent work, and adopting policies of equal pay for work of comparable worth. For single mothers not to fall into poverty, the government also must provide additional income support, such as unemployment insurance and temporary disability insurance, that provides coverage for the types of earning losses common to single mothers, who sometimes must take time off from work to care for children or other family members. According to Spalter-Roth and Hartmann (1994), such strategies could mean that a larger number of women would qualify for earned income credit benefits and for unemployment insurance.

Because women generally have low wages, and because many must work part-time to care for their children, many mothers, particularly single mothers, need to have their incomes from paid work supplemented by basic income guarantees. Family allowances, for example, would reduce the inordinately high levels of child poverty found in the United States (Bergmann, 1994). Many other democracies provide support through universal governmental transfers such as child allowances or basic income guarantees. Spalter-Roth and Hartmann (1994) favor a strategy of "income packaging" for single mothers, which would help mothers combine income sources from work, from ex-husbands or ex-partners, and from the state. Finally, mothers need health care for themselves and their children because many jobs do not offer health care benefits, a situation that leaves some women and children vulnerable to the high cost of medical care and, as noted earlier, forces others to turn to welfare, now itself increasingly unreliable.

Those who do not favor state support of families will find many grounds on which to oppose these proposals. First, such reforms are costly. While refusing to support care work and the needs of single mothers, opponents of increased government support for families fail to acknowledge that middle-class families receive many direct and indirect government subsidies that remain hidden from view such as marriage and inheritance laws, insurance and benefit regulations, probate laws, and zoning ordinances (Coontz, 1992). Middle-class families also benefit from tax deductions for children that are worth much more than the child allowances given to mothers on welfare who have more children (Fineman, 1995a). Furthermore, middle-class taxpayers benefit from deductions such as interest paid on mortgage debt and some child care expenses and from the fact that employer contributions to health and life insurance policies frequently are not counted as income (Fineman, 1995b).

Those with business interests also undoubtedly would object to more generous family benefits and would see family leave, higher wages, and greater support for part-time work as unjustifiably cutting into their profits. Thus the reforms just proposed will not be easy to achieve and will only be enacted after considerable political struggle. Despite the denials of those who oppose social supports for families, however, our society does have the money to fund generous social welfare programs. According to Coontz (1992),

> Redistributing just 1 percent of the income of America's richest 5 percent would life one million people above the poverty line. A 1 percent tax on the net wealth of the richest 2 percent of American families would allow us to double federal spending on education and still have almost $20 billion left to spend somewhere else. . . . One commission has recently suggested that it would be possible to restructure the military to transfer $125 billion a year to other uses over the next ten years. (p. 286).

In addition to objections to cost, some, especially supporters of the traditional family, will oppose reforms that support the family because they fear that increased wages for women, as well as increased support for family health care and day care, will enable women to live more easily outside of marriage (Popenoe, 1988). This may well be true. We must question why we designate the marital unit as the primary site for distributing family resources, however, and consider other ways in which to distribute these resources. In our social policies, we should consider privileging relationships that care for and sustain dependents and distributing benefits through such relationships. This will require developing new models for providing care and resources to families.

Many writers have examined our current conceptions of care and have pointed to their failure to acknowledge that dependency is a permanent feature of life for everyone, whether during childhood, old age, or periods of illness—not an aberrant, deviant condition (DeVault, 1991: Fraser & Gordon, 1994; Hochschild, 1995). Fineman (1995b) and Kittay (1995) argued that there is an additional kind of dependency that affects a major sector of the population: the dependency that results when one is responsible for the care of others. Fineman made a distinction between what she calls "inevitable dependency," the dependency that results from not being able to take care of oneself, and "derivative dependency," the dependency that results from taking care of others. Fineman argued that derivative dependency needs must be given social support. Similarly, Kittay referred to the need to "care for those who care" and urges that our social institutions be responsive both to dependents and to those who attend to dependents (Kittay, 1995).

We must rethink our current ideas about what is "private" and what is "public" and acknowledge that what we currently think of as private often has very public dimensions. For example, Kittay (1995) pointed out that even though we think of many dependency decisions, such as marital decisions, as private, in fact we grant many social privileges on the basis of marriage and we require employers, landlords, hospitals, insurance agencies, and the Internal Revenue Service, for example, to take certain actions on behalf of married people. She argued that, similarly, the private decision to take on the work of dependency and to support a dependent outside of marriage also should include third-party obligations to support the dependency worker. To avoid overinstitutionalizing or overprivatizing family care, Hochschild (1995) called for a balanced sharing of public and private responsibility for care work.

CONCLUSION

Viewing the family from a feminist perspective, I have identified what I see as two of the major problems of the contemporary family: the double burden faced by mothers who do the majority of family care work while also working in the paid economy, and the poverty and stress faced by single-mother heads of household and their children. Many scholars and other analysts have failed to see these problems because they have not understood how inequities based on gender and race bias are built into the structure of the family. Policymakers similarly fail to take into account factors of gender and race when making family policy. For example, one of the primary solutions proposed to address the current crises of the family has been to promote greater father participation in family life to do caretaking work and provide more financial assistance for single-mother families. Although father involvement should be seriously promoted, and at some point in the future there may be equitable father participation in the social and financial lives of all families, for the time being father participation is not a solution to family problems. Furthermore, in some cases, such as when fathers have been violent, great care must be taken to determine whether fathers should participate in family life. What is required is for our society to assume more collective social responsibility for families through stronger social supports.

Current government policies toward the family are inadequate and flawed. Although there have been government initiatives in the provision of day care and family leave, the government's approach to supporting families continues to assign priority to the traditional nuclear family model. We persist in viewing the welfare state, increasingly associated with AFDC (or welfare), with *dependency*, a stigmatizing and derogatory term (Fraser & Gordon, 1994). Our culture, looking to eradicate dependency, identifies the problems of poverty and hardship as rooted in the characteristics of vulnerable groups such as women, particularly minority women. Racism feeds this tendency to view any kind of family help as dependency. Instead of passing the legislation necessary to reduce poverty, legislators have rushed to restrict benefits for women on AFDC.

We must replace this reliance on outdated views with new thinking about the concepts of *dependency* and *independence*. Our social policies must reflect the reality that family members all are highly interdependent and that all families require social supports. We need to provide all types of families with many more universal, non-means-tested, nonstigmatizing benefits including day care and health care. These measures are necessary for women but also for their children. We need a new model of social welfare that includes not handouts but rather universal benefits for all such as the right to shelter, income supports for working parents, an expanded earned income tax program, and family allowances. The goal of such benefits would be to integrate beneficiaries into the social mainstream, unlike current social welfare programs, which segregate them.

NOTE

1. There are two types of joint custody: legal, in which a parent has the right to share in important decisions about a child's life (e.g., educational and religious training), and physical, in which the child resides with a parent. Research shows that the enactment of laws creating joint legal custody does not lead to increased contact between a father and his children.

REFERENCES

Abramowitz, M. (1988). Regulating the lives of women: Social welfare policy from colonial times to the present. Boston: South End.

Arendell, T. (1986). *Mothers and divorce: Legal, economic, and social dilemmas.* Berkeley: University of California Press.

Arendell, T. (1995). Fathers and divorce. Thousand Oaks: Sage.

Arendell, T. (1996, January). *Co-parenting: A review of the literature.* Unpublished manuscript commissioned by the National Center on Fathers and Families, Philadelphia.

Babcock, B. A., Copelon, R., Freedman, A., Norton, E. H., Ross, S., Taub, N., & Williams, W. (1996). *Sex discrimination and the law: Causes and remedies* (2nd ed.). Boston: Little, Brown.

Barnett, R. C. & Rivers, C. (1996). She works, he works. How two-income families are happier, healthier, and better-off. San Francisco: Harper.

Bergmann, B. (1994, May). Childcare: The key to ending child poverty. Paper presented at the conference on Social Policies for Children, Princeton, NJ.

Bergmann, B. (1986). *The economic emergence of women.* New York: Basic Books.

Blair, S., & Johnson, M. (1992). Wives' perceptions of the fairness of the division of labor: The intersection of housework and ideology. *Journal of Marriage and the Family, 5,* 570–582.

Blankenhorn, D. (1995). *Fatherless america: Confronting our most urgent social problem.* New York: Basic Books.

Blau, F. D., & Ferber, M. A. (1992). *The economics of women, men, and work,* Englewood Cliffs, NJ: Prentice Hall.

Coltrane, S. (1996). *Family man: Fatherhood, housework, and gender equity.* New York: Oxford University Press.

Coontz, S. (1992). *The way we never were.* New York: Basic Books.

Czapanskiy, K. (1989). Child support and visitation: Rethinking the connections. *Rutger's Law Journal, 20,* 619–665.

DeVault, M. L. (1991). *Feeding the family: The social organization of caring as gendered work.* Chicago: University of Chicago Press.

Fineman, M. A. (1995a). Masking dependency: The political role of family rhetoric. Virginia Law Review, 81, 501–534.

Fineman, M. A. (1995b). *The neutered mother, the sexual family and other twentieth century tragedies.* New York: Routledge.

Fineman, M. A. (1996). The nature of dependencies and welfare "reform." *Santa Clara Law Review, 36,* 1401–1425.

Fraser, N., & Gordon, L. (1994). "Dependency" demystified: Inscriptions of power in a keyword of the welfare state." *Social Politics, 1*(1), 4–31.

Furstenberg, F. F., Jr., & Cherlin, A. (1991). *Divided families.* Cambridge, MA: Harvard University Press.

Garfinkel, I. (1992). *Assuring child support: An extension of social security.* New York: Russell Sage.

Gerson, K. (1993). No man's land: Men's changing commitments to family and work. New York: Basic Books.

Glazer, N. (1987). Servants to capital: Unpaid domestic labor and paid work. In N. Gerstel & H. Gross (Eds.), *Families and work: Towards reconceptualization* (pp. 236–255). Philadelphia: Temple University Press.

Gordon, L. (1994). Pitied but not entitled: Single mothers and the history of welfare, 1890–1935. New York: Free Press.

Hartmann, H. (1981). The family as the locus of gender, class, and political struggle: The example of housework. *Signs: Journal of Women in Society and Culture, 6*, 366–394.

Herz, D. E., & Wootton, B. H. (1996). Women in the workforce: An overview. In C. Costello & B. Krimgold (Eds.), *The American woman 1996–97* (pp. 44–78). New York: Norton.

Hochschild, A. (1995). The culture of politics: Traditional, postmodern, cold-modern, and warm-modern ideals of care. *Social Politics, 2*, 331–346.

Hochschild, A., with Machung, A. (1989). *The second shift: Working parents and the revolution at home.* New York: Viking.

Hoffman, L. W. (1989). Effects of maternal employment in the two-parent family. *American Psychologist, 44,* 283–292.

Kittay, E. F. (1995). Taking dependency seriously: The Family and Medical Leave Act considered in light of the social organization of dependency work and gender equality." *Hypatia, 10*(1), 7–29.

Kurz, D. (1995). For richer, for poorer: Mothers confront divorce. New York: Routledge.

Kurz, D. (1996). Separation, divorce, and woman abuse. *Violence Against Women, 2*(1), 63–81.

Luxton, M. (1980). *More than a labour of love: Three generations of women's work in the home.* Toronto: Women's Press.

Malveaux, J. (1985). The economic interests of black and white women: Are they similar. *Review of Black Political Economy, 14*, 5–27.

McLanahan, S., & Adams, J. (1987). Parenthood and psychological well-being. *Annual Review of Sociology, 13*, 237–257.

Mirowsky, J., & Ross, C. E. (1989). *Social causes of psychological distress.* New York: Aldine de Gruyter.

Murray, C. (1993, October 29). The coming white underclass. *Wall Street Journal,* p. A14.

Pearce, D. (1993). The feminization of poverty: Update. In A. Jaggar & P. Rothenberg (Eds.) *Feminist frameworks* (3rd ed., pp. 290–296). New York: McGraw-Hill.

Pleck, J. H. (1996, June). *Paternal involvement: Levels, sources, and consequences.* Paper presented at the Co-Parenting Roundtable of the Fathers and Families Roundtable Series sponsored by the National Center on Fathers and Families, Philadelphia.

Popenoe, David. (1988). *Disturbing the nest: Family change and decline in modern societies.* New York: Aldine de Gruyter.

Population Reference Bureau. (1993). *What the 1990 census tells us about women: A state factbook* (Vol. 20). Washington, DC: Author.

Raphael, J. (1995). Domestic violence and welfare reform. *Poverty and Race, 4*(1), 19–29.

Reskin, B., & Padavic, I. (1994). *Women and men at work.* Thousand Oaks, CA: Pine Forge.

Roberts, D. (1995). Racism and patriarchy in the meaning of motherhood. In M. Fineman & I. Karpin (Eds.), *Mothers in law* (pp. 224–249). New York: Columbia University Press.

Roberts, P. G. (1994). *Ending poverty as we know it: The case for child support enforcement and assurance.* Washington, DC: Center for Law and Social Policy.

Sandquist, K. (1987). Swedish family policy and the attempt to change paternal roles. In C. O'Brien & M. O'Brien (Eds.), *Reassessing fatherhood* (pp. 1444–1460). London: Sage.

Seltzer, J. (1991). Relationships between fathers and children who live apart: The father's role after separation." *Journal of Marriage and the Family, 53*, 79–101.

Sidel, R. (1996). Keeping women and children last. New York: Penguin Books.

Silverstein, L. B. (1996). Fathering is a feminist issue. *Psychology of Women Quarterly, 20*, 3–37.

Smith, D. (1987). Women's inequality and the family. In N. Gerstl & H. Gross (Eds.), *Families and work* (pp. 23–54). Philadelphia: Temple University Press.

Spalter-Roth, R. M., & Hartmann, H. (1994). AFDC recipients as care-givers and workers: A feminist approach to income security policy for American women. *Social Politics, 1*, 190–210.

Steil, J. M. (1994). Equality and entitlement in marriage. In M. Lerner & G. Mikula (Eds.), *Entitlement and the affectional bond: Justice in close relationships* (pp. 229–258). New York: Plenum.

Thoits, P. (1986). Multiple identities: Examining gender and marital status differences in distress. *American Sociological Review, 51*, 259–272.

U.S. Bureau of the Census. (1993). Poverty in the United States: 1992. In *Current Population Reports* (Series P60, No. 185), Washington, DC: Government Printing Office.

U.S. Bureau of the Census. (1994). Who's minding the kids? Child care arrangements: Fall 1991. In *Current Population Reports* (Series P70, No. 36). Washington, DC: Government Printing Office.

U.S. Bureau of the Census. (1995). Dynamics of economic well-being: Program participation 1990–1992. In Current Population Reports (Series P70, No. 41). Washington, D.C.: Government Printing Office.

U.S. Department of Health and Human Services, Administration of Children and Families, Office of Child Support Enforcement. (1994). Seventeenth annual report to Congress. Washington, D.C.: Government Printing Office.

U.S. Department of Labor, Bureau of Labor Statistics. (1989). Labor force statistics 1948-1987. In *Current Population Survey* (Bulletin 22307). Washington, DC: Government Printing Office.

Weitzman, L. J. (1985). *The divorce revolution.* New York: Free Press.

Wilkerson, M. B., & Gresham, J. H. (1993). The racialization of poverty. In A. Jagger & P. Rothberg (Eds.), *Feminist frameworks: Alternative theoretical relations between men and women* (3rd ed., pp. 297–303). New York: McGraw-Hill.

Wyche, K. F. (1993). Psychology and African-American women: Findings from applied research. *Applied and Preventive Psychology, 2*, 115–121.

DISCUSSION QUESTIONS

1. What does the research find about men's participation in the work of family life? What consequences does this have for women?

2. What is social speedup and how is it affecting the experiences of women, men, and children in contemporary families? What specific changes would you recommend to address these problems?

INFOTRAC COLLEGE EDITION

You can use your access to InfoTrac College Edition to learn more about the subjects covered in this essay. Some suggested search terms include:

child care
domestic labor (or domestic work)
fatherhood
housework

motherhood
parenting
social speedup

55

"Male-Order" Brides:

Immigrant Women, Domestic Violence and Immigration Law

UMA NARAYAN

Domestic violence has been widely identified as resulting from the power dynamics within families. Here, Uma Narayan discusses the threat of violence for immigrant women. She argues that women's status as immigrants can both increase their vulnerability to violence and limit the options they have available in trying to leave violent relationships. Her discussion reminds us of the unique impact that race, class, ethnicity, and national origin can have in shaping people's family experiences.

Despite increased awareness of the need to attend to the experiences of women who are on the margins of society because of their class, race, ethnicity, and sexual orientation, little attention has been paid . . . to problems that particularly affect *immigrant* women, many of whom are women from Third World countries. The experience of immigration is often a difficult one for both men and women, involving moving great distances from the familiar contexts of one's homeland to the rigors of life in a foreign country, where they face not only the disempowering unfamiliarities of the new context, but also prejudice and discrimination. For many women immigrants in particular, the shift to a new context only exacerbates their gender-linked vulnerabilities and powerlessness.

This paper analyzes one set of problems that confront many immigrant women to the United States. The majority of immigrants to the United States today are women (Housten et al. 1984). Although some of these women immigrate in their own right, others enter the United States with an immigration status dependent on marriage to a man who is either a U.S. citizen or a Legal Permanent Resident (LPR). Their ability to acquire citizenship or LPR status is dependent on the goodwill of their husbands and on the survival of their marriages. I wish to focus on the problems faced by women with such "dependent immigration status" if they face abuse and violence within their marriage.[1] Although recent events have given issues of domestic violence extensive media coverage, there has been little attention to battered immigrant women, who are not only disempowered by all the factors that affect battered women who are citizens, but who also confront legal prohibitions against seeking employment and the threat of deportation if they leave abusive marriages.

From: *Hypatia* 10, no. 1 (Winter 1995): 104–119. © by Uma Narayan.

For many immigrants, moving to a new country often precedes acquiring the legal status of a citizen or of a permanent resident of that country by several years. In a context of increasing global immigration, we need to attend to the implications of citizenship as a status that, for many immigrants, is obtained via the mediation of complex legal rules, which are often insensitive to predicaments faced by immigrant women. When immigration rules render women legally dependent on their husbands in a manner that is oblivious to problems of domestic violence or make legal provisions to help battered immigrant women that assume immigrant women to have the knowledge, resources and choices of the sort enjoyed by mainstream male citizens, these rules *exacerbate* immigrant women's lack of autonomy instead of helping to enhance their autonomy. Such immigration rules seem more concerned with "policing the borders" between noncitizens and citizens than with helping to make empowered citizens of immigrant women who are in the process of legally negotiating these borders in order to acquire citizenship. . . .

IMMIGRANT WOMEN AND INCREASED VULNERABILITIES TO BATTERING

Although data on battered women are difficult to obtain and evaluate, empirical evidence suggests that women whose immigration status depends on their husbands are more at risk for battery than women in general. There are wide variations in the studies that seek to ascertain the percentage of all married women who experience some form of violence within their marriages. Estimates range from 12 percent to 50 percent (see Sigler 1989, 12–13). A recent report by the American Medical Association reported that one in three women will be assaulted by a domestic partner in her lifetime.[2] The estimated rates of battery for immigrant women seem considerably higher. One study reports that 77 percent of women with dependent immigrant status are battered. Of the victims of domestic violence at the Victim's Services Agency in Jackson Heights, Queens, 90 percent were immigrants (see Anderson 1993, n. 9).

What factors contribute to such heightened vulnerability to abuse? Women with dependent immigration status are often more economically, psychologically and linguistically dependent on their spouses than wives in general. Dependent immigration status legally prohibits them from seeking employment. Many lack fluency in English, a factor that impedes their ability to negotiate the routines of everyday life without their husbands' assistance. The language barrier impedes social relationships except with those who share their linguistic background, but members of their linguistic community are often people connected to the husband, unlikely to assist against the husband's abusive conduct (Hogeland and Rosen 1990). Contact with members of one's linguistic community can be entirely lacking for foreign wives whose husbands are Americans who do not share their ethnic background. Community norms within immigrant communities also work to disempower battered women. Citing the director of Everywoman's Shelter in Los Angeles who points out that in many Asian communities saving the honor of the family from shame is a priority that deters immigrant women from reporting domestic violence (Rimonte 1991), Kimberlé Crenshaw remarks, "Unfortunately, this priority tends to be interpreted as obliging women not to scream rather than obliging men not to hit" (Crenshaw, N.d.).

There are additional factors that heighten vulnerability to domestic violence for *particular groups* of women with dependent immigration status. Due to U.S. military presence overseas, since World War II more than 200,000 women, mostly Asian, have married U.S. servicemen and immigrated to the United States as "military brides" (Anderson 1993, 1406). Studies reveal that *rates* as well as the *severity* of domestic violence, are greater in military families than in civilian families. One study indicated that military men used weapons on their wives twice as often as civilians, and that three-fourths of the cases were life-threatening, compared to one-third of cases involving civilians (Anderson 1993, 1406). Researchers conclude that "the worst of the civilian cases were the norm for military cases" (Shupe et al. 1987, 67–70). A number of factors account for this heightened degree of violence. Periods of extended separation when soldiers are stationed away from home and the social isolation due to the transient nature of military postings are thought to increase the stresses on military family life; but the most significant factor is believed to be the aggressive values indoctrinated into soldiers that carry over into domestic contexts (Anderson 1993, 1406). If military wives are at greater risk for domestic violence than civilian wives, and if women who are dependent immigrants are at greater risk than other women, the possibility that dependent immigrants who are military brides face substantial risks of violence is frighteningly high indeed.

A number of women immigrate to the United States to marry virtual strangers. Approximately 2,000 to 3,500 American men annually marry "mail-order wives," according to current estimates (Kadohata 1990). The typical man who seeks a "mail-order bride" is described as an older, politically conservative, college-educated white man, with higher than average income, who has had bitter experiences with divorce or breakups.[3] The vast majority of women who are available as "mail-order brides" are young Asian women from poorer Southeast Asian countries. However, women from Eastern European countries and from areas that were formerly part of the Soviet Union seem to have recently joined these lists in significant numbers (Henneberger 1992). Many of these women come from backgrounds marked by grinding poverty, unemployment, and political turmoil. Men who seek these women as marriage partners have motivations in which sexist and racist stereotypes play significant roles. Men who marry "mail-order brides" want women who will be totally dependent on them; they are disenchanted with changing gender roles and often blame the women's movement for their inability to find locally the sort of woman they wish to marry (Joseph 1984). They hope to find "beautiful, traditional, faithful" Asian wives, who will not seek to work outside the home, in contrast to local women whom they deem "overly liberated," and not devoted enough to their husbands and families (Villipando 1989, 321). The "mail-order bride" businesses exploit stereotypes of Asian women, tapping into existing views of Asian women as Lotus Blossoms and

Geisha Girls, devoted and deferential to men, the "feminine" and "delicate" counterparts of their loud, independent western sisters (Tajima 1989).

The "mail-order bride" phenomenon illuminates interesting details of the nexus between race and gender stereotypes prevalent in the United States. Economically vulnerable women from African countries, for instance, are not in demand as wives. Submissiveness and deference do not seem to be stereotypical qualities attributed to women of African descent. White "mail-order brides" have increased in number as a result of the recent economic and political upheavals in the former Soviet bloc, and have become quickly "popular," judging by the catalogues. This suggests that many men prefer white "mail-order brides" over Asian women, when available. I was unable to find any full-scale study of such "mail-order" marriages, but newspaper reports suggest that "mail-order brides" are frequently subject to violent assault and abandonment.[4]

Many immigrant men settled in the United States marry women from their countries of origin. I focus, for reasons of familiarity, on marriages between Indian men settled abroad and women from India. There are similarities between the situations of "mail-order brides" and the situations confronted by these Indian women, since both are entering into arranged marriages to relative strangers and are vulnerable to domestic abuse for many of the reasons previously discussed. Some immigrant Indian women confront particular problems in leaving an abusive marriage, problems that might affect women in other immigrant communities as well. Leaving an abusive marriage to return home, even if economically feasible (which it frequently is not), often results in social stigma for the woman and her family, and leaves her with few hopes for re-marriage. In India, men who are U.S. citizens or LPRs are regarded as very attractive matrimonial prospects, which puts them in a position to demand very high dowries from the families of the women they marry. Paying a large dowry constitutes a tremendous economic sacrifice on the part of a woman's family, resulting in great pressure on daughters to "make the marriage work" (see Narayan 1993, 159–70).

Quite a few expatriate Indian men seek to marry women raised in India because of a conviction that such women will make better wives, be more "traditional," "family oriented," and less independent or assertive than their Indian counterparts "corrupted" by being raised in western contexts. The stereotypes held by Indian men looking for "really Indian" wives have an astonishing resemblance to those held by western men looking for Asian "mail-order brides." The same insidious and stereotypical contrast between traditional, home-loving, faithful, self-sacrificing nonwestern women and independent, aggressive, sexually promiscuous "Westernized" women operates in both sorts of marriages. The gender stereotypes that are culturally chauvinistic oddly mirror those that are racist. The qualities that constitute a "good wife" seem to cut across a plurality of cultural landscapes. What all these men are looking for is not only a wife with the appropriately subservient attitudes toward her husband, but a wife who is materially and socially disempowered in ways that will prevent her from challenging their authority. They are looking for a genuine "male-order" bride, in fact. The very dependencies that make these women "attractive wives" ensure their relative powerlessness to confront violence within their marriages.

Even many women who are citizens endure domestic violence within their marriages because of economic dependence, fear of losing custody of their children, social isolation, and ignorance about institutions that might offer shelter or assistance. If marriage is not an empowering institution for many women who are citizens, the disempowering nature of the institution is only exacerbated when a woman's ability to remain in the country depends on the continuation of the marriage. Dependent immigration status is often exploited by husbands, who use the threat of deportation to ensure that their wives do not leave or seek assistance when abused. Since many western and nonwestern men who marry immigrant women are explicitly looking for dependent and subservient wives, they are unlikely to have qualms about using the conjunction of violence and threat of deportation to control their wives.

It is not surprising that women whose dependence on their marriage is high and who have few resources with which to negotiate changes in their husbands' behavior are more susceptible to prolonged physical abuse. While women from all categories and walks of life are susceptible to domestic violence, the *degree* of this susceptibility is clearly affected by material and social structures that disempower particular groups of women. Unfortunately, a variety of factors collude against calling attention to the fact that particular groups of marginalized women are often more vulnerable to domestic violence than their more privileged sisters. These factors contribute to a lack of awareness about the specific problems that confront battered immigrant women. I will discuss some of these factors.

Sexism and cultural chauvinism often collaborate to create tremendous resistance to acknowledging the extent of domestic violence within immigrant communities. I have encountered resistance, among expatriate Indians as well as Indians back in India, to confronting the degree of abuse women face within arranged marriages. Attempts by Indian feminists to call attention to issues of violence against women, such as rape, harassment, dowry-murder and dowry-related marital abuse were often condemned by sections of the Indian intelligentsia as an imposition of irrelevant "Western" agendas (Katzenstein 1991–92, 7). Among Indian immigrants as well, there is a sexist and culturally chauvinistic insistence that "our traditions" guarantee respectful treatment of women and that "our families" do not suffer from problems perceived as endemic to "Western" marriages. Lower rates of divorce among Indians are often uncritically offered as proof that women do not suffer abuse within the institution of arranged marriage. Variants of these attitudes might be prevalent in other immigrant communities as well.

While such attitudes constitute a willful denial of domestic violence within these immigrant communities, other elements also impede acknowledgment of factors that may heighten vulnerability to domestic violence on the part of women of color and immigrant women. Kimberlé Crenshaw points out an odd and detrimental collaboration between the politics of U.S. feminist groups addressing domestic violence and the cultural politics of many minority and immigrant communities. Crenshaw reveals that the Los Angeles Police Department would not release statistics that would show the correlation between arrests for domestic violence and racial group, because of conjoint pressures from domestic violence activists who feared the statistics would be used to dismiss domestic violence as a "minority problem" and by representatives from minority communities who worried that the data would reinforce racist stereotypes. Crenshaw concludes, "This account sharply illustrates how women of color can be erased by the strategic silences of antiracism and feminism" (Crenshaw, N.d.), making it difficult to address the specific problems confronting victims of domestic violence who are women of color.

A genuinely feminist and antiracist approach to domestic violence should recognize that such "strategic silences" betray the interests of women of color, and instead advocate sensitivity to the ways in which different groups of battered women are differently impeded in their abilities to secure assistance. Immigrant women are more vulnerable to domestic violence because of economic, social, and legal factors—such as lacking legal standing to work, and the threat of deportation if they leave their marriage—that do not burden women who are citizens. Worries that acknowledging immigrant women's heightened susceptibility to domestic violence will reflect badly on immigrant communities or immigrant men can, I argue, be countered by emphasizing that immigrant women who marry mainstream U.S. citizens (as is the case with most "military brides" and "mail-order brides") also face greater obstacles to leaving abusive marriages.

The "strategic silences" Crenshaw discusses contribute to policies that are blind to the specific problems that confront battered immigrant women. When immigrant community leaders, who are most often men, deny the domestic abuse prevalent in their communities, they make it harder to create, publicize, and maintain accessible community-based sources of assistance that would most effectively help battered immigrant women. Such "strategic silences" deflect attention from the *particular* problems that affect battered immigrant women since they contribute to public unawareness of the extent to which the policies of domestic-violence support services and shelters themselves often disempower immigrant or minority women. Crenshaw discusses the case of a non-English-speaking battered Latina woman fleeing murder threats from her husband, who was denied accommodation at a shelter with only English-speaking staff, on grounds that the shelter's policies required the battered woman to call the shelter *herself* for screening, and to take part in support group sessions which a woman lacking English proficiency could not participate in (Crenshaw, N.d.). Lack of attention to the special vulnerabilities of battered immigrant women also results in inadequate immigration policies. . . .

NOTES

1. That "undocumented women" suffer similar problems with respect to domestic violence is borne out by Hogeland and Rosen, (1990). However, I limit the scope of my discussion to women who enter the United States legally, and not on women who are illegal immigrants.

2. When violence hits home, *Time Magazine*, 4 July 1994, 12–13.

3. This profile is based on a 1983 survey of 265 American men "actively seeking a

partner from the Orient," Mates by mail: This couple catalogues affairs of the heart, *Chicago Sun Times*, 12 August 1984.

4. See Deanna Hodgin, 'Mail-order' brides marry pain to get green cards, *Washington Times*, 16 April 1991, E1; James Leung, Many mail-order brides find intimidation, abuse: Marriages made in China for U.S. citizenship, *San Francisco Chronicle*, 4 September 1990, A9; and Kalinga Seneylatne, Australia; Filipino mail-order brides end up being murdered, *Inter Press Service*, 20 July 1991.

REFERENCES

Anderson, Michelle J. 1993. License to abuse: The impact of conditional status on female immigrants. *Yale Law Journal* 102 (6): 1401–30.

Crenshaw, Kimberlé. N.d. Mapping the margins: Intersectionality, identity politics and violence against women of color. *Stanford Law Review.* Forthcoming.

Henneberger, Melinda. 1992. Well, the Ukraine girls really knock them out. *New York Times,* 15 November 1992, E6.

Hogeland, Chris and Karen Rosen. 1990. *Dreams lost, Dreams found: undocumented women in the land of opportunity.* Booklet published by the Coalition for Immigrant and Refugee Rights and Services, Immigrant Women's Task Force.

Housten, Marion F. 1984. Female predominance in immigration to the United States since 1930: A first look. *International Migration Review* 18(1): 902–25.

Joseph, Raymond A. 1984. American men find Asian brides fill the unliberated bill: Mail-order firms help them look for the ideal woman they didn't find at home. *Wall Street Journal*, 25 January 1984.

Kadohata, Cynthia. 1990. More than he bargained for. *New York Times,* 7 January 1990, 15, book reviews.

Katzenstein, Mary F. 1991–92. Getting women's issues onto the public agenda: Body politics in India. *Samya Shakti* 6: 1–16.

Narayan, Uma. 1993. Paying the price of change: Women, modernization, and arranged marriages in India. In *Women's lives and public policy: The international experience, ed.* Meredeth Turshen and Briavel Holcomb. Westport, CT: Greenwood Press.

Rimonte, Nilda. 1991. A question of culture: Cultural approval of violence against women in the Pacific-Asian community and the cultural defense. *Stanford Law Review* 43(6): 1311–26.

Shupe, Anson D., William A. Stacey, and Lonnie R. Hazlewood. 1987. *Violent men, violent couples: The dynamics of domestic violence.* Lexington, MA: Lexington Books.

Sigler, Robert T. 1989. *Domestic violence in context: An assessment of community attitudes.* Lexington, MA: Lexington Books.

Tajima, Renee E. 1989. Lotus blossoms don't bleed: Images of Asian women. In *Making waves: An anthology of writing by and about Asian-American women,* ed. Asian Women United of California. Boston: Beacon Press.

Villipando, Venny. 1989. The business of selling mail-order brides. In *Making waves: An anthology of writing by and about Asian-American women,* ed. Asian Women United of California. Boston: Beacon Press.

DISCUSSION QUESTIONS

1. What reasons does Narayan give for immigrant women being particularly vulnerable to domestic violence? What implications does this have for changes in immigration policy?

2. In what ways does the practice of "mail-order" brides leave women susceptible to domestic violence? How does such a practice develop from race and gender stereotypes?

INFOTRAC COLLEGE EDITION

You can use your access to InfoTrac College Edition to learn more about the subjects covered in this essay. Some suggested search terms include:

battered women's shelters
domestic violence
immigration

immigration law
mail-order brides
marital rape (or wife rape)

56

Clique Dynamics

PATRICIA ADLER AND PETER ADLER

Patricia Adler and Peter Adler take a look at clique formation and friendship groupings in schools. In their study of children's friendship groups, they analyze how cliques can generate tremendous power and influence over clique members.

A dominant feature of children's lives is the clique structure that organizes their social world. The fabric of their relationships with others, their levels and types of activity, their participation in friendships, and their feelings about themselves are tied to their involvement in, around, or outside the cliques organizing their social landscape. Cliques are, at their base, friendship circles, whose members tend to identify each other as mutually connected. Yet they are more than that; cliques have a hierarchical structure, being dominated by leaders, and are exclusive in nature, so that not all individuals who desire membership are accepted. They function as bodies of power within grades, incorporating the most popular individuals, offering the most exciting social lives, and commanding the most interest and attention from classmates. . . . As such they represent a vibrant component of the preadolescent experience, mobilizing powerful forces that produce important effects on individuals.

The research on cliques is cast within the broader literature on elementary school children's friendship groups. A first group of such works examines independent variables that can have an influence on the character of children's friendship groups. A second group looks at the features of children's inter-and intragroup relations. A third group concentrates on the behavioral dynamics specifically associated with cliques. Although these studies are diverse in their focus, they identify several features as central to clique functioning without thoroughly investigating their role and interrelation: boundary maintenance and definitions of membership (exclusivity); a hierarchy of popularity (status stratification and differential power), and relations between in-groups and out-groups (cohesion and integration).

In this [essay] we look at these dynamics and their association, at the way clique leaders generate and maintain their power and authority (leadership, power/dominance), and at what it is that influences followers to comply so readily with clique leaders' demands (submission). These interactional dynamics are not intended to apply to all children's friendship groups, only those (populated by

From: Patricia Adler and Peter Adler. 1998. *Peer Power: Preadolescent Culture and Identity.* New Brunswick, NJ: Rutgers University Press, pp. 56–69. Reprinted with permission.

one-quarter to one-half of the children) that embody the exclusive and stratified character of cliques.

TECHNIQUES OF INCLUSION

The critical way that cliques maintained exclusivity was through careful membership screening. Not static entities, cliques irregularly shifted and evolved their membership, as individuals moved away or were ejected from the group and others took their place. In addition, cliques were characterized by frequent group activities designed to foster some individuals' inclusion (while excluding others). Cliques had embedded, although often unarticulated, modes for considering and accepting (or rejecting) potential new members. These modes were linked to the critical power of leaders in making vital group decisions. Leaders derived power through their popularity and then used it to influence membership and social stratification within the group. This stratification manifested itself in tiers and subgroups within cliques composed of people who were hierarchically ranked into levels of leaders, followers, and wannabes. Cliques embodied systems of dominance, whereby individuals with more status and power exerted control over others' lives.

Recruitment

Initial entry into cliques often occurred at the invitation or solicitation of clique members. . . . Those at the center of clique leadership were the most influential over this process, casting their votes for which individuals would be acceptable or unacceptable as members and then having other members of the group go along with them. If clique leaders decided they liked someone, the mere act of their friendship with that person would accord them group status and membership. . . .

Potential members could also be brought to the group by established members who had met and liked them. The leaders then decided whether these individuals would be granted a probationary period of acceptance during which they could be informally evaluated. If the members liked them, the newcomers would be allowed to remain in the friendship circle, but if they rejected them, they would be forced to leave.

Tiffany, a popular, dominant girl, reflected on the boundary maintenance she and her best friend Diane, two clique leaders, had exercised in fifth grade:

Q: *Who defines the boundaries of who's in or who's out?*
TIFFANY: Probably the leader. If one person might like them, they might introduce them, but if one or two people didn't like them, then they'd start to get everyone up. Like in fifth grade, there was Dawn Bolton and she was new. And the girls in her class that were in our clique liked her, but Diane and I didn't like her, so we kicked her out. So then she went to the other clique, the Emily clique. . . .

Application

A second way for individuals to gain initial membership into a clique occurred through their actively seeking entry. . . . Several factors influenced the likelihood that a person would be accepted as a candidate for inclusion, as Darla, a popular fourth-grade girl described: "Coming in, it's really hard coming in, it's like really hard, even if you are the coolest person, they're still like, 'What is *she* doing [exasperated]?' You can't be too pushy, and like I don't know, it's really hard to get in, even if you can. You just got to be there at the right time, when they're nice, in a nice mood."

According to Rick, a fifth-grade boy who was in the popular clique but not a central member, application for clique entry was more easily accomplished by individuals than groups. He described the way individuals found routes into cliques: "It can happen any way. Just you get respected by someone, you do something nice, they start to like you, you start doing stuff with them. It's like you just kind of follow another person who is in the clique back to the clique, and he says, 'Could this person play?' So you kind of go out with the clique for a while and you start doing stuff with them, and then they almost like invite you in. And then soon after, like a week or so, you're actually in. It all depends. . . . But you can't bring your whole group with you, if you have one. You have to leave them behind and just go in on your own."

Successful membership applicants often experienced a flurry of immediate popularity. Because their entry required clique leaders' approval, they gained associational status.

Friendship Realignment

Status and power in a clique were related to stratification, and people who remained more closely tied to the leaders were more popular. Individuals who wanted to be included in the clique's inner echelons often had to work regularly to maintain or improve their position.

Like initial entry, this was sometimes accomplished by people striving on their own for upward mobility. In fourth grade, Danny was brought into the clique by Mark, a longtime member, who went out of his way to befriend him. After joining the clique, however, Danny soon abandoned Mark when Brad, the clique leader, took an interest in him. Mark discussed the feelings of hurt and abandonment this experience left him with: "I felt really bad, because I made friends with him when nobody knew him and nobody liked him, and I put all my friends to the side for him, and I brought him into the group, and then he dumped me. He was my friend first, but then Brad wanted him. . . . He moved up and left me behind, like I wasn't good enough anymore."

The hierarchical structure of cliques, and the shifts in position and relationships within them, caused friendship loyalties within these groups to be less reliable than they might have been in other groups. People looked toward those above them and were more susceptible to being wooed into friendship with individuals more popular than they. When courted by a higher-up, they could easily drop their less popular friends. . . .

Ingratiation

Currying favor with people in the group, like previous inclusionary endeavors, can be directed either upward (supplication) or downward (manipulation). . . . Note that children often begin their attempts at entry into groups with low-risk tactics; they first try to become accepted by more peripheral members, and only later do they direct their gaze and inclusion attempts toward those with higher status. The children we observed did this as well, making friendly overtures toward clique followers and hoping to be drawn by them into the center.

The more predominant behavior among group members, however, involved currying favor with the leader to enhance their popularity and attain greater respect from other group members. One way they did this was by imitating the style and interests of the group leader. Marcus and Adam, two fifth-grade boys, described the way borderline people would fawn on their clique and its leader to try to gain inclusion:

> MARCUS: Some people would just follow as around and say, "Oh yeah, whatever he says, yeah, whatever his favorite kind of music is, is my favorite kind of music."
> ADAM: They're probably in a position then they want to be more in because if they like what we like, then they think more people will probably respect them. Because if some people in the clique think this person likes their favorite groups, say it's REM, or whatever, so it's say Bud's [the clique leader's], this person must know what we like in music and what's good and what's not, so let's tell him that he can come up and join us after school and do something.

Fawning on more popular people not only was done by outsiders and peripherals but was common practice among regular clique members, even those with high standing. Darla, a second-tier fourth-grade girl,. . . described how, in fear, she used to follow the clique leader and parrot her opinions: "I was never mean to the people in my grade because I thought Denise might like them and then I'd be screwed. Because there were some people that I hated that she liked and I acted like I loved them, and so I would just be mean to the younger kids, and if she would even say, 'Oh she's nice,' I'd say, 'Oh yeah, she's really nice!'" Clique members, then, had to stay abreast of the leader's shifting tastes and whims if they were to maintain status and position in the group. Part of their membership work involved a regular awareness of the leader's fads and fashions, so that they would accurately align their actions and opinions with the current trends in timely manner. . . .

TECHNIQUES OF EXCLUSION

Although inclusionary techniques reinforced individuals' popularity and prestige while maintaining the group's exclusivity and stratification, they failed to contribute to other, essential, clique features such as cohesion and integration, the management of in-group and out-group relationships, and submission to clique leadership. These features are rooted, along with further sources of domination and power, in cliques' exclusionary dynamics.

Out-Group Subjugation

When they were not being nice to try to keep outsiders from straying too far from their realm of influence, clique members predominantly subjected outsiders to exclusion and rejection. They found sport in picking on these lower-status individuals. As one clique follower remarked, "One of the main things is to keep picking on unpopular kids because it's just fun to do." Eder. . . notes that this kind of ridicule, where the targets are excluded and not enjoined to participate in the laughter, contrasts with teasing, where friends make fun of each other in a more lighthearted manner but permit the targets to remain included in the group by also jokingly making fun of themselves. Diane, a clique leader in fourth grade, described the way she acted toward outsiders: "Me and my friends would be mean to the people outside of our clique. Like, Eleanor Dawson, she would always try to be friends with us, and we would be like, 'Get away, ugly.'"

Interactionally sophisticated clique members not only treated outsiders badly but managed to turn others in the clique against them. Parker and Gottman. . . observe that one of the ways people do this is through gossip. Diane recalled the way she turned all the members of her class, boys as well as girls, against an outsider: "I was always mean to people outside my group like Crystal, and Sally Jones; they both moved schools. . . I had this gummy bear necklace, with pearls around it and gummy bears. She [Crystal] came up to me one day and pulled my necklace off. I'm like, 'It was my favorite necklace,' and I got all of my friends, and all the guys even in the class, to revolt against her. No one liked her. That's why she moved schools, because she tore my gummy bear necklace off and everyone hated her. They were like, 'That was mean. She didn't deserve that. We hate you.'". . .

In-Group Subjugation

Picking on people within the clique's confines was another way to exert dominance. More central clique members commonly harassed and were mean to those with weaker standing. Many of the same factors prompting the ill treatment of outsiders motivated high-level insiders to pick on less powerful insiders. Rick, a fifth-grade clique follower, articulated the systematic organization of downward harassment: "Basically the people who are the most popular, their life outside in the playground is picking on other people who aren't as popular, but are in the group. But the people just want to be more popular so they stay in the group, they just kind of stick with it, get made fun of, take it. . . . They come back everyday, you do more ridicule, more ridicule, more ridicule, and they just keep taking it because they want to be more popular, and they actually like you but you don't like them. That goes on a lot, that's the main thing in the group. You make fun of someone, you get more popular, because insults is what they like, they like insults."

The finger of ridicule could be pointed at any individual but the leader. It might be a person who did something worthy of insult, it might be someone who the clique leader felt had become an interpersonal threat, or it might be someone singled out for no apparent reason. . . . Darla, the second tier fourth grader discussed earlier, described the ridicule she encountered and her feelings of mortification when the clique leader derided her hair: "Like I remember, she embarrassed me so bad one day. Oh my God, I wanted to kill her! We were in music class and we were standing there and she goes, 'Ew! what's all that shit in your hair?' in front of the whole class. I was so embarrassed, 'cause, I guess I had dandruff or something."

Often, derision against insiders followed a pattern, where leaders started a trend and everyone followed it. This intensified the sting of the mockery by compounding it with multiple force. Rick analogized the way people in cliques behaved to the links on a chain: "Like it's a chain reaction, you get in a fight with the main person, then the person right under him will not like you, and the person under him won't like you, and et cetera, and the whole group will take turns against you. A few people will still like you because they will do their own thing, but most people will do what the person in front of them says to do, so it would be like a chain reaction. It's like a chain; one chain turns, and the other chain has to turn with them or else it will tangle."

Compliance

Going along with the derisive behavior of leaders or other high-status clique members could entail either active or passive participation. Active participation occurred when instigators enticed other clique members to pick on their friends. For example, leaders would often come up with the idea of placing phony phone calls to others and would persuade their followers to do the dirty work. They might start the phone call and then place followers on the line to finish it, or they might pressure others to make the entire call, thus keeping one step distant from becoming implicated, should the victim's parents complain.

Passive participation involved going along when leaders were mean and manipulative, as when Trevor submissively acquiesced in Brad's scheme to convince Larry that Rick had stolen his money. Trevor knew that Brad was hiding the money the whole time, but he watched while Brad whipped Larry into a frenzy, pressing him to deride Rick, destroy Rick's room and possessions, and threaten to expose Rick's alleged theft to others. It was only when Rick's mother came home, interrupting the bedlam, that she uncovered the money and stopped Larry's onslaught. The following day at school, Brad and Trevor could scarcely contain their glee. As noted earlier, Rick was demolished by the incident and cast out by the clique; Trevor was elevated to the status of Brad's best friend by his coconspiracy in the scheme. . . .

Stigmatization

Beyond individual incidents of derision, clique insiders were often made the focus of stigmatization for longer periods of time. Unlike outsiders who commanded less enduring interest, clique members were much more involved in picking on their friends, whose discomfort more readily held their attention.

Rick noted that the duration of this negative attention was highly variable: "Usually at certain times, it's just a certain person you will pick on all the time, if they do something wrong. I've been picked on for a month at a time, or a week, or a day, or just a couple of minutes, and then they will just come to respect you again." When people became the focus of stigmatization, as happened to Rick, they were rejected by all their friends. The entire clique rejoiced in celebrating their disempowerment. They would be made to feel alone whenever possible. Their former friends might join hands and walk past them through the play yard at recess, physically demonstrating their union and the discarded individual's aloneness.

Worse than being ignored was being taunted. Taunts ranged from verbal insults to put-downs to singsong chants. Anyone who could create a taunt was favored with attention and imitated by everyone. Even outsiders, who would not normally be privileged to pick on a clique member, were able to elevate themselves by joining in on such taunting. . . .

The ultimate degradation was physical. Although girls generally held themselves to verbal humiliation of their members, the culture of masculinity gave credence to boys' injuring each other. . . . Fights would occasionally break out in which boys were punched in the ribs or stomach, kicked, or given black eyes. When this happened at school, adults were quick to intervene. But after hours or on the school bus boys could be hurt. Physical abuse was also heaped on people's homes or possessions. People spit on each other or others' books or toys, threw eggs at their family's cars, and smashed pumpkins in front of their house.

Expulsion

While most people returned to a state of acceptance following a period of severe derision. . . this was not always the case. Some people became permanently excommunicated from the clique. Others could be cast out directly, without under going a transitional phase of relative exclusion. Clique members from any stratum of the group could suffer such a fate, although it was more common among people with lower status.

When Davey, mentioned earlier, was in sixth grade, he described how expulsion could occur as a natural result of the hierarchical ranking, where a person at the bottom rung of the system of popularity was pushed off. He described the ordinary dynamics of clique behavior:

Q: *How do clique members decide who they are going to insult that day?*
DAVEY: It's just basically everyone making fun of everyone. The small people making fun of smaller people, the big people making fun of the small people. Nobody is really making fun of people bigger than them because they can get rejected, because then they can say, "Oh yes, he did this and that, this and that, and we shouldn't like him anymore." And everybody else says, "Yeah, yeah, yeah," 'cause all the lower people like him, but all the higher people don't. So the lowercase people just follow the highercase people. If one person is doing something wrong, then they will say, "Oh yeah, get out, good-bye.". . .

DISCUSSION QUESTIONS

1. Take a look at your own friendship group in school. Which of the processes of both inclusion and exclusion do you observe?
2. What forms of negative sanction, or punishment, do the more powerful high status clique members deliver to others? List some, noting how they differ in severity.

INFOTRAC COLLEGE EDITION

You can use your access to InfoTrac College Edition to learn more about the subjects covered in this essay. Some suggested search terms include:

clique   stigma
friendship group

57

Ghetto Schooling

JEAN ANYON

The plight of today's urban ghetto schools is detailed here. Also discussed are the many difficulties of school reform.

It is "World Day" on the Rutgers University campus. The colleges and universities in downtown Newark have joined together to celebrate the diversity of people who make up the city and the world. Jesse Jackson's son is to speak. Computer screens and science laboratories are temporarily empty as college students and faculty mingle among the banners and booths on the decorated street. I drive from my office and this bustling scene to the school where I am working with Newark public school teachers.

The street on which Marcy School stands is empty. Stretching for many blocks is a desolate area of closed factories, housing projects, and small, tired-looking homes. I see no people or cars, no trees or other signs of life. I feel as if I have left one country and entered another. I have left the world of progress

From: Jean Anyon. 1997. *Ghetto Schooling: A Political Economy of Urban Educational Reform.* New York: Teachers College Press, pp. 3–13. Reprinted with permission.

and technology on the Rutgers campus, and even though less than two miles away, feel as if I am on a continent in which a holocaust has wiped out a civilization, leaving only a few disoriented survivors. The dilapidated buildings and empty lots breathe a postholocaust weariness and devastation. The celebration of diversity, the computers and science labs, the arrival of Jesse Jackson's son, are not known to the inhabitants of this place. (Field notes, April 8, 1992).

The contrast between the university and the neighborhood around Marcy School illustrates the extreme isolation and poverty of many residents not only in Newark, New Jersey, but in the majority of our large American cities. In this [essay] I set the stage for the analyses to follow by describing this current social milieu of isolation and poverty, then illustrate how these conditions affect urban schools. I use Newark as a specific example in each section. I then analyze the visions driving current educational reform, and end by questioning whether these visions are capable of overpowering the effects of the urban environment on education.

THE RACIAL ISOLATION
AND POVERTY OF CENTRAL CITIES

Central cities now hold only 29% of the nation's population and comprise less than 12% of the national electorate. About 48% of the U.S. population lived in suburbs when the 1990 census was taken. Because they turn out for elections at a higher rate than do rural and central city voters, suburban voters cast a majority of the votes in most state elections, as they did in the 1992 presidential election. Large cities cast less than half the votes in most states: For example, New York City casts 31% of the state's vote, Chicago's share of the popular vote in Illinois is 22%, St. Louis's share of Missouri's vote is 6%, and the six largest cities in New Jersey together produce under 10% of the state's vote. . . . Inner city residents thus elect proportionately very few representatives in state and national contests; most state and national politicians represent suburban constituencies. . . .

Most residents of large cities are African American or Latino. Of the nation's eight largest cities with a population of one million or more in 1990, only two, Philadelphia and San Diego, were less than half minority. Of these eight largest cities, racial ethnic minorities (mostly black and Latino, with percentages of Asian and other minorities averaging 5.7%) are 57% in New York City, 62% in Chicago, 63% in Los Angeles, 70% in Atlanta, 79% in Detroit, and 88% in Miami. Among the 14 cities with between 500,000 and a million inhabitants, 8—Baltimore, Cleveland, El Paso, Memphis, San Antonio, San Francisco, San Jose, and Washington, DC—are also more than half African American, Latino, and Asian. . . .

However, the percentage of blacks in suburbia is miniscule. In 1990, African Americans made up only 8.7% of suburban residents in 12 large standard metropolitan statistical areas studied—New York, Los Angeles-Long Beach, Chicago, Philadelphia, Detroit, San Francisco-Oakland, Boston, Pittsburgh, St. Louis, Washington, DC, Cleveland, and Baltimore. . . .

This political isolation of American cities—and their minority populations—is accompanied by the isolation of poorer urban residents from the economic mainstream of middle-class jobs. Most central city industries have closed or dispersed. As jobs moved to the suburbs, residential segregation prevented most black families from following. Almost all new jobs created in cities are now either low-wage or top-tier (technical, professional) positions . . . Since 53% of adults living in extreme poverty tracts in American cities have not completed high school. . . they are automatically precluded from participation in the high-wage sectors of the economy. . . .

THE PLIGHT OF URBAN SCHOOLS

Total public school enrollment in the United States is about 38 million. Of these, 10.4 million students are in urban, 16.8 million in suburban, and 10.5 million in rural schools. City demographics are reflected in the enrollments in urban schools. Most (approximately 76%) of the students in America's central city schools are African American or Latino. In school year 1992–93, 50 of the Great City School districts enrolled 5.7 million students, including 36.1% of our nation's African American students, 29.8% of our Hispanic students, and only 4.8% of our white students. These districts enrolled 13.5% of the nation's students, but 22.1% of the nation's school poor and 35.9% of the nation's students with limited English proficiency.

Nationally, 42% of urban students are eligible to receive subsidized school lunches, and 40% attend schools defined by the U.S. Department of Education as high-poverty schools, in which more than 40% of students receive free or reduced-price lunch. Against these figures, only 10% of suburban students and 25% of rural students attend high-poverty schools. If present trends continue, the United States will, in 25 years, have a majority of "minority" students in its public schools, enrolling most of the black and Hispanic students in the large cities, with more than half of them living in poverty. . . .

Less than half of the ninth graders entering high schools in our large city systems typically graduate in 4 years; . . . Urban drop-out rates for low-income African American and Hispanic students, already high, increased between 1990 and 1993. Less than half of urban students are above national achievement norms. . . The large percentages of students needing special services or programs strain city school budgets, in some cases accounting for up to one quarter of expenses. . . .

Old school buildings, many dating from the nineteenth and early twentieth centuries, have not been well maintained. Classrooms typically have few instructional supplies and little equipment. . . . Oakes found . . . that students in schools in central cities tend to have less access to science and math resources, programs, and teachers with science or math backgrounds than do those in more advantaged schools. Moreover, math and science teachers in central city schools rate themselves as less confident about their science and math teaching than do teachers in advantaged schools. . . . According to Darling-Hammond and

Sclan, students in urban schools have only a 50% chance of being taught by a certified mathematics or science teacher.

Research has shown that instruction in inner city schools is often based on cognitively low-level, unchallenging, rote material. . . . Members of the Washington, DC-based Education Trust relate their dismay at witnessing "English classrooms [in urban schools] where 14-year-olds were assigned to *color* the definitions to a list of vocabulary words and required to recite—over and over again—the parts of speech". . . . Although high percentages of city students need supplementary academic instruction, there is a 50% higher shortage of teachers in cities than the national average. . . .

Why, with such a large percentage of urban districts reporting restructuring activity, has the picture of education in inner cities not brightened considerably? In the last 15 years Boston, Chicago, Cincinnati, Cleveland, Dade County (Miami), Detroit, Hammond (Indiana), Louisville, New York, Philadelphia, Pittsburgh, and Rochester, New York—among other cities—have been in the news for their attempts to make city schools more successful for low-income students by using a variety of restructuring activities. This most recent wave of reform, of course, follows close on the heels of several decades of other educational reforms in urban districts: federal programs in the late 1960s and early 1970s, and the decades of accountability and standards reforms in the 1970s, 1980s, and 1990s. . . . The latest wave of reform, of which restructuring is a component, is referred to by advocates as "systemic reform." Activity to produce systemic change is an attempt to pull together disparate types of reform from the past several decades, and to overcome contradictory initiatives generated by different levels of the educational system.

Given the dismal state of schooling in most of our central cities, however, it seems clear that this recent wave of educational reform has not succeeded there. I believe it is imperative, therefore, to examine the adequacy of assumptions underlying these efforts. How do educational reformers define school, for example, and what is their vision of change?

THE INADEQUACY OF CURRENT VISIONS OF EDUCATIONAL REFORM

A basic tenet of reform efforts has been that schools are complex, "loosely coupled" organizations in which all parts affect each other, in which control is vested at the top, but in which change cannot be dictated from above. . . . These loosely coupled organizations are enmeshed in systems of district and state regulations. . . . Participants in the school are the key to reform. As longtime authority on school reform Seymour Sarason, says, "salvation for our schools will not come from without but from within". . . .

Reform thus depends upon alteration of the work patterns of school actors. The regularities of their daily practice need to be changed. . . . This bottom-up change needs to be supported by district and state rules, professional and—if possible—community support. The alteration of the regularities of schools will be accomplished by creating new mechanisms, or new structures, for the way school actors participate in the daily educational enterprise. . . . The goal of the new practices and structures is to transform the whole school rather than individual projects or classrooms. . . .

Examples of the new organizational forms for changing the regularities of daily life in school are joint decision making among teachers and administrators, the teaming of teachers, flexible scheduling, multiage groups, and core planning in individual schools. . . .

In the case of Chicago, an altered structure produced by the 1988 reform legislation resulted in the creation of local councils in all schools—assigning unprecedented power to parents and teachers. . . . (In 1995, however, a new law brought about the reorganization of authority, returning the preponderance of power back to a chief executive officer and board of trustees.) In New York City, system-wide decentralization, several districts with school choice, and a number of alternative schools within the public system have involved many of the restructuring movement's most cherished reforms: small, personalized schools; cooperative learning; integration of curriculum; longer class periods; deeper curriculum study and fewer topics; parental involvement; joint management by teachers and administrators; and new forms of assessment, wherein students show both their knowledge and their capabilities in a comfortable setting through demonstrations and portfolios of work rather than through pencil-and-paper tests. . . .

In Philadelphia, small, self-contained learning communities (charters), within large comprehensive high schools and closely linked to local universities and national networks of other restructured schools, represent a district goal of decentralization of administration to support democratic governance of schools, school-based decision making, and management of resources. The task in Philadelphia has been to "reinvigorate intellectually and professionally the educators," "to reengage the students," and "to organize the parents" in order to achieve what Michelle Fine has called "radical, systemic reform of the school system". . . .

The current emphasis on restructuring decision making in schools has been informed by models for reforming complex business organizations in pursuit of higher productivity. One source of the corporate vision is the idea, publicized by Xerox vice president John Seely Brown and business consultant W. Edwards Deming, that participants in an organization are more productive when they are involved in decisions affecting their efforts and when authority is decentralized. . . . Deming goes further. His approach to quality control in organizations entails, rather than blaming the workers when problems arise, instituting new organizational structures—such as "quality control circles"—so that groups of employees can work together to solve the problems. Producing a quality product is everyone's job, and collaborative planning by constituents from various strata of an organization results in solutions. . . .

When the corporate model is applied to schools, principals, teachers, parents, citizens, and sometimes students are supposed to come together to discuss and decide on the educationally best practice for their school. Rather than blaming the students for educational failure, adults in the school should examine and change their own behavior. Superintendents are key, as, in this model, they will need to give up absolute power in deference to others in the system. Principals are also [reluctant] to give up total control, while at the same time retaining their instructional leadership and motivational roles, long associated with school success. . . .

In the view of reformers, teachers are perhaps the most important element, because it is they on whom change ultimately depends: The reforms are, in the final analysis, classroom reforms. Democratization of school governance and organization serves the purpose of improving classroom teaching. An important goal of systemic reform, for example, moving away from teaching by rote drill and toward "active student learning" and use of higher order thinking skills, cannot be met if teachers do not enact it.

However, classroom change is also in great measure dependent on changes in the culture of the school. . . . This involves changing the attitudes of people in all parts of the system—from teachers and students in classrooms, to district administrators, to legislators in state capitals. . . . The involvement of all levels of educational decision making gives the movement its name— "systemic reform". . . . Educational change also needs to be assisted by professional networks and university researchers, and by consultants, business executives, and foundations. . . .

DISCUSSION QUESTIONS

1. Did you attend an inner-city school—either elementary, junior high (middle school) or high school? How accurately does this essay describe your school?

2. The author lists the kinds of reforms that are necessary in city schools. Can you think of any additional kinds of reforms that she might have left out?

INFOTRAC COLLEGE EDITION

You can use your access to InfoTrac College Edition to learn more about the subjects covered in this essay. Some suggested search terms include:

desegregation   school reform
ghetto education   urban schools [education]
school financing

58

Intelligence

HOWARD F. TAYLOR

Howard Taylor reviews the origins and validity of intelligence testing. Also summarized is the matter of racial-ethnic cultural differences and the issue of whether intelligence is primarily due to genes (nature) or to social, familial, and educational environment (nurture).

Intelligence is a hypothetical concept . . . . on which individuals differ or vary from each other. The concept of intelligence is not itself directly observable or measurable. Intelligence, an abstract continuum, is distinct from its nonabstract presumed indicators or measures; "intelligence" (concept) is distinct from "intelligence test" (indicator). The accuracy of the indicator—that is, its validity and reliability—is a matter of how accurately it measures observable phenomena presumed to reflect differing amounts of intelligence. The IQ or "intelligence quotient" is thus an operational, observable, presumed indicator of the concept, intelligence, and not the concept itself.

Various definitions of intelligence have appeared since the idea of a unitary coordinating "mental" faculty was used by the ancient Greek philosophers. The principal formulation of the modern notion of intelligence as some kind of mental capacity began in 1850 with Herbert Spencer and Sir Francis Galton. Since then, definitions of intelligence have varied but generally involve a notion of "potential," "ability," or "capacity" as distinct from actual achievement, attainment, or accomplishment. It is a cognitive disposition as distinct from an affect or emotion. Researchers have defined intelligence as "the ability to carry on abstract thinking" (Terman in 1916, inventor in 1916 of the still popular Stanford–Binet IQ test); "innate general cognitive ability" (Burt); and "the ability to adapt to the environment" (Thorndike . . .).

. . . The development of psychological testing for intelligence began in the early 1900s with A. Binet in France, who invented the notion of IQ as the ratio (quotient) of mental age to chronological age; J. M. Cattell in the United States; and C. Spearman and C. L. Burt in England. This inaugurated the continuing controversy of whether intelligence is reducible to some common set of highly correlated abilities or capacities (the unidimensional theory) or whether it is a matter of several, or many, relatively uncorrelated or independent (nonoverlapping) abilities or capacities (the multidimensional theory).

Measurement accuracy—or, conversely, measurement error—is the degree to which presumed indicators of intelligence (intelligence tests such as the Stanford-Binet or the Wechsler Adult Intelligence Scales [WAIS], or the "ability" portion,

From: *Encyclopedia of Sociology*, Vol 2. Edited by E. F. Borgatta and M. L. Borgatta. New York: Macmillan Publishing Co., 1992. pp. 941–949. Reprinted with permission.

as distinct from the achievement sections, of the popular Scholastic Aptitude Test [SAT]) accurately reflect or measure the extent to which an individual possesses intellectual ability. The degree of measurement accuracy is called the validity and reliability of the indicator(s). Validity is the degree to which the score on an indicator reflects the unknown, unobserved score on the concept "intelligence." Reliability refers to the stability or consistency of an indicator across, for example, different times (test—retest reliability), different forms or wordings of the same test (equivalent forms reliability), or different researchers. Hence, measurement accuracy (and its complement, measurement error) is the relationship or "epistemic" correlation between indicator and concept. Since the true score on the concept is not directly observable or measurable, validity is assessed indirectly, as by observing the relationships among several presumed indicators themselves.

A closely allied issue of measurement is whether the concept intelligence is unidimensional or multidimensional. In 1904 Spearman maintained that there was a statistical dimension or factor (called simply g) for "general intelligence," and that this factor accounted for, and correlated very highly with, specific separate abilities or performances on tests. This view—that intelligence is fundamentally reducible to one basic, though general, master capacity—is still somewhat popular today. This would mean that persons who score high on one type of ability (say, mathematics) would tend to score high on certain other abilities as well (such as reading comprehension, vocabulary, verbal analogies, and so on); and that persons who score low on one ability would also tend to score low on others. In fact, modern-day factor analysis, a set of techniques for assessing the extent to which many indicators may be parsimoniously accounted for by their relationship to (correlations with) a fewer number of variables called factors, began with Spearman's analyses.

This unidimensional formulation was challenged later by L. L. Thurstone's multidimensional principle, which argued that intelligence consists of six or seven largely uncorrelated or nonoverlapping skills or capacities, called "primary abilities," such as spatial, quantitative, verbal, and inductive. Today, the multidimensional view of intelligence predominates over the strict unidimensional view, and it is argued that intelligence probably reflects both some unidimensionality as well as quite a bit of multidimensionality.

From 1960 to about 1975, important contributions to the multidimensional view were made by J. P. Guilford, who theorized that intelligence was so multidimensional that it could be broken down into as many as 120 specific abilities. A number of these abilities encompass what has come to be called creative intelligence. While the unidimensional view stressed individual abilities at inductive and also deductive reasoning, creative intelligence stresses "divergent" reasoning, or the ability to draw new and unanticipated conclusions or inferences. Debates in the professional literature now center on the question of whether creativity is a separate trait itself or whether it bears some relationship or overlap to unidimensional general intelligence. . . . . Finally, a recent multidimensional formulation is that of H. Gardner, who posits seven independent abilities: logical—mathematical, linguistic, spatial, musical, interpersonal, intrapersonal, and body-kinesthetic.

Reliability of measures of intelligence have been assessed mainly by examining how stable one's intelligence scores for different types of abilities remain as a function of increasing age. Most studies show moderate to high correlations . . . between IQ at age five and IQ during the late teenage years. . . .

Attempts to alter or increase IQ by means of specific coaching or environment-manipulation programs have met with some limited success. In general, such attempts to increase IQ are more successful for individuals in the normal ranges of IQ than for individuals who score either very high or very low in IQ. Programs geared to specific abilities (as increasing math skills) are more likely to result in an increase in that particular ability rather than in other abilities not addressed by the program. Recent evidence shows that such attempts to increase IQ among black and Hispanic youths can result in gains of up to 20 IQ points and that such gains do persist at least for several years. Finally, studies of the effect of coaching to increase SAT scores show that after removing the effect of test experience—that is, subtracting out the average difference between first and subsequent test scores for matched control, individuals who have not had an intervening coaching seminar—(most people score higher the second time they take the SAT than on their first attempt), some increase in SAT scores results, depending on the length and type of coaching program. There is no evidence that coaching has differing effects on males versus females, or upon white, black, or Asian groups.

Validity of measures of intelligence has been most often assessed through what is called predictive, or criterion, validation—finding the degree of relationship . . . between some measure of intelligence, taken at one point in time, and some other "criterion" measure taken later, such as grades obtained later on in college or the type of occupation chosen after graduation. The higher the correlation or slope or both between test score and criterion, the higher the predictive validity of the test. . . .

GROUP AND CULTURAL DIFFERENCES

Predictive validity has been assessed in many studies for different racial and ethnic groups (white, black, Hispanic, Asian). It is generally found that whites tend to score *on the average* about one standard deviation higher than blacks and Hispanics on IQ tests (about 15 IQ points higher) and on ability tests such as the SAT (about 50 to 100 SAT points higher, on a scale of 200 to 800). Since the 1980s, Asians have scored on the average slightly higher than or the same as whites on the quantitative portions of such tests, and either the same as or somewhat lower than whites on the verbal portions.

These average differences are regarded as being environmental in origin, reflecting group differences in education, socioeconomic status, childhood socialization, language, nutrition, and cultural advantages. There is no evidence whatever that such *between-group* differences are in any way genetic in origin. (*Within-group* differences in IQ that may be genetic in origin are discussed below under "Nature versus Nurture.")

Average differences in test scores reflect not only differences in social environment, but they may reflect lack of equivalence between groups in the predictive validities of the test. A test may be more predictively valid for one group than for another; namely, a test may have less measurement error for one group than for another. This is determined by assessing the relationship between the test score (X) at one point in time and some criterion measure (Y) such as college grades at a later point in time for the two groups being compared. . . .

A large number of studies show roughly equivalent test-to-grade predictions for whites and blacks. . . . This evidence suggests that one can, to a modest extent, predict later grades (such as first-year college grades) from ability test scores (SAT scores) and that these predictions are roughly the same for whites and blacks.

However, a significant number of studies (about one-fourth of those done) show some differences in . . . correlation . . . for whites compared to blacks and Hispanics, with the . . . correlations being greater for whites, with blacks and Hispanics being roughly the same. This means that, at least for these studies, the test is less predictively valid for blacks and Hispanics than for whites. This is evidence for relatively more measurement error (thus measurement bias) in the case of blacks and Hispanics. Considerably less data are available on Asians; some studies show equal predictive validities, a few show less, and very few show slightly higher predictive validities for some Asian groups.

Apart from the question of validity, average differences in the test score itself exist on the basis of race and ethnicity, of socioeconomic status (social class and the linguistic and cultural differences associated with class differences), and of gender.

In almost every multiracial nation in the world, race differences are strongly confounded with social class differences. Thus, it is difficult to separate the effects of race on IQ from the effects of social class. In studies of race differences in intelligence in countries where white culture is the dominant culture, as in the United States, it has been consistently found that groups classified as Negro, black, or colored are disproportionately represented among the lower classes, and they also score lower on intelligence tests than those classified as white. These are *average* differences, as the distributions of scores for the two groups overlap considerably. In most cases, individuals were tested by white examiners, and there is evidence that one performs somewhat better on a test when examiner and examinee are of the same race. . . .

Gender differences in intelligence, like race differences, are confounded by the differences in social status that men and women occupy in society. In all known societies, gender is correlated with social status; in the vast majority of societies and particularly the United States, women have been traditionally forced to occupy lower average social and economic status than men. Hence, like the minority-to-white comparison, the female-to-male comparison reveals some differences in IQ, with females scoring less on abilities defined as valuable (such as quantitative ability) by society. It used to be thought that the variability in IQ scores is greater among men than among women, with proportionately more men than women in both the gifted as well as the severely retarded range. More recent studies have shown this not to be the case and that with adequate sampling designs (absent in many past studies) the extremes of the distribution are equal for males and for females, as are the middle ranges. While recent studies show average IQ scores to be roughly equal for males and females, differences do arise on specific abilities: Men score higher, on average, in numerical reasoning, gross motor skills, spatial perception, and mechanical aptitude; women score higher, on average, in perception of detail, verbal facility, and memory. These average differences may reflect differences between the sexes in childhood socialization and differences in societal expectations or norms pertaining to men and women. Finally, there is some limited evidence that sex-typing (gender bias) in certain test questions, even in the quantitative sections, is present and thus may account for a portion of the gender difference.

NATURE VERSUS NURTURE

Since the mid-nineteenth century a controversy has raged in the social, behavioral, and natural sciences: Are human differences in intelligence influenced more by biological heredity (nature), by social environment (nurture), or primarily by combinations (interactions) of both? The contemporary approach to the issue has been to attempt to estimate empirically what is called the broad heritability coefficent, defined as the proportion of the total differences (total variance) in intelligence in a population that is causally attributable to genetic factors. Equivalently, it is the proportion of variance in intelligence that is accounted for by the genetic similarity between pairs of biological relatives. Consequently, if heritability in a population is 30 percent, then statistically speaking 30 percent of the differences in intelligence among individuals in that population are due to genetic factors, assuming that intelligence is validly and reliably measured. A 30 percent heritability would mean that 70 percent of the differences in intelligence in the population would be due to environment or to combinations or interactions . . . of genes and enviornment.

The heritability estimate is not generalizable from one population to another; thus, for example, heritability estimates on whites cannot be generalized to blacks or any other racial or ethnic group. Nor can the heritability coefficient be used to draw conclusions about between-group average differences. Heritability has been variously estimated to be as low as 20 percent and as high as 80 percent, depending on the methodological accuracy of the study, the statistical asumptions used, types of biological relatives studied, and a host of other things. Some argue convincingly that heritability is impossible to estimate in any reliable way. The most recent reliable studies tend to center on a heritability of 40 to 50 percent.

How then is the heritability of intelligence actually calculated or estimated? . . . Identical (monozygotic) twins are genetically identical to each other; they are genetic clones of one another. It is noted that identical twins raised together in the same family are very similar in IQ and in other traits suspected of high genetic causation, such as height. Their average IQ pair correlation [for height] is about 0.90, which is very high. This correlation reflects both their genetic similarity and their environmental similarity, since they were raised together. . . .

There have been several procedures used to attempt to disentangle the effects of environmental similarity from the effects of genetic similarity for pairs of relatives. One of these is widely regarded as the best single–kinship procedure for estimating heritability: the study of identical twins who have been raised separately. In such instances, if the twins were separated virtually at birth (because of the death of a parent or other family problems) and raised in randomly differing environments and scattered over a wide range of environments, then any remaining similarity in their IQs would be caused genetically, since all they have in common would be their genes (and since the effects of prenatal environment are assumed to be slight). Also, with these givens, the magnitude of the resulting IQ correlation is a direct estimate of the magnitude of the heritability of IQ.

Given the extreme rarity of separated identical twins, there have been only four studies to date of pairs of separated identical twins (Newman, Freeman, and Holzinger . . . ; Bouchard et al. . . . ; Shields . . . ; Juel-Nielsen . . . ; a fifth study, . . . that of Burt has been thoroughly discredited as having fabricated and falsified data on twins). These studies find the IQ similarity (correlation) of separated twins to be between 0.62 and 0.77. This would suggest strong genetic causation for the IQ similarity.

Methodological critics, however, discovered that quite a few of the twin pairs in three out of these four studies were not actually raised separately but were instead raised in different branches of the same family, or were separated not at the birth but later during the teenage years, or were raised in social enviornments that were very similar in many respects. In quite a few cases the twins had actually attended the same schools for the same number of years before being tested for IQ. When the truly separated twins are singled out for analysis, their pair correlation, and thus the heritability estimate, falls to 0.30 or 0.40. . . .

REFERENCE

Terman, L. M. 1916 *The Measurement of Intelligence.* Boston: Houghton Mifflin.

DISCUSSION QUESTIONS

1. In your opinion, is intelligence determined more by genes, environment, or some combination of the two? What evidence do you have for your point of view?

2. List some of the varied reasons that people of color get on the average lower scores than Whites (and some Asians) on standardized cognitive ability tests such as the SAT (Scholastic Assessment Test).

INFOTRAC COLLEGE EDITION

You can use your access to InfoTrac College Edition to learn more about the
subjects covered in this essay. Some suggested search terms include:

IQ testing

multidimensional theory
  of intelligence

nature versus nurture

predictive validity

race and intelligence testing

59

School Girls

PEGGY ORENSTEIN

*Peggy Orenstein gives a concise summary of the many ways that schools shortchange and
cheat girls in the educational system from kindergarten through high school.*

The bell rings, as it always does, at 8:30 sharp. Twenty-eight sixth graders
file into their classroom at Everett Middle School in San Francisco, strag-
gling a bit since this is the first warm day in months—warm enough for
shorts and cutoffs, warm enough for Stüssy T-shirts.

The students take their seats. Heidi, who wears bright green Converse sneak-
ers and a matching cap, pulls off her backpack and shouts, "Did everyone bring
their permission slips? You have to bring them so we can have the pizza party."

"Pizza," moans Carrie, who has brown bangs and a permanently bored expression. "That has milk. I'm allergic to milk."

Heidi looks stunned. "You can't eat pizza?"

The drama is interrupted as Judy Logan, a comfortably built woman with
gray-flecked hair and oversized glasses, steps to the front of the classroom. She
tapes two four-foot lengths of butcher paper to the chalkboard. Across the top of
one she writes: "MALES," across the other: "FEMALES."

Ms. Logan is about to begin the lesson from which her entire middle school
curriculum flows, the exercise that explains why she makes her students bother
to learn about women, why the bookshelves in her room are brimful with
women's biographies, why her walls are covered with posters that tout women's
achievements and draped with quilts that depict women through history and
women in the students' own lives.

It's time for the gender journey.

From: Peggy Orenstein. 1994. *School Girls: Young Women, Self-Esteem and the Confidence Gap.*
New York: Anchor Doubleday, pp. xi–xix. Reprinted with permission.

"Ladies and gentlemen," Ms. Logan says, turning toward the children and clasping her hands. "I'd like you to put your heads down and close your eyes. We're going to take a journey back in time."

Ms. Logan's already soothing voice turns soft and dreamy. "Go back," she tells her students. "Forget about everything around you and go back to fifth grade. Imagine yourself in your classroom, at your desk, sitting in your chair. Notice who your teacher is, what you have on, who's sitting around you, who your friends are.

"Continue your journey backward in time to third grade. Picture your third-grade teacher, your place in class. Imagine yourself in your room at home. What do you like to do when you have free time? What kind of toys do you play with? What books are you reading?"

The children go further back, to the first magical day of kindergarten, then further still, remembering preschool, remembering their discovery of language, remembering their first toddling steps. Then Ms. Logan asks her students to recall the moment of their birth, to imagine the excitement of their parents. And then, when the great moment arrives . . .

They are each born the opposite sex.

The class gasps.

"Gross," offers Jonathan with great enthusiasm.

"Yuck," adds Carrie. "That's worse than being allergic to milk."

"You are born the opposite sex," Ms. Logan repeats firmly, and then asks her students to imagine moving forward through their lives again, exactly as they were and are, except for that one crucial detail.

Again they imagine themselves walking on tiny, uncertain feet. Again they imagine speaking, entering kindergarten. Again they envision their clothes as third graders, their toys, their books, their friends.

"I can't do this," says Jonathan, who has braces and short blond hair "I just picture myself like I am now except in a pink dress."

"This is stupid," agrees Carrie. "It's too hard."

"Just try," says Ms. Logan. "Try to imagine yourself in fourth grade, in fifth grade." By the time thirty minutes have elapsed, the students are back in their classroom, safe and sound and relieved to find their own personal anatomies intact.

"Without talking," Ms. Logan says, "I'd like you to make a list, your own personal list, not to turn in, of everything that would be different if you were the opposite sex."

The students write eagerly, with only occasional giggles. When there is more horseplay than wordplay, Ms. Logan asks them to share items from their lists with the class. The offerings go up on the butcher paper.

"I wouldn't play baseball because I'd worry about breaking a nail," says Mark, who wears a San Jose Sharks jersey.

"My father would feel more responsibility for me, he'd be more in my life," says Dayna, a soft-spoken African American girl.

Luke virtually spits his idea. "My room would be *pink* and I'd think everything would be *cute*."

"I'd have my own room," says a girl.

"I wouldn't care how I look or if my clothes matched," offers another.

"I'd have to spend lots of time in the bathroom on my hair and stuff," says a boy whose own hair is conspicuously mussed. The other boys groan in agreement.

"I could stay out later," ventures a girl.

"I'd have to help my mom cook," says a boy.

"I'd get to play a lot more sports," says Annie, a freckled, red-haired girl who looks uncomfortable with the entire proposition. Many of the students are, in fact, unsettled by this exercise. Nearly a third opt to pass when their turns come, keeping their lists to themselves.

"I'd have to stand around at recess instead of getting to play basketball," says George, sneering. "And I'd worry about getting pregnant."

Raoul offers the final, if most obvious comment, which cracks up the crowd. "I'd have to sit down to go to the bathroom."

At this point the bell rings, although it is not the end of the lesson. The students will return after a short break to assess the accuracy of their images of one another. But while they're gone, I scrutinize the two butcher paper lists. Almost all of the boys' observations about gender swapping involve disparaging "have to"s, whereas the girls seem wistful with longing. By sixth grade, it is clear that both girls and boys have learned to equate maleness with opportunity and femininity with constraint.

It was a pattern I'd see again and again as I undertook my own gender journey, spending a year observing eighth-grade girls in two other Northern California middle schools. The girls I spoke with were from vastly different family structures and economic classes, and they had achieved varying degrees of academic success. Yet all of them, even those enjoying every conceivable advantage, saw their gender as a liability.

Sitting with groups of five or six girls, I'd ask a variation on Ms. Logan's theme: what did they think was lucky about being a girl? The question was invariably followed by a pause, a silence. Then answers such as "Nothing, really. All kinds of bad things happen to girls, like getting your period. Or getting pregnant."

Marta, a fourteen-year-old Latina girl, was blunt. "There's nothing lucky about being a girl," she told me one afternoon in her school's cafeteria. "I wish I was a boy."

SHORTCHANGING GIRLS: WHAT THE AAUW
SURVEY REVEALS

Like many people, I first saw the results of the American Association of University Women's report *Shortchanging Girls, Shortchanging America* in my daily newspaper. The headline unfurled across the front page of the San Francisco *Examiner*: "Girls' Low Self-Esteem Slows Their Progress," and *The New York Times* proclaimed: "Girls' Self-Esteem Is Lost on the Way to Adolescence." And, like many people, as I read further, I felt my stomach sink.

This was the most extensive national survey on gender and self-esteem ever conducted, the articles said: three thousand boys and girls between the ages of nine and fifteen were polled on their attitudes toward self, school, family, and friends. As part of the project the students were asked to respond to multiple-choice questions, provide comments, and in some cases, were interviewed in focus groups. The results confirmed something that many women already knew too well. For a girl, the passage into adolescence is not just marked by menarche or a few new curves. It is marked by a loss of confidence in herself and her abilities, especially in math and science. It is marked by a scathingly critical attitude toward her body and a blossoming sense of personal inadequacy.

In spite of the changes in women's roles in society, in spite of the changes in their own mothers' lives, many of today's girls fall into traditional patterns of low self-image, self-doubt, and self-censorship of their creative and intellectual potential. Although all children experience confusion and a faltering sense of self at adolescence, girls' self-regard drops further than boys' and never catches up. They emerge from their teenage years with reduced expectations and have less confidence in themselves and their abilities than do boys. Teenage girls are more vulnerable to feelings of depression and hopelessness and are four times more likely to attempt suicide.

The AAUW discovered that the most dramatic gender gap in self-esteem is centered in the area of competence. Boys are more likely than girls to say they are "pretty good at a lot of things" and are twice as likely to name their talents as the thing they like most about themselves. Girls, meanwhile, cite an aspect of their physical appearance. Unsurprisingly, then, teenage girls are much more likely than boys to say they are "not smart enough" or "not good enough" to achieve their dreams.

The education system is supposed to provide our young people with opportunity, to encourage their intellectual growth and prepare them as citizens. Yet students in the AAUW survey reported gender bias in the classroom—and illustrated its effects—with the canniness of investigative reporters. Both boys and girls believed that teachers encouraged more assertive behavior in boys, and that, overall, boys receive the majority of their teachers' attention. The result is that boys will speak out in class more readily, and are more willing to "argue with my teachers when I think I'm right."

Meanwhile, girls show a more precipitous drop in their interest in math and science as they advance through school. Even girls who like the subjects are, by age fifteen, only half as likely as boys to feel competent in them. These findings are key: researchers have long understood that a loss of confidence in math usually *precedes* a drop in achievement, rather than vice versa. A confidence gap, rather than an ability gap, may help explain why the numbers of female physical and computer scientists actually went down during the 1980s. The AAUW also discovered a circular relationship between math confidence and overall self-confidence, as well as a link between liking math and aspiring to professional careers—a correlation that is stronger for girls than boys. Apparently girls who can resist gender-role stereotypes in the classroom resist them elsewhere more effectively as well.

Among its most intriguing findings, the AAUW survey revealed that, although all girls report consistently lower self-esteem than boys, the severity and the nature of that reduced self-worth vary among ethnic groups. Far more African American girls retain their overall self-esteem during adolescence than white or Latina girls, maintaining a stronger sense of both personal and familial importance. They are about twice as likely to be "happy with the way I am" than girls of other groups and report feeling "pretty good at a lot of things" at nearly the rate of white boys. The one exception for African American girls is their feelings about school: black girls are more pessimistic about both their teachers and their schoolwork than other girls. Meanwhile, Latina girls' self-esteem crisis is in many ways the most profound. Between the ages of nine and fifteen, the number of Latina girls who are "happy with the way I am" plunges by 38 percentage points, compared with a 33 percent drop for white girls and a 7 percent drop for black girls. Family disappears as a source of positive self-worth for Latina teens, and academic confidence, belief in one's talents, and a sense of personal importance all plummet. During the year in which *Shortchanging Girls, Shortchanging America* was conducted, urban Latinas left school at a greater rate than any other group, male or female.

DISCUSSION QUESTIONS

1. As in Orenstein's exercise, imagine that you were born the opposite gender, and then begin to list the many differences in your life in school that would likely result.
2. How would you change education in the United States to help reduce the gender gap in certain areas of academic achievement?

INFOTRAC COLLEGE EDITION

You can use your access to InfoTrac College Edition to learn more about the subjects covered in this essay. Some suggested search terms include:

AAUW Report (American Association   gender bias in education
   of University Women)   single-sex schools
gender and self-esteem

60

The Protestant Ethic
and the Spirit of Capitalism

MAX WEBER

Max Weber's classic analysis of the Protestant Ethic and the spirit of capitalism shows how cultural belief systems, such as a religious ethic, can support the development of specific economic institutions. His multidimensional analysis shows how capitalism became morally defined as something more than pursuing monetary interests and, instead, has been culturally defined as a moral calling because of its consistency with Protestant values.

The impulse to acquisition, pursuit of gain, of money, of the greatest possible amount of money, has in itself nothing to do with capitalism. This impulse exists and has existed among waiters, physicians, coachmen, artists, prostitutes, dishonest officials, soldiers, nobles, crusaders, gamblers, and beggars. One may say that it has been common to all sorts and conditions of men at all times and in all countries of the earth, wherever the objective possibility of it is or has been given. It should be taught in the kindergarten of cultural history that this naïve idea of capitalism must be given up once and for all. Unlimited greed for gain is not in the least identical with capitalism, and is still less its spirit. Capitalism may even be identical with the restraint, or at least a rational tempering, of this irrational impulse. But capitalism is identical with the pursuit of profit, and forever renewed profit, by means of continuous, rational, capitalistic enterprise. . . .

If any inner relationship between certain expressions of the old Protestant spirit and modern capitalistic culture is to be found, we must attempt to find it, for better or worse, not in its alleged more or less materialistic or at least antiascetic joy of living, but in its purely religious characteristics. . . .

In the title of this study is used the somewhat pretentious phrase, the *spirit* of capitalism. What is to be understood by it? The attempt to give anything like a definition of it brings out certain difficulties which are in the very nature of this type of investigation.

If any object can be found to which this term can be applied with any understandable meaning, it can only be an historical individual, i.e. a complex of elements associated in historical reality which we unite into a conceptual whole from the standpoint of their cultural significance. . . .

From: Max Weber. 1958. *The Protestant Ethic and the Spirit of Capitalism*, translated by Talcott Parsons. New York: Scribner, pp. 17–27, 44–83, 157–183.

"Remember, that *time* is money. He that can earn ten shillings a day by his labour, and goes abroad, or sits idle, one half of that day, though he spends but sixpence during his diversion or idleness, ought not to reckon *that* the only expense; he has really spent, or rather thrown away, five shillings besides.

"Remember, that *credit* is money. If a man lets his money lie in my hands after it is due, he gives me the interest, or so much as I can make of it during that time. This amounts to a considerable sum where a man has good and large credit, and makes good use of it. . . .

"The most trifling actions that affect a man's credit are to be regarded. The sound of your hammer at five in the morning, or eight at night, heard by a creditor, makes him easy six months longer; but if he sees you at a billiard-table, or hears your voice at a tavern, when you should be at work, he sends for his money the next day; demands it, before he can receive it, in a lump.". . .

Truly what is here preached is not simply a means of making one's way in the world, but a peculiar ethic. The infraction of its rules is treated not as foolishness but as forgetfulness of duty. That is the essence of the matter. It is not mere business astuteness, that sort of thing is common enough, it is an ethos. *This* is the quality which interests us.

When Jacob Fugger, in speaking to a business associate who had retired and who wanted to persuade him to do the same, since he had made enough money and should let others have a chance, rejected that as pusillanimity and answered that "he (Fugger) thought otherwise, he wanted to make money as long as he could," the spirit of his statement is evidently quite different from that of Franklin.[1] What in the former case was an expression of commercial daring and a personal inclination morally neutral, in the latter takes on the character of an ethically coloured maxim for the conduct of life. The concept spirit of capitalism is here used in this specific sense, it is the spirit of modern capitalism. For that we are here dealing only with Western European and American capitalism is obvious from the way in which the problem was stated. Capitalism existed in China, India, Babylon, in the classic world, and in the Middle Ages. But in all these cases, as we shall see, this particular ethos was lacking. . . .

And in truth this peculiar idea, so familiar to us today, but in reality so little a matter of course, of one's duty in a calling, is what is most characteristic of the social ethic of capitalistic culture, and is in a sense the fundamental basis of it. It is an obligation which the individual is supposed to feel and does feel towards the content of his professional activity, no matter in what it consists, in particular no matter whether it appears on the surface as a utilization of his personal powers, or only of his material possessions (as capital). . . .

Rationalism is an historical concept which covers a whole world of different things. It will be our task to find out whose intellectual child the particular concrete form of rational thought was, from which the idea of a calling and the devotion to labour in the calling has grown, which is, as we have seen, so irrational from the standpoint of purely eudæmonistic self-interest, but which has been and still is one of the most characteristic elements of our capitalistic culture. We are here particularly interested in the origin of precisely the irrational element which lies in this, as in every conception of a calling. . . .

. . . Like the meaning of the word, the idea is new, a product of the Reformation. This may be assumed as generally known. It is true that certain suggestions of the positive valuation of routine activity in the world, which is contained in this conception of the calling, had already existed in the Middle Ages, and even in late Hellenistic antiquity. We shall speak of that later. But at least one thing was unquestionably new: the valuation of the fulfilment of duty in worldly affairs as the highest form which the moral activity of the individual could assume. This it was which inevitably gave every-day worldly activity a religious significance, and which first created the conception of a calling in this sense. . . . late Scholasticism, is, from a capitalistic view-point, definitely backward. Especially, of course, the doctrine of the sterility of money which Anthony of Florence had already refuted.

. . . For, above all, the consequences of the conception of the calling in the religious sense for worldly conduct were susceptible to quite different interpretations. The effect of the Reformation as such was only that, as compared with the Catholic attitude, the moral emphasis on and the religious sanction of, organized worldly labour in a calling was mightly increased. . . .

The real moral objection is to relaxation in the security of possession, the enjoyment of wealth with the consequence of idleness and the temptations of the flesh, above all of distraction from the pursuit of a righteous life. In fact, it is only because possession involves this danger of relaxation that it is objectionable at all. For the saints' everlasting rest is in the next world; on earth man must, to be certain of his state of grace, "do the works of him who sent him, as long as it is yet day." Not leisure and enjoyment, but only activity serves to increase the glory of God, according to the definite manifestations of His will.

Waste of time is thus the first and in principle the deadliest of sins. The span of human life is infinitely short and precious to make sure of one's own election. Loss of time through sociability, idle talk, luxury, even more sleep than is necessary for health, six to at most eight hours, is worthy of absolute moral condemnation. It does not yet hold, with Franklin, that time is money, but the proposition is true in a certain spiritual sense. It is infinitely valuable because every hour lost is lost to labour for the glory of God. Thus inactive contemplation is also valueless, or even directly reprehensible if it is at the expense of one's daily work. . . .

It is true that the usefulness of a calling, and thus its favour in the sight of God, is measured primarily in moral terms, and thus in terms of the importance of the goods produced in it for the community. But a further, and, above all, in practice the most important, criterion is found in private profitableness. For if that God, whose hand the Puritan sees in all the occurrences of life, shows one of His elect a chance of profit, he must do it with a purpose. Hence the faithful Christian must follow the call by taking advantage of the opportunity. "If God show you a way in which you may lawfully get more than in another way (without wrong to your soul or to any other), if you refuse this, and choose the less gainful way, you cross one of the ends of your calling, and you refuse to be God's steward, and to accept His gifts and use them for Him when He requireth it: you may labour to be rich for God, though not for the flesh and sin."

Wealth is thus bad ethically only in so far as it is a temptation to idleness and sinful enjoyment of life, and its acquisition is bad only when it is with the purpose of later living merrily and without care. But as a performance of duty in a calling it is not only morally permissible, but actually enjoined. . . .

Let us now try to clarify the points in which the Puritan idea of the calling and the premium it placed upon ascetic conduct was bound directly to influence the development of a capitalistic way of life. As we have seen, this asceticism turned with all its force against one thing: the spontaneous enjoyment of life and all it had to offer. . . .

On the side of the production of private wealth, asceticism condemned both dishonesty and impulsive avarice. What was condemned as covetousness, Mammonism, etc., was the pursuit of riches for their own sake. For wealth in itself was a temptation. But here asceticism was the power "which ever seeks the good but ever creates evil"; what was evil in its sense was possession and its temptations. For, in conformity with the Old Testament and in analogy to the ethical valuation of good works, asceticism looked upon the pursuit of wealth as an end in itself as highly reprehensible; but the attainment of it as a fruit of labour in a calling was a sign of God's blessing. And even more important: the religious valuation of restless, continuous, systematic work in a worldly calling, as the highest means to asceticism, and at the same time the surest and most evident proof of rebirth and genuine faith, must have been the most powerful conceivable lever for the expansion of that attitude toward life which we have here called the spirit of capitalism.

When the limitation of consumption is combined with this release of acquisitive activity, the inevitable practical result is obvious: accumulation of capital through ascetic compulsion to save. The restraints which were imposed upon the consumption of wealth naturally served to increase it by making possible the productive investment of capital. . . .

One of the fundamental elements of the spirit of modern capitalism, and not only of that but of all modern culture: rational conduct on the basis of the idea of the calling, was born—that is what this discussion has sought to demonstrate—from the spirit of Christian asceticism. . . .

The Puritan wanted to work in a calling; we are forced to do so. For when asceticism was carried out of monastic cells into everyday life, and began to dominate worldly morality, it did its part in building the tremendous cosmos of the modern economic order. This order is now bound to the technical and economic conditions of machine production which to-day determine the lives of all the individuals who are born into this mechanism, not only those directly concerned with economic acquisition, with irresistible force. . . .

Since asceticism undertook to remodel the world and to work out its ideals in the world, material goods have gained an increasing and finally an inexorable power over the lives of men as at no previous period in history. To-day the spirit of religious asceticism—whether finally, who knows? has escaped from the cage. But victorious capitalism, since it rests on mechanical foundations, needs its support no longer. The rosy blush of its laughing heir, the Enlightenment, seems also to be irretrievably fading, and the idea of duty in one's calling prowls about in our lives like the ghost of dead religious beliefs. Where the fulfilment of the calling cannot directly be related to the highest spiritual and cultural values, or when, on the other hand, it need not be felt simply as economic compulsion, the individual generally abandons the attempt to justify it at all. In the field of its highest development, in the United States, the pursuit of wealth, stripped of its religious and ethical meaning, tends to become associated with purely mundane passions, which often actually give it the character of sport.

No one knows who will live in this cage in the future, or whether at the end of this tremendous development entirely new prophets will arise, or there will be a great rebirth of old ideas and ideals, or, if neither, mechanized petrification, embellished with a sort of convulsive self-importance. For of the last stage of this cultural development, it might well be truly said: "Specialists without spirit, sensualists without heart; this nullity imagines that it has attained a level of civilization never before achieved." . . .

The modern man is in general, even with the best will, unable to give religious ideas a significance for culture and national character which they deserve. But it is, of course, not my aim to substitute for a one-sided materialistic an equally one-sided spiritualistic causal interpretation of culture and of history. Each is equally possible, but each, if it does not serve as the preparation, but as the conclusion of an investigation, accomplishes equally little in the interest of historical truth.

NOTE

1. The quotations are attributed to Benjamin Franklin.

DISCUSSION QUESTIONS

1. Weber is known for developing a multidimensional view of human society. What role does he see the Protestant ethic as playing in the development of capitalism?

2. Weber's analysis sees western capitalists as not pursuing money just for the sake of money, but because of the moral calling invoked by the Protestant ethic. Given the place of consumerism in contemporary society, how do you think Weber might modify his argument were he writing now? In other words, are there still remnants of the Protestant ethic in our beliefs about stratification? If so, how do they fit with contemporary capitalist values?

INFOTRAC COLLEGE EDITION

You can use your access to InfoTrac College Edition to learn more about the subjects covered in this essay. Some suggested search terms include:

American dream
capitalism and religious values
Max Weber

Protestant ethic
work ethic

61

Growing Up Religious

ROBERT WUTHNOW

Using different case studies of people growing up Christian and others growing up Jewish, Robert Wuthnow shows how religion is transmitted through socialization practices in the family. His analysis suggests that to understand religious belief we must analyze the practices of religious upbringing that are part of family and community traditions.

GROWING UP JEWISH

Ninety percent of Hal Meyerson's extended family died in the Holocaust. Growing up in the United States after World War II, Hal never remembers any of his family talking about it. "They just buried it," he says. "They tried to fit in and live like everyone else." But he is grateful that his parents took him to temple regularly, taught him Hebrew and Yiddish, and observed religious customs at home. "It gave me stability," he says. "I wasn't a ship just drifting in any direction. It was tradition. It had gone on since King David and it was going to go on forever. There was no question that this would go on forever." . . .

As a child, Hal spent a great deal of his time at the synagogue his parents and maternal grandparents attended. It was a Conservative temple. Both sides of the family had been Orthodox for as long as anyone could remember. But Samuel had gone away to college during the Depression, served in the Coast Guard during World War II, and then become a prominent government contractor. He joined the Conservative temple, Hal thinks, "because it was more appropriate to society and his lifestyle."

If they accommodated themselves to society in this way, Hal's parents nevertheless made sure he received an abundance of religious training. They sent him to Hebrew school three afternoons a week for an hour and a half after regular school. On Friday evenings the family lit candles and said the traditional blessings at dinner. "My father would say the hamotzi over the bread and my mother would say the blessing over the candles, my father over the wine," Hal recalls. Saturday mornings were always spent at the temple, as were Sundays. Hal especially remembers the Sunday ritual. It always began with a father-and-son breakfast of chocolate milk, bagels, and lox, and this was followed by Hebrew classes. . . .

Hal attributes much of his current interest in spirituality to the fact that his father took time to explain what everything meant while he was growing up.

From: Robert Wuthnow. 1999. *Growing Up Religious.* Boston: Beacon Press, pp. i–xl. Reprinted with permission.

Like prayers. Hal says it would have been easy to read them in Hebrew and know the words but not understand why the prayers were important. For example, in reading the Amidah, Hal's father would say, " 'Do you know what this means? Do you know why the martyrs said it while they were being tortured to death by the Spanish Inquisition or by the Russian pogroms? Do you understand what the mourner's Kaddish means?' " Hal says, "When I was a kid I used to think, 'Well, this is a prayer for dead people.' But it's not. It doesn't even mention the dead. So he would explain those things to me."

Religious upbringing was indistinguishable from family rituals and ethnic customs. Hal's parents both spoke Yiddish fluently and talked a lot about the stories they had heard of ancestors who had been rabbis or who had been persecuted in Russia during the pogroms. Hal's mother cooked ethnic foods, including ample bowls of chicken soup whenever he was sick. She made her own pickles, served chopped chicken liver on Jewish rye bread, and on special occasions brought out some of her mother's homemade gefilte fish.

The seders were especially important because they drew the wider family together. Hal recalls: "The seders were all celebrated with my father's family. And they were seders! They started at sunset and didn't finish until three or four o'clock in the morning. I remember how we used to meet in Aunt Minnie's apartment. There was a living room and a dining room, and the table stretched from the wall of the living room through both rooms to the wall of the dining room. Then there was a bathroom and two bedrooms. The coats were all piled in the bedrooms and literally nobody could move. Thirty or forty of us. All the aunts and uncles and cousins."

Growing up religious was thus a way to retain one's ties to the past. Many of the outward customs that had characterized Jewish ghettos in the past might be deemphasized, but religion could be performed in the relative safety of the family and at temple. . . .

The more he learns about the Holocaust, the more Hal believes it is essential for Jews to take their identity seriously. He thinks his ancestors probably took their faith for granted before the Holocaust but had to become more self-conscious about it after World War II. "The Holocaust sent an electric shock through the Diaspora," he says. "It's one thing to read in the books that this guy hated Jews or that in the sixteenth century there was a Spanish Inquisition. But to see the pictures of starving bodies of people just piled on top of each other and know that they were killed only because they were related to you has to have made a tremendous effect on the whole community." People need to understand their tradition and remember what happened.

SAINT PATRICK'S GIRL

When Mary Shannon was in sixth grade, she thought everyone in the world was Irish. She and her eight brothers (four older and four younger) had grown up within a few blocks of Saint Patrick's Church, attended parochial school since kindergarten, and lived exclusively in an Irish neighborhood. Besides the Malloys (her maiden name), there were lots of Kellys, Kennedys, Brennans, and O'Sheas. Two of the families at church had different-sounding names—Morelli and Cuneo—but Mary assumed they were Irish, too. It was a close neighborhood in many ways. People lived in row houses that all looked alike. Everybody shopped at Morelli's grocery store, just down the block. And Mary remembers running in and out of her friends' houses, all of which seemed to be the same: crucifixes on the walls, holy water, little statues of Mary, shamrocks.

A year later, Mary's innocence was shattered. Her father suffered a massive heart attack, causing him to lose his job and forcing the family to move to another community. The neighbors were now German, Polish, and Russian. Most were Jewish. "It was like being sucked out of something that had been your whole life and put into something that was totally foreign," she explains. Mary attended public school that year. The strain was almost too much. "I felt scared and there was a lot of sadness. I didn't feel accepted. I missed what had been such an integral part of our life. I felt very different, and up to that point I just felt like I was just like everybody else."

The following year the family moved again, this time to another Irish Catholic community. Mary went back to attending parochial school. After high school graduation, she attended a Catholic college. There she met Tom Shannon, whom she married at age twenty-two. By the time she was twenty-five, she had given birth to two daughters and a son. A few years later, she started teaching and commuted to another Catholic college in the area for a master's degree in school administration. Since then, she has taught for eight years at the elementary level and served for six years as a principal, all in Catholic schools.

Being Catholic is thus an integral fact of Mary's identity. As she reflects on the values she was taught as a child, she says instantly that "the church was the center of our life." She means this quite literally. Her paternal grandparents were Irish immigrants who lived nearby and made sure their grandchildren were receiving proper religious instruction. Her father had been forced to quit school in eighth grade in order to help support the family. Over the years he worked his way up the ladder at a local manufacturing company and was active in the union.

Her mother had worked before marriage and got a job at a bakery when the children were old enough to be in school. But with nine children, the family always struggled financially. They lived in a row house at first and eventually moved into a run-down twin house that her father rented cheaply in return for doing maintenance work. Mary remembers that the church pitched in on a number of occasions to bring food to the family and to supply presents for the children at Christmas. "The community was very tightly knit," she remembers. "When some were struggling, the others would help." . . .

TARRYING WITH JESUS

Jess Hartley has known few of the material advantages that Hal Meyerson and Mary Shannon have enjoyed. Jess grew up in a low-income housing development; his family was supported by the meager wages his father earned as a clerk at a toy store and, later, as an independent contractor doing janitorial work and polishing hardwood floors. Like Mary, Jess is also the fifth of nine children. He is a former high-jump champion and soccer star who enjoys singing in the church choir, eating his mother's fried chicken, and taking his daughter on outings to Chuck E Cheese. At thirty-seven, Jess is a muscular man with medium brown skin, hazel eyes, and a neatly trimmed beard, who stands six feet two and weighs about 180 pounds.

He has followed in his father's footsteps, taking over the floor polishing business when his dad went into semiretirement a decade ago. For the first ten years of their marriage, Jess and his wife lived in a "semi" (or duplex) located in an inner-city neighborhood just three blocks from an area infested with drug dealers. They considered themselves fortunate because they had purchased the duplex with a mortgage of $300 a month, which they paid by renting out the other half. Their daughter is now five years old and they have recently moved to a working-class community a few miles away in order to be in a better school district. Naturally gregarious, Jess enjoys meeting his customers, often engaging them in long conversations while he waits for a section of freshly waxed hardwood to dry.

When asked what sort of values he learned from his parents, Jess spontaneously emphasizes his religious upbringing. "I can remember we always had prayer in the home. One of the things we always learned was that God was the head of our life. It wasn't just that you have to do this or that. We always were reminded that whatever we had, it was because of God. God was the one that allowed us to have these things."

Jess says he learned from an early age the story about Christ's coming into the world and dying on the cross. He knew the story long before he had any idea what it meant. Gradually he learned that it was a privilege to receive Christ and that a person had to seek Christ with utmost sincerity for this to happen. . . .

Jess's family belonged to the United Holy Church of America, a small denomination founded in North Carolina, by revivalist Isaac Cheshier, in 1886. By the time Jess was a teenager, the denomination numbered some sixty-five thousand members in nearly seven hundred congregations scattered from Georgia to New England, including approximately two hundred churches that split off to form their own denomination a few years later. The "tarrying" Jess experienced as a child was common to most of the holiness and Pentecostal denominations that emerged around the turn of the century and spread rapidly among both African Americans and European Americans in working-class communities. . . .

The church was a pervasive influence in Jess's life from the time he was little. The Hartleys went to church every week, despite one or two of the children's being sick on any given Sunday, simply loading everyone into the battered family station wagon and driving the few miles to their house of worship. Generally the service began at eleven o'clock and ran until two-thirty, sometimes longer "if the spirit started moving." The Hartleys also attended Sunday school before the morning service, an afternoon service, and an evening service. . . .

Conscience may not be totally effective in corralling adventuresome youth, but the inner turmoil it creates can leave a powerful impression. As Jess grew older, he realized increasingly that his religious upbringing made him different from other young people. Especially in relating to the opposite sex, he felt inadequate and inexperienced. Despite the fact that he played three sports and was vice president of the student council, Jess had little interaction with girls during his first three years in high school. Then in his senior year he realized what he was missing and started dating a girl named Wanda.

Time came for the senior prom and Jess asked Wanda to go. The only problem was that Jess's church did not allow dancing, other than the gospel dancing and shouting that took place during the services. So Jess arranged to pick up his tuxedo at a friend's house and told his parents he and Wanda were just going to dinner. The plan worked: Jess and Wanda went to the prom, barely avoiding being seen by some people in the neighborhood who knew his parents, and then went to an after-prom party at another friend's house.

Unknown to Jess, however, his parents had discovered the truth by interrogating his sister, and when he arrived home at dawn the front door was locked and chained. . . .

Using his prom experience as an example, Jess says he learned to forgive himself and move on. "You forgive and forget because that's what Jesus did for us. He forgave our sins. You put them in the sea of forgetfulness and go on from there."

For Jess, going "on from there" has meant staying involved in his denomination and continuing to pray, like his mother did, even when things don't go as planned. He and his wife suffered the anguish of a miscarriage and the death of a newborn, and they have struggled continuously to make ends meet. For the past five years they have attended a congregation of about 150 people who support each other in good times and bad. . . .

THE SIGNIFICANCE OF THE PAST

Growing up religious has a profound influence on many people's lives. It is the source of fond memories for some and of painful experiences for others. When people meet others who have stories to tell about the congregations in which they grew up, the verses they memorized, and the special holidays they celebrated, there is an instant bond. People who have been raised in the same tradition feel the connection acutely, but religious upbringings also bridge traditions, creating a subculture of common understanding.

For many people, remembering and telling their stories is a way of making sense of their lives and of their continuing quest for the sacred. Childhood religiosity is not simply the experience of children but of adults as well. Memories ripen with age, especially as they are refined in an ongoing dialogue with one's experiences. . . .

Effective religious socialization comes about through embedded practices; that is, through specific, deliberate religious activities that are firmly intertwined with the daily habits of family routines, of eating and sleeping, of having conversations, of adorning the spaces in which people live, of celebrating the holidays, and of being part of a community. Compared with these practices, the formal teachings of religious leaders often pale in significance. Yet when such practices are present, formal teachings also become more important.

The past is not static. It is a remembered past and thus one that people are continually revising, making sense of, and reinterpreting. Many people of course have little in their childhood to remember about religion. But those who grew up in a religious household continue to have a very substantial impact on the character of American religion. They are far more likely than other people to be active members of a congregation as adults and to want their own children to receive religious training. Many of them sense that society is declining as a result of forgetting its religious heritage. Yet they are often painfully aware of the need for change, having undergone much change themselves. Understanding how they have changed is a way of gaining a clearer sense of how America itself is changing. . . .

The first lesson to be learned from people's accounts of their religious up-bringing is thus that particular events and experiences were significant enough to be memorable. These of course varied from person to person; indeed, it was their particularity that helped define each person's individuality. The common feature of these events and experiences, nevertheless, is the fact that they consisted of embedded practices. . . .

The second lesson to be learned as we listen to people talk about their religious upbringing is that it is part of the continuing experience of adults rather than an event occurring only in childhood. The conversations people carry on with their past transform growing up religious into a living memory; their reflections are an act of anamnesis, an antidote to amnesia, to forgetting. Those who had reflected most deeply on their religious upbringing felt they were able to live a more fully integrated life as a result. . . .

Anamnesis is facilitated by family gatherings, such as holiday celebrations, when people sit for long hours with parents and siblings and with members of their congregations, retelling and reinterpreting the stories of their past. Such occasions are themselves memorable, serving as crucibles in which to reconsider the stories told by previous generations. When they are absent, they are missed; as one of our older respondents says in talking about her parents: "I should have forced them to tell me their stories!" These occasions are also becoming fragile because people are being geographically dispersed and living in the interstices among multiple communities.

Apart from the lessons about personal spirituality, growing up religious also teaches an important lesson about how to live in an increasingly diverse society. Those who have been raised in intensely religious families often recognize an affinity with one another, and this affinity frequently includes people whose particular traditions have been quite different from their own. They know what it was like to make long preparations for a religious service, even though the services were different, for example. They may also regard one another as kindred spirits because of having been taught a certain appreciation of God or of particular moral values. Because of their loyalty to their own tradition and their awareness of other traditions, they often have an appreciation of what it means to live in a pluralistic society.

Living in a multicultural society requires an ability to forge connections with people who are similar in some respects and different in others. Jews and Christians, Catholics and Protestants, European Americans and African Americans who were able to craft such links with one another through some understanding of their common religious experiences have much to tell us about the possibilities of cultural harmony. The pluralistic character of American religion, encouraging loyalty to particular communities of faith and yet elevating certain unifying ethical standards, also has much to teach about these possibilities. . . .

DISCUSSION QUESTIONS

1. Robert Wuthnow describes religious socialization as occurring through specific cultural practices, many of which are part of family routines. Think of two people you know who have grown up in different religious faiths and compare the religious socialization practices in their families.

2. What significance does Wuthnow give to memory as a part of the process of religious socialization?

INFOTRAC COLLEGE EDITION

You can use your access to InfoTrac College Edition to learn more about the subjects covered in this essay. Some suggested search terms include:

religiosity   religious traditions
religious rituals   sociology of religion
religious socialization

62

Growing Up American: The Complexity of Ethnic Involvement

MIN ZHOU

Religion and ethnicity are often intertwined. Based on a community study of Vietnamese immigrants in New Orleans, Min Zhou explains how participation in religion is an important part of building identification with an ethnic community. In this and other ethnic communities, religious participation can be an important part of ethnic involvement.

Vietnamese communities across the United States are characterized by multiple levels of ethnic involvement. In one way or another, the common refugee experience and cultural heritage have brought almost all Vietnamese refugees and most of their offspring into a dense social system made up not merely of family and friendship ties but also of connections of ethnic religious and work organizations. These networks help Vietnamese gain access to material support as well as to intangible support and thus actively engage both adult and young members of the community. Over time, these connections become increasingly closed and complex in form. . . .

Religious participation provides yet another form of ethnic involvement. Buddhism and Catholicism are the most important religions among Vietnamese refugees in the United States, and for both Buddhists and Catholics, the religious institution is much more than a house of worship; it is a place where they can share feelings and emotions, engage collectively in the struggle to reestablish their lives, and transmit the ancestral language and culture to the younger generation. Thich Thien Chi, the chief monk of Phap Hoa Temple in a small Vietnamese community in New York City, said to us, "The point of this temple is to have people come together and teach them how to be a good person. To guide them out of their suffering."

The temples and churches are sites for regular worship and other formal religious practices; they also function as social service organizations, operating a wide range of programs such as after-school programs for children, youth programs, summer camps, festival celebrations, and family counseling. One Buddhist center in Little Saigon, for example, conducts courses for Vietnamese children on Sun-

From: Min Zhou. 1998. *Growing Up American: How Vietnamese Children Adapt to Life in America.* New York: Russell Sage Foundation, pp. 96–107.

day mornings, when roughly two hundred children aged 7 to 18 are taught Vietnamese and hear talks on Vietnamese culture. Similar programs for both children and adults can be found in the Catholic church.

Versailles Village is heavily Catholic, and the Catholic church is the single most important ethnic institution. Of the Vietnamese high school students we surveyed, 87 percent told us that they were Catholic; 10 percent were Buddhists, and the rest fell into a range of other denominations, such as Baptists and *Cao Dai* (an indigenous Vietnamese religion). There were no Vietnamese Buddhist temples in the immediate vicinity of Versailles Village, but there was a Vietnamese temple in New Orleans, easily reached by car. The neighborhood contained two Vietnamese churches: a small Baptist church, with a tiny but almost exclusively Vietnamese congregation, and a large Catholic church, Mary Queen of Vietnam Church, whose location at the geographical center of the neighborhood reflected its institutional centrality to the community, as all the Catholics were parishioners of that church. The pastor of the Catholic church maintained close contact with monks at the nearby Buddhist temple, and the religious leaders often coordinated activities that concern all the Vietnamese in New Orleans. Despite the theological differences between the Catholics and the small number of Buddhists, we observed little difference in the social functions of Vietnamese Catholic churches and Vietnamese Buddhist temples in New Orleans, which corresponded to the anthropologist Paul J. Rutledge's findings of a Vietnamese community in Oklahoma City (1985).

Like the religious institutions in Little Saigon, Mary Queen of Vietnam Church in Versailles Village ran after-school classes for young people. While these classes were organized by religious personnel, such as Brother John Nhon, the volunteer teachers were frequently Vietnamese public school teachers or assistant teachers. The church concerned itself not just with religious teachings but with broader educational issues as well. For instance, when leaders in the Vietnamese community attempted to initiate the teaching of Vietnamese as an elective in a public high school in 1991, and when Vietnamese educators and concerned citizens were protesting the elimination of the Office of ESL–Bilingual Education in 1992, community leaders met with an official of the Orleans Parish ESL–Bilingual Education Section in Monsignor Luong's office at the church (personal communications with Jesse Nash, Sept. 1, 1993, and Charlotte Stever, Sept. 23, 1993).

Given its centrality, the church serves as a primary mechanism for integrating young people into the community's system of ethnic relations. Theoretically, we thus argue that the involvement with ethnic religious institutions can strengthen ethnic identification while also reaffirming ethnic affiliation. To validate this theoretical argument, we asked in our 1994 survey whether young people who participated more often in Vietnamese religious organizations were more likely than others to describe themselves unequivocally as "Vietnamese" rather than as "Vietnamese American" or "American." Since most young people in this community showed some level of religious participation, we looked at differences among those who attend their church or temple once a month or less, about once a week, and more than once a week.

It should be noted that religious participation appears to be extremely intense. As is shown in table 4.1, about 43 percent of the Vietnamese high school

**Table 4.1   Ethnic Self-Identification of Vietnamese Youths,
by Frequency of Church or Temple Attendance**

	Church or Temple Attendance			
	Once a Month or Less	About Once a Week	More Than Once a Week	Row Total (*N*)
American (%)	10.3	3.4	0.0	3.3 (13)
Vietnamese American (%)	38.5	29.5	33.9	33.2 (131)
Vietnamese (%)	51.3	67.1	66.1	63.5 (251)
Column total (%) (*N*)	19.7 (78)	37.7 (149)	42.5 (168)	100.0 (395)

Source: The Versailles Village Survey of 1994.

Chi-Square = 20.63; $p < .01$

students whom we surveyed went to church or temple more than once a week; another 38 percent went to church or temple about once a week, and only 20 percent of them were infrequent churchgoers or nonparticipants in ethnic religious institutions. Those who went once a week or more than once a week were more likely to describe themselves as "Vietnamese" than the infrequent churchgoers or nonparticipants. None of those who attended more than once a week chose "American" as a self-description. By contrast, over 10 percent of the infrequent churchgoers or nonparticipants preferred "American" as a self-description.

Self-description, of course, is only one aspect of involvement with an ethnic group: one can describe onself as belonging to a group and yet maintain no day-to-day contact with coethnic members. Moreover, our argument establishes a linkage between religious involvement and ethnic affiliations. We measured ethnic affiliations by ethnic preferences for coethnic friendship and marital partners. One can judge the extent to which individuals find themselves enmeshed in an ethnic network by the degree to which ethnicity defines their friendship circles. To explore this issue, we looked at the relationship between church or temple attendance and the proportion of the respondents' friends who were coethnic (see table 4.2). Here there was a linear relationship, with striking differences between frequent and infrequent religious participants. Nearly 70 percent of the students who attended church or temple more than weekly responded that either all or almost all of their friends were Vietnamese, and only 4 percent of them reported that they had some or very few Vietnamese friends. In contrast, only 42 percent of those who showed little or no participation in ethnic religious institutions said most of their friends were Vietnamese, while well over a fifth (22 percent) of them reported that they had some or very few Vietnamese friends.

We then asked about preference for marital partners. As table 4.3 shows, young Vietnamese who attended a religious institution often were more likely

Table 4.2   **Proportion of Coethnic Friends Among Vietnamese Youths, by Frequency of Church or Temple Attendance**

	Church or Temple Attendance			
	Once a Month or Less	About Once a Week	More Than Once a Week	Row Total (N)
None (%)	3.8	1.3	0.0	1.3
				(5)
Very few (%)	7.7	5.4	2.4	4.6
				(18)
Some (%)	10.3	3.4	1.8	4.1
				(16)
About half (%)	15.4	8.1	9.5	10.1
				(40)
Most (%)	20.5	34.2	17.8	24.6
				(97)
Almost all or all (%)	42.3	47.6	68.5	55.3
				(219)
Column Total (%)	19.8	37.7	42.5	100.0
(N)	(78)	(149)	(168)	(395)

Source: The Versailles Village Survey of 1994.

Chi-square = 40.88; p < .01

to prefer a Vietnamese spouse. Of the two groups of respondents who attended once a week or more, almost two-thirds said that they would "prefer" or "definitely want" to marry someone who was Vietnamese. Marrying within the group was widely desired and widely expected. Those who attended church or temple infrequently or not at all, however, were more likely to say that they did not care whether they married a Vietnamese person or not (46 percent, compared with 38 percent for weekly participants and 28 percent for those who attended more than weekly), or even that they definitely did not want or preferred not to marry someone who was Vietnamese (9 percent, compared with 2 percent of those who attended weekly and less than 1 percent—one individual—of those who attended more than weekly). Moreover, commitment to endogamy is not simply associated with participation in ethnic religious institutions. The relationship is linear: the greater the participation, the greater the commitment to marrying within the group. Of course, the theoretical choice of a marriage partner is highly speculative, and the actual choice may be limited by the lack of contact. For example, one of the young Vietnamese men that we interviewed remarked, "I don't have anything against other (non-Vietnamese) girls. I just don't think they would completely understand what I have to say, even if I say it in English." Nonetheless, endogamy indicates intense ethnic involvement.

Table 4.3 Commitment to Endogamy Among Vietnamese Youths,
by Frequency of Church or Temple Attendance

	Church or Temple Attendance			
	Once a Month or Less	About Once a Week	More Than Once a Week	Row Total (*N*)
Definitely do not want Vietnamese spouse (%)	5.1	1.3	0.6	1.8 (7)
Prefer non-Vietnamese spouse (%)	3.8	0.7	0.0	1.0 (4)
Do not care (%)	46.2	37.6	28.0	35.2 (139)
Prefer Vietnamese spouse (%)	28.2	40.9	41.6	38.7 (153)
Definitely want Vietnamese spouse (%)	16.7	19.5	29.8	23.3 (92)
Column Total (%)	19.8	37.7	42.5	100.0
(*N*)	(78)	(149)	(168)	(395)

Source: The Versailles Village Survey of 1994.

Chi-square = 28.09; $p < .01$

COMMUNITY-BASED ORGANIZATIONS

Closely connected to the religious institutions are various secular social organizations. In Versailles Village, for example, the formalized, well-established, and influential organizations include the Vietnamese-American Voters' Association, the Political Prisoner Veterans Union, the Versailles Neighborhood Association, the *Dung Lac* (a youth program), and the Vietnamese Educational Association. The most important organization affecting young people directly is the Vietnamese Educational Association. This association runs two major projects—after-school classes at the Child Development Center and an annual awards ceremony in honor of Vietnamese students who have excelled in the public school system. Although both projects are held on the grounds of the Catholic church, the educational association is not an exclusively Catholic organization. A Buddhist monk sits on its board of directors, and Buddhist as well as Catholic children may participate in after-school classes and receive awards at the annual ceremony. The after-school classes, offered on a voluntary basis to elementary and high school students, emphasize language instruction in both English and Vietnamese, although other academic subjects are also offered from time to time. English language classes serve the needs of relatively new arrivals and others whose English skills are weak; the Vietnamese language classes serve the native-born and those who have lived in the United States since early childhood. The Vietnamese language classes, which are taught by a Vietnamese priest attached to the church, place a heavy emphasis on reading and writing skills, since many young people who learn to speak their parental language in the home have never had an opportunity to develop literacy.

Vietnamese social workers run the *Dung Lac* (named after a Vietnamese religious martyr), which was set up by a Vietnamese priest in 1991 to cope with the growing problems of troubled youths. This organization, which features weekend retreats, evening sports events, and service projects such as cleaning up the neighborhood, seeks to involve troubled youths in productive activities and eventually get them into "life planning" courses that provide counseling and access to jobs.

Virtually all community organizations and activities have a church connection. For example, the local Vietnamese Voters Association, which helps to prepare eligible community members for the test for U.S. citizenship, holds all its meetings on the grounds of the church, and a priest serves as its advisor. The church also provides the site for community meetings at irregular intervals to discuss problems and goals. Every Saturday morning, the church grounds become an open-air market, where all Vietnamese in the Versailles neighborhood can sell their goods. The entire church parish divides itself into zones, each of which has a "zone leader," an influential person, who represents zone residents at meetings held at the church to decide both secular and religious activities and policies. All these continuing church-centered activities provide ample opportunities for ethnic interaction and thus help strengthen ties among members while also reinforcing the leadership roles of religious institutions and community-based organizations. . . .

The community and family networks of Versailles Village create a distinctive set of social relations based on "respect" according to social roles, surround young people with a complex system of ethnic involvements including economic, religious, and psychological elements, and hold young people to a system of norms and values directing them toward constructive patterns of behavior. Paradoxically, intense ethnic involvement increases rather than decreases the probability that young people will gain entry into the world beyond the ethnic community. . . .

REFERENCES

Rutledge, Paul J. 1985. *The Role of Religion in Ethnic Self-Identity: A Vietnamese Community.* Lanham, Md.: University Press of America.

DISCUSSION QUESTIONS

1. Sociologists who work from a functionalist perspective often see social institutions as fulfilling purposes in addition to those for which they are organized. What function does religion serve in the Vietnamese immigrant community that Min Zhou studies?

2. How are religion and ethnicity intermixed in the community that Zhou describes? Compare and contrast this with the religion of other ethnic groups with which you are familiar.

INFOTRAC COLLEGE EDITION

You can use your access to InfoTrac College Edition to learn more about the subjects covered in this essay. Some suggested search terms include:

ethnicity (ethnic group)   interfaith marriage
immigrant communities   religious participation
immigrant religion   Vietnamese immigrants

63

Insane Therapy

MARYBETH F. AYELLA

People have been fascinated with the study of those who join religious cults. Converts are often assumed to be "brainwashed" or to have some sort of psychological problem. By describing her experience as a participant observer in the Unification Church (otherwise known as the "Moonies"), sociologist Marybeth Ayella shows how normal processes of social influence explain how people become members of cults.

My interest in "cults" began with a chance encounter with a "Moonie" in Sproul Plaza, Berkeley, in August 1975. In response to what I thought was a pick-up attempt by a man standing behind me at a sandwich vendor, I began a conversation with a man about my age. We wound up sitting and talking as I ate my lunch. He described a wonderful communal group he lived in, in which there was never any conflict. This intrigued me—no fights over who does dishes? over who hogs the bathroom? He invited me to come and see for myself; in fact, I could come home with him that night, since they were having a special Friday night dinner.

From: Marybeth F. Ayella. 1998. *Insane Therapy Portrait of a Psychotherapy Cult*. Philadelphia: Temple University Press, pp. 1–14.

I was twenty-three years old, just beginning a doctoral program at the University of California at Berkeley. Since I knew no one in town I did not have dinner plans, so I accepted the invitation. Jack was taking public transportation home, and I felt safe in accompanying him. Before going, I asked if the group he belonged to was religious, since I was not interested in religious communes. He said no.

Jack and I arrived at the Oakland house, located next to a Roman Catholic church, together with a second group member, a woman we had met on the bus. We all took off our shoes at the entrance, and I was invited into the kitchen to meet others who were making dinner. This was the first thing that favorably impressed me—the members preparing dinner were all men, yet they seemed to know just what to do. I had known several male housemates and plenty of male friends who were unable to prepare anything but frozen foods (according to directions); these men were not only washing and chopping vegetables but doing so happily. I joined in, and everyone was very friendly and warm to me. The apparently changed gender roles fit my feminist orientation.

Everyone seemed to be about my age, and I was curious as to how they earned their living. I had just arrived from Boston, and one of the members came over to talk when he heard that. He had just dropped out of the University of Massachusetts at Amherst, only a year before graduation. I asked why. He said that since he had met the group, he felt like he wanted to stay and give communal living a try. I asked what he did to support himself, and he said he worked in a group-owned business. This made no sense to me—why not finish up his last year, come back to the group, and earn a living in journalism, as he'd planned? His explanation that he had quit because he had never met such a great collection of people made no more sense. I was mystified; others in the house told similar stories of being in college, meeting the group, and deciding to drop out to work in businesses with group members. What was it about this group that would lead a person to decide to change career and life plans so abruptly, all for a "great" group of people?

Dinnertime came, and it was obvious that I was not the only guest. Our hosts and hostesses were clearly solicitous of our comfort from the moment we walked in. Before we started dinner, a man sitting at the head of the long table began a lengthy grace. At this I turned to Jack and asked why say a grace if the group was not religious? He replied that some of the members were religious, and whoever led the dinner could begin with grace if she or he wanted.

Later, we went into the living room to hear more in a lecture. It was very vague, describing the group's interest in such things as living in peace and harmony, first on a small-group scale, then nationally and internationally. I did not find myself disagreeing with anything that was said, but I had no idea how the ideals were put into practice. After the lecture, we saw color slides of the group at their Boonville farm, north of San Francisco. It looked like summer camp, with a large number of people clowning around with each other, mugging for the camera.

We had dessert afterwards. We sat around on the floor, in small groups, getting to know each other. As the conversation was winding down, we were all invited to come for a weekend seminar in "group living" for twelve dollars. I was intrigued, and I said I would like to go.

It seemed as though the bulk of the guests were also deciding to go. Group members brought out consent forms for us to sign, in the event we were hurt. I asked what kind of injury they anticipated, and I was told "sprained ankles," because the farm's terrain was hilly and rocky. That didn't seem too bad, so I signed. One other thing had aroused my curiosity—the letterhead said Unification Church. I asked whether the group was a church; I was told no and given some answer about simply using their letterhead paper, but not being connected in any way. The impression I got was that they were cheap, using the excess letterhead paper of some church.

I was told I could drive up with Jack and some other members in one of the group's cars that night. We stopped at my apartment so that I could pack a suitcase, then we got on the road about 11:30 P.M. When we got into the car, I suddenly realized I was traveling alone with three men, and I had a momentary qualm as I remembered my mother's warnings about not talking to, let alone traveling with, "strange men." I wondered what I was doing in a car with three men, at midnight, going to a farm outside a town I had never heard of until that day.

Shortly after we turned onto the northbound highway my companions produced songbooks and suggested we sing. I refused; it seemed too corny and reminded me of those musicals where people going about routine activities burst into song. Further into the trip, when I noticed Jack, the driver, dozing off, it was I who suggested we sing, fearful we would be in a car crash. I also suggested we get coffee and something to eat. We stopped at a roadside restaurant, and I was the first in line to order. I was also the only one to order—suddenly the others were not hungry, and they didn't want coffee, not even Jack. Thinking they had no money, I offered to buy them food, but they refused, insisting they were just fine. On our way out, we came across another carload of people also headed to Boonville. I began to wonder how many of them there were.

We at last reached Boonville, which appeared to me to be a one-street town in the middle of nowhere, and we turned off the main street down a winding road, with steep inclines at various spots. Jack was still nodding off and I was seated within elbow's reach. I jogged his arm every time his head dipped forward, and I sang or asked questions the rest of the ride. I was really scared for the last half hour or so, because the road was so dimly lit, winding, and especially precipitous around curves. Jack had refused all my offers to drive, so I sat with my heart in my throat.

When we finally arrived at the farm, we approached a gate topped with barbed wire and a barbed-wire fence, with what appeared to be a guard at the gate. When I asked about the barbed wire, Jack told me people were always trying to sneak in. Soon after we arrived and parked, a car full of women from the restaurant pulled in; Jack asked one of them to help me find a sleeping bag and get settled. Sleeping quarters were segregated. We entered a mobile trailer that was wall-to-wall bodies in sleeping bags. As I crawled into the bag I was given to use, I had an acute anxiety attack: what am I doing here? what kind of people are they? what if they're like Charles Manson's Family? are they into drugs? violence?

I didn't sleep that night. The last time I looked at my watch it was 7:00 A.M., and people had been coming in all night. An hour or so later, I was awakened by people playing guitars and singing "when the red red robin comes bob-bob-bob-bin' along." At this novel wake-up call; everyone seemed to wake up, jump up, and roll up their sleeping bags, except me. I had decided I was tired and would prefer to miss the morning seminar. Of course, after being asked repeatedly to get up so as not to miss any of the day's exciting activities I finally realized I was wide awake and not likely to fall back to sleep. So I got up.

Wandering out after dressing and washing, I stood on the sidelines observing the group exercises. Again, I was repeatedly asked to take part, and I finally gave in to "This guy is so shy, and you're so friendly and outgoing, would you be his exercise partner?" During my stay, I frequently wished I had asked more questions in my methods class in college, as I confronted a myriad of situations in which I did not know how to act. Key was how much to participate. Get up or sleep in? exercise or watch? were just the first of many choices I had to make.

When we were put into small groups after the exercises, Noah, the lead lecturer for the weekend, instructed us on how to behave in order to "get the most out of the weekend." Should I do exactly what Noah said, for example by not talking to other newcomers? I did not feel constrained to go along with group directives completely, because I felt I could not get a full picture of the persons involved: how could I find out why these others were thinking of living communally with this group if I did not ask? But I wanted to fit in, I did not want to stick out like a sore thumb, and to have members act normally, I did not want always to point up my research interests.

I ate breakfast with the small group I had been assigned to. Introductions were made, and we began to eat. Abruptly I realized that none of the people I identified as members of the group (including Jack) ate any food—suddenly thinking that the food was poisoned, I stopped eating, and decided to eat only what they ate. I had to wait until lunch to eat, and I carefully observed the members' choices.

The remainder of that day was a comedy of misperception on my part. My definition of the situation—that I was participating in a "weekend seminar in communal living"—was rudely shattered during the first lecture, which dealt with Adam and Eve's fall from grace. There I realized the definition of the situation on their part—conversion. This was obviously a religious group (something I had repeatedly been told it was not). Anger at having been lied to was the first feeling I shared with my small group after the lecture. We were asked, "What do you think of the lecture?" I said, "You lied, this is a religious group." The response was, "Would you have come if we said it was a religious group?" When I replied no, I was told that was precisely why they lied, to get people who otherwise would not have come to come, so that they could appreciate the organization's obvious merit.

My uncertainty, always a part of field research, increased because I did not know what kind of a religious group this was. I did not know how to act, I had been lied to about what I considered a major thing, and so I wondered what if anything I could believe. As I learned over time, this was one example of "Heavenly Deception," which was explained to me as justifying various instances of deceptiveness that I saw members engage in during different aspects of recruitment.

I had been clear about my identity from the start. I felt betrayed by the group, my naïve trust was destroyed, but I still felt compelled to be truthful with them in answering questions about my reaction to the group and my research interests. How much to be revealed of self, and where, were two key questions I had. How open to be in lectures and groups (public settings) as compared to casual talks with members and other prospects (some of which were private settings)? I later found out that I had also misperceived the nature of public and private settings. I discovered this when I learned that information I had confided to only one other person was suddenly revealed by someone else and that this information was being used to persuade me to stay.

Other occasions caused me to question how much to participate: Afternoon dodge-ball game—watch or play, cheer while playing or not? Composing a skit about life in the "family" for Saturday night, telling how one came to group— how to express my research interests as all around me members told dramatic before-and-after stories. During lectures, deciding how much to participate was difficult—I kept questioning the lecturer, because he asked for questions and the lecture seemed so flawed, so one-sided. "Not everyone." I might say, "would agree that Richard Nixon was on the side of good/God and feminists are on the side of bad/Satan." I couldn't help but try to get answers, since I was really curious, and more so after getting the impression that this group really believed we were in the "last days." Moreover, many "questions" other participants offered were in the form of comments praising the group— "I really loved that lecture, everything is so clear now." Soon, the lecturer began to ignore my raised hand, even when there were no other hands raised.

I learned when I went to the bathroom before returning to the lecture that a woman I had been introduced to in my small group, who sat next to me at breakfast and at the first lecture, was intent on sticking to me like glue. When she accompanied me the short distance to the bathroom, and came into the two-stall building, I did not know what to make of her. Overly friendly? lesbian? weird? That caused me to see how everyone who was new seemed to have the same kind of "buddy." It soon became clear that I was expected to confide all my reactions to this buddy. My problem was that I just did not like this woman, so I kept trying to change our conversations around to elicit information on why she had joined the group.

Early on Sunday members began questioning us newcomers about whether we wanted to stay for the "advanced" set of lectures. My choice as I saw it was to stay and study a millenarian religious group or leave. I still did not know the organization was the Unification Church (UC), but at that time the name would not have meant much to me. The chief attraction of staying was the chance to study a group dramatically different from what I was familiar with, churchgoing Roman Catholicism. I was sure someone would want to publish an analysis of the community because it seemed so interesting in its strangeness. I was puzzled at how people who seemed similar to me could want to become or remain members. Some of my students can not understand this initial decision: "Why not get out when the going's good? They're a bunch of crackpots! You must have been crazy to stay, with a bunch of strangers, of religious fanatics, in an isolated location." Looking back at the twenty-three-year-old, I too wonder at her staying, but not at her interest.

When I made the choice to remain for the next set of lectures, I felt fairly certain that the group was not violent and did not use drugs, my two biggest fears. Their beliefs seemed opposed to both. I was intensely curious, I had time on my hands before the school semester began, and I had dreams of glory (my master's thesis had just been accepted for publication).

Weighing against these pros were the cons: twinges of anxiety aroused by certain things, such as that no one knew where I was and the one public pay phone bore an out-of-order sign. I wondered how much I could trust my new acquaintances. I questioned whether they were as they appeared to be, a peaceful group; if so, why the barbed wire? What happened to the guy from my group, the one other person seeming to have doubts, who disappeared Saturday night? Why didn't Jack and the others eat at first, either at the roadside diner or at Saturday breakfast?

I was heavily pressured by various members to stay, and I agreed to do so on the one condition that I be allowed to write about them. The members agreed, and even introduced me to a woman who they said had also come to study the organization and who had remained.[1] This only increased my anxiety: I did not want to become like her, I did not want to lead this kind of life.

I filled the requirements for what the Moonies were looking for then in a convert—young, unattached, idealistic. I did not know this at first, but very few of the members or prospects I encountered were older than their late twenties or early thirties. This made it more difficult for me to be objective. Members seemed to be very similar to me, and one of my initial questions was why they were taking such a different path. I could not believe anyone would decide to join such an organization virtually overnight, with dropping out of school and not returning to one's home state or native country (the United Kingdom and Australia, for the people I met) being among the consequences. But, the people in my group seemed to have made this choice. So one thing that differentiated "us" was impulsiveness—no listing of pros and cons, no talking to people to weigh options (a major reason being that we were on the farm and couldn't contact anyone).

After the weekend I was given a phone number and an address, and I included the number in a letter to friends. Later, I learned that those who tried to call me were told "no one lives here with that name." This was not necessarily a case of malicious withholding by the UC; it seemed, rather, to be a result of not knowing where all prospects were at all times. However, my friends did wonder where I was, and I wondered why no one wrote or called me during my three-week stay. Thus, I was effectively isolated from all but members and prospects during my participant observation.[2]

My account differs from Barker's (1984) in that she well knew what the UC was when she finally went to the northern California farm. Older than I, with a family, and an established sociologist of religion, she was approached by the UC to do research (intended by the church to counter other, negative reports). Barker made her visit with the realization that this was the organization's most effective, and most deceptive, branch. This deceptiveness may have contributed to the effectiveness of the northern California branch of the UC. For example, it was only on the tenth day of my stay that I learned the church was headed by the Reverend Sun Myung Moon. Accordingly, the view I present is more or less similar to that of the ordinary newcomer to the UC.

Thus began three of the most interesting weeks of my life. This participant observation resulted in two graduate-school papers on Moonie conversion attempts and left me with an abiding interest in "cults."

When I was writing my two papers, very little scholarly analysis of conversion attempts had been published. Much of what appeared in the media assumed that people who became Moonies (or Hare Krishnas, etc.) had been "brainwashed."[3] This seemed too simplistic to me. My analysis of the Moonie attempts at conversion I had observed persuaded me that interaction between members and prospects was key to prospects' staying. The information presented on the group seemed secondary in making initial decisions to remain. What was it about the interaction that was most effective? "Love bombing" seemed essential, with prospects being bombarded with loving attention at all times.[4] This style of interaction helped to develop strong personal ties between members of the group and prospective members.

My analysis of Moonie conversion attempts also emphasized that the prospective convert was not a passive recipient throughout the process. Rather, I pointed to ways in which individuals participated in acquiring the belief system and becoming a committed member. I concluded:

> In sum, it is not a matter of an individual being assaulted by information and losing "self" in a one-sided process of conversion, but a far more complicated process in which interaction figures most prominently, wherein the individual gradually "chooses," for various reasons, the new belief system. (Ayella, 1981:4)

NORMALITY, INFLUENCE, AND DEVIANCE

A dissatisfaction with popular explanations available at the time of my meeting the moonies in 1975, implying that these were essentially "crackpots" who were involved in cults, has remained with me. In the past ten years of discussing my research with college students, their chief question has been. "What kind of people are they?" The students insist that "they must be nuts," and thus dramatically unlike "us." Such thinking reflects the "kind of person" explanation of deviant behavior, which emphasizes that "deviants" can be clearly differentiated from "normals."

Similarly, the psychologist Margaret Singer (Singer with Lalich, 1995:15–16) mentions the "not me" attitude she believes the general public has about who joins cults. Yet, much of the by now burgeoning research on persons who enter cults has emphasized the essential normality of such people. In Lofland's (1966) early work on the Moonies, the clearly incompetent could not last in the group because they could not make a commitment; Barker's (1984) later research shows that participants are normal. Levine's (1984) main conclusion after looking in depth at individuals who joined a variety of cults is that they show no serious psychopathology and are "normal" adolescents. Singer's more recent book, based on more than three thousand interviews with current and former members of cults, indicates that the "majority of adolescents and adults in cults come from middle-class backgrounds, are fairly well educated, and are not seriously disturbed prior to joining" (Singer with Lalich, 1995:17).

This does not mean that behavioral and attitudinal changes, a "before" and "after," do not occur in individuals, but rather that both can be accounted for by reference to social processes and interaction, rather than by a "they're crazy" psychopathological explanation.

Although the Unification Church, with its seemingly "overnight" transformations of young, middle-class kids into "Moonies," was the subject of some noteworthy media stories (e.g., mass marriages performed in Madison Square Garden), it was the 1978 mass suicide-homicide of 912 persons in Jonestown, Guyana, that was most important in bringing "cults" to national awareness. Investigation of how the People's Temple could have come to this terrible end revealed many strange practices within the group, among them extreme disciplinary methods and the regulation of sexual and affectional matters so as to promote allegiance to Jim Jones alone (Coser and Coser, 1979; Hall, 1979, 1987; Reiterman and Jacobs, 1982; Richardson, 1982).

Not long after, we read about the "Synanon Horrors", . . . which included the forced shaving of heads (of members and nonmembers), sterilization of all males within the group (except the leader, Chuck Dederich), coerced abortions, forced breakups of marriages and other relationships, beatings and other attacks, and harassment of past members and nonmembers. Perhaps best known was the rattlesnake attack on Paul Morantz (the snake was put in his mailbox, minus its rattles), a lawyer who had just won a judgment against Synanon. Before information about the "horrors" emerged, Synanon had received great acclaim as an unprecedentedly effective drug-rehabilitation group, generating many imitators (Ofshe, 1980).

Two other groups labeled as cults that got lots of media attention were the Hare Krishnas and Scientology. The full title of a book about the former, *Monkey on a Stick: Murder, Madness, and the Hare Krishnas* (Hubner and Gruson, 1988), gives a hint of the deviant practices (arms buying, drug running, child abuse, murder) within parts of the movement; these attracted great negative coverage of the group in the 1980s and 1990s.

Scientology has been controversial since its founding an the 1950s. It achieved more notoriety through a series of 1990 *Los Angeles Times* articles and with the publication of the journalist Richard Behar's highly critical article "The Thriving Cult of Greed and Power" (*Time,* May 6, 1991). Behar described the cost of "enlightenment," estimated at $200,000 to $400,000 for the average person; deceptive attempts to attract mainstream members through an array of "front groups and financial scams"; the commission of federal crimes; and harassment of critics, especially through litigation paying "an estimated $20 million annually to more than 100 lawyers," according to Behar). This harassment extended to Behar himself, who said that "for the Time story, at least 10 attorneys and six private detectives were unleashed by Scientology and its followers in an effort to threaten, harass and discredit me."

Many others have also discovered that being critical of cults can bring various forms of retribution. A whole chapter of Singer and Lalich's *Cults in Our Midst* (1995), "The Threat of Intimidation," details a multitude of instances of cults' harassment of their critics. Singer herself uses assumed names because of cult scrutiny. Among many other things her office has been broken into, video and audio interviews of erstwhile cult members have been stolen, a woman from a cult posed as a student and "helped" in her office, her trash has been repeatedly stolen, and two dozen large brown rats were put into her home through a duct to the attic (Singer with Lalich, 1995:239–42).

Some later newsworthy examples of cults include David Koresh's Branch Davidians in Waco, Texas, which group met a fiery end on April 19, 1993, killing Koresh and seventy-seven followers (*CQ Researcher,* 1993:396). In October 1994, fifty-three members of the Order of the Solar Temple died in Canada and Switzerland "of various combinations of bullets, fire, stabbings, plastic bags over their heads, and injected drugs" (Singer with Lalich, 1995:339). Members of this group were affluent people, who retained their jobs while following Luc Jouret, a "forty-six-year-old Belgian homeopathic doctor," who was one of the dead (ibid.).

In 1995 a Japanese cult, Aum Shinrikyo (Supreme Truth), got great media attention when it used homemade nerve gas to poison subway riders in Tokyo. And last, the March 28, 1997, suicide of thirty-nine members of the group Heaven's Gate called to mind the 1978 Jonestown suicides. . . . In addition, the revelation of the castration of some adherents recalled the sterilization of men belonging to Synanon.

There are many other stories of persons interested in change or self-improvement joining cultlike groups. And today we are more aware of a proliferation of cults. Margaret Singer (Singer with Lalich, 1995:5) estimates that there are from three to five thousand cults in our society and that "between two and five million Americans are involved in cults at any one time" (12). Singer describes as the fastest growing "cultic groups" the ones "centered around New Age thinking and certain personal improvement training, life-styles, or prosperity programs" (13). As we approach the millennium, she estimates that there are "more than 1,100 end-of-time groups" (quoted in McCullough, 1997).

All kinds of strange happenings in cults have become known to us through the news media. To many, these experiences may seem so bizarre that we cannot imagine people like ourselves being involved. But in fact we are all vulnerable to influence, and it is only through influence that we change, for good or for ill . . .

Extreme groups are part of a social process that sociologists term "deviance defining." By this they mean that such groups, with their (in many ways) deviant beliefs and practices cause us as individuals and as a society to grapple with the question of where to draw the line between "normal" and "deviant" behavior among individuals and groups. These organizations also illustrate the role of social influence in shaping individuals' decisions to join, remain, and leave groups.

NOTES

1. This turned out to be Barbara Underwood, who later left and wrote (with her mother, Betty Underwood) *Hostage to Heaven* (1979).

2. At no point during this time did I become a Moonie. I left at the end of the three-week period that I had free, before joining my parents in San Francisco for their visit.

3. I maintained at the time that "brainwashing" in these accounts is a one-time phenomenon, with effects on the individual monolithic and all-binding. One is programmed to a particular, deviant belief system as if one were a robot. The "programming" involved most often seems to be a very simplistic, stimulus-response notion of how learning occurs, a repetitive process of providing informaton, rewarding appropriate behavior and belief, again and again and again. To reverse the process one "deprograms" the individual, and provides alternative information in the same repetitious manner. (Ayella, 1981:2)

4. Both John Lofland's (1977) and David Taylor's (1982) articles emphasize the importance of "love bombing."

REFERENCES

Ayella, Marybeth F. 1981. "An Analysis of Current Conversion Practices of Followers of Reverend Sun Myung Moon." Unpublished manuscript.

Barker, Eileen. 1984. *The Making of a Moonie.* New York: Basil Blackwell.

Behar, Richard. 1991. "The Thriving Cult of Greed and Power." *Time,* May 6.

Coser, Rose Laub, and Lewis Coser. 1979. "Jonestown as a Perverse Utopia." *Dissent* (Spring): 158–63.

CQ Researcher. 1993. "Cults in America," May 7.

Hall, John R. 1982, "Apocalypse at Jonestown." pp. 35–54 in *Violence and Religious Commitment,* ed. Ken Levi. University Park: Pennsylvania State University Press.

Hubner, John, and Lindsey Gruson. 1988. *Monkey on a Stick: Murder, Madness, and the Hare Krishna.* New York, Penguin Books USA.

Levine, Saul, 1984. *Radical Departures: Desparate Detours to Growing Up.* New York: Harcourt Brace Jovanovich.

Lofland, John. 1977. "Becoming a World Saver' Revisited." *American Behavioral Scientist* 20(6): 862–75.

McCullough, Marie. 1997. "Cults' influence Is Growing, Say Two Researchers." *Philadelphia Inquirer,* May 30.

Ofshe, Richard. 1980. "The Social Development of the Synanon Cult: The Managerial Strategy of Organizational Transformation." *Sociological Analysis* 41(2): 109–27.

Reiterman, Tim, and John Jacobs. 1982. *Raven.* New York: E. P. Dutton.

Richardson, James T. 1982. "A Comparison between Jonestown and Other Cults." pp. 21–34 in *Violence and Religious Commitment,* ed. Ken Levi. University Park: Pennsylvania State University Press.

Singer, Margaret, with Janja Lalich. 1995. *Cults in Our Midst.* San Francisco: Jossey-Bass Publishers.

Taylor, David. 1982. "Becoming New People: The Recruitment of Young Americans into the Unification Church." Pp. 177–230 in *Millennialism and Charisma,* ed. Roy Wallis. Belfast, U.K.: Queen's University.

Underwood, Barbara, and Betty Underwood. 1979. *Hostage to Heaven.* New York: Potter.

DISCUSSION QUESTIONS

1. How does Marybeth Ayella's experience with the Unification Church illustrate the concept of resocialization? What specific techniques do cults use to incorporate new members?

2. Why does Ayella suggest that it is wrong to understand cult members as somehow abnormal or deviant?

INFOTRAC COLLEGE EDITION

You can use your access to InfoTrac College Edition to learn more about the subjects covered in this essay. Some suggested search terms include:

brainwashing
conversion
cults
degradation ceremonies

new religious movements
resocialization
Unification Church (Moonies)

64

The Service Society and the Changing Experience of Work

CAMERON LYNNE MACDONALD AND CARMEN SIRIANNI

The U.S. economy has changed from being based primarily on manufacturing to being based on service industries. The transition to more "service work" has changed the character of workplace control. The service economy is embedded in systems of race, gender, and class stratification that are revealed in patterns of employment and perceptions of who is most fit for particular jobs.

We live and work in a service society. Employment in the service sector currently accounts for 79 percent of nonagricultural jobs in the United States (U.S. Department of Labor 1994: 83). More important, 90 percent of the new jobs projected to be created by the year 2000 will be in service occupations, while the number of goods-producing jobs is projected to decline (Kutscher 1987: 5). Since the mid-nineteenth century the U.S. economy has been gradually transformed from an agriculture-based economy to a manufacturing-based economy to a service-based economy. Near the turn of the century, employment distribution among the three major economic sectors was equally divided at roughly one-third each. Since then, agriculture's labor market share has declined rapidly, now accounting for only about 3 percent of U.S. jobs, while the service sector provides over 70 percent and the goods-producing sector about 25 percent.

The decrease in proportion of manufacturing jobs occurred not because U.S. corporations manufacture fewer goods, but primarily because they use fewer workers to make the goods they produce (Albrecht and Zemke 1990). They use fewer workers due to increasing levels of automation and the exportation of manufacturing functions to low-wage job markets overseas. In addition, the feminization of the work force has created a self-fulfilling cycle in which the entrance of more women into the work force has led to increased demand for those consumer services once provided gratis by housewives (cleaning, cooking, child care, etc.), which in turn has produced more service jobs that are predominantly filled by women.

Still, these trends fail to account fully for the dominance of service work in the U.S. economy, since companies outside of the service sector also contain service occupations. For example, 13.2 percent of the employees in the manufacturing sector work in service occupations such as clerical work, customer service,

From: Cameron Lynne Macdonald and Carmen Sirianni, eds. 1996. *Working in the Service Society*. Philadelphia: Temple University Press, pp. 1–24.

telemarketing, and transportation (Kutscher 1987). Further, manufacturing and technical occupations are comprised increasingly of service components as U.S. firms adopt Total Quality Management (TQM) and other customer-focused strategies to generate a competitive edge in the global economy. When production efficiency and quality are maximized, the critical variable in the struggle for economic dominance is the quality of interactions with customers. As one business school professor remarks, "Sooner or later, new technology becomes available to everyone. Customer-oriented employees are a lot harder to copy or buy" (Schlesinger and Heskett 1991: 81). So whether one believes that U.S. manufacturing is going to Mexico, to automation, or to the dogs, it is clear that the United States is increasingly becoming a service society and that service work is here to stay.

What do we mean when we speak of "service work?" By definition, a service is intangible; it is produced and consumed simultaneously, and the customer generally participates in its production (Packham 1992). . . . . Service work includes jobs in which face-to-face or voice-to-voice interaction is a fundamental element of the work. "Interactive service work" (Leidner 1993) generally requires some form of what Arlie Hochschild (1983) has termed "emotional labor," meaning the conscious manipulation of the workers' self-presentation either to display feeling states and/or to create feeling states in others. In addition, the guidelines, or "feeling rules," for this emotional labor are created by management and conveyed to the worker as a critical aspect of the job. . . . .

Much managerial and professional work also entails emotional labor. For example, doctors are expected to display an appropriate "bedside manner," lawyers are expert actors in and out of the courtroom, and managers, at the most fundamental level, try to instill feeling states and thus promote action in others. However, there remains a critical distinction between white-collar work and work in the emotional proletariat: in management and in the professions, guidelines for emotional labor are generated collegially and, to a great extent, are self-supervised. In front-line service jobs, workers are given very explicit instructions concerning what to say and how to act, and both consumers and managers watch to ensure that these instructions are carried out. However, one could argue that even those in higher ranking positions increasingly experience the kinds of monitoring of their interactive labor encountered by those lower on the occupational ladder, be it by customers, supervisors, or employees. . . . .

Given the rising dominance of service occupations in the labor force, what are the special difficulties and opportunities that workers encounter in a service society? A key problem seems to be how to inhabit the job. In the past there was a clear distinction between *careers,* which required a level of personalization, emotion management, authenticity in interaction, and general integration of personal and workplace identities, and *jobs,* which required the active engagement of the body and parts of the mind while the spirit and soul of the worker might be elsewhere. Workers in service occupations are asked to inhabit jobs in ways that were formerly limited to managers and professionals alone. They are required to bring some level of personal identity and self-expression into their work, even if it is only at the level of basic interactions, and even if the job itself is only temporary.

The assembly-line worker could openly hate his job, despise his supervisor, and even dislike his co-workers, and while this might be an unpleasant state of affairs, if he completed his assigned tasks efficiently, his attitude was his own problem. For the service worker, inhabiting the job means, at the very least, pretending to like it, and, at most, actually bringing his whole self into the job, liking it, and genuinely caring about the people with whom he interacts.

This demand has several implications: who will be asked to fill what jobs, how they are expected to perform, and how they will respond to those demands. Because personal interaction is a primary component of all service occupations, managers continually strive to find ways to oversee and control those interactions, and worker responses to these attempts vary along a continuum from enthusiastic compliance to outright refusal. Hiring, control of the work process, and the stresses of bringing one's emotions to work are all shaped by the characteristics of the worker and the nature of the work. The self-presentation and other personal characteristics of the worker make up the work process and the work product, and are increasingly the domain of management-worker struggles (see Leidner 1993). In addition, because much of the labor itself is invisible, contests over control of the labor process are often more implicit than explicit. . . .

There are three trends emanating from the rising dominance of service work. First, the need to supervise the production of an intangible, good service, has given rise to particularly invasive forms of workplace control and has led managers to attempt to oversee areas of workers' personal and psychic lives that have heretofore been considered off-limits. Second, the fact that workers' personal characteristics are so firmly linked to their "suitability" for certain service occupations continues to lead to increasing levels of stratification within the service *labor* force. Finally, . . . how [do] workers respond to these and other aspects of working in the service society, and how they might build autonomy and dignity into their work, ensuring that service work does not equal servitude? . . .

GENDER, RACE, AND STRATIFICATION
IN THE SERVICE SECTOR

Service industries tend to produce two kinds of jobs: large numbers of low-skill, low-pay jobs and a smaller number of high-skill, high-income jobs, with very few jobs that could be classified in the middle. As Joel Nelson (1994) notes, "Service workers are more likely than manufacturing workers to have lower incomes, fewer opportunities for full-time employment, and greater inequality in earnings" (p. 240). A typical example of this kind of highly stratified work force can be found in fast food industries. These firms tend to operate with a small core of managers and administrators and a large, predominantly part-time work force who possess few skills and therefore are considered expendable (Woody 1989).

As a result, service jobs fall into two broad categories: those likely to be production-line jobs and those likely to be empowered jobs. This distinction not only refers to the level of responsibility and autonomy expected of workers, but also to wages, benefits, job security, and potential for advancement. While empowered service jobs are associated with full-time work, decent wages and benefits, and internal job ladders, production-line jobs offer none of these. Some researchers have described the distinction between empowered and production-line service jobs as one between "core" and "periphery" jobs in the service economy (Hirschhorn 1988; Walsh 1990; Wood 1989). . .

The core/periphery distinction may be a misleading characterization of functions in service industries, however. In many firms contingent workers perform functions essential to the operation of the firm and can comprise up to two-thirds of a firm's labor force while "core" workers perform nonessential functions (Walsh 1990). For example, a majority of key functions in industries such as hospitality, food service, and retail sales are performed by workers who, based on their level of benefits, pay, and job security, would be considered periphery workers. In low-skill service positions, job tenure has no relation to output or productivity. Therefore, employers can rely on contingent workers to provide high-quality service at low costs. As T. J. Walsh (1990: 527) points out, it is therefore likely that the poor compensation afforded these workers is due not to their productivity level but to the perceptions of their needs, level of commitment, and availability.

Given the proliferation of service jobs in the United States, key questions for labor analysts are what kinds of jobs are service industries producing, and who is likely to fill them? At the high end, service industries demand educated workers who can rapidly adapt to changing economic conditions. This means that employers may demand a college degree or better for occupations that formerly required only a high school diploma,

> even though many of the job-holders' activities have not changed, or appear relatively simple, because they want workers to be more responsive to the general situation in which they are working and the broader purposes of their work. (Hirschhorn 1988: 35)

In addition, core workers are frequently expected to take on more responsibilities, work longer hours, and intensify their output.

At the low end of the spectrum of service occupations are periphery workers who are frequently classified as part-time, temporary, contract, or contingent. These flexible-use workers act as a safety valve for service firms, allowing managers to redeploy labor costs in response to market conditions. In addition, they allow managers to minimize overhead because they rarely qualify for benefits and generally receive low wages. In 1988, 86 percent of all part-time workers worked in service sector industries, and this trend has continued since (Tilly 1992: p 30). Labor analysts argue that the bulk of the expansion in part-time and contingent work is due to the expansion of the service sector, which has always used shift and part-time workers as cost-control mechanisms. . . . .

Service sector expansion has also sparked the rapid growth of the temporary help industry. Over the past decade, the number of temporary workers has tripled (Kilborn 1995). As the chairman of Manpower, Inc., the largest single employer in the United States, remarked, "The U.S. is going from just-in-time manufacturing to just-in-time employment. The employer tells us, 'I want them delivered exactly when I want them, as many as I need, and when I don't need them, I don't want them here'" (quoted in Castro 1993: 44). Like part-time workers, temporary workers carry no "overhead" costs in terms of taxes and benefits, and they are on call as needed. Since they experience little workplace continuity, they are less likely than full-time, continuous employees to organize or advocate changes in working conditions.

Women, youth, and minorities comprise the bulk of the part-time and contingent work force in the service sector. For example, Karen Brodkin Sacks (1990) has noted that the health care industry is "so stratified by race and gender that the uniforms worn to distinguish the jobs and statuses of health care workers are largely redundant" (p. 188). Patients respond to the signals implicitly transmitted via gender and race and act accordingly, offering deference to some workers and expecting it from others. Likewise in domestic service, race and gender determine who gets which jobs. White American and European women are most likely to be hired for domestic jobs defined primarily as child care, while women of color predominate in those defined as house cleaning (Rollins 1985), regardless of what the actual allocation of work might be. The same kinds of stratification can be found in secretarial work, food service, hotels, and sales occupations (Hochschild 1983; Leidner 1991). In all of these occupational groups, demographic characteristics of the worker determine the job title and thus other factors such as status, pay, benefits, and degree of autonomy on the job.

As a result, the shift to services has had a differential impact on various sectors of the labor market, increasing stratification between the well employed and the underemployed. The service sector work force is highly feminized, especially at the bottom. Within personal and business services, for example, Bette Woody (1989) found that "men are concentrated in high-ticket, high commission sales jobs, and women in retail and food service" (p. 57). And although the decline in manufacturing forced male workers to move into service industries, Jon Lorence (1992: 150) notes that within the service sector, occupations are highly stratified by gender. From 1950 to 1990, 60 percent of all new service sector employment and 74 percent of all new low-skill jobs were filled by women. . . .

Service sector employment is equally stratified by race. For example, Woody (1989) finds that the shift to services has affected black women in two important ways. First, it has meant higher rates of unemployment and underemployment for black men, which increased pressure on black women to be the primary breadwinners for their families and contributed to the overall reduction in black family income. Second, although the increase in service sector jobs has meant greater opportunity for black women and has allowed them to move out of domestic service into the formal economy, as Evelyn Nakano Glenn (1996) also notes, they have remained at the lowest rung of the service employment ladder. Unlike some white women who "moved up" to male-intensive occupations with the shift to services, black women "moved over" to sectors traditionally employing white women (Woody 1989: 54). These traditionally male occupations are not only low-security, low-pay jobs, but they lack internal career ladders. As Ruth Needleman and Anne Nelson (1988) note, "There is no progression from nurse to physician, from secretary to manager" (p. 297).

Service work differs most radically from manufacturing, construction, or agricultural work in the relationship between worker characteristics and the job. Even though discrimination in hiring, differential treatment, differential pay, and other forms of stratification exist in all labor markets, service occupations are the only ones in which the producer in some sense equals the product. In no other area of wage labor are the personal characteristics of the workers so strongly associated with the nature of work. Because at least part of the job in all service occupations is to "manufacture social relations" (Filby 1992; 37), traits such as gender, race, age, and sexuality serve a signaling function, indicating to the customer/employer important cues about the tone of the interaction. Women are expected to be more nurturing and empathetic than men and to tolerate more offensive behavior from customers (Hochschild 1983; Leidner 1991; Pierce 1996, Sutton and Rafaeli 1989). Similarly, both women and men of color are expected to be deferential and to take on more demeaning tasks (Rollins 1985; Woody 1989). In addition, a given task may be viewed as more or less demeaning depending on who is doing it.

These occupations are so stratified that worker characteristics such as race and gender determine not only who is considered desirable or even eligible to fill certain jobs, but also who will want to fill certain jobs and how the job itself is performed. Worker characteristics shape what is expected of a worker by management and customers, how that worker adapts to the job, and what aspects of the job he or she will resist or embrace. The strategies workers use to adapt to the demands of service jobs are likely to differ according to gender and other characteristics. Women are more likely to embrace the emotional demands (e.g., nurturing, care giving) of certain types of service jobs because these demands generally fit their notion of gender-appropriate behavior. As Pierce notes, heterosexual men tend to resist these demands because they find "feminine" emotional labor demeaning; in response they either reframe the nature of the job to emphasize traditional masculine qualities or distance themselves by providing service by rote, making it clear that they are acting under duress.

All of these interconnections between worker, work, and product result in tendencies toward very specific types of labor market stratification. A long and heated debate has raged concerning the ultimate impact of deindustrialization on the structure of the labor market. On one side are those who argue that a shift to a service-based economy will produce skill upgrading and a leveling of job hierarchies as information and communications technologies reshape the labor market (see, for example, Bell 1973). Others take a more pessimistic view, arguing that the shift to services will give rise to two trends: "towards polarization and towards the proliferation of low-wage jobs" (Bluestone and Harrison 1988: 126). In a sense, both positions are correct but for different segments of the labor market.

Overall, the transition to a service-based economy will likely mean a more stratified work force in which more part-time and contingent jobs are filled predominantly by women, minorities, and workers without college degrees. In jobs lacking internal career ladders, these workers have little chance for upward mobility. Contingent workers are also less likely to organize successfully due to their tenuous attachment to specific employers and to the labor force in general. At the opposite end of the service sector occupational spectrum are highly educated managerial, professional, and paraprofessional workers, who will have equally weak attachments to employers, but who, due to their highly marketable skills, will move with relative ease from one well-paying job to another. Given this divided and economically segregated work force, what are the opportunities for workers to advocate, collectively or individually, for greater security, better working conditions, and a voice in shaping their work?. . .

REFERENCES

Albrecht, Karl, and Ron Zemke. 1990. *Service America!: Doing Business in the New Economy.* New York: Warner Books.

Bell, Daniel. 1973. *The Coming of Post-Industrial Society: A Venture in Social Forecasting.* New York: Basic Books.

Bluestone, Barry, and Bennett Harrison. 1988. *The Great U-Turn: Corporate Restructuring and the Polarizing of America.* New York: Basic Books.

Castro, Janice. 1993. "Disposable Workers." *Time Magazine,* March 29, pp. 43–57.

Filby, M. P. 1992, "'The Figures, the Personality, and the Bums': Service Work and Sexuality." *Work, Employment, and Society* 6 (March): 23–42.

Glenn, Evelyn Nakano. 1996. "From Servitude to Service Work: Historical Continuities in the Racial Division of Paid Reproductive Labor." In Cameron Lynne Macdonald and Carmen Sirianni, eds., *Working in the Service Society,* pp. 115–156. Philadelphia: Temple University Press.

Hirschhorn, Larry. 1988. "The Post-Industrial Economy: Labour, Skills, and the New Mode of Production." *The Service Industries Journal* 8: 19–38.

Hochschild, Arlie Russell. 1983. *The Managed Heart: Commercialization of Human Feeling.* Berkeley: University of California Press.

Kilborn, Peter T. 1995. "In New Work World, Employers Call All the Shots: Job Insecurity, a Special Report." *New York Times,* July 3, p. A1.

Kutscher, Ronald E. 1987. "Projections 2000: Overview and Implications of the Projections to 2000." *Monthly Labor Review* (September): 3–9.

Leidner, Robin. 1993. *Fast Food, Fast Talk: Service Work and the Routinization of Everyday Life.* Berkeley: University of California Press.

———. 1991. "Serving Hamburgers and Selling Insurance: Gender, Work, and Identity in Interactive Service Jobs." *Gender & Society* 5: 154–77.

Lorence, Jon. 1992. "Service Sector Growth and Metropolitan Occupational Sex Segregation." *Work and Occupations* 19:128–56.

Needleman, Ruth, and Anne Nelson. 1988. "Policy Implications: The Worth of Women's Work." In Anne Starham, Eleanor M. Miller, and Hans O. Mauksch, eds., *The Worth of Women's Work: A Qualitative Synthesis* (pp. 293–308). Albany: SUNY Press.

Nelson, Joel I. 1994. "Work and Benefits: The Multiple Problems of Service Sector Employment." *Social Problems* 41: 240–55.

Packham, John. 1992. "The Organization of Work on the Service-Sector Shop Floor." Unpublished paper.

Pierce, Jennifer L. 1996. *Gender Trials: Emotional Lives in Contemporary Law Firms.* Berkeley: University of California Press.

Rollins, Judith. 1985. *Between Women: Domestics and Their Employers.* Philadelphia: Temple University Press.

Sacks, Karen Brodkin. 1990. "Does It Pay to Care?" In Emily K. Abel and Margaret K. Nelson, eds., *Circles of Care: Work and Identity in Women's Lives* (pp. 188–206). Albany: SUNY Press.

Schlesinger, Leonard, and James Heskett. 1991. "The Service-Driven Service

Company." *Harvard Business Review* (September–October): 71–81.

Sutton, Robert I., and Anat Rafaeli. 1989. "The Expression of Emotion in Organizational Life." *Research in Organizational Behavior* 11: 1–42.

Tilly, Chris. 1992. "Short Hours, Short Shrift: The Causes and Consequences of Part-Time Employment." In Virginia L. Du Rivage, ed., *New Policies for the Part-Time and Contingent Work-Force* (pp. 15–43). Armonk, NY: M. E. Sharpe, Economic Policy Institute Series.

U.S. Department of Labor, Bureau of Labor Statistics. 1994. *Monthly Labor Review* (July): 74–83.

Walsh, T. J. 1990. "Flexible Labour Utilisation in the Private Service Sector." *Work, Employment, and Society* 4: 517–30.

Wood, Stephen, ed. 1989. *The Transformation of Work? Skill, Flexibility, and the Labour process.* London: Unwin Hyman.

Woody, Bette. 1989. "Black Women in the Emerging Services Economy." *Sex Roles* 21: 45–67.

DISCUSSION QUESTIONS

1. How has the transition to a service-based economy changed employment patterns in the contemporary economy? What implications does this have for the relationship between management and labor?

2. How do race and gender stratification influence the perceptions of some workers as suitable for particular forms of service work? What evidence have you witnessed of this phenomenon in your experiences in a service-based society?

INFOTRAC COLLEGE EDITION

You can use your access to InfoTrac College Edition to learn more about the subjects covered in this essay. Some suggested search terms include:

automation
contingent work
dual labor market
emotional labor

gender division of labor
service sector
service work

65

Toward a 24-Hour Economy

HARRIET B. PRESSER

Analysis of employment data indicates that few in the United States are working a "typical" forty-hour work week. This change is being driven by the changing economy, the changing demography of the workplace, and changing technology. Harriet Presser discusses the effects of the increase in working hours on families.

Americans are moving toward a 24-hour, 7-day-a-week economy. Two-fifths of all employed Americans work mostly during the evenings or nights, on rotating shifts, or on weekends. Much more attention has been given to the number of hours Americans work[1,2] than to the issue of which hours—or days—Americans work. Yet the widespread prevalence of nonstandard work schedules is a significant social phenomenon, with important implications for the health and well-being of individuals and their families and for the implementation of social policies. Here I discuss recent national data on the widespread prevalence of nonstandard work schedules, explain why this has come about, and highlight some of the important social implications.

PREVALENCE

As of 1997, only 29.1% of employed U.S. citizens worked a "standard work week," defined as 35 to 40 hours a week, Monday through Friday, on a fixed daytime schedule. For employed men, the proportion is 26.5%; for employed women, 32.8%. Only 54.4%—a bare majority—regularly work a fixed daytime schedule, all five weekdays, for any number of hours.

These figures are derived from the May 1997 Current Population Survey (CPS), a representative sample of about 48,000 U.S. households. I selected for further study a subset of about 50,000 employed Americans ages 18 and over in these households with nonagricultural occupations and who reported on their specific work hours and/or work days.

Of the people in this group, one in five work other than on a fixed daytime schedule, and one in three work on weekends (and, for most, on weekdays as well). Men and women are similar in their prevalence of evening employment,

From: *Science* 284 (June 1999): 1778–1779. Reprinted with permission.

but a somewhat higher proportion of men than women work fixed nights, rotating and variable hours, and weekends. The most marked differences are between those working full time and part-time. More part-timers work other than a fixed day (29.6%) than do full-timers (17.0%); evening employment is especially high among part timers. The difference between full- and part-timers is less marked for weekend employment (30.7% and 34.7%, respectively).

For the modal U.S. family—the two-earner couple–the prevalence of nonstandard work schedules is especially high, because either the husband or wife may be working nondays or weekends. (Rarely do both work the same nonstandard schedules.) Among two-earner couples, 27.8% include at least one spouse who works other than a fixed daytime schedule, and 54.6% include at least one spouse working weekends. When children under age 14 are in the household the respective percentages are 31.1 and 46.8%. Indeed of all two-earner couples with children, those with both spouses working fixed daytime schedules and weekdays are a minority; 57.3% do not fit this description. Thus, the temporal context in which millions of American couples are raising their children today is diverse and is likely to become even more so in the future.

ORIGINS AND CAUSES

At least three interrelated factors are increasing the demand for Americans to work late hours and weekends: a changing economy, changing demography, and changing technology. With regard to the changing economy, an important aspect is the growth of the service sector with its high prevalence of nonstandard work schedules relative to the goods-producing sector. In the 1960s, employees in manufacturing greatly exceeded those in service industries, whereas by 1995 the percentage was about twice as high in services as in manufacturing[3]. In particular, there is an interaction between the growth of women's employment and the growth of the service sector because there is a disproportionately high percentage of female occupations in this sector. In turn, the increasing participation of women in the labor force contributes to the growth of the service economy. For example, the decline in full-time homemaking has generated an increase in family members eating out and purchasing other services. Moreover, women's increasing daytime labor force participation has generated a demand for services during nondaytime hours and weekends[4].

Demographic changes also have contributed. The postponement of marriage, along with the rise in real family income resulting from two earners, has increased the demand for recreation and entertainment during late hours and weekends. The aging of the population has increased the demand for medical services over a 24-hour day, 7 days a week.

Finally, technological change, along with reduced costs, has moved us to a global 24-hour economy. The ability to be "on call" at all hours of the day and night to others around the world at low cost generates a need to do so. For example, the rise of multinational corporations, along with the use of computers, faxes, and other forms of rapid communication, increases the demand for branch offices to operate at the same time that corporate headquarters are open. Similarly, international financial markets are expanding their hours of operation. Express mailing companies such as United Parcel Service require round-the-clock workers all days of the week.

We do not have precise national estimates of the amount of growth over recent decades in the prevalence of nonstandard work schedules as a consequence of these changes. Questions on work hours have been asked differently by the Bureau of Labor Statistics in each of the CPSs since 1980; questions on work days were not even asked until 1991.

Most of the top 10 occupations projected by the Bureau of Labor Statistics to have the largest job growth between 1996 and 2006 are service occupations[5]. Using the May 1997 CPS data. I calculated the percentages in the top growth occupations for which nonstandard schedules are prevalent and considered their gender and racial composition.

The data suggest that not only will future job growth generate an increase in employment during nonstandard hours and weekends, but also that this increase will be experienced disproportionately by females and blacks. Many of the top growth occupations that tend to have nonstandard work schedules also have high percentages of female workers: cashiers, registered nurses, retail salespersons, nurses' aides, orderlies, and attendants combined with home health aides. The top growth occupations that disproportionately include blacks and tend to have nonstandard work schedules are cashiers, truck drivers, nurses' aides, orderlies, and attendants combined with home health aides.

Although nonstandard work schedules are pervasive throughout the occupational structure, such schedules are disproportionately concentrated in jobs low in the occupational hierarchy[6]. This fact, combined with the expectation that women and blacks will disproportionately increase their participation in nonstandard work schedules, suggests that this phenomenon will increasingly affect the working poor.

Effects on Families

The physical consequences of working non-standard hours, particularly night and rotating hours, have been well documented[7]. Such work schedules alter one's circadian rhythms, often leading to sleep disturbances, gastrointestinal disorders, and chronic malaise. The social consequences of such employment have received less attention, although working nonstandard schedules may be significantly altering the structure and stability of family life. Some of the consequences can be viewed as positive, others negative, and both may vary by family member. Moreover, short-term benefits may be offset by long-term costs and vice versa.

Consider, for example, the care of children among dual-earner couples. As noted above, one-third of such couples with preschool-aged children are split-shift couples with one spouse working days and the other evenings, nights, or rotating schedules. A national study of American couples with preschool-aged children showed that in virtually all cases in which mothers and fathers are employed different hours and neither are on rotating schedules, fathers are the primary caregivers of children when their wives are employed[8]. Insofar as we view the greater involvement of fathers in child care as desirable, and considering the economic benefits to the family of reduced child care expenses resulting from this arrangement, such split-shift parenting may be a positive outcome.

However, these gains may be more than offset by the longer term costs to the marriage. New research shows that among couples with children, when men work nights (and are married less than 5 years) the likelihood of separation or divorce 5 years later is some six times that when men work days. When women work nights (and are married more than 5 years) the odds of divorce or separation are three times as high. Moreover, the data suggest that the increased tendency for divorce is not because spouses in troubled marriages are more likely to opt for night work; the causality seems in the opposite direction[9].

Single as well as married mothers often engage in a split-shift caregiving arrangement with grandmothers. More than one-third of grandmothers who provide care for preschool-aged children are otherwise employed[10]. Here, too, there may be both positive and negative aspects of such arrangements, but this has not been studied. The observation that single mothers are more likely than married mothers to work long as well as nonstandard hours and are more likely to be among the working poor [11,12] suggests that the problems of managing time and money are especially stressful for such mothers.

Policymakers and scholars must take a more realistic view of the temporal nature of family life among Americans. With regard to welfare reform, for example, close to half (43.3%) of employed mothers with a high school education or less, ages 18 to 34, work other than a fixed daytime schedule, weekdays only[13]. If mothers on welfare are to move into jobs similar to these mothers, a key policy issue is how to improve the fit between the availability of child care and these working mothers' schedules. Expanding day care alone will not be satisfactory.

The movement toward a 24-hour economy is well underway, and will continue into the next century. Although driven by factors external to individual families, it will affect the lives of family members in profound ways. The home-time structure of families is becoming temporally very complex. We need to change our conception of family life to include such complexities. This should help to improve social policies that seek to ease the economic and social tensions that often result from the dual demands of work and family, particularly among the working poor.

REFERENCES AND NOTES

1. J. Schor, *The Overworked American* (Basic Books, New York, 1991).
2. J. P. Robinson and G. Godbey. *Time for Life: The Surprising Ways Americans Use Their Time* (Pennsylvania State Univ. Press, University Park, PA, 1997).
3. J. R. Meisenheimer II, *Mon. Labor Rev.* **121, 22** (February 1998).
4. H. B. Presser, *Demography* **26,** 523 (1989).
5. G. T. Silvestri, *Mon. Labor Rev.* **120,** 58 (November 1997).
6. H. B. Presser, *Demography* **32,** 577 (1995).
7. *Biological Rhythms: Implications for the Worker* (OTA-BA-463, Office of

Technological Assessment, Washington, DC, 1991).

8. H. B. Presser, *J. Marr. Fam.* **50,** 133 (1988).

9. ——— ibid., in press.

10. ——— ibid., **51,** 581 (1989).

11. A. G. Cox, thesis, University of Maryland (1994).

12. ——— and H. B. Presser, in *Work and Family: Research Informing Policy.* T. Parcel and D. B. Cornfield, Eds. (Sage. Thousand Oaks, CA, in press.).

13. H. B. Presser and A. G. Cor, *Mon. Labor Rev.* **120,** 25 (April 1997).

DISCUSSION QUESTIONS

1. What factors does Presser identify as leading to increases in the working hours of employed people? What impact are these changes likely to have on the labor force experiences of two-earner households?

2. Sociologists have often argued that work and family structures are mutually interdependent. How is this evidenced by the effects of increasing work hours on families' experiences? What evidence of this have you seen in your own life?

INFOTRAC COLLEGE EDITION

You can use your access to InfoTrac College edition to learn more about the subjects covered in this essay. Some suggested search terms include:

flex-time
part-time work
time-budget studies
women's labor force participation

work hours
worker stress
working poor

66

The Smile Factory:
Work at Disneyland

JOHN VAN MAANEN

Walt Disney Enterprises epitomizes a work environment where strong mechanisms of social control dictate every aspect of workers' activities, including dress, demeanor, and "emotion management." John Van Maanen analyzes how corporate control is established through worker training and supervision. In doing so, he debunks the image of Disneyland as the "Happiest Place on Earth."

Part of Walt Disney Enterprises includes the theme park Disneyland. In its pioneering form in Anaheim, California, this amusement center has been a consistent money maker since the gates were first opened in 1955. Apart from its sociological charm, it has, of late, become something of an exemplar for culture vultures and has been held up for public acclaim in several best-selling publications as one of America's top companies. . . . To outsiders, the cheerful demeanor of its employees, the seemingly inexhaustible repeat business it generates from its customers, the immaculate condition of park grounds, and, more generally, the intricate physical and social order of the business itself appear wondrous.

Disneyland as the self-proclaimed "Happiest Place on Earth" certainly occupies an enviable position in the amusement and entertainment worlds as well as the commercial world in general. Its product, it seems, is emotion—"laughter and well-being." Insiders are not bashful about promoting the product. Bill Ross, a Disneyland executive, summarizes the corporate position nicely by noting that "although we focus our attention on profit and loss, day-in and day-out we cannot lose sight of the fact that this is a feeling business and we make our profits from that."

The "feeling business" does not operate, however, by management decree alone. Whatever services Disneyland executives believe they are providing to the 60 to 70 thousand visitors per day that flow through the park during its peak summer season, employees at the bottom of the organization are the ones who most provide them. The work-a-day practices that employees adopt to amplify or dampen customer spirits are therefore a core concern of this feeling business. The happiness trade is an interactional one. It rests partly on the symbolic resources put into place by history and park design but it also rests on an animated work force that is more or less eager to greet the guests, pack the trams, push the but-

From: John Van Maanen. 1991. "The Smile Factory." pp. 58–75 in *Reframing Organizational Culture*, edited by Peter Frost. Thousand Oaks, CA: Sage. Reprinted with permission.

tons, deliver the food, dump the garbage, clean the streets, and, in general, marshal the will to meet and perhaps exceed customer expectations. False moves, rude words, careless disregard, detected insincerity, or a sleepy and bored presence can all undermine the enterprise and ruin a sale. The smile factory has its rules.

IT'S A SMALL WORLD

This rendition is of course abbreviated and selective. I focus primarily on such matters as the stock appearance (vanilla), status order (rigid), and social life (full), and swiftly learned codes of conduct (formal and informal) that are associated with Disneyland ride operators. These employees comprise the largest category of hourly workers on the payroll. During the summer months, they number close to four thousand and run the 60-odd rides and attractions in the park.

They are also a well-screened bunch. There is—among insiders and outsiders alike—a rather fixed view about the social attributes carried by the standard-make Disneyland ride operator. Single, white males and females in their early twenties, without facial blemish, of above average height and below average weight, with straight teeth, conservative grooming standards, and a chin-up, shoulder-back posture radiating the sort of good health suggestive of a recent history in sports are typical of these social identifiers. There are representative minorities on the payroll but because ethnic displays are sternly discouraged by management, minority employees are rather close copies of the standard model Disneylander, albeit in different colors.

This Disneyland look is often a source of some amusement to employees who delight in pointing out that even the patron saint, Walt himself, could not be hired today without shaving off his trademark pencil-thin mustache. But, to get a job in Disneyland and keep it means conforming to a rather exacting set of appearance rules. These rules are put forth in a handbook on the Disney image in which readers learn, for example, that facial hair or long hair is banned for men as are aviator glasses and earrings and that women must not tease their hair, wear fancy jewelry, or apply more than a modest dab of makeup. Both men and women are to look neat and prim, keep their uniforms fresh, polish their shoes, and maintain an upbeat countenance and light dignity to complement their appearance—no low spirits or cornball raffishness at Disneyland.

The legendary "people skills" of park employees, so often mentioned in Disneyland publicity and training materials, do not amount to very much according to ride operators. Most tasks require little interaction with customers and are physically designed to practically insure that is the case. The contact that does occur typically is fleeting and swift, a matter usually of only a few seconds. In the rare event sustained interaction with customers might be required, employees are taught to deflect potential exchanges to area supervisors or security. A Training Manual offers the proper procedure: "On misunderstandings, guests should be told to call City Hall. . . . In everything from damaged cameras to physical injuries, don't discuss anything with guests . . . there will always be one of us nearby." Employees learn quickly that security is hidden but everywhere. On Main Street security cops are Keystone Kops; in Frontierland, they are Town Marshalls; on Tom Sawyer's Island, they are Cavalry Officers, and so on.

Occasionally, what employees call "line talk" or "crowd control" is required of them to explain delays, answer direct questions, or provide directions that go beyond the endless stream of recorded messages coming from virtually every nook and cranny of the park. Because such tasks are so simple, consisting of little more than keeping the crowd informed and moving, it is perhaps obvious why management considers the sharp appearance and wide smile of employees so vital to park operations. There is little more they could ask of ride operators whose main interactive tasks with visitors consist of being, in their own terms, "information booths," "line signs," "pretty props," "shepherds," and "talking statues."

A few employees do go out of their way to initiate contact with Disneyland customers but, as a rule, most do not and consider those who do to be a bit odd. In general, one need do little more than exercise common courtesy while looking reasonably alert and pleasant. Interactive skills that are advanced by the job have less to do with making customers feel warm and welcome than they do with keeping each other amused and happy. This is, of course, a more complex matter.

Employees bring to the job personal badges of status that are of more than passing interest to peers. In rough order, these include: good looks, college affiliation, career aspirations, past achievements, age (directly related to status up to about age 23 or 24 and inversely related thereafter), and assorted other idiosyncratic matters. Nested closely alongside these imported status badges are organizational ones that are also of concern and value to employees.

Where one works in the park carries much social weight. Postings are consequential because the ride and area a person is assigned provide rewards and benefits beyond those of wages. In-the-park stature for ride operators turns partly on whether or not unique skills are required. Disneyland neatly complements labor market theorizing on this dimension because employees with the most differentiated skills find themselves at the top of the internal status ladder, thus making their loyalties to the organization more predictable.

Ride operators, as a large but distinctly middle-class group of hourly employees on the floor of the organization, compete for status not only with each other but also with other employee groupings whose members are hired for the season from the same applicant pool. A loose approximation of the rank ordering among these groups can be constructed as follows:

1. The upper-class prestigious Disneyland Ambassadors and Tour Guides (bilingual young women in charge of ushering—some say rushing—little bands of tourists through the park);

2. Ride operators performing coveted "skilled work" (such as live narrations or tricky transportation tasks like those who symbolically control customer access to the park and drive the costly entry vehicles such as the antique trains, horse-drawn carriages, and Monorail);

3. All other ride operators;

4. The proletarian Sweepers (keepers of the concrete grounds);

5. The sub-prole or peasant status Food and Concession workers (whose park sobriquets reflect their lowly social worth—"pancake ladies," "peanut pushers," "coke blokes," "suds divers," and the seemingly irreplaceable "soda jerks").

Pay differentials are slight among these employee groups. The collective status adheres, as it does internally for ride operators, to assignment or functional distinctions. As the rank order suggests, most employee status goes to those who work jobs that require higher degrees of special skill, relative freedom from constant and direct supervision, and provide the opportunity to organize and direct customer desires and behavior rather than to merely respond to them as spontaneously expressed.

The basis for sorting individuals into these various broad bands of job categories is often unknown to employees—a sort of deep, dark secret of the casting directors in personnel. When prospective employees are interviewed, they interview for "a job at Disneyland," not a specific one. Personnel decides what particular job they will eventually occupy. Personal contacts are considered by employees as crucial in this job-assignment process as they are in the hiring decision. Some employees, especially those who wind up in the lower ranking jobs, are quite disappointed with their assignments as is the case when, for example, a would-be Adventureland guide is posted to a New Orleans Square restaurant as a pot scrubber. Although many of the outside acquaintances of our pot scrubber may know only that he works at Disneyland, rest assured, insiders will know immediately where he works and judge him accordingly.

Uniforms are crucial in this regard for they provide instant communication about the social merits or demerits of the wearer within the little world of Disneyland workers. Uniforms also correspond to a wider status ranking that casts a significant shadow on employees of all types. Male ride operators on the Autopia wear, for example, untailored jump-suits similar to pit mechanics and consequently generate about as much respect from peers as the grease-stained outfits worn by pump jockeys generate from real motorists in gas stations. The ill-fitting and homogeneous "whites" worn by Sweepers signify lowly institutional work tinged, perhaps, with a reminder of hospital orderlies rather than street cleanup crews. On the other hand, for males, the crisp, officer-like Monorail operator stands alongside the swashbuckling Pirate of the Caribbean, the casual cowpoke of Big Thunder Mountain, or the smartly vested Riverboat pilot as carriers of valued symbols in and outside the park. Employees lust for these higher status positions and the rights to small advantages such uniforms provide. A lively internal labor market exists wherein there is much scheming for the more prestigious assignments.

For women, a similar market exists although the perceived "sexiness" of uniforms, rather than social rank, seems to play a larger role. To wit, the rather heated antagonisms that developed years ago when the ride "It's a Small World" first opened and began outfitting the ride operators with what were felt to be the shortest skirts and most revealing blouses in the park. Tour Guides, who traditionally headed the fashion vanguard at Disneyland in their above-the-knee kilts, knee socks, tailored vests, black English hats, and smart riding crops were apparently appalled at being upstaged by their social inferiors and lobbied actively (and, judging by the results, successfully) to lower the skirts, raise the necklines, and generally remake their Small World rivals.

Important, also, to ride operators are the break schedules followed on the various rides. The more the better. Work teams develop inventive ways to increase the number of "time-outs" they take during the work day. Most rides are organized on a rotational basis (e.g., the operator moving from a break, to queue monitor, to turnstile overseer, to unit loader, to traffic controller, to driver, and, again, to a break). The number of break men or women on a rotation (or ride) varies by the number of employees on duty and by the number of units on line. Supervisors, foremen, and operators also vary as to what they regard as appropriate break standards (and, more importantly, as to the value of the many situational factors that can enter the calculation of break rituals—crowd size, condition of ride, accidents, breakdowns, heat, operator absences, special occasions, and so forth). Self-monitoring teams with sleepy supervisors and lax (or savvy) foremen can sometimes manage a shift comprised of 15 minutes on and 45 minutes off each hour. They are envied by others, and rides that have such a potential are eyed hungrily by others who feel trapped by their more rigid (and observed) circumstances.

Movement across jobs is not encouraged by park management, but some does occur (mostly within an area and job category). Employees claim that a sort of "once a sweeper, always a sweeper" rule obtains but all know of at least a few exceptions to prove the rule. The exceptions offer some (not much) hope for those working at the social margins of the park and perhaps keep them on the job longer than might otherwise be expected. Dishwashers can dream of becoming Pirates, and with persistence and a little help from their friends, such dreams just might come true next season (or the next).

These examples are precious, perhaps, but they are also important. There is an intricate pecking order among very similar categories of employees. Attributes of reward and status tend to cluster, and there is intense concern about the cluster to which one belongs (or would like to belong). To a degree, form follows function in Disneyland because the jobs requiring the most abilities and offering the most interest also offer the most status and social reward. Interaction patterns reflect and sustain this order. Few Ambassadors or Tour Guides, for instance, will stoop to speak at length with Sweepers who speak mostly among themselves or to Food workers. Ride operators, between the poles, line up in ways referred to above with only ride proximity (i.e., sharing a break area) representing a potentially significant intervening variable in the interaction calculation. . . .

Paid employment at Disneyland begins with the much renowned University of Disneyland whose faculty runs a day-long orientation program (Traditions I) as part of a 40-hour apprenticeship program, most of which takes place on the rides. In the classroom, however, newly hired ride operators are given a very thorough introduction to matters of managerial concern and are tested on their absorption of famous Disneyland fact, lore, and procedure. Employee demeanor is governed, for example, by three rules:

First, we practice the friendly smile.
Second, we use only friendly and courteous phrases.
Third, we are not stuffy—the only Misters in Disneyland are Mr. Toad and Mr. Smee.

Employees learn too that the Disneyland culture is officially defined. The employee handbook put it in this format:

Dis-ney Cor-po-rate Cul-ture (diz'ne kor'pr'it kul'cher) *n* 1. Of or pertaining to the Disney organization, as a: the philosophy underlying all business decisions; b: the commitment of top leadership and management to that philosophy; c: the actions taken by individual cast members that reinforce the image.

Language is also a central feature of university life, and new employees are schooled in its proper use. Customers at Disneyland are, for instance, never referred to as such, they are "guests." There are no rides at Disneyland, only "attractions." Disneyland itself is a "Park," not an amusement center, and it is divided into "back-stage," "on-stage," and "staging" regions. Law enforcement personnel hired by the park are not policemen, but "security hosts." Employees do not wear uniforms but check out fresh "costumes" each working day from "wardrobe." And, of course, there are no accidents at Disneyland, only "incidents." . . .

The university curriculum also anticipates probable questions ride operators may someday face from customers, and they are taught the approved public response. A sample:

Question (posed by trainer): What do you tell a guest who requests a rain check?
Answer (in three parts): We don't offer rain checks at Disneyland because (1) the main attractions are all indoors; (2) we would go broke if we offered passes; and (3) sunny days would be too crowded if we gave passes.

Shrewd trainees readily note that such an answer blissfully disregards the fact that waiting areas of Disneyland are mostly outdoors and that there are no subways in the park to carry guests from land to land. Nor do they miss the economic assumption concerning the apparent frequency of Southern California rains. They discuss such matters together, of course, but rarely raise them in the training classroom. In most respects, these are recruits who easily take the role of good student.

Classes are organized and designed by professional Disneyland trainers who also instruct a well-screened group of representative hourly employees straight from park operations on the approved newcomer training methods and materials. New-hires seldom see professional trainers in class but are brought on board by enthusiastic peers who concentrate on those aspects of park procedure thought highly general matters to be learned by all employees. Particular skill training (and "reality shock") is reserved for the second wave of socialization occurring on the rides themselves as operators are taught, for example, how and when to send a mock bobsled caroming down the track or, more delicately, the proper ways to stuff an obese adult customer into the midst of children riding the Mon-key car on the Casey Jones Circus Train or, most problematically, what exactly to tell an irate customer standing in the rain who, in no uncertain terms, wants his or her money back and wants it back now.

During orientation, considerable concern is placed on particular values the Disney organization considers central to its operations. These values range from the "customer is king" verities to the more or less unique kind, of which "every-one is a child at heart when at Disneyland" is a decent example. This latter piety is one few employees fail to recognize as also attaching to everyone's mind as well after a few months of work experience. Elaborate checklists of appearance stan-dards are learned and gone over in the classroom and great efforts are spent try-ing to bring employee emotional responses in line with such standards. Employees are told repeatedly that if they are happy and cheerful at work, so, too, will the guests at play. Inspirational films, hearty pep talks, family imagery, and exemplars of corporate performance are all representative of the strong symbolic stuff of these training rites. . . .

In general, Disneyland employees are remarkable for their forbearance and polite good manners even under trying conditions. They are taught, and some come to believe, for a while at least, that they are really "on-stage" at work. And, as noted, surveillance by supervisory personnel certainly fades in light of the un-ceasing glances an employee receives from the paying guests who tromp daily through the park in the summer. Disneyland employees know well that they are part of the product being sold and learn to check their more discriminating man-ners in favor of the generalized countenance of a cheerful lad or lassie whose en-thusiasm and dedication is obvious to all.

At times, the emotional resources of employees appear awesome. When the going gets tough and the park is jammed, the nerves of all employees are frayed and sorely tested by the crowd, din, sweltering sun, and eyeburning smog. Cus-tomers wait in what employees call "bullpens" (and park officials call "reception areas") for up to several hours for a 3½ minute ride that operators are some-times hell-bent on cutting to 2½ minutes. Surely a monument to the human ability to suppress feelings has been created when both users and providers alike can maintain their composure and seeming regard for one another when in such a fix.

It is in this domain where corporate culture and the order it helps to sustain must be given its due. Perhaps the depth of a culture is visible only when its members are under the gun. The orderliness—a good part of the Disney formula for financial success—is an accomplishment based not only on physical design and elaborate procedures, but also on the low-level, part-time employees who, in the final analysis, must be willing, even eager, to keep the show afloat. The ease with which employees glide into their kindly and smiling roles is, in large mea-sure, a feat of social engineering. Disneyland does not pay well; its supervision is arbitrary and skin-close; its working conditions are chaotic; its jobs require mini-mal amounts of intelligence or judgment; and asks a kind of sacrifice and loyalty of its employees that is almost fanatical. Yet, it attracts a particularly able work-force whose personal backgrounds suggest abilities far exceeding those required of a Disneyland traffic cop, people stuffer, queue or line manager, and button pusher. As I have suggested, not all of Disneyland is covered by the culture put forth by management. There are numerous pockets of resistance and various degrees of autonomy maintained by employees. Nonetheless, adherence and support for the organization are remarkable. And, like swallows returning to Capistrano, many part-timers look forward to their migration back to the park for several seasons.

THE DISNEY WAY

Four features alluded to in this unofficial guide to Disneyland seem to account for a good deal of the social order that obtains within the park. First, socialization, although costly, is of a most selective, collective, intensive, serial, sequential, and closed sort. These tactics are notable for their penetration into the private spheres of individual thought and feeling. . . . Incoming identities are not so much dismantled as they are set aside as employees are schooled in the use of new identities of the situational sort. Many of these are symbolically powerful and, for some, laden with social approval. It is hardly surprising that some of the more problematic positions in terms of turnover during the summer occur in the food and concession domains where employees apparently find little to identify with on the job. Cowpokes on Big Thunder Mountain, Jet Pilots, Storybook Princesses, Tour Guides, Space Cadets, Jungle Boat Skippers, or Southern Belles of New Orleans Square have less difficulty on this score. Disneyland, by design, bestows identity through a process carefully set up to strip away the job relevance of other sources of identity and learned response and replace them with others of organizational relevance. It works.

Second, this is a work culture whose designers have left little room for individual experimentation. Supervisors, as apparent in their focused wandering and attentive looks, keep very close tabs on what is going on at any moment in all the lands. Every bush, rock, and tree in Disneyland is numbered and checked continually as to the part it is playing in the park. So too are employees. Discretion of a personal sort is quite limited while employees are "on-stage." Even "back-stage" and certain "off-stage" domains have their corporate monitors. Employees are indeed aware that their "off-stage" life beyond the picnics, parties, and softball games is subject to some scrutiny, for police checks are made on potential and current employees. Nor do all employees discount the rumors that park officials make periodic inquiries on their own as to a person's habits concerning sex and drugs. Moreover, the sheer number of rules and regulations is striking, thus making the grounds for dismissal a matter of multiple choice for supervisors who discover a target for the use of such grounds. The feeling of being watched is, unsurprisingly, a rather prevalent complaint among Disneyland people, and it is one that employees must live with if they are to remain at Disneyland.

Third, emotional management occurs in the park in a number of quite distinct ways. From the instructors at the university who beseech recruits to "wish every guest a pleasant good day," to the foremen who plead with their charges to, "say thank you when you herd them through the gate," to the impish customer who seductively licks her lips and asks, "what does Tom Sawyer want for Christmas?" appearance, demeanor, and etiquette have special meanings at Disneyland. Because these are prized personal attributes over which we normally feel in control, making them commodities can be unnerving. Much self-monitoring is involved, of course, but even here self-management has an organizational side. Consider ride operators who may complain of being "too tired to smile" but, at the same time, feel a little guilty for uttering such a confession. Ride operators who have worked an early morning shift on the Matterhorn (or other popular rides) tell of a queasy feeling they get when the park is opened for business and they suddenly feel the ground begin to shake under their feet and hear the low thunder of the hordes of customers coming at them, oblivious of civil restraint and the small children who might be among them. Consider, too, the discomforting pressures of being "on-stage" all day and the cumulative annoyance of having adults ask permission to leave a line to go to the bathroom, whether the water in the lagoon is real, where the well-marked entrances might be, where Walt Disney's cryogenic tomb is to be found, or—the real clincher—whether or not one is "really real."

The mere fact that so much operator discourse concerns the handling of bothersome guests suggests that these little emotional disturbances have costs. There are, for instance, times in all employee careers when they put themselves on "automatic pilot," "go robot," "can't feel a thing," "lapse into a dream," "go into a trance," or otherwise "check out" while still on duty. Despite a crafty supervisor's (or curious visitor's) attempt to measure the glimmer in an employee's eye, this sort of willed emotional numbness is common to many of the "on-stage" Disneyland personnel. Much of this numbness is, of course, beyond the knowledge of supervisors and guests because most employees have little trouble appearing as if they are present even when they are not. It is, in a sense, a passive form of resistance that suggests there still is a sacred preserve of individuality left among employees in the park.

Finally, taking these three points together, it seems that even when people are trained, paid, and told to be nice, it is hard for them to do so all of the time. But, when efforts to be nice have succeeded to the degree that is true of Disneyland, it appears as a rather towering (if not always admirable) achievement. It works at the collective level by virtue of elaborate direction. Employees—at all ranks—are stage-managed by higher ranking employees who, having come through themselves, hire, train, and closely supervise those who have replaced them below. Expression rules are laid out in corporate manuals. Employee timeouts intensify work experience. Social exchanges are forced into narrow bands of interacting groups. Training and retraining programs are continual. Hiding places are few. Although little sore spots and irritations remain for each individual, it is difficult to imagine work roles being more defined (and accepted) than those at Disneyland. Here, it seems, is a work culture worthy of the name.

DISCUSSION QUESTIONS

1. How is the corporate culture of Disney reproduced through the training that workers undergo? What effect is being achieved and whose interests does this support?
2. Identify a work organization with which you are familiar. Using an analysis similar to that done by Van Maanen, how would you describe the culture of this work place? How does this culture affect the experience of workers in the organization? of managers? on customers?

INFOTRAC COLLEGE EDITION

You can use your access to InfoTrac College Edition to learn more about the subjects covered in this essay. Some suggested search terms include:

corporate culture
cultures of resistance
feeling rules
hospitality industry

total quality management
worker control
worker surveillance

67

Delicate Transactions: Gender, Home, and Employment among Hispanic Women

M. PATRICIA FERNÁNDEZ KELLY

Comparing the work experience of Mexican women workers in southern California and Cuban women workers in Miami, M. Patricia Fernández Kelly identifies differences and commonalities in the connections between work and family experience for women workers. Her analysis shows how the class experience of women also constructs their gender experience and she identifies the variations in work and family patterns that can be found within racial and ethnic groups.

From: Faye Genslurg and Anna Lowenhaupt Tsing. 1990. *Uncertain Terms. Negotiating Gender in American Culture*. Boston: Beacon Press, pp. 182–195. Reprinted with permission.

Although there are many studies comparing minorities and whites in the U.S., there have been few attempts to look at variations of experience *within* ethnic groups. This is true for Hispanics in general and for Hispanic women in particular; yet contrasts abound. For example, Mexicans comprise more than half of all Hispanics between eighteen and sixty-four years of age living in the U.S. Of these, approximately 70% were born in this country. Average levels of educational attainment are quite low with less than 50% having graduated from high school. In contrast, Cubans represent about 7% of the Hispanic population. They are mostly foreign-born; 58% of Cubans have 12 or more years of formal schooling.[1]

Both in Southern California and in Southern Florida most direct production workers in the garment industry are Hispanic. In Los Angeles most apparel firm operatives are Mexican women, in Miami, Cuban women.[2] The labor force participation rates of Mexican and Cuban women dispel the widespread notion that work outside the home is a rare experience for Hispanic women.[3] Yet the Los Angeles and Miami communities differ in a number of important respects. One can begin with contrasts in the garment industry in each area.

The two sites differ in the timing of the industry, its evolution, maturity, and restructuring. In Los Angeles, garment production emerged in the latter part of the nineteenth-century and expanded in the 1920s, stimulated in part by the arrival of runaway shops evading unionization drives in New York. The Great Depression sent the Los Angeles garment industry into a period of turmoil, but soon fresh opportunities for the production of inexpensive women's sportswear developed, as the rise of cinema established new guidelines for fashion. During the 1970s and 1980s the industry reorganized in response to foreign imports; small manufacturing shops have proliferated, as has home production. In contrast, the apparel industry in Miami has had a shorter and more uniform history. Most of the industry grew up since the 1960s, when retired manufacturers from New York saw the advantage of opening new businesses and hiring exiles from the Cuban Revolution.

The expansion of the Los Angeles clothing industry resulted from capitalists' ability to rely on continuing waves of Mexican immigrants, many of whom were undocumented. Mexican migration over the last century ensured a steady supply of workers for the apparel industry; from the very beginning, Mexican women were employed in nearly all positions in the industry.[4] By contrast, the expansion of garment production in Miami was due to an unprecedented influx of exiles ejected by a unique political event. Cubans working in the Florida apparel industry arrived in the United States as refugees under a protected and relatively privileged status. Exile was filled with uncertainty and the possibility of dislocation but not, as in the case of undocumented Mexican aliens, with the probability of harassment, detention, and deportation.

Mexican and Cuban workers differ strikingly in social class. For more than a century, the majority of Mexican immigrants have had a markedly proletarian background. Until the 1970s, the majority had rural roots, although in more recent times there has been a growing number of urban immigrants.[5] In sharp contrast, Cuban waves of migration have included a larger proportion of professionals, mid-level service providers, and various types of entrepreneurs ranging from those with previous experience in large companies to those qualified to start small family enterprises. Entrepreneurial experience among Cubans and reliance on their own ethnic networks accounts, to a large extent, for Cuban success in business formation and appropriation in Miami.[6] Thus, while Mexican migration has been characterized by relative homogeneity regarding class background, Cuban exile resulted in the transposition of an almost intact class structure containing investors and professionals as well as unskilled, semiskilled, and skilled workers.

In addition to disparate class compositions, the two groups differ in the degree of their homogeneity by place of birth. Besides the sizable undocumented contingent mentioned earlier, the Los Angeles garment industry also employs U.S.-born citizens of Mexican heritage. First-hand reports and anecdotal evidence indicate that the fragmentation between "Chicana" and "Mexicana" workers causes an unresolved tension and animosity within the labor force. Cubans, on the other hand, were a highly cohesive population until the early 1980s, when the arrival of the so-called "Marielitos" resulted in a potentially disruptive polarization of the community.

Perhaps the most important difference between Mexicans in Los Angeles and Cubans in Florida is related to their distinctive labor market insertion patterns. Historically, Mexicans have arrived in the U.S. labor market in a highly individuated and dispersed manner. As a result, they have been extremely dependent on labor market supply and demand forces entirely beyond their control. Their working-class background and stigma attached to their frequent undocumented status has accentuated even further their vulnerability vis-à-vis employers. By contrast, Cubans have been able to consolidate an economic enclave formed by immigrant businesses, which hire workers of a common cultural and national background. The economic enclave partly operates as a buffer zone separating and often shielding members of the same ethnic group from the market forces at work in the larger society. The existence of an economic enclave does not preclude exploitation on the basis of class; indeed, it is predicated upon the existence of a highly diversified immigrant class structure. However, commonalities of culture, national background, and language between immigrant employers and workers can become a mechanism for collective improvement of income levels and standards of living. As a result, differences in labor market insertion patterns among Mexicans and Cubans have led to varying social profiles and a dissimilar potential for socioeconomic attainment.

THE WOMEN GARMENT WORKERS

These differences between the two Hispanic communities have led to important differences between the two groups of women who work in the garment industry. For Mexican women in Southern California, employment in garment production is the consequence of long-term economic need. Wives and daughters choose to work outside the home in order to meet the survival requirements of their families in the absence of satisfactory earnings by men. Some female heads of household join the labor force after losing male support through illness, death, and, more often, desertion. In many of these instances, women opt for industrial homework in order to reconcile child care and the need for wage employment. They are particularly vulnerable members of an economically marginal ethnic group.

By contrast, Cuban women who arrived in Southern Florida during the 1960s saw jobs in garment assembly as an opportunity to recover or attain middle-class status. The consolidation of an economic enclave in Miami, which accounts for much of the prosperity of Cubans, was largely dependent upon the incorporation of women into the labor force. While they toiled in factories, men entered business or were self-employed. Their vulnerability was tempered by shared goals of upward mobility in a foreign country.

Despite their different nationalities, migratory histories, and class backgrounds, Mexicans and Cubans share many perceptions and expectations. In both cases, patriarchal norms of reciprocity are favored; marriage, motherhood, and devotion to family are high priorities among women, while men are expected to hold authority, to be good providers, and to be loyal to their wives and children. However, the divergent economic and political conditions surrounding Mexicans in Southern California and Cubans in Southern Florida have had a differing impact upon each group's ability to uphold these values. Mexican women are often thrust into financial "autonomy" as a result of men's inability to fulfill their socially assigned role. Among Cubans, by contrast, men have been economically more successful. Indeed, ideological notions of patriarchal responsibility have served to maintain group cohesion; that offers women an advantage in getting and keeping jobs within the ethnic enclave.

Cuban and Mexican women both face barriers stemming from their subordination in the family and their status as low-skilled workers in highly competitive industries. Nevertheless, their varying class backgrounds and modes of incorporation into local labor markets entail distinctive political and socioeconomic effects. How women view their identities as women is especially affected. Among Mexican garment workers disillusion about the economic viability of men becomes a desire for individual emancipation, mobility, and financial independence as women. However, these ideals and ambitions for advancement are most often frustrated by poverty and the stigmas attached to ethnic and gender status.

Cuban women, on the other hand, tend to see no contradiction between personal fulfillment and a strong commitment to patriarchal standards. Their incorporation and subsequent withdrawal from the labor force are both influenced by their acceptance of hierarchical patterns of authority and the sexual division of labor. As in the case of Mexicans in Southern California, Cuban women's involvement in industrial homework is an option bridging domestic and income-generating needs. However, it differs in that homework among them was brought about by relative prosperity and expanding rather than diminishing options. Women's garment work at home does not contradict patriarchal ideals of women's place at the same time as it allows women to contribute to the economic success that confirms gender stratification.

The stories of particular women show the contrasts in how women in each of these two groups negotiate the links among household, gender, and employment arrangements. Some of the conditions surrounding Mexican home workers in Southern California are illustrated by the experience of Amelia Ruíz.[7] She was born into a family of six children in El Cerrito, Los Angeles County. Her mother, a descendant of Native American Indians, married at a young age the son of Mexican immigrants. Among Amelia's memories are the fragmentary stories of her paternal grandparents working in the fields and, occasionally, in canneries. Her father, however was not a stoop laborer but a trained upholsterer. Her mother was always a homemaker. Amelia grew up with a distinct sense of the contradictions that plague the relationships between men and women:

> All the while I was a child, I had this feeling that my parents weren't happy. My mother was smart but she could never make much of herself. Her parents taught her that the fate of woman is to be a wife and mother; they advised her to find a good man and marry him. And that she did. My father was reliable and I think he was faithful but he was also distant; he lived in his own world. He would come home and expect to be served hand and foot. My mother would wait on him but she was always angry about it. I never took marriage for granted.

After getting her high school diploma, Amelia found odd jobs in all the predictable places: as a counter clerk in a dress shop, as a cashier in a fast-food establishment, and as a waitress in two restaurants. When she was 20, she met Miguel—Mike as he was known outside the barrio. He was a consummate survivor, having worked in the construction field, as a truck driver, and even as an English as a Second Language instructor. Despite her misgivings about marriage, Amelia was struck by Mike's penchant for adventure:

> He was different from the men in my family. He loved fun and was said to have had many women. He was a challenge. We were married when I was 21 and he 25. For a while I kept my job but when I became pregnant, Miguel didn't want me to work any more. Two more children followed and then, little by little, Miguel became abusive. He wanted to have total authority over me and the children. He said a man should know how to take care of a family and get respect, but it was hard to take him seriously when he kept changing jobs and when the money he brought home was barely enough to keep ends together.

After the birth of her second child, Amelia started work at Shirley's, a women's wear factory in the area. Miguel was opposed to the idea. For Amelia, work outside the home was an evident need prompted by financial stress. At first, it was also a means to escape growing disenchantment:

> I saw myself turning into my mother and I started thinking that to be free of men was best for women. Maybe if Miguel had had a better job, maybe if he had kept the one he had, things would have been different, but he didn't. . . . We started drifting apart.

Tension at home mounted over the following months. Amelia had worked at Shirley's for almost a year when, one late afternoon after collecting the three children from her parents' house, she returned to an empty home. She knew, as soon as she stepped inside, that something was amiss. In muted shock, she confirmed the obvious: Miguel had left, taking with him all personal possessions; even the wedding picture in the living room had been removed. No explanations had been left behind. Amelia was then 28 years of age, alone, and the mother of three small children.

As a result of these changes, employment became even more desirable, but the difficulty of reconciling home responsibilities with wage work persisted. Amelia was well regarded at Shirley's, and her condition struck a sympathetic cord among the other factory women. In a casual conversation, her supervisor described how other women were leasing industrial sewing machines from the local Singer distributor and were doing piecework at home. By combining factory work and home assembly, she could earn more money without further neglecting the children. Mr. Driscoll, Shirley's owner and general manager, made regular use of home workers, most of whom were former employees. That had allowed him to retain a stable core of about 20 factory seamstresses and to depend on approximately 10 home workers during peak seasons.

Between 1979, the year of her desertion, and 1985, when I met her, Amelia had struggled hard, working most of the time and making some progress. Her combined earnings before taxes fluctuated between $950 and $1,150 a month. Almost half of her income went to rent for the two-bedroom apartment which she shared with the children. She was in debt and used to working at least 12 hours a day. On the other hand, she had bought a double-needle sewing machine and was thinking of leasing another one to share additional sewing with a neighbor. She had high hopes:

> Maybe some day I'll have my own business; I'll be a liberated woman. . . I won't have to take orders from a man. Maybe Miguel did me a favor when he left after all. . .

With understandable variations, Amelia's life history is shared by many garment workers in Southern California. Three aspects are salient in this experience. First, marriage and a stable family life are perceived as desirable goals which are, nonetheless, fraught with ambivalent feelings and burdensome responsibilities.

Second, tensions between men and women result from contradictions between the intent to fulfill gender definitions and the absence of the economic base necessary for their implementation. The very definition of manhood includes the right to hold authority and power over wives and children, as well as the responsibility of providing adequately for them. The difficulties in implementing those goals in the Mexican communities I studied are felt equally by men and women but expressed differently by each. Bent on restoring their power, men attempt to control women in abusive ways. Women often resist their husbands' arbitrary or unrealistic impositions. Both reactions are eminently political phenomena.

Third, personal conflict regarding the proper behavior of men and women may be tempered by negotiation. It can also result in the breach of established agreements, as in the case of separation or divorce. Both paths are related to the construction of alternative discourses and the redefinition of gender roles. Women may seek personal emancipation, driven partly by economic need and partly by dissatisfaction with men's performance as providers. In general, individuals talk about economic and political conflict as a personal matter occurring in their own homes. Broader contextual factors are less commonly discussed.

The absence of economic underpinnings for the implementation of patriarchal standards may bring about more equitable exchanges between men and women, and may stimulate women's search for individual well-being and personal autonomy as women. However, in the case at hand, such ideals remain elusive. Mexican garment workers, especially those who are heads of households, face great disadvantages in the labor market. They are targeted for jobs that offer the lowest wages paid to industrial workers in the United States; they also have among the lowest unionization rates in the country. Ironically, the breakdown of patriarchal norms in the household draws from labor market segmentation that reproduces patriarchal (and ethnic) stratification.

Experiences like the ones related are also found among Cuban and Central American women in Miami. However, a large proportion have had a different trajectory. Elvira Gómez's life in the U.S. is a case in point. She was 34 when she arrived in Miami with her four children, ages three to twelve. The year was 1961.

> Leaving Havana was the most painful thing that ever happened to us. We loved our country. We would have never left willingly. Cuba was not like Mexico: we didn't have immigrants in large numbers. But Castro betrayed us and we had to join the exodus. We became exiles. My husband left Cuba three months before I did and there were moments when I doubted I would ever see him again. Then, after we got together, we realized we would have to forge ahead without looking back.
>
> We lost everything. Even my mother's china had to be left behind. We arrived in this country as they say, "covering our nakedness with our bare hands" (*una mano delante y otra detrás*). My husband had had a good position in a bank. To think that he would have to take any old job in Miami was more than I could take; a man of his stature having to beg for a job in a hotel or in a factory? It wasn't right!

Elvira had worked briefly before her marriage as a secretary. As a middle-class wife and mother, she was used to hiring at least one maid. Coming to the United States changed all that:

> Something had to be done to keep the family together. So I looked around and finally found a job in a shirt factory in Hialeah. Manolo (her husband) joined a childhood friend and got a loan to start an export-import business. All the time they were building the firm, I was sewing. There were times when we wouldn't have been able to pay the bills without the money I brought in.

Elvira's experience was shared by thousands of women in Miami. Among the first waves of Cuban refugees there were many who worked tirelessly to raise the standards of living of their families to the same levels or higher than those they had been familiar with in their country of origin. The consolidation of an ethnic enclave allowed many Cuban men to become entrepreneurs. While their wives found un-skilled and semi-skilled jobs, they became businessmen. Eventually, they purchased homes, put their children through school, and achieved comfort. At that point, many Cuban men pressed their wives to stop working outside of the home; they had only allowed them to have a job, in the first place, out of economic necessity. . .

This discussion partly shows that decisions made at the level of the household can remove workers, actively sought and preferred by employers, from the mar-ketplace. This, in turn, can threaten certain types of production. In those cases, loyalty to familial values can mitigate against the interests of capitalist firms. Inter-views with Cuban women involved in homework confirm the general accuracy of this interpretation. After leaving factory employment, many put their experi-ence to good use by becoming subcontractors and employing neighbors or friends. They also transformed so-called "Florida rooms" (the covered porches in their houses) into sewing shops. It was in one of them that Elvira Gómez was first interviewed. In her case, working outside the home was justified only as a way to maintain the integrity of her family and as a means to support her hus-band's early incursions into the business world:

> For many long years I worked in the factory but when things got better financially, Manolo asked me to quit the job. He felt bad that I couldn't be at home all the time with the children. But it had to be done. There's no reason for women not to earn a living when it's necessary; they should have as many opportunities and responsibilities as men. But I also tell my daughters that the strength of a family rests on the intelligence and work of women. It is foolish to give up your place as a mother and a wife only to go take orders from men who aren't even part of your family. What's so liberated about that? It is better to see your husband succeed and to know you have supported one another.

Perhaps the most important point here is the unambiguous acceptance of pa-triarchal mores as a legitimate guideline for the behavior of men and women. Exile did not eliminate these values; rather, it extended them in telling ways. The high labor force participation rates of Cuban women in the United States have been mentioned before. Yet, it should be remembered that, prior to their migra-tion, only a small number of Cuban women had worked outside the home for any length of time. It was the need to maintain the integrity of their families and to achieve class-related ambitions that precipitated their entrance into the labor force of a foreign country.

In descriptions of their experience in exile, Cuban women often make clear that part of the motivation in their search for jobs was the preservation of known definitions of manhood and womanhood. Whereas Mexican women worked as a response to what they saw as a failure of patriarchal arrangements, Cuban women worked in the name of dedication to their husbands and children, and in order to preserve the status and authority of the former. Husbands gave them "permis-sion" to work outside the home, and only as a result of necessity and temporary economic strife. In the same vein, it was a ritual yielding to masculine privilege that led women to abandon factory employment. Conversely, men "felt bad" that their wives had to work for a wage and welcomed the opportunity to remove them from the marketplace when economic conditions improved.

As with Mexicans in Southern California, Cuban women in Miami earned low wages in low- and semi-skilled jobs. They too worked in environments devoid of the benefits derived from unionization. Nevertheless, the outcome of their experience as well as the perceptions are markedly different. Many Cuban women interpret their subordination at home as part of a viable option ensuring economic and emotional benefits. They are bewildered by feminist goals of equality and fulfillment in the job market. Yet, the same women have had among the highest rates of participation in the U.S. labor force.

CONCLUSIONS

For Mexican women in Southern California, proletarianization is related to a high number of female-headed households, as well as households where the earnings provided by women are indispensable for maintaining standards of modest subsistence. In contrast, Cuban women's employment in Southern Florida was a strategy for raising standards of living in a new environment. These contrasts in the relationship between households and the labor market occurred despite shared values regarding the family among Mexicans and Cubans. Both groups partake of similar mores regarding the roles of men and women; nevertheless, their actual experience has differed significantly. Contrasting features of class, educational background, and immigration history have created divergent gender and family dilemmas for each group.

This analysis underscores the impact of class on gender. Definitions of manhood and womanhood are implicated in the very process of class formation. At the same time, the norms of reciprocity sanctioned by patriarchal ideologies can operate as a form of social adhesive consolidating class membership. For poor men and women, the issue is not only the presence of the sexual division of labor and the persistence of patriarchal ideologies but the difficulties of upholding either.

Thus, too, the meaning of women's participation in the labor force remains plagued by paradox. For Mexican women in Southern California, paid employment responds to and increases women's desires for greater personal autonomy and financial independence. Ideally, this should have a favorable impact upon women's capacity to negotiate an equitable position within their homes and in the labor market. Yet these women's search for paid employment is most often the consequence of severe economic need; it expresses vulnerability not strength within homes and in the marketplace. Indeed, in some cases, women's entry into the labor force signals the collapse of reciprocal exchanges between men and women. Women deserted by their husbands are generally too economically marginal to translate their goals of gender equality and autonomy into socially powerful arrangements. Conversely, Cuban women in Southern Florida have more economic power, but this only strengthens their allegiance to patriarchal standards. The conjugal "partnership for survival" Elvira Gómez describes is not predicated on the existence of a just social world, but rather an ideological universe entailing differentiated and stratified benefits and obligations for men and women.

NOTES

1. Frank D. Bean and Marta Tienda, *The Hispanic Population of the United States* (New York: Russell Sage Foundation, 1987). There are almost twenty million Hispanics in the United States, that is, 14.6% of the total population.

2. Approximately 75% and 67% of operatives in Los Angeles and Miami apparel firms are Mexican and Cuban women respectively.

3. Note 54.2% of native-born and 47.5% of foreign-born Mexican women were employed outside the home in 1980. The equivalent figure for the mostly foreign-born Cuban women was almost 65%. Non-Hispanic white women's labor force participation in 1980 was assessed at 57.9% (U.S. Census of Population, 1980).

4. Peter S. Taylor, "Mexican Women in Los Angeles Industry in 1928," *Aztlán: International Journal of Chicano Studies Research*, **11**, 1 (Spring, 1980): 99–129.

5. Alejandro Portes and Robert L. Bach, *Latin Journey: Cuban and Mexican Immigrants in the United States* (Berkeley: University of California Press, 1985), 67.

6. Alejandro Portes, "The Social Origins of the Cuban Enclave Economy of Miami," *Pacific Sociological Review*, Special Issue on the Ethnic Economy, 30, 4 (October, 1987): 340–372. See also Lisandro Perez, "Immigrant Economic Adjustment and Family Organization: The Cuban Success Story Reexamined," *International Migration Review*, 20 (1986): 4–20.

7. The following descriptions are chosen from a sample of 25 Mexican and 10 Cuban women garment workers interviewed in Los Angeles and Miami Counties. The names of people interviewed, and some identifying characteristics, have been changed.

DISCUSSION QUESTIONS

1. How do the different experiences of the Mexican and Cuban American women in M. Patricia Fernández Kelly's study illustrate how the gender division of labor intersects with the class system?

2. Using the examples given in this article, describe how the distinct labor market experiences of different groups are intertwined with family and gender relations.

INFOTRAC COLLEGE EDITION

You can use your access to InfoTrac College Edition to learn more about the subjects covered in this essay. Some suggested search terms include:

garment workers   proletarianization
home work   refugee labor
migrant workers   women's wage labor
migration

68

The Power Elite

C. WRIGHT MILLS

C. Wright Mills' classic book, The Power Elite, *first published in 1956, remains an important analysis of the system of power in the United States. He argues that national power is located in three particular institutions: the economy, politics, and the military. An important point in his article is that the power of elites is derived from their institutional location, not their individual attributes.*

The powers of ordinary men are circumscribed by the everyday worlds in which they live, yet even in these rounds of job, family, and neighborhood they often seem driven by forces they can neither understand nor govern. 'Great changes' are beyond their control, but affect their conduct and outlook none the less. The very framework of modern society confines them to projects not their own, but from every side, such changes now press upon the men and women of the mass society, who accordingly feel that they are without purpose in an epoch in which they are without power.

But not all men are in this sense ordinary. As the means of information and of power are centralized, some men come to occupy positions in American society from which they can look down upon, so to speak, and by their decisions mightily affect, the everyday worlds of ordinary men and women. They are not made by their jobs; they set up and break down jobs for thousands of others; they are not confined by simple family responsibilities; they can escape. They may live in many hotels and houses, but they are bound by no one community. They need not merely 'meet the demands of the day and hour'; in some part, they create these demands, and cause others to meet them. Whether or not they profess their power, their technical and political experience of it far transcends that of the underlying population. What Jacob Burckhardt said of 'great men,' most Americans might well say of their elite: 'They are all that we are not.'

The power elite is composed of men whose positions enable them to transcend the ordinary environments of ordinary men and women; they are in positions to make decisions having major consequences. Whether they do or do not make such decisions is less important than the fact that they do occupy such pivotal positions: their failure to act, their failure to make decisions, is itself an act that is often of greater consequence than the decisions they do make. For they are in command of the major hierarchies and organizations of modern society. They rule the big corporations. They run the machinery of the state and claim its

From: C. Wright Mills. 1956. *The Power Elite.* New York: Oxford University Press, pp. 3–29, 269–297.

prerogatives. They direct the military establishment. They occupy the strategic command posts of the social structure, in which are now centered the effective means of the power and the wealth and the celebrity which they enjoy.

The power elite are not solitary rulers. Advisers and consultants, spokesmen and opinion-makers are often the captains of their higher thought and decision. Immediately below the elite are the professional politicians of the middle levels of power, in the Congress and in the pressure groups, as well as among the new and old upper classes of town and city and region. Mingling with them, in curious ways which we shall explore, are those professional celebrities who live by being continually displayed but are never, so long as they remain celebrities, displayed enough. If such celebrities are not at the head of any dominating hierarchy, they do often have the power to distract the attention of the public or afford sensations to the masses, or, more directly, to gain the ear of those who do occupy positions of direct power. More or less unattached, as critics of morality and technicians of power, as spokesmen of God and creators of mass sensibility, such celebrities and consultants are part of the immediate scene in which the drama of the elite is enacted. But that drama itself is centered in the command posts of the major institutional hierarchies.

The truth about the nature and the power of the elite is not some secret which men of affairs know but will not tell. Such men hold quite various theories about their own roles in the sequence of event and decision. Often they are uncertain about their roles, and even more often they allow their fears and their hopes to affect their assessment of their own power. No matter how great their actual power, they tend to be less acutely aware of it than of the resistances of others to its use. Moreover, most American men of affairs have learned well the rhetoric of public relations, in some cases even to the point of using it when they are alone, and thus coming to believe it. The personal awareness of the actors is only one of the several sources one must examine in order to understand the higher circles. Yet many who believe that there is no elite, or at any rate none of any consequence, rest their argument upon what men of affairs believe about themselves, or at least assert in public.

There is, however, another view: those who feel, even if vaguely, that a compact and powerful elite of great importance does now prevail in America often base that feeling upon the historical trend of our time. They have felt, for example, the domination of the military event, and from this they infer that generals and admirals, as well as other men of decision influenced by them, must be enormously powerful. They hear that the Congress has again abdicated to a handful of men decisions clearly related to the issue of war or peace. They know that the bomb was dropped over Japan in the name of the United States of America, although they were at no time consulted about the matter. They feel that they live in a time of big decisions; they know that they are not making any. Accordingly, as they consider the present as history, they infer that at its center, making decisions or failing to make them, there must be an elite of power.

On the one hand, those who share this feeling about big historical events assume that there is an elite and that its power is great. On the other hand, those who listen carefully to the reports of men apparently involved in the great decisions often do not believe that there is an elite whose powers are of decisive consequence.

Both views must be taken into account, but neither is adequate. The way to understand the power of the American elite lies neither solely in recognizing the historic scale of events nor in accepting the personal awareness reported by men of apparent decision. Behind such men and behind the events of history, linking the two, are the major institutions of modern society. These hierarchies of state and corporation and army constitute the means of power; as such they are now of a consequence not before equaled in human history—and at their summits, there are now those command posts of modern society which offer us the sociological key to an understanding of the role of the higher circles in America.

Within American society, major national power now resides in the economic, the political, and the military domains. Other institutions seem off to the side of modern history, and, on occasion, duly subordinated to these. No family is as directly powerful in national affairs as any major corporation; no church is as directly powerful in the external biographies of young men in America today as the military establishment; no college is as powerful in the shaping of momentous events as the National Security Council. Religious, educational, and family institutions are not autonomous centers of national power; on the contrary, these decentralized areas are increasingly shaped by the big three, in which developments of decisive and immediate consequence now occur.

Families and churches and schools adapt to modern life; governments and armies and corporations shape it; and, as they do so, they turn these lesser institutions into means for their ends. Religious institutions provide chaplains to the armed forces where they are used as a means of increasing the effectiveness of its morale to kill. Schools select and train men for their jobs in corporations and their specialized tasks in the armed forces. The extended family has, of course, long been broken up by the industrial revolution, and now the son and the father are removed from the family, by compulsion if need be, whenever the army of the state sends out the call. And the symbols of all these lesser institutions are used to legitimate the power and the decisions of the big three.

The life-fate of the modern individual depends not only upon the family into which he was born or which he enters by marriage, but increasingly upon the corporation in which he spends the most alert hours of his best years; not only upon the school where he is educated as a child and adolescent, but also upon the state which touches him throughout his life; not only upon the church in which on occasion he hears the word of God, but also upon the army in which he is disciplined.

If the centralized state could not rely upon the inculcation of nationalist loyalties in public and private schools, its leaders would promptly seek to modify the decentralized educational system. If the bankruptcy rate among the top five hundred corporations were as high as the general divorce rate among the thirty-seven million married couples, there would be economic catastrophe on an international scale. If members of armies gave to them no more of their lives than do believers to the churches to which they belong, there would be a military crisis.

Within each of the big three, the typical institutional unit has become enlarged, has become administrative, and, in the power of its decisions, has become centralized. Behind these developments there is a fabulous technology, for as institutions, they have incorporated this technology and guide it, even as it shapes and paces their developments.

The economy—once a great scatter of small productive units in autonomous balance—has become dominated by two or three hundred giant corporations, administratively and politically interrelated, which together hold the keys to economic decisions.

The political order, once a decentralized set of several dozen states with a weak spinal cord, has become a centralized, executive establishment which has taken up into itself many powers previously scattered, and now enters into each and every cranny of the social structure.

The military order, once a slim establishment in a context of distrust fed by state militia, has become the largest and most expensive feature of government, and, although well versed in smiling public relations, now has all the grim and clumsy efficiency of a sprawling bureaucratic domain.

In each of these institutional areas, the means of power at the disposal of decision makers have increased enormously; their central executive powers have been enhanced; within each of them modern administrative routines have been elaborated and tightened up.

As each of these domains becomes enlarged and centralized, the consequences of its activities become greater, and its traffic with the others increases. The decisions of a handful of corporations bear upon military and political as well as upon economic developments around the world. The decisions of the military establishment rest upon and grievously affect political life as well as the very level of economic activity. The decisions made within the political domain determine economic activities and military programs. There is no longer, on the one hand, an economy, and, on the other hand, a political order containing a military establishment unimportant to politics and to money-making. There is a political economy linked, in a thousand ways, with military institutions and decisions. On each side of the world-split running through central Europe and around the Asiatic rimlands, there is an ever-increasing interlocking of economic, military, and political structures. If there is government intervention in the corporate economy, so is there corporate intervention in the governmental process. In the structural sense, this triangle of power is the source of the *interlocking directorate* that is most important for the historical structure of the present. . . .

At the pinnacle of each of the three enlarged and centralized domains, there have arisen those higher circles which make up the economic, the political, and the military elites. At the top of the economy, among the corporate rich, there are the chief executives; at the top of the political order, the members of the political directorate; at the top of the military establishment, the elite of soldier-statesmen clustered in and around the Joint Chiefs of Staff and the upper echelon. As each of these domains has coincided with the others, as decisions tend to become total in their consequence, the leading men in each of the three domains of power—the warlords, the corporation chieftains, the political directorate—tend to come together, to form the power elite of America. . . .

By the powerful we mean, of course, those who are able to realize their will, even if others resist it. No one, accordingly, can be truly powerful unless he has access to the command of major institutions, for it is over these institutional means of power that the truly powerful are, in the first instance, powerful. Higher politicians and key officials of government command such institutional power; so do admirals and generals, and so do the major owners and executives of the larger corporations. Not all power, it is true, is anchored in and exercised by means of such institutions, but only within and through them can power be more or less continuous and important.

Wealth also is acquired and held in and through institutions. The pyramid of wealth cannot be understood merely in terms of the very rich; for the great inheriting families, as we shall see, are now supplemented by the corporate institutions of modern society: every one of the very rich families has been and is closely connected—always legally and frequently managerially as well—with one of the multi-million dollar corporations.

The modern corporation is the prime source of wealth, but, in latter-day capitalism, the political apparatus also opens and closes many avenues to wealth. The amount as well as the source of income, the power over consumer's goods as well as over productive capital, are determined by position within the political economy. If our interest in the very rich goes beyond their lavish or their miserly consumption, we must examine their relations to modern forms of corporate property as well as to the state; for such relations now determine the chances of men to secure big property and to receive high income. . . .

If we took the one hundred most powerful men in America, the one hundred wealthiest, and the one hundred most celebrated away from the institutional positions they now occupy, away from their resources of men and women and money, away from the media of mass communication that are now focused upon them—then they would be powerless and poor and uncelebrated. For power is not of a man. Wealth does not center in the person of the wealthy. Celebrity is not inherent in any personality. To be celebrated, to be wealthy, to have power requires access to major institutions, for the institutional positions men occupy determine in large part their chances to have and to hold these valued experiences. . . .

DISCUSSION QUESTIONS

1. What evidence do you see of the presence of the power elite in today's economic, political, and military institutions? Suppose that Mills were writing his book today; what might he change about his essay?

2. Mills argues that the power elite use institutions such as religion, education, and the family as the means to their ends. Find an example of this from the daily news and explain how Mills would see this institution as being shaped by the power elite.

INFOTRAC COLLEGE EDITION

You can use your access to InfoTrac College Edition to learn more about the subjects covered in this essay. Some suggested search terms include:

billionnaires

corporate power

Forbes Four Hundred

military-industrial complex

nouveau riche

power elite

69

Diversity in the Power Elite

RICHARD ZWEIGENHAFT AND G. WILLIAM DOMHOFF

Richard Zweigenhaft and G. William Domhoff ask here whether the power elite has changed by incorporating more diverse groups into these higher circles. They find limited evidence that women, Jews, African Americans, Latinos, Asian Americans, and gays and lesbians have entered the power elite. Those who have tend to share the perspectives and values of those already in power.

Since the 1870s the refrain about the new diversity of the governing circles has been closely intertwined with a staple of American culture created by Horatio Alger Jr., whose name has become synonymous with upward mobility in America. Born in 1832 to a patrician family—Alger's father was a Harvard graduate, a Unitarian minister, and a Massachusetts state senator—Alger graduated from Harvard at the age of nineteen. There followed a series of unsuccessful efforts to establish himself in various careers. Finally, in 1864 Alger was hired as a Unitarian minister in Brewster, Massachusetts. Fifteen months later, he was dismissed from this position for homosexual acts with boys in the congregation.

Alger returned to New York, where he soon began to spend a great deal of time at the Newsboys' Lodging House, founded in 1853 for footloose youngsters between the ages of twelve and sixteen and home to many youths who had been mustered out of the Union Army after serving as drummer boys. At the Newsboys' Lodging House Alger found his literary niche and his subsequent claim to fame: writing books in which poor boys make good. His books sold by the hundreds of thousands in the last third of the nineteenth century, and by 1910 they were enjoying annual sales of more than one million in paperback.[1]

From: Richard Zweigenhaft and G. William Domhoff. 1998. *Diversity in the Power Elite: Have Women and Minorities Reached the Top?* New Haven, CT: Yale University Press, pp. 1–7, 192–194.

The deck is not stacked against the poor, according to Horatio Alger. When they simply show a bit of gumption, work hard, and thereby catch a break or two, they can become part of the American elite. The persistence of this theme, reinforced by the annual Horatio Alger Awards to such well-known personalities as Ronald Reagan, Bob Hope, and Billy Graham (who might not have been so eager to accept them if they had known of Alger's shadowed past), suggests that we may be dealing once again with a cultural myth. In its early versions, of course, the story concerned the great opportunities available for poor white boys willing to work their way to the top. More recently, the story has featured black Horatio Algers who started in the ghetto, Latino Horatio Algers who started in the barrio, Asian-American Horatio Algers whose parents were immigrants, and female Horatio Algers who seem to have no class backgrounds—all of whom now sit on the boards of the country's largest corporations.

But is any of this true? Can anecdotes and self-serving autobiographical accounts about diversity, meritocracy, and upward social mobility survive a more systematic analysis? Have very many women and previously excluded minorities made it to the top? Has class lost its importance in shaping life chances?

. . . We address these and related questions within the framework provided by the iconoclastic sociologist C. Wright Mills in his hard-hitting classic *The Power Elite*, published in 1956 when the media were in the midst of what Mills called the Great American Celebration. In spite of the Depression of the 1930s, Americans had pulled together to win World War II, and the country was both prosperous at home and influential abroad. Most of all, according to enthusiasts, the United States had become a relatively classless and pluralistic society, where power belonged to the people through their political parties and public opinion. Some groups certainly had more power than others, but no group or class had too much. The New Deal and World War II had forever transformed the corporate-based power structure of earlier decades.

Mills challenged this celebration of pluralism by studying the social backgrounds and career paths of the people who occupied the highest positions in what he saw as the three major institutional hierarchies in postwar America—the corporations, the executive branch of the federal government, and the military. He found that almost all the members of this leadership group, which he called the power elite, were white Christian males who came from "at most, the upper third of the income and occupational pyramids," despite the many Horatio Algeresque claims to the contrary.[2] A majority came from an even narrower stratum, the 11 percent of U.S. families headed by businesspeople or highly educated professionals like physicians and lawyers. Mills concluded that power in the United States in the 1950s was just about as concentrated as it had been since the rise of the large corporations, although he stressed that the New Deal and World War II had given political appointees and military chieftains more authority than they had exercised previously.

It is our purpose, therefore, to take a detailed look at the social, educational, and occupational backgrounds of the leaders of these three institutional hierarchies to see whether they have become more diverse in terms of gender, race, ethnicity, and sexual orientation, and also in terms of socioeconomic origins. Unlike Mills, we think the power elite is more than a set of institutional leaders. It is also the leadership group for the small upper class of owners and managers of large income-producing properties, the 1 percent of Americans who in 1992 possessed 37.2% of all net worth.[3] But that theoretical difference is not of great moment here. The important commonality is the great wealth and power embodied in these institutional hierarchies and the people who lead them. . . .

In addition to studying the extent to which women and minorities have risen in the system, we focus on whether they have followed different avenues to the top than their predecessors did, and on any special roles they may play. Are they in the innermost circles of the power elite, or are they more likely to serve as buffers and go-betweens? Do they go just so far and no farther? What obstacles does each group face?

We also examine whether or not the presence of women and minorities affects the power elite itself. Do those women and minorities who become part of the power elite influence it in a more liberal direction, or do they end up endorsing traditional conservative positions, such as opposition to trade unions, taxes, and government regulation of business? In addition, . . . we consider the possibility that the diversity forced on the power elite has had the ironic effect of strengthening it, at least in the short run, by providing it with people who can reach out to the previously excluded groups and by showing that the American system can deliver on its most important promise, an equal opportunity for every individual.

These are not simple issues, and the answers to some of the questions we ask vary greatly depending on which previously disadvantaged group we are talking about. Nonetheless, in the course of our research, a few general patterns emerged. . . .

1. The power elite now shows considerable diversity, at least as compared with its state in the 1950s, but its core group continues to be wealthy white Christian males, most of whom are still from the upper third of the social ladder. They have been filtered through a handful of elite schools in law, business, public policy, and international relations.

2. In spite of the increased diversity of the power elite, high social origins continue to be a distinct advantage in making it to the top. There are relatively few rags-to-riches stories in the groups we studied, and those we did find tended to come through the electoral process, usually within the Democratic Party. In general, it still takes at least three generations to rise from the bottom to the top in the United States.

3. The new diversity within the power elite is transcended by common values and a sense of hard-earned class privilege. The newcomers to the power elite have found ways to signal that they are willing to join the game as it has always been played, assuring the old guard that they will call for no more than relatively minor adjustments, if that. There are few liberals and fewer crusaders in the power elite, despite its new multiculturalism. Class backgrounds, current roles, and future aspirations are more powerful in shaping behavior in the power elite than gender, ethnicity, or race.

4. Not all the groups we studied have been equally successful in contributing to the new diversity in the power elite. Women, blacks, Latinos, Asian Americans, and openly homosexual men and women are all underrepresented, but to varying degrees and with different rates of increasing representation. . . .

5. Although the corporate, political, and military elites accepted diversity only in response to pressure from minority activists and feminists, these elites have benefited from the presence of new members. Some serve either a buffer or a liaison function with such groups and institutions as consumers, angry neighborhoods, government agencies, and wealthy foreign entrepreneurs.

6. There is greater diversity in Congress than in the power elite, and the majority of the female and minority elected officals are Democrats. . . .

The power elite has been strengthened because diversity has been achieved primarily by the selection of women and minorities who share the prevailing perspectives and values of those already in power. The power elite is not "multicultural" in any full sense of the concept, but only in terms of ethnic or racial origins. This process has been helped along by those who have called for the inclusion of women and minorities without any consideration of criteria other than sex, race, or ethnicity. Because the demand was strictly for a woman on the Supreme Court, President Reagan could comply by choosing a conservative upper-class corporate lawyer, Sandra Day O'Connor. When pressure mounted to have more black justices, President Bush could respond by appointing Clarence Thomas, a conservative black Republican with a law degree from Yale University. It is yet another irony that appointments like these served to undercut the liberal social movements that caused them to happen.[4]

It is not surprising, therefore, that when we look at the business practices of the women and minorities who have risen to the top of the corporate world, we find that their perspectives and values do not differ markedly from those of their white male counterparts. When Linda Wachner, one of the few women to become CEO of a *Fortune*-level company, the Warnaco Group, concluded that one of Warnaco's many holdings, the Hathaway Shirt Company, was unprofitable, she decided to stop making Hathaway shirts and to sell or close down the factory. It did not matter to Wachner that Hathaway, which started making shirts in 1837, was one of the oldest companies in Maine, that almost all of the five hundred employees at the factory were working-class women, or even that the workers had given up a pay raise to hire consultants to teach them to work more effectively and, as a result, had doubled their productivity. The bottomline issue was that the company was considered unprofitable, and the average wage of the Hathaway workers, $7.50 an hour, was thought to be too high. (In 1995 Wachner was paid $10 million in salary and stock, and Warnaco had a net income of $46.5 million.) "We did need to do the right thing for the company and the stockholders," explained Wachner.[5]

Nor did ethnic background matter to Thomas Fuentes, a senior vice president at a consulting firm in Orange County, California, a director of Fleetwood Enterprises, and chairman of the Orange County Republican Party. Fuentes targeted fellow Latinos who happened to be Democrats when he sent uniformed security guards to twenty polling places in 1988 "carrying signs in Spanish and

English warning people not to vote if they were not U.S. citizens." The security firm ended up paying $60,000 in damages when it lost a lawsuit stemming from this intimidation.[6]

We also recall that the Fanjuls, the Cuban-American sugar barons, have had no problem ignoring labor laws in dealing with their migrant labor force, and that the Sakioka family illegally gave short-handled hoes to its migrant farm workers. These people were acting as employers, not as members of ethnic groups. That is, members of the power elite of both genders and all ethnicities have practiced class politics, making it possible for the power structure to weather the challenge created by the social movements that began in the 1960s.

Those who challenged Christian white male homogeneity in the power structure during the 1960s not only sought to create civil rights and new job opportunities for men and women who had previously been mistreated, important though these goals were. They also hoped that new perspectives in the boardrooms and the halls of government would bring greater openness throughout the society. The idea was both to diversify the power elite and to shift some of its power to previously excluded groups and social classes. The social movements of the 1960s were strikingly successful in increasing the individual rights and freedoms available to all Americans, especially African Americans. As we have shown, they also created pressures that led to openings at the top for individuals from groups that had previously been excluded.

But as the concerns of social movements, political leaders, and the courts came to focus more and more on individual rights, the emphasis on social class and "distributive justice" was lost. The age-old American commitment to individualism, reinforced at every turn by members of the power elite, won out over the commitment to greater equality of income and wealth that had been one strand of New Deal liberalism and a major emphasis of left-wing activists in the 1960s.

We therefore have to conclude on the basis of our findings that the diversification of the power elite did not generate any changes in an underlying class system in which the top 1 percent have 45.6 percent of all financial wealth, the next 19 percent have 46.7 percent, and the bottom 80 percent have 7.8 percent.[7] The values of liberal individualism embedded in the Declaration of Independence, the Bill of Rights, and the civic culture were renewed by vigorous and courageous activists, but despite their efforts the class structure remains a major obstacle to individual fulfillment for the overwhelming majority of Americans. This fact is more than an irony. It is a dilemma. It combines with the dilemma of race to create a nation that celebrates equal opportunity but is, in reality, a bastion of class privilege and conservatism.

NOTES

1. See Richard M. Huber, *The American Idea of Success* (New York: McGraw Hill, 1971), 44–46; Gary Scharnhorst, *Horatio Alger, Jr.* (Boston: Twayne, 1980), 24, 29, 141.

2. C. Wright Mills, *The Power Elite* (New York: Oxford University Press, 1956),

279. For Mills's specific findings, see 104–105, 128–129, 180–181, 393–394, and 400–401.

3. Edward N. Wolff, *Top Heavy* (New York: New Press, 1996), 67.

4. In addition, evidence from experimental work in social psychology suggests that

tokenism has the effect of undercutting the impetus for collective action by the excluded group. See, for example, Stephen C. Wright, Donald M. Taylor, and Fathali M. Moghaddam, "Responding to Membership in a Disadvantaged Group: From Acceptance to Collective Protest," *Journal of Personality and Social Psychology* 58, no. 6 (1990), 994–1003. See also Bruce R. Hare, "On the Desegregation of the Visible Elite; or, Beware of the Emperor's New Helpers: He or She May Look Like You or Me," *Sociological Forum* 10, no. 4 (1995), 673–678.

5. Sara Rimer, "Fall of a Shirtmaking Legend Shakes its Maine Hometown," *New York Times,* May 15, 1996. See also Floyd Norris, "Market Place," *New York Times,* June 7, 1996; Stephanie Strom, "Double Trouble at Linda Wachner's Twin Companies," *New York Times,* August 4, 1996. Strom's article reveals that Hathaway Shirts "got a reprieve" when an investor group stepped in to save it.

6. Claudia Luther and Steven Churm, "GOP Official Says He OK'd Observers at Polls," *Los Angeles Times,* November 12, 1988; Jeffrey Perlman, "Firm Will Pay $60,000 in Suit Over Guards at Polls," *Los Angeles Times,* May 31, 1989.

7. Edward N. Wolff, *Top Heavy* (New York: New press, 1996), 67.

DISCUSSION QUESTIONS

1. To what extent do Zweigenhaft and Domhoff see diverse groups as becoming a part of the power elite? What factors do they identify as important in gaining entrance to the power elite and how does this challenge the myth of upward mobility typically symbolized by the Horatio Alger story?

2. At the heart of Zweigenhaft and Domhoff's analysis is the question, "Does having more women and people of color in positions of power change institutions?" Using empirical evidence, how would you answer this question?

INFOTRAC COLLEGE EDITION

You can use your access to InfoTrac College Edition to learn more about the subjects covered in this essay. Some suggested search terms include:

black corporate leaders
diversity training
glass ceiling
Hispanics in corporations

interlocking directorates
White Anglo-Saxon Protestants (WASPs)
women in corporations

70

Power and Class
in the United States

G. WILLIAM DOMHOFF

G. William Domhoff shows how the corporate power elite dominate political interests in the federal government. They do this through the formation of coalitions formed to defend their interests. Using the example of Bohemian Grove, an annual recreational retreat of conservative upper-class politicians, academic, and corporate leaders, Domhoff also shows how the upper class bonds together around their common interests.

*P*ower and *class* are terms that make Americans a little uneasy, and concepts like *power elite* and *dominant class* immediately put people on guard. The idea that a relatively fixed group of privileged people might shape the economy and government for their own benefit goes against the American grain. Nevertheless, . . . the owners and top-level managers in large income-producing properties are far and away the dominant power figures in the United States. Their corporations, banks, and agribusinesses come together as a *corporate community* that dominates the federal government in Washington. Their real estate, construction, and land development companies form *growth coalitions* that dominate most local governments. Granted, there is competition within both the corporate community and the local growth coalitions for profits and investment opportunities, and there are sometimes tensions between national corporations and local growth coalitions, but both are cohesive on policy issues affecting their general welfare, and in the face of demands by organized workers, liberals, environmentalists, and neighborhoods.

As a result of their ability to organize and defend their interests, the owners and managers of large income-producing properties have a very great share of all income and wealth in the United States, greater than in any other industrial democracy. Making up at best 1 percent of the total population, by the early 1990s they earned 15.7 percent of the nation's yearly income and owned 37.2 percent of all privately held wealth, including 49.6 percent of all corporate stocks and 62.4 percent of all bonds. Due to their wealth and the lifestyle it makes possible, these owners and managers draw closer as a common social group. They belong to the same exclusive social clubs, frequent the same summer and winter resorts, and send their children to a relative handful of private schools. Members of the

From: G. William Domhoff. 1998. *Who Rules America? Power & Politics in the Year 2000.* Mountain View, CA: Mayfield Publishing Co., pp. 1–4, 88–92.

corporate community thereby become a *corporate rich* who create a nationwide *social upper class* through their social interaction. . . . Members of the growth coalitions, on the other hand, are *place entrepreneurs,* people who sell locations and buildings. They come together as local upper classes in their respective cities and sometimes mingle with the corporate rich in educational or resort settings.

The corporate rich and the growth entrepreneurs supplement their small numbers by developing and directing a wide variety of nonprofit organizations, the most important of which are a set of taxfree charitable foundations, think tanks, and policy-discussion groups. These specialized nonprofit groups constitute a *policy-formation network* at the national level. Chambers of commerce and policy groups affiliated with them form similar policy-formation networks at the local level, aided by a few national-level city development organizations that are available for local consulting.

Those corporate owners who have the interest and ability to take part in general governance join with top-level executives in the corporate community and the policy-formation network to form the *power elite*, which is the leadership group for the corporate rich as a whole. The concept of a power elite makes clear that not all members of the upper class are involved in governance; some of them simply enjoy the lifestyle that their great wealth affords them. At the same time, the focus on a leadership group allows for the fact that not all those in the power elite are members of the upper class; many of them are high-level employees in profit and nonprofit organizations controlled by the corporate rich. . . . The power elite, in other words, is based in both ownership and in organizational positions. . . .

The power elite is not united on all issues because it includes both moderate conservatives and ultraconservatives. Although both factions favor minimal reliance on government on all domestic issues, the moderate conservatives sometimes agree to legislation advocated by liberal elements of the society, especially in times of social upheaval like the Great Depression of the 1930s and the Civil Rights Movement of the early 1960s. Except on defense spending, ultraconservatives are characterized by a complete distaste for any kind of government programs under any circumstances—even to the point of opposing government support for corporations on some issues. Moderate conservatives often favor foreign aid, working through the United Nations, and making attempts to win over foreign enemies through patient diplomacy, treaties, and trade agreements. Historically, ultraconservatives have opposed most forms of foreign involvement, although they have become more tolerant of foreign trade agreements over the past thirty or forty years. At the same time, their hostility to the United Nations continues unabated.

Members of the power elite enter into the electoral arena as the leaders within a *corporate-conservative coalition,* where they are aided by a wide variety of patriotic, antitax, and other single-issue organizations. These conservative advocacy organizations are funded in varying degrees by the corporate rich, direct-mail appeals, and middle-class conservatives. This coalition has played a large role in both political parties at the presidential level and usually succeeds in electing a conservative majority to both houses of Congress. Historically, the conservative majority in Congress was made up of most Northern Republicans and most Southern Democrats, but that arrangement has been changing gradually since the 1960s as the conservative Democrats of the South are replaced by even more conservative Southern Republicans. The corporate-conservative coalition also has access to the federal government in Washington through lobbying and the appointment of its members to top positions in the executive branch.

During the past twenty-five years the corporate-conservative coalition has formed an uneasy alliance within the Republican Party with what is sometimes called the "New Right" or "New Christian Right," which consists for the most part of middle-level religious groups concerned with a wide range of "social issues," such as teenage sexual and drinking behavior, abortion, and prayer in school. I describe the alliance as an "uneasy" one because the power elite and the New Right do not have quite the same priorities, except for a general hostility to government and liberalism, and because it is not completely certain that the New Right is helping the corporate-conservative coalition as much as its publicists and fund-raisers claim. Nevertheless, ultraconservatives within the power elite help to finance some of the single-issue organizations and publications of the New Right.

Despite their preponderant power within the federal government and the many useful policies it carries out for them, members of the power elite are constantly critical of government as an alleged enemy of freedom and economic growth. Although their wariness toward government is expressed in terms of a dislike for taxes and government regulations, I believe their underlying concern is that government could change the power relations in the private sphere by aiding average Americans through a number of different avenues: (1) creating government jobs for the unemployed; (2) making health, unemployment, and welfare benefits more generous; (3) helping employees gain greater workplace rights and protections; and (4) helping workers organize unions. All of these initiatives are opposed by members of the power elite because they would increase wages and taxes, but the deepest opposition is toward any government support for unions because unions are a potential organizational base for advocating the whole range of issues opposed by the corporate rich. . . .

THE BOHEMIAN GROVE AS A MICROCOSM

The Bohemian Club is the most unusual and widely known club of the upper class. Its annual two-week retreat seventy-five miles north of San Francisco brings together members of the upper class, corporate leaders, celebrities, and government officials for relaxation and entertainment. They are joined by several hundred "associate" members who pay lower dues in exchange for producing plays, skits, artwork and other forms of entertainment. Fifty to 100 professors and university administrators, most of them from Stanford University and campuses of the University of California, are also included in the associate category. The encampment provides the best possible insight into the role of clubs in uniting the corporate community and the upper class. It is a microcosm of the world of the corporate rich.

The 2,700-acre pristine forest setting called the Bohemian Grove was purchased by the club in the 1890s after twenty years of holding the retreat in rented quarters. Bohemians and their guests number anywhere from 1,500 to 2,500 for the three weekends in the encampment, which is always held during the last two weeks in July. However, there may be as few as 400 men in residence in the middle of the week because most return to their homes and jobs after the weekends. During their stay the campers are treated to plays, symphonies, concerts, lectures, and political commentaries by entertainers, musicians, scholars, corporate executives, and government officials. They also trapshoot, canoe, swim, drop by the Grove art gallery, and take guided tours into the outer fringe of the mountain forest. But a stay at the Bohemian Grove is mostly a time for relaxation and drinking in the modest lodges, bunkhouses, and even teepees that fit unobtrusively into the landscape along the two or three dirt roads that join the few "developed" acres within the Grove. It is like a summer camp for the power elite and their entertainers.

The men gather in little camps of from ten to thirty members during their stay—although the camps for associate members are often larger. Each of the approximately 120 camps has its own pet name, such as Sons of Toil, Cave Man, Mandalay, Toyland, Owl's Nest, Hill Billies, and Parsonage. A group of men from Los Angeles named their camp Lost Angels, and the men in the Bohemian chorus call their camp Aviary. Some camps are noted for special drinking parties, brunches, or luncheons to which they invite members from other camps. The camps are a fraternity system within the larger fraternity.

There are many traditional events during the encampment, including plays called the High Jinx and the Low Jinx. The most memorable event, however, is an elaborate ceremonial ritual called the Cremation of Care, which is held the first Saturday night. It takes place at the base of the forty-foot Owl Shrine constructed out of poured concrete and made even more resplendent by the mottled forest mosses that cover much of it. The Owl Shrine is only one of many owl symbols and insignias to be found in the Grove and the downtown clubhouse. The owl was adopted early in the club's history as its mascot, or totem animal. According to the club's librarian—who is also a historian at a large university—the event "incorporates druidical ceremonies, elements of medieval Christian liturgy, sequences directly inspired by the Book of Common Prayer, traces of Shakespearean drama and the seventeenth-century masque, and late nineteenth-century American lodge rites." Bohemians are proud that the ceremony had been carried out 125 consecutive years as of 1997.

The opening ceremony is called the Cremation of Care because it involves the burning of an effigy named Dull Care, who symbolizes the burdens and responsibilities that these busy Bohemians now wish to shed temporarily. More than 250 Bohemians take part in the ceremony as priests, elders, acolytes, shore patrols, brazier bearers, boatmen, and woodland voices. After many flowery speeches and a long conversation with Dull Care, the high priest lights the fire with the flame from the Lamp of Fellowship, located on the "Altar of Bohemia" at the base of the shrine. The ceremony, which has the same initiatory functions as those of any fraternal or tribal group, ends with fireworks, shouting, and the playing of "There'll Be a Hot Time in the Old Town Tonight." The attempt to create a sense of cohesion and ingroup solidarity among the assembled is complete. The laughter, drinking, and storytelling can now begin.

But the retreat sometimes provides an occasion for more than fun and merriment. Although business is rarely discussed except in an informal way in groups of two or three, the retreat provides members with an opportunity to introduce their friends to politicians and hear formal noontime speeches (called Lakeside Talks because they take place across the lake from the Owl Shrine) from political candidates. Every Republican president of the twentieth century has been a member or guest at the Bohemian Grove. President Herbert Hoover (1929–1933) was the first Republican president to be a member, which gave him the honor of giving the final Lakeside Talk from the 1930s until his death in 1964. He was a member of Cave Man Camp, as was President Nixon. President Ford is in Mandalay, President Reagan in Owl's Nest, and President Bush in Hill Billies.

In 1995, House Speaker Newt Gingrich delivered the Lakeside Talk on the middle Saturday of the encampment and President Bush gave it on the final Saturday. The featured Saturday speakers in 1996 were the Republican governor of California and a former Republican secretary of state. Perhaps the most striking change in the Lakeside Talks in the 1990s is the absence of any leading Democrats. Although a Democratic president has never been a member or guest at the Grove, cabinet members from the Kennedy, Johnson, and Carter Administrations were prominent guests and Lakeside speakers in the past.

An exhaustive analysis of the members and guests at the Bohemian Grove in 1970 and 1980 demonstrates the way in which one club intertwines the upper class with the entire corporate community. In 1970, 29 percent of the top 800 corporations had at least one officer or director at the Bohemian Grove festivities; in 1980, the figure was 30 percent. As might be expected, the overlap was especially great among the largest corporations, with twenty-three of the top twenty-five industrials represented in 1970, fifteen of twenty-five in 1980. Twenty of the twenty-five largest banks had at least one officer or director in attendance in both 1970 and 1980. Other business sectors were represented somewhat less.

An even more intensive study by sociologist Peter Phillips, which includes participant-observation and interviews as well as membership network analysis, extends the sociological understanding of the Bohemian Grove into the 1990s. Using a list of 1,144 corporations—well beyond the 800 used in the studies for 1970 and 1980—Phillips nonetheless found that 24 percent of these companies had at least one director who was a member or guest in 1993. For the top 100 corporations outside of California, the figure was 42 percent, compared to 64 percent in 1971. . . .

As the case of the Bohemian Grove and its theatrical performances rather dramatically illustrates, there seems to be a great deal of truth to the earlier-cited suggestion by Crane Brinton that clubs may function within the upper class the way that the clan or brotherhood does in tribal societies. With their restrictive membership policies, initiatory rituals, private ceremonials, and great emphasis on tradition, clubs carry on the heritage of primitive secret societies. They create among their members an attitude of prideful exclusiveness that contributes greatly to an in-group feeling and a sense of fraternity within the upper class. . . .

DISCUSSION QUESTIONS

1. What does Domhoff mean when he writes of a *corporate-conservative coalition*? What interests do these two groups share in common and where do they differ?

2. Domhoff describes Bohemia Grove as a place where the power elite engage in rituals that promote an in-group feeling. How is this achieved and what significance does this have for the formation and durability of the upper class and its role in politics?

INFOTRAC COLLEGE EDITION

You can use your access to InfoTrac College Edition to learn more about the subjects covered in this essay. Some suggested search terms include:

Bohemian Club   political coalitions
New Christian Right   think tanks
political action committees   upper class

71

The First Americans:

American Indians

C. MATTHEW SNIPP

The history of American Indians is marked by subordination by the U.S. government, as well as by Indian resistance to domination. Here, C. Matthew Snipp identifies several different periods that characterize relations between the federal government and American Indians, depicting this history as one where American Indians have strived for self-sufficiency, even in the face of genocide and extermination.

By the end of the nineteenth century, many observers predicted that American Indians were destined for extinction. Within a few generations, disease, warfare, famine, and outright genocide had reduced their numbers from millions to less than 250,000 in 1890. Once a self-governing, self-sufficient

From: Silvia Pedraza, and Rubén G. Rumbaut, eds., 1996, *Origins and Destinies: Immigration, Race, and Ethnicity in America*. Belmont, CA: Wadsworth, pp. 390–403. Reprinted by permission.

people, American Indians were forced to give up their homes and their land and to subordinate themselves to an alien culture. The forced resettlement to reservation lands or the Indian Territory (now Oklahoma) frequently meant a life of destitution, hunger, and complete dependency on the federal government for material needs.

Today, American Indians are more numerous than they have been for several centuries. While still one of the most destitute groups in American society, tribes have more autonomy and are now more self-sufficient than at any time since the last century. In cities, modern pan-Indian organizations have been successful in making the presence of American Indians known to the larger community, and have mobilized to meet the needs of their people (Cornell 1988; Nagel 1989; Weibel-Orlando 1991). In many rural areas, American Indians and especially tribal governments have become increasingly more important and increasingly more visible by virtue of their growing political and economic power. The balance of this [reading] is devoted to explaining their unique place in American society.

The current political and economic status of American Indians is the result of the process by which they were incorporated into Euro-American society (Hall 1989). This amounts to a long history of efforts aimed at subordinating an otherwise self-governing and self-sufficient people that eventually culminated in widespread economic dependency. The role of the U.S. government in this process can be seen in the five major historical periods of federal Indian relations: removal, assimilation, the Indian New Deal, termination and relocation, and self-determination.

REMOVAL

In the early nineteenth century, the population of the United States expanded rapidly at the same time that the federal government increased its political and military capabilities. The character of Indian-American relations changed after the War of 1812. The federal government increasingly pressured tribes settled east of the Appalachian Mountains to move west to the territory acquired in the Louisiana Purchase. Numerous treaties were negotiated by which the tribes relinquished most of their land and eventually were forced to move west.

Initially the federal government used bargaining and negotiation to accomplish removal, but many tribes resisted (Prucha 1984). However, the election of Andrew Jackson by a frontier constituency signaled the beginning of more forceful measures to accomplish removal. In 1830 Congress passed the Indian Removal Act, which mandated the eventual removal of the eastern tribes to points west of the Mississippi River, in an area which was to become the Indian Territory and is now the state of Oklahoma. Dozens of tribes were forcibly removed from the eastern half of the United States to the Indian Territory and newly created reservations in the west, a long process ridden with conflict and bloodshed.

As the nation expanded beyond the Mississippi River, tribes of the plains, southwest, and west coast were forcibly settled and quarantined on isolated reservations.

This was accompanied by the so-called Indian Wars—a bloody chapter in the history of Indian-White relations (Prucha 1984; Utley 1984). This period in American history is especially remarkable because the U.S. government was responsible for what is unquestionably one of the largest forced migrations in history.

The actual process of removal spanned more than a half-century and affected nearly every tribe east of the Mississippi River. Removal often meant extreme hardships for American Indians, and in some cases this hardship reached legendary proportions. For example, the Cherokee removal has become known as the "Trail of Tears." In 1838, nearly 17,000 Cherokees were ordered to leave their homes and assemble in military stockades (Thornton 1987, p. 117). The march to the Indian Territory began in October and continued through the winter months. As many as 8,000 Cherokees died from cold weather and diseases such as influenza (Thornton 1987, p. 118).

According to William Hagan (1979), removal also caused the Creeks to suffer dearly as their society underwent a profound disintegration. The contractors who forcibly removed them from their homes refused to do anything for "the large number who had nothing but a cotton garment to protect them from the sleet storms and no shoes between them and the frozen ground of the last stages of their hegira. About half of the Creek nation did not survive the migration and the difficult early years in the West" (Hagan 1979, p. 77–81). In the West, a band of Nez Perce men, women, and children, under the leadership of Chief Joseph, resisted resettlement in 1877. Heavily outnumbered, they were pursued by cavalry troops from the Wallowa valley in eastern Oregon and finally captured in Montana near the Canadian border. Although the Nez Perce were eventually captured and moved to the Indian Territory, and later to Idaho, their resistance to resettlement has been described by one historian as "one of the great military movements in history" (Prucha 1984, p. 541).

ASSIMILATION

Near the end of the nineteenth century, the goal of isolating American Indians on reservations and the Indian Territory was finally achieved. The Indian population also was near extinction. Their numbers had declined steadily throughout the nineteenth century, leading most observers to predict their disappearance (Hoxie 1984). Reformers urged the federal government to adopt measures that would humanely ease American Indians into extinction. The federal government responded by creating boarding schools and the allotment acts—both were intended to "civilize" and assimilate American Indians into American society by Christianizing them, educating them, introducing them to private property, and making them into farmers. American Indian boarding schools sought to accomplish this task by indoctrinating Indian children with the belief that tribal culture was an inferior relic of the past and that Euro-American culture was vastly superior and preferable. Indian children were forbidden to wear their native attire, to eat their native foods, to speak their native language, or to practice their traditional religion. Instead, they were issued Euro-American clothes, and expected to speak English and become Christians. Indian children who did not relinquish their culture were punished by school authorities. The curriculum of these schools taught vocational arts along with "civilization" courses.

The impact of allotment policies is still evident today. The 1887 General Allotment Act (the Dawes Severalty Act) and subsequent legislation mandated that tribal lands were to be allotted to individual American Indians in fee simple title, and the surplus lands left over from allotment were to be sold on the open market. Indians who received allotted tribal lands also received citizenship, farm implements, and encouragement from Indian agents to adopt farming as a livelihood (Hoxie 1984, Prucha 1984).

For a variety of reasons, Indian lands were not completely liquidated by allotment, many Indians did not receive allotments, and relatively few changed their lifestyles to become farmers. Nonetheless, the allotment era was a disaster because a significant number of allotees eventually lost their land. Through tax foreclosures, real estate fraud, and their own need for cash, many American Indians lost what for most of them was their last remaining asset (Hoxie 1984).

Allotment took a heavy toll on Indian lands. It caused about 90 million acres of Indian land to be lost, approximately two-thirds of the land that had belonged to tribes in 1887 (O'Brien 1990). This created another problem that continues to vex many reservations: "checkerboarding." Reservations that were subjected to allotment are typically a crazy quilt composed of tribal lands, privately owned "fee" land, and trust land belonging to individual Indian families. Checkerboarding presents reservation officials with enormous administrative problems when trying to develop land use management plans, zoning ordinances, or economic development projects that require the construction of physical infrastructure such as roads or bridges.

THE INDIAN NEW DEAL

The Indian New Deal was short-lived but profoundly important. Implemented in the early 1930s along with the other New Deal programs of the Roosevelt administration, the Indian New Deal was important for at least three reasons. First, signaling the end of the disastrous allotment era as well as a new respect for American Indian tribal culture, the Indian New Deal repudiated allotment as a policy. Instead of continuing its futile efforts to detribalize American Indians, the federal government acknowledged that tribal culture was worthy of respect. Much of this change was due to John Collier, a long-time Indian rights advocate appointed by Franklin Roosevelt to serve as Commissioner of Indian Affairs (Prucha 1984).

Like other New Deal policies, the Indian New Deal also offered some relief from the Great Depression and brought essential infrastructure development to many reservations, such as projects to control soil erosion and to build hydroelectric dams, roads, and other public facilities. These projects created jobs in New Deal programs such as the Civilian Conservation Corps and the Works Progress Administration.

An especially important and enduring legacy of the Indian New Deal was the passage of the Indian Reorganization Act (IRA) of 1934. Until then, Indian self-government had been forbidden by law. This act allowed tribal governments, for the first time in decades, to reconstitute themselves for the purpose of over-seeing their own affairs on the reservation. Critics charge that this law imposed an alien form of government, representative democracy, on traditional tribal authority. On some reservations, this has been an on-going source of conflict (O'Brien 1990). Some reservations rejected the IRA for this reason, but now have tribal governments authorized under different legislation.

TERMINATION AND RELOCATION

After World War II, the federal government moved to terminate its long-standing relationship with Indian tribes by settling the tribes' outstanding legal claims, by terminating the special status of reservations, and by helping reservation Indians relocate to urban areas (Fixico 1986). The Indian Claims Commission was a special tribunal created in 1946 to hasten the settlement of legal claims that tribes had brought against the federal government. In fact, the Indian Claims Commission became bogged down with prolonged cases, and in 1978 the commission was dissolved by Congress. At that time, there were 133 claims still unresolved out of an original 617 that were first heard by the commission three decades earlier (Fixico 1986, p. 186). The unresolved claims that were still pending were transferred to the Federal Court of Claims.

Congress also moved to terminate the federal government's relationship with Indian tribes. House Concurrent Resolution (HCR) 108, passed in 1953, called for steps that eventually would abolish all reservations and abolish all special programs serving American Indians. It also established a priority list of reservations slated for immediate termination. However, this bill and subsequent attempts to abolish reservations were vigorously opposed by Indian advocacy groups such as the National Congress of American Indians. Only two reservations were actually terminated, the Klamath in Oregon and the Menominee in Wisconsin. The Menominee reservation regained its trust status in 1975 and the Klamath reservation was restored in 1986.

The Bureau of Indian Affairs (BIA) also encouraged reservation Indians to relocate and seek work in urban job markets. This was prompted partly by the desperate economic prospects on most reservations, and partly because of the federal government's desire to "get out of the Indian business." The BIA's relocation programs aided reservation Indians in moving to designated cities, such as Los Angeles and Chicago, where they also assisted them in finding housing and employment. Between 1952 and 1972, the BIA relocated more than 100,000 American Indians (Sorkin 1978). However, many Indians returned to their reservations (Fixico 1986). For some American Indians, the return to the reservation was only temporary—for example, during periods when seasonal employment such as construction work was hard to find.

SELF-DETERMINATION

Many of the policies enacted during the termination and relocation era were steadfastly opposed by American Indian leaders and their supporters. As these programs became stalled, critics attacked them for being harmful, ineffective, or both. By the mid–1960s, these policies had very little serious support. Perhaps inspired by the gains of the Civil Rights movement, American Indian leaders and their supporters made "self-determination" the first priority on their political agendas. For these activists, self-determination meant that Indian people would have the autonomy to control their own affairs, free from the paternalism of the federal government.

The idea of self-determination was well received by members of Congress sympathetic to American Indians. It also was consistent with the "New Federalism" of the Nixon administration. Thus, the policies of termination and relocation were repudiated in a process that culminated in 1975 with the passage of the American Indian Self-Determination and Education Assistance Act, a profound shift in federal Indian policy. For the first time since this nation's founding, American Indians were authorized to oversee the affairs of their own communities, free of federal intervention. In practice, the Self-Determination Act established measures that would allow tribal governments to assume a larger role in reservation administration of programs for welfare assistance, housing, job training, education, natural resource conservation, and the maintenance of reservation roads and bridges (Snipp and Summers 1991). Some reservations also have their own police forces and game wardens, and can issue licenses and levy taxes. The Onondaga tribe in upstate New York have taken their sovereignty one step further by issuing passports that are internationally recognized. Yet there is a great deal of variability in terms of how much autonomy tribes have over reservation affairs. Some tribes, especially those on large and well-organized reservations have nearly complete control over their reservations, while smaller reservations with limited resources often depend heavily on BIA services. . . .

CONCLUSION

Though small in number, American Indians have an enduring place in American society. Growing numbers of American Indians occupy reservation and other trust lands, and equally important has been the revitalization of tribal governments. Tribal governments now have a larger role in reservation affairs than ever in the past. Another significant development has been the urbanization of American Indians. Since 1950, the proportion of American Indians in cities has grown rapidly. These American Indians have in common with reservation Indians many of the same problems and disadvantages, but they also face other challenges unique to city life.

The challenges facing tribal governments are daunting. American Indians are among the poorest groups in the nation. Reservation Indians have substantial needs for improved housing, adequate health care, educational opportunities, and employment, as well as developing and maintaining reservation infrastructure. In the face of declining federal assistance, tribal governments are assuming an ever-larger burden. On a handful of reservations, tribal governments have assumed completely the tasks once performed by the BIA.

As tribes have taken greater responsibility for their communities, they also have struggled with the problems of raising revenues and providing economic opportunities for their people. Reservation land bases provide many reservations with resources for development. However, these resources are not always abundant, much less unlimited, and they have not always been well managed. It will be yet another challenge for tribes to explore ways of efficiently managing their existing resources. Legal challenges also face tribes seeking to exploit unconventional resources such as gambling revenues. Their success depends on many complicated legal and political contingencies.

Urban American Indians have few of the resources found on reservations, and they face other difficult problems. Preserving their culture and identity is an especially pressing concern. However, urban Indians have successfully adapted to city environments in ways that preserve valued customs and activities—-powwows, for example, are an important event in all cities where there is a large Indian community. In addition, pan-Indianism has helped urban Indians set aside tribal differences and forge alliances for the betterment of urban Indian communities.

These alliances are essential, because unlike reservation Indians, urban American Indians do not have their own form of self-government. Tribal governments do not have jurisdiction over urban Indians. For this reason, urban Indians must depend on other strategies for ensuring that the needs of their community are met, especially for those new to city life. Coping with the transition to urban life poses a multitude of difficult challenges for many American Indians. Some succumb to these problems, especially the hardships of unemployment, economic deprivation, and related maladies such as substance abuse, crime, and violence. But most successfully overcome these difficulties, often with help from other members of the urban Indian community.

Perhaps the greatest strength of American Indians has been their ability to find creative ways for dealing with adversity, whether in cities or on reservations. In the past, this quality enabled them to survive centuries of oppression and persecution. Today this is reflected in the practice of cultural traditions that Indian people are proud to embrace. The resilience of American Indians is an abiding quality that will no doubt ensure that they will remain part of the ethnic mosaic of American society throughout the twenty-first century and beyond.

REFERENCES

Cornell, Stephen. 1988. *The Return of the Native: American Indian Political Resurgence.* New York: Oxford University Press.

Fixico, Donald L. 1986. *Termination and Relocation: Federal Indian Policy, 1945–1950.* Albuquerque, NM: University of New Mexico Press.

Hagan, William T. 1979. *American Indians.* Chicago, IL: University of Chicago Press.

Hall, Thomas D. 1989. *Social Change in the Southwest, 1350–1880.* Lawrence, KS: University Press of Kansas.

Hoxie, Frederick E. 1984. *A Final Promise: The Campaign to Assimilate the Indians, 1880–1920.* Lincoln, NE: University of Nebraska Press.

Nagel, Joanne. 1989. "American Indian Repertoires of Contention." Paper presented at the annual meeting of the American Sociological Association, San Francisco, CA.

O'Brien, Sharon. 1990. *American Indian Tribal Governments.* Norman, OK: University of Oklahoma Press.

Prucha, Francis Paul. 1984. *The Great Father.* Lincoln, NE: University of Nebraska Press.

Snipp, C. Matthew, and Gene F. Summers. 1991. "American Indian Development Policies." Pp. 166–180 in *Rural Policies*

for the 1990s, ed. Cornelia B. Flora and James A. Christensen. Boulder, CO: Westview Press.

Sorkin, Alan L. 1978. *The Urban American Indian.* Lexington, MA: Lexington Books.

Thornton, Russell. 1987. *American Indian Holocaust and Survival: A Population History Since 1492.* Norman, OK: University of Oklahoma Press.

Utley, Robert. 1984. *The Indian Frontier of the American West, 1846–1890.* Albuquerque, NM: University of New Mexico Press.

Weibel-Orlando, Joan. 1991. *Indian Country, L.A.* Urbana, IL: University of Illinois Press.

DISCUSSION QUESTIONS

1. Snipp identifies five periods of relations between the U.S. federal government and Native Americans. What are these five periods and how do the events in each shape relations between the government and American Indians?

2. The history of Native Americans can be read as a constant struggle between subordination and the desire for self-sufficiency. How have Native Americans resisted subordination and how is this shaping contemporary Indian experience?

INFOTRAC COLLEGE EDITION

You can use your access to InfoTrac College Edition to learn more about the subjects covered in this essay. Some suggested search terms include:

American Indian Movement (AIM)
Bureau of Indian Affairs
Dawes Severalty Act

Indian New Deal
reclamation of land
tribal government

72

Medical Students' Contacts
with the Living and the Dead

ALLEN C. SMITH III AND SHERRYL KLEINMAN

Smith and Kleinman examine how medical students, and the institutions that train them, deal with the issue of physical intimacy in the treatment of patients. They argue that medical students are taught to manage their emotions in order to remain scientifically objective, in keeping with the culture of Western medicine. Their strategies for maintaining affective neutrality include dehumanizing the patient, accentuating scientific learning, joking, and avoiding contact.

The ideology of affective neutrality is strong in medicine; yet no courses in the medical curriculum deal directly with emotion management, specifically learning to change or eliminate inappropriate feelings (Hochschild 1979). Rather, two years of participant observation in a medical school revealed that discussion of the students' feelings is taboo; their development toward emotional neutrality remains part of the hidden curriculum. Under great pressure to prove themselves worthy of entering the profession, students are afraid to admit that they have uncomfortable feelings about patients or procedures, and hide those feelings behind a "cloak of competence" (Haas and Shaffir 1977, 1982). Beneath their surface presentations, how do students deal with the "unprofessional" feelings they bring over from the personal realm? Because faculty members do not address the problem, students are left with an individualistic outlook: they expect to get control of themselves through sheer willpower.

Despite the silence surrounding this topic, the faculty, the curriculum, and the organization of medical school do provide students with resources for dealing with their problem. The culture of medicine that informs teaching and provides the feeling rules also offers unspoken supports for dealing with unwanted emotions. Students draw on aspects of their experience in medical school to manage their emotions. Their strategies include transforming the patient or the procedure into an analytic object or event, accentuating the comfortable feelings that come from learning and practicing "real medicine," blaming patients, empathizing with patients, joking, and avoiding sensitive contact. . . .

From: *Social Psychology Quarterly* 52 (1989):56–69. Reprinted with permission.

THE STUDENTS' PROBLEM

As they encounter the human body, students experience a variety of uncomfortable feelings including embarrassment, disgust, and arousal. Medical school, however, offers a barrier against these feelings by providing the anesthetic effect of long hours and academic pressure.

> You know the story. On call every third night, and stay in the hospital late most other evenings. I don't know how you're supposed to think when you're that tired, but you do, plod through the day insensitive to everything (Third-year male).

Well before entering medical school, students learn that their training will involve constant pressure and continuing fatigue. Popular stories prepare them for social isolation, the impossibility of learning everything, long hours, test anxiety, and the fact that medical school will permeate their lives (Becker, Geer, Hughes, and Strauss 1961). These difficulties and the sacrifices that they entail legitimate the special status of the profession the students are entering. They also blunt the students' emotional responses.

Yet uncomfortable feelings break through. Throughout the program, students face provocative situations—some predictable, others surprising. They find parts of their training, particularly dissection and the autopsy, bizarre or immoral. . . .

Much of the students' discomfort is based on the fact that the bodies they have contact with are or were *people*. Suddenly students feel uncertain about the relationship of the person to the body, a relationship they had previously taken for granted.

> It felt tough when we had to turn the whole body over from time to time (during dissection). It felt like real people (First-year female). . . .

When the person is somehow reconnected to the body, such as when data about the living patient who died is brought into the autopsy room, students feel less confident and more uneasy.

Students find contact with the sexual body particularly stressful. In the anatomy lab, in practice sessions with other students, and in examining patients, students find it difficult to feel neutral as contact approaches the sexual parts of the body.

> When you listen to the heart you have to work around the breast, and move it to listen to one spot. I tried to do it with minimum contact, without staring at her tit. . . .breast. . . .The different words (pause) shows I was feeling both things at once (Second-year male).

Though they are rarely aroused, students worry that they will be. They feel guilty, knowing that sexuality is proscribed in medicine, and they feel embarrassed. Most contact involves some feelings, but contact with the sexual body presents a bigger problem. . . .

Students also feel disgust. They see feces, smell vomit, touch wounds, and hear bone saws, encountering many repulsive details with all of their senses.

One patient was really gross! He had something that kept him standing, and coughing all the time. Coughing phlegm, and that really bothers me. Gross! Just something I don't like. Some smelled real bad. I didn't want to examine their axillae. Stinking armpits! It was just not something I wanted to do (Second-year female).

When the ugliness is tied to living patients, the aesthetic problem is especially difficult. On opening the bowels of the cadaver, for example, students permit themselves some silent expressions of discomfort, but even a wince is unacceptable with repugnant living patients.

To make matters worse, students learn early on that they are not supposed to talk about their feelings with faculty members or other students. Feelings remain private. The silence encourages students to think about their problem as an individual matter, extraneous to the "real work" of medical school. They speak of "screwing up your courage," "getting control of yourself," "being tough enough," and "putting feelings aside." They worry that the faculty would consider them incompetent and unprofessional if they admitted their problem.

I would be embarrassed to talk about it. You're supposed to be professional here. Like there's an unwritten rule about how to talk (First-year female). . . .

The "unwritten rule" is relaxed enough sometimes to permit discussion, but the privacy that surrounds these rare occasions suggests the degree to which the taboo exists. At times, students signal their uncomfortable feelings—rolling their eyes, turning away, and sweating—but such confirmation is limited. Exemplifying pluralistic ignorance, each student feels unrealistically inadequate in comparison with peers (yet another uncomfortable feeling). Believing that other students are handling the problem better than they are, each student manages his or her feelings privately, only vaguely aware that all students face the same problem. . . .

EMOTION MANAGEMENT STRATEGIES

How do students manage their uncomfortable and "inappropriate" feelings? The deafening silence surrounding the issue keeps them from defining the problem as shared, or from working out common solutions. They cannot develop strategies collectively, but their solutions are not individual. Rather, students use the *same* basic emotion management strategies because social norms, faculty models, curricular priorities, and official and unofficial expectations provide them with uniform guidelines and resources for managing their feelings.

Transforming the Contact

Students feel uncomfortable because they are making physical contact with people in ways they would usually define as appropriate only in a personal context, or as inappropriate in any context. Their most common solution to this problem is cognitive (Hochschild 1979; Thoits 1985).

Mentally they transform the body and their contact with it into something entirely different from the contacts they have in their personal lives. Students transform the person into a set of esoteric body parts and change their intimate contact with the body into a mechanical or analytic problem.

> I just told myself, "OK, doc, you're here to find out what's wrong, and that includes the axillae (armpits)." And I detach a little, reduce the person for a moment. . . .Focus real hard on the detail at hand, the fact, or the procedure or the question. Like with the cadaver. Focus on a vessel. Isolate down to whatever you're doing (Second-year female). . . .

Students also transform the moment of contact into a complex intellectual puzzle, the kind of challenge they faced successfully during previous years of schooling. They interpret details according to logical patterns and algorithms, and find answers as they master the rules. . . .

> The patient is really like a math word problem. You break it down into little pieces and put them together. The facts you get from a history and physical, from the labs and chart. They fit together, once you begin to see how to do it. . . . It's an intellectual challenge (Third-year female).

Defining contact as a part of scientific medicine makes the students feel safe. They are familiar with and confident about science, they feel supported by its cultural and curricular legitimacy, and they enjoy rewards for demonstrating their scientific know-how. In effect, science itself is an emotion management strategy. By competing for years for the highest grades, these students have learned to separate their feelings from the substance of their classes and to concentrate on the impersonal facts of the subject matter. In medical school they use these "educational skills" not only for academic success but also for emotion management. . . .

The scientific, clinical language that the students learn also supports intellectualization. It is complex, esoteric, and devoid of personal meanings. "Palpating the abdomen" is less personal than "feeling the belly." . . .

Further, the structure of the language, as in the standard format for the presentation of a case helps the students to think and speak impersonally. Second-year students learn that there is a routine, acceptable way to summarize a patient: chief complaint, history of present illness, past medical history, family history, social history, review of systems, physical findings, list of problems, medical plan. In many situations they must reduce the sequence to a two- or three-minute summary. Faculty members praise the students for their ability to present the details quickly. Medical language labels and conveys clinical information, and it leads the students away from their emotions.

Transformation sometimes involves changing the body into a nonhuman object. Students think of the body as a machine or as an animal specimen, and recall earlier, comfortable experiences in working on that kind of object. The body is no longer provocative because it is no longer a body. . . .

> You can't tell what's wrong without looking under the hood. It's different when I'm talking with a patient. But when I'm examining them it's like an

automobile engine. There's a bad connotation with that, but it's literally what I mean (Third-year male). . . .

The curriculum supports these dehumanizing transformations by eliminating the person in most of the students' contact with the body. Contact is usually indirect, based on photographs, X-rays (and several newer technologies), clinical records, diagrams, and written words. . . .

Accentuating the Positive

As we hinted in the previous section, transforming body contact into an analytic event does not merely rid students of their uncomfortable feelings, producing neutrality. It often gives them opportunities to have *good* feelings about what they are doing. Their comfortable feelings include the excitement of practicing "real medicine," the satisfaction of learning, and the pride of living up to medical ideals.

Students identify much of their contact with the body as "real medicine," asserting that such contact separates medicine from other professions. As contact begins in dissection and continues through the third-year clinical clerkships, students feel excited about their progress. . . .

> This (dissection) is the part that is really medical school. Not like any other school. It feels like an initiation rite, something like when I joined a fraternity. We were really going to work on people (First-year male). . . .

Eventually students see contact as their responsibility and their right, and forget the sense of privilege they felt at the beginning. Still, some excitement returns as they take on clinical responsibility in the third year. All of these feelings can displace the discomfort which also attends most contact.

Contact also provides a compelling basis for several kinds of learning, all of which the students value. They sense that they learn something important in contact, something richer than the "dry facts" of textbooks and lectures. Physicians, they believe, rely on touch, not on text. . . .

Laughing About It

Students can find or create humor in the situations that provoke their discomfort. Humor is an acceptable way for people to acknowledge a problem and to relieve tension without having to confess weaknesses. In this case, joking also lets other students know that they are not alone with the problem. . . .

By redefining the situation as at least partially humorous, students reassure themselves that they can handle the challenge. They believe that the problem can't be so serious if there is a funny side to it. Joking also allows them to relax a little and to set ideals aside for a time.

Where do students learn to joke in this way? The faculty, including the residents (who are the real teachers on the clinical teams), participate freely, teaching the students that humor is an acceptable way to talk about uncomfortable encounters in medicine. . . .

Unlike the students' other strategies, joking occurs primarily when they are alone with other medical professionals. Jokes are acceptable in the hallways, over coffee, or in physicians' workrooms, but usually are unacceptable when outsiders might overhear. Joking is backstage behavior. Early in their training, students sometimes make jokes in public, perhaps to strengthen their identity as "medical student," but most humor is in-house, reserved for those who share the problem and have a sense of humor about it.

Avoiding the Contact

Students sometimes avoid the kinds of contact that give rise to unwanted emotions. They control the visual field during contact, and eliminate or abbreviate particular kinds of contact. . . .

Keeping personal body parts covered in the lab and in examinations prevents mold, maintains a sterile field, and protects the patient's modesty. Covers also eliminate disturbing sites and protect students from their feelings. Such nonprofessional purposes are sometimes most important. Some students, for example, examine the breasts by reaching under the patient's gown, bypassing the visual examination emphasized in training. . . .

CONCLUSION

Medical students sometimes feel attracted to or disgusted by the human body. They want to do something about these feelings, but they find that the topic is taboo. Even among themselves, students generally refrain from talking about their problem. Yet despite the silence, the culture and the organization of medical school provide students with supports and guidelines for managing their emotions. Affective socialization proceeds with no deliberate control, but with profound effect. . . .

The emotion management strategies used by the students illustrate the culture of modern Western medicine. In relying on these strategies, the students reproduce that culture (Foucault 1973), creating a new generation of physicians who will support the biomedical model of medicine and the kind of doctor-patient relationship in which the patient is too frequently dehumanized. Students sometimes criticize their teachers for an apparent insensitivity to their patients, but they turn to desensitizing strategies themselves in their effort to control the emotions that medical situations provoke. These strategies exclude the patient's feelings, values, and social context, the important psychosocial aspects of medicine (Engel 1977; Gorlin and Zucker 1983). Contradicting their previous values, students reinforce biomedicine as they rely on its emotion management effects. . . .

It would be unfair to conclude that medical training is uniquely responsible for the specific character of the students' emotion management problem and for its unspoken solution. The basic features of the culture of medicine are consistent with the wider cultural context in which medicine exists. Biomedicine fits with the emphasis in Western culture on rationality and scientific "objectivity." In Western societies the mind is defined as superior to the body, and thoughts are defined as superior to feelings (Mills and Kleinman 1988; Tuan 1982; Turner 1984). Not surprisingly, students know the feeling rules of professional life before they arrive at medical school. Childhood socialization and formal education teach them to set aside their feelings in public, to master "the facts," and to present themselves in intellectually defensible ways (Bowers 1984). Medical situations provide vivid challenges, but students come equipped with emotion management skills that they need only to strengthen. . . .

REFERENCES

Becker, H., B. Geer, E. Hughes and A. Strauss. 1961. *Boys in White*. New Brunswick, NJ: Transaction.

Bowers, C. 1984. *The Promise of Theory: Education and the Politics of Cultural Change*. New York: Longmans.

Engel, G. 1977. "The Need for a New Medical Model: A Challenge for Biomedicine." *Science* 196(4286): 129–36.

Foucault, M. 1973. *The Birth of the Clinic: An Archaeology of Medical Perception*. New York: Pantheon.

Gorlin, R. and H. Zucker. 1983. "Physicians' Reactions to Patients: A Key to Teaching Humanistic Medicine." *New England Journal of Medicine* 308(18):1059–63.

Haas, J. and W. Shaffir. 1977. "The Professionalization of Medical Students: Developing Competence and a Cloak of Competence." *Symbolic Interaction* 1:71–88.

———. 1982. "Taking on the Role of Doctor: A Dramaturgical Analysis of Professionalization." *Symbolic Interaction* 5:187–203.

Hochschild, A. 1979. "Emotion Work, Feeling Rules, and Social Structure." *American Journal of Sociology* 85(3):551–75.

Mills, T. and S. Kleinman, 1988. "Emotions, Reflexivity, and Action: An Interactionist Analysis." *Social Forces* 66(4):1009–27.

Thoits, P. 1985. "Self-Labeling Processes in Mental Illness: The Role of Emotional Deviance." *American Journal of Sociology* 91:221–49.

Tuan, Y.-F. 1982. *Segmented Worlds and Self: Group Life and Individual Consciousness*. Minneapolis: University of Minnesota Press.

Turner, B. 1984. *The Body and Society*. New York: Basil Blackwell.

DISCUSSION QUESTIONS

1. How do the emotion management strategies outlined in this essay contribute to the notion that Western medicine treats the problem and not the person?

2. What would be the consequence of medical students who do not learn to manage their emotions when treating patients? How would the practice of medicine be changed?

INFOTRAC COLLEGE EDITION

You can use your access to InfoTrac College Edition to learn more about the subjects covered in this essay. Some suggested search terms include:

affective neutrality

emotion management

medical school

physician–patient relationship

73

Do Doctors Eat Brains?

ANN FADIMAN

In this essay, Ann Fadiman summarizes the differences between Western practices of medicine and the healing techniques of the Hmong people in East Asia. Her account illustrates both misperceptions of Western medicine and the significant differences in treatment ideology between the two cultures. Overall, the essay reveals that Western medicine traditionally ignores the connection between body and soul.

In 1982, Mao Thao, a Hmong woman from Laos who had resettled in St. Paul, Minnesota, visited Ban Vinai, the refugee camp in Thailand where she had lived for a year after her escape from Laos in 1975. She was the first Hmong-American ever to return there, and when an officer of the United Nations High Commissioner for Refugees, which administered the camp, asked her to speak about life in the United States, 15,000 Hmong, more than a third of the population of Ban Vinai, assembled in a soccer field and questioned her for nearly four hours. Some of the questions they asked her were: Is it forbidden to use a *txiv neeb* to heal an illness in the United States? Why do American doctors take so much blood from their patients? After you die, why do American doctors try to open up your head and take out your brains? Do American doctors eat the livers, kidneys, and brains of Hmong patients? When Hmong people die in the United States, is it true that they are cut into pieces and put in tin cans and sold as food?

The general drift of these questions suggests that the accounts of the American health care system that had filtered back to Asia were not exactly enthusiastic. The limited contact the Hmong had already had with Western medicine in the camp hospitals and clinics had done little to instill confidence, especially when compared to the experiences with shamanistic healing to which they were accustomed. A *txiv neeb* might spend as much as eight hours in a sick person's home; doctors forced their patients, no matter how weak they were, to come to the hospital, and then might spend only twenty minutes at their bedsides. *Txiv neebs* were polite and never needed to ask questions; doctors asked many rude and intimate questions about patients' lives, right down to their sexual and excretory habits. *Txiv neebs* could render an immediate diagnosis; doctors often demanded samples of blood (or even urine or feces, which they liked to keep in little bottles), took X rays, and waited for days for the results to come back from the laboratory—and then, after all that, sometimes they were unable to identify the cause of the

From: Ann Fadiman, 1997. *The Spirit Catches You and Then You Fall Down: A Hmong Child, Her American Doctors, and the Clash of Two Cultures* New York: Farrar, Straus, and Giroux, pp. 32–35.

problem. *Txiv neebs* never undressed their patients; doctors asked patients to take off all their clothes, and sometimes dared to put their fingers inside women's vaginas. *Txiv neebs* knew that to treat the body without treating the soul was an act of patent folly; doctors never even mentioned the soul. *Txiv neebs* could preserve unblemished reputations even if their patients didn't get well, since the blame was laid on the intransigence of the spirits rather than the competence of the negotiators, whose stock might even rise if they had had to do battle with particularly dangerous opponents; when doctors failed to heal, it was their own fault.

To add injury to insult, some of the doctors' procedures actually seemed more likely to threaten their patients' health than to restore it. Most Hmong believe that the body contains a finite amount of blood that it is unable to replenish, so repeated blood sampling, especially from small children, may be fatal. When people are unconscious, their souls are at large, so anesthesia may lead to illness or death. If the body is cut or disfigured, or if it loses any of its parts, it will remain in a condition of perpetual imbalance, and the damaged person not only will become frequently ill but may be physically incomplete during the next incarnation; so surgery is taboo. If people lose their vital organs after death, their souls cannot be reborn into new bodies and may take revenge on living relatives; so autopsies and embalming are also taboo. (Some of the questions on the Ban Vinai soccer field were obviously inspired by reports of the widespread practice of autopsy and embalming in the United States. To make the leap from hearing that doctors removed organs to believing that they ate them was probably no crazier than to assume, as did American doctors, that the Hmong ate human placentas—but it was certainly scarier.)

The only form of medical treatment that was gratefully accepted by at least some of the Hmong in the Thai camps was antibiotic therapy, either oral or by injection. Most Hmong have little fear of needles, perhaps because some of their own healers (not *txiv neebs,* who never touch their patients) attempt to release fevers and toxicity through acupuncture and other forms of dermal treatment, such as massage; pinching; scraping the skin with coins, spoons, silver jewelry, or pieces of bamboo; applying a heated cup to the skin; or burning the skin with a sheaf of grass or a wad of cotton wool. An antibiotic shot that could heal an infection almost overnight was welcomed. A shot to immunize someone against a disease he did not yet have was something else again. In his book *Les naufragés de la liberté,* the French physician Jean-Pierre Willem, who worked as a volunteer in the hospital at the Nam Yao camp, related how during a typhoid epidemic, the Hmong refugees refused to be vaccinated until they were told that only those who got shots would receive their usual allotments of rice—whereupon 14,000 people showed up at the hospital, including at least a thousand who came twice in order to get seconds. . . .

Wendy Walker-Moffat, an educational consultant who spent three years teaching and working on nutritional and agricultural projects in Phanat Nikhom and Ban Vinai, suggests that one reason the Hmong avoided the camp hospitals is that so many of the medical staff members were excessively zealous volunteers from Christian charitable organizations. "They were there to provide medical aid, but they were also there—though not overtly—to convert people," Walker-Moffat told me. "And part of becoming converted was believing in Western medicine. I'll never forget one conversation I overheard when I was working in the hospital area at Ban Vinai. A group of doctors and nurses were talking to a Hmong man whom they had converted and ordained as a Protestant minister. They had decided that in order to get the Hmong to come into the hospital they were going to allow a traditional healer, a shaman, to practice there. I knew they all thought shamanism was witch-doctoring. So I heard them tell this Hmong minister that if they let a shaman work in the medical center he could only give out herbs, and not perform any actual work with the spirits. At this point they asked the poor Hmong minister, 'Now *you* never go to a shaman, do you?' He was a Christian convert, he knew you cannot tell a lie, so he said, 'Well, yes, I do.' But then their reaction was so shocked that he said, 'No, no, no, I've never been. I've just heard that *other* people go.' What they didn't realize was that—to my knowledge, at least—no Hmong is ever fully converted.". . . .

DISCUSSION QUESTIONS

1. What are some stereotypical images Americans have of Eastern medicine? How has this essay educated you with regard to those stereotypes?

2. Has there been a recent change in Western medicine that allows for some of these Eastern treatment practices? How does society respond to the incorporation of non-Western treatment approaches in U.S. medicine?

INFOTRAC COLLEGE EDITION

You can use your access to InfoTrac College Edition to learn more about the subjects covered in this essay. Some suggested search terms include:

Hmong   reincarnation
non–Western medicine   shamanism

74

Corporatization and the Social Transformation of Doctoring

JOHN B. MCKINLAY AND JOHN D. STOECKLE

In this essay, John McKinlay and John Stoeckle outline changes in the medical profession from highly skilled independent physicians to a managed care approach to delivering medical treatment. They outline Marx's and Weber's theory as it applies to the medical profession and suggest proletarianization as a useful theory to explain the change process.

We are witnessing a transformation of the health care systems of developed countries that is without parallel in modern times. This dramatic change has implications for patients and, without exception, affects the entire division of labor in health care. What are some of these changes and how are they manifesting themselves with respect to doctoring?

THE CHANGES

Over the last few years especially, many multinational corporations, with highly diverse activities, have become involved in all facets of the generally profitable business of medical care, from medical manufacturing and the ownership of treatment institutions to the financing and purchase of services in preferred provider organizations and health maintenance organizations (HMOs). Conglomerates such as General Electric, AT&T, and IBM, among many others, now have large medical manufacturing enterprises within their corporate divisions. Aerospace companies are involved in everything from computerized medical information systems to life support systems. Even tobacco companies and transportation enterprises have moved into the medical arena. In addition to this industrial or manufacturing capital, even larger financial capital institutions (e.g., commercial banks, life insurance companies, mutual and pension funds, and diversified financial organizations) are also stepping up their involvement in medical care and experiencing phenomenal success.

Besides corporate investments in health care, corporate mergers of treatment organizations and industrial corporations are also taking place. Privately owned

From: John B. McKinley, and John D. Stoeckle. 1998. "Corporatization and the Social Transformation of Doctoring." *International Journal of Health Services* 18 (1998): 191–205.

hospital chains, controlled by larger corporations, evidence continuing rapid growth. Much of this growth comes from buying up local, municipal, and voluntary community hospitals, many of which were "going under" as a result of cutbacks in government programs and regulations on hospital use and payment. By 1990, about 30 percent of general hospital beds will be managed by investor hospital chains. Because the purpose of an investor-owned organization is to make money, there is understandable concern over the willingness of such organizations to provide care to the 35 million people who lack adequate insurance coverage and who are not eligible for public programs.

RESPONSES

Regulations

Confronted with an ever deepening fiscal crisis, the state continues to cast around for regulatory solutions—one of the latest of which is diagnosis related groups (DRGs) for Medicare patients, which reimburse hospitals by diagnosis and with rates determined by government. If the actual cost of treatment is less than the allowable payment, then the hospital makes a profit; if treatment costs are more, then the hospital faces a loss, even bankruptcy, especially since an average of 40 percent of hospital revenues come from Medicare patients. This probably ineffective measure follows many well-documented policy failures (e.g., Professional Standards Review Organizations), and its consequences for the health professions are profound. These regulatory efforts, corporate mergers, investor-owned hospital chains, federally mandated cost-containment measures, among many other changes, are transforming the shape, content, and even the moral basis of health care. How are these institutional changes affecting the everyday work of the doctor?

New Management

By all accounts, hospitals are being managed by a new breed of physician administrators, whom Alford aptly terms "corporate rationalizers"[1]. While some have medical qualifications, most are trained in the field of hospital administration, which emphasizes, among other things, rationalization, productivity, and cost efficiency. Doctors used to occupy a privileged position at the top of the medical hierarchy. Displaced by administrators, doctors have slipped down to the position of middle management where their prerogatives are also challenged or encroached upon by other health workers. Clearly, managerial imperatives often compete or conflict with physicians' usual mode of practice. Increasingly, it seems, administrators, while permitting medical staff to retain ever narrower control of technical aspects of care, are organizing the necessary coordination for collaborative work, the work schedules of staff, the recruitment of patients to the practice, and the contacts with third-party purchasers, and are determining the fiscal rewards.

Some argue that many administrators are medically qualified, and thus act so as to protect the traditional professional prerogatives. This view confuses the usual distinction between status and role. As many hospital and HMO doctors will attest, a physician who is a full-time administrator is understandably concerned to protect the bottom line, not the prerogatives of the profession. When these interests diverge, as they increasingly must, it becomes clear where the physician/administrator's divided loyalty really resides. One recent survey of doctors shows that a majority do not believe that their medical directors represent the interests of the medical staff. As a result, the American Medical Association (AMA) has concluded that "as hospital employees . . . medical directors may align their loyalty more with hospitals than with medical staff interests." To counteract these trends, it has been seriously suggested that "physicians should be trained in organization theory . . . to act as liaisons among all those with an interest in medicine, including patients, health care providers, insurers, politicians, economists, and administrators."

Specialization-Deskilling

Specialization in medicine, while deepening knowledge in a particular area, is also circumscribing the work that doctors may legitimately perform. Specialization can, with task delegation, reduce the hospital's dependence on its highly trained medical staff. Other health workers (e.g., physicians' assistants and nurse practitioners) with less training, more narrowly skilled, and obviously cheaper can be hired. Doctors, while believing that specialization is invariably a good thing, are being "deskilled"—a term employed by Braverman[2] to describe the transfer of skills from highly trained personnel to more narrowly qualified specialists. Many new health occupations (physicians' assistants, nurse practitioners, certified nurses) have emerged over the last several decades to assume some of the work that doctors used to perform. Not only is work deskilled but it is increasingly conducted without M.D.s' control as other professional groups and workers seek their own autonomy. These processes receive support from administrators constantly searching for cheaper labor, quite apart from the controlled trials revealing that "allied health professionals" can, in many circumstances, do the same work just as effectively and efficiently for those patients who must use them. Preference for the term "allied health professional" rather than "physician extender" or "physician assistant" reflects the promotion of this occupational division of labor.

Just over a decade ago, Victor Fuchs[3] viewed the physician as "captain of the team." Around that time, doctors (usually male) were the unquestioned masters and other health workers (usually female), especially nurses, worked "under the doctor" to carry out his orders. That subordination is disappearing. Nowadays, physicians are required to work alongside other professionals on the "health care team." The ideology of *team work* is a leveler in the hierarchical division of health care labor. Other health workers—for example, physiotherapists, pharmacists, medical social workers, inhalation therapists, podiatrists, and even nurses in general—may have more knowledge of specific fields than physicians, who are increasingly required to defer to other workers, now providing some of the technical and humane tasks of doctoring. While some M.D.s continue to resist these trends, and have publicly complained about "the progressive exclusion of doctors from nursing affairs," still others have accommodated to the changing scene captured in the title of a recent article: "At This Hospital, 'The Captain of the Ship' is Dead."

DOCTOR OVERSUPPLY

The growing oversupply of doctors in developed countries reinforces these trends in medical work and professional power by intensifying intraprofessional competition and devaluing their position in the job market. During the 1970s, the supply of physicians increased 36 percent, while the population grew only 8 percent. Medical schools in the United States continue to pump 17,000 physicians into the system annually. One report projected an . . . excess of 150,000 by the year 2000 . . . This level of intensity, obviously much higher in the northeast and on the west coast, renders fee-for-service solo practice economically less feasible. Again, the changes occurring are captured in the title of a recent article, "Doctor, the Patient Will see You Now." There are reliable reports that doctors are unemployed in a number of countries and increasingly underemployed in quite a few others. Doctors have apparently received unemployment payments in Scandinavian countries, Canada, and Australia. Official recognition of physician oversupply exists in Belgium, which is restricting specialty training, and the Netherlands, which is reducing both medical school intake and specialty training. . . .

Anecdotal reports from older doctors indicate that medicine today is not like "the good old days." The malpractice crisis, DRGs, the likelihood of fixed fees, and shrinking incomes (projected at a 30 percent decline over the next decade) all combine to remove whatever "fun" there was in medical practice. Some wonder aloud whether they would choose medicine if, with the benefit of hindsight, they had to do it all over again. While doctors used to want their children to follow in their footsteps, many report that they would not recommend medicine today. Recent graduates have doubts of other kinds. They fear that their debts will force them into specialities on the basis of anticipated earnings rather than intrinsic interest. College advisors may dissuade the highly talented students they counsel from choosing medicine because its job market looks so bleak. . . .

Unionization—A Harbinger of a Trend?

There are reports from across the United States. . . that physicians are rebelling against the continuing challenges to their authority and attempts to cut their incomes by HMOs and other corporate-like means of organizing profitable production of medical care. One recent manifestation of doctors' frustration with the profound changes already described is increased interest in unionization. Several unions have been or are being formed in different areas of the country to represent doctors working as full-time employees of state and local government and in HMOs. The largest is the Union of American Physicians and Dentists, based in Oakland, California, with a membership of around 43,000 in 17 states. . . . The

HMOs' organizational structure and that of other similar prepaid health care plans appears to generate disgruntlement among their salaried physician employees who were socialized to expect considerably more status and professional autonomy than the HMOs permit. . .

. . . For understandable reasons, many physicians and the lay public recoil at the thought of and disparage unionized doctors. Only a decade ago unionization among physicians was unthinkable, a movement commonly considered to be working class. . .

THEORIES OF CHANGE

Some of the forces transforming medical care and the work of doctors have been described. How does one *explain* what is occurring? *Why* is it happening?

Probably the best account of the stage-by-stage transformation of the labor process under capitalism is provided by Karl Marx. Although not concerned with health care, his thesis is applicable. During the precapitalist period, small-scale independent craftsmen (solo practitioners) operated domestic workshops, sold their products on the free market, and controlled the production of goods. Over time, capitalists steered many of these skilled workers into their factories (hospitals) where they were able to continue traditional crafts semi-autonomously in exchange for wages. Eventually, the owners of production (investors) began to rationalize the production process in their factories by encouraging specialization, allocating certain tasks to cheaper workers, and enlisting managers to coordinate the increasingly complex division of labor that developed. Rationalization was completed during the final stage when production was largely performed by engineering systems and machines, with the assistance of unskilled human machine-minders. The worker's autonomy and control over work and the workplace diminished, while the rate of exploitation increased with each successive stage in the transformation of production.

Weber's account of the same process (bureaucratization) is strikingly similar. According to Weber, bureaucracy is characterized by the following: (*a*) a hierarchical organization; (*b*) a strict chain of command from top to bottom; (*c*) an elaborate division of labor; (*d*) assigning specialized tasks to qualified individuals; (*e*) detailed rules and regulations governing work; (*f*) hiring personnel based on competence, specialized training, and qualifications (as opposed to family ties, political power, or tradition); and (*g*) expectations of a life-time career from officials. He described how workers were increasingly "separated from ownership of the means of production or administration." Bureaucratic workers became specialists performing circumscribed duties within a hierarchical structure subject to the authority of superiors and to established rules and procedures. According to Weber, bureaucratic employees are "subject to strict and systematic discipline and control in the conduct of the office" they occupy. For Weber, the bureaucratic form of work was present not only in the area of manufacturing but also in churches, schools, and government organizations. It is noteworthy that he also included hospitals: "this type of bureaucracy is found in private clinics, as well as in endowed hospitals or the hospitals maintained by religious orders." While Weber viewed bureaucracy as the most rational and efficient mode of organizing work, he also saw the accompanying degradation of working life as inevitable.

It is argued that the process outlined by Marx and Weber with respect to a different group of workers, during a different historical era, is directly applicable to the changing situation of doctors today, now that the "industrial revolution has finally caught up with medicine" (George Rosen). Whereas, generally speaking, most other workers have been quickly and easily corporatized, physicians have been able to postpone or minimize this process in their own case. Now, primarily as a result of the bureaucratization that has been forced on medical practice, physicians are being severely reduced in function and their formerly self-interested activities subordinated to the requirements of the highly profitable production of medical care. . . .

TOWARD PROLETARIANIZATION

The healthy debate over the changing position of doctors within the rapidly changing health care system is likely to continue for some time. Along with others in Britain, Australia, Canada, Scandinavia, and the United States, we have elaborated one viewpoint (proletarianization), and have presented as much data as can be easily mustered. . . . The theory of proletarianization seeks to explain *the process by which an occupational category is divested of control over certain prerogatives relating to the location, content, and essentiality of its task activities, thereby subordinating it to the broader requirements of production under advanced capitalism.* That is admittedly and necessarily a general definition. However, in order to provide operational specificity, and to facilitate the collection of the evidence that everyone desires, seven specific professional prerogatives, that are lost or curtailed through the process of proletarianization, are identified as follows:

1. *The criteria for entrance* (e.g., the credentialing system and membership requirements);

2. *The content of training* (e.g., the scope and content of the medical curriculum);

3. *Autonomy regarding the terms and content of work* (e.g., the ways in which what must be done is accomplished);

4. *The objects of labor* (e.g., commodities produced or the clients served);

5. *The tools of labor* (e.g., machinery, biotechnology, chemical apparatus);

6. *The means of labor* (e.g., hospital buildings, clinic facilities, laboratory services); and

7. *The amount and rate of remuneration for labor* (e.g., wage and salary levels, fee schedules). . . .

It is thus our argument that the industrial revolution has fully caught up with medicine. We are beginning to see the same phenomena in this sphere of work. From the preceding description, it is clear that we view the theory of proletarianization as a useful explanation of a process under development, *not* a state that has been or is just about to be achieved. The process described will most likely continue for a considerable period of time. . . . The term "proletarianization" denotes a *process*. Use of the preposition "towards" was intended to indicate that the process is still continuing.

NOTES

1. Alford, R. *Health Care Politics: Ideological and Interest Group Barriers to Reform.* University of Chicago Press, Chicago, 1995.

2. Braverman, H. *Labor and Monopoly Capital.* Monthly Review Press, New York, 1974.

3. Fuchs, V. *Who Shall Live?* Basic Books, New York, 1975.

DISCUSSION QUESTIONS

1. What are the consequences of corporate run medicine for patient care? How do the changes described in this essay affect the doctor-patient relationships?

2. What are the consequences of corporate-run medicine for doctors? Have we seen a change in social status among physicians? Are they considered less prestigious than other professionals?

INFOTRAC COLLEGE EDITION

You can use your access to InfoTrac College Edition to learn more about the subjects covered in this essay. Some suggested search terms include:

bureaucratization of medicine
health maintenance organization (HMO)
managed care

specialization
universal health care

75

Urbanism as a Way of Life

LOUIS WIRTH

In 1938 Louis Wirth defined the nature of urbanism in terms of population size, density, and heterogeneity. While this formulation has been refined over the years, it nonetheless remains as a fundamental treatment and definition of urbanism.

A SOCIOLOGICAL DEFINITION OF THE CITY

Despite the preponderant significance of the city in our civilization, our knowledge of the nature of urbanism and the process of urbanization is meager, notwithstanding many attempts to isolate the distinguishing characteristics of urban life. Geographers, historians, economists, and political scientists have incorporated the points of view of their respective disciplines into diverse definitions of the city. While in no sense intended to supersede these, the formulation of a sociological approach to the city may incidentally serve to call attention to the interrelations between them by emphasizing the peculiar characteristics of the city as a particular form of human association. A sociologically significant definition of the city seeks to select those elements of urbanism which mark it as a distinctive mode of human group life. . . .

For sociological purposes a city may be defined as a relatively large, dense, and permanent settlement of socially heterogeneous individuals. On the basis of the postulates which this minimal definition suggests, a theory of urbanism may be formulated in the light of existing knowledge concerning social groups.

A THEORY OF URBANISM

Given a limited number of identifying characteristics of the city, I can better assay the consequences or further characteristics of them in the light of general sociological theory and empirical research. I hope in this manner to arrive at the essential propositions comprising a theory of urbanism. Some of these propositions can be supported by a considerable body of already available research materials; others may be accepted as hypotheses for which a certain amount of presumptive

From: *American Journal of Sociology,* 44 (July 1938): 1–24. Copyright © 1938 by The University of Chicago. All rights reserved. Reprinted with permission.

evidence exists, but for which more ample and exact verification would be required. At least such a procedure will, it is hoped, show what in the way of systematic knowledge of the city we now have and what are the crucial and fruitful hypotheses for future research.

The central problem of the sociologist of the city is to discover the forms of social action and organization that typically emerge in relatively permanent, compact settlements of large numbers of heterogeneous individuals. We must also infer that urbanism will assume its most characteristic and extreme form in the measure in which the conditions with which it is congruent are present. Thus the larger, the more densely populated, and the more heterogeneous a community, the more accentuated the characteristics associated with urbanism will be. . . .

Some justification may be in order for the choice of the principal terms comprising our definition of the city, a definition which ought to be as inclusive and at the same time as denotative as possible without unnecessary assumptions. To say that large numbers are necessary to constitute a city means, of course, large numbers in relation to a restricted area or high density of settlement. There are, nevertheless, good reasons for treating large numbers and density as separate factors, because each may be connected with significantly different social consequences. Similarly the need for adding heterogeneity to numbers of population as a necessary and distinct criterion of urbanism might be questioned, since we should expect the range of differences to increase with numbers. In defense, it may be said that the city shows a kind and degree of heterogeneity of population which cannot be wholly accounted for by the law of large numbers or adequately represented by means of a normal distribution curve. Because the population of the city does not reproduce itself, it must recruit its migrants from other cities, the countryside and—in the United States until recently—from other countries. The city has thus historically been the melting-pot of races, peoples, and cultures, and a most favorable breeding-ground of new biological and cultural hybrids. It has not only tolerated but rewarded individual differences. It has brought together people from the ends of the earth because they are different and thus useful to one another, rather than because they are homogeneous and like-minded.

A number of sociological propositions concerning the relationship between (a) numbers of population, (b) density of settlement, (c) heterogeneity of inhabitants and group life can be formulated on the basis of observation and research.

Size of the Population Aggregate

Ever since Aristotle's *Politics,* it has been recognized that increasing the number of inhabitants in a settlement beyond a certain limit will affect the relationships between them and the character of the city. Large numbers involve, as has been pointed out, a greater range of individual variation. Furthermore, the greater the number of individuals participating in a process of interaction, the greater is the *potential* differentiation between them. The personal traits, the occupations, the cultural life, and the ideas of the members of an urban community may, therefore, be expected to range between more widely separated poles than those of rural inhabitants.

That such variations should give rise to the spatial segregation of individuals according to color, ethnic heritage, economic and social status, tastes and preferences, may readily be inferred. The bonds of kinship, of neighborliness, and the sentiments arising out of living together for generations under a common folk tradition are likely to be absent or at best, relatively weak in an aggregate the members of which have such diverse origins and backgrounds. Under such circumstances competition and formal control mechanisms furnish the substitutes for the bonds of solidarity that are relied upon to hold a folk society together.

Increase in the number of inhabitants of a community beyond a few hundred is bound to limit the possibility of each member of the community knowing all the others personally. Max Weber, in recognizing the social significance of this fact, explained that from a sociological point of view large numbers of inhabitants and density of settlement mean a lack of that mutual acquaintanceship which ordinarily inheres between the inhabitants in a neighborhood. The increase in numbers thus involves a changed character of the social relationships. As Georg Simmel points out: "[If] the unceasing external contact of numbers of persons in the city should be met by the same number of inner reactions as in the small town, in which one knows almost every person he meets and to each of whom he has a positive relationship, one would be completely atomized internally and would fall into an unthinkable mental condition." The multiplication of persons in a state of interaction under conditions which make their contact as full personalities impossible produces that segmentalization of human relationships which has sometimes been seized upon by students of the mental life of the cities as an explanation for the "schizoid" character of urban personality. This is not to say that the urban inhabitants have fewer acquaintances than rural inhabitants, for the reverse may actually be true; it means rather that in relation to the number of people whom they see and with whom they rub elbows in the course of daily life, they know a smaller proportion, and of these they have less intensive knowledge.

Characteristically, urbanites meet one another in highly segmental roles. They are, to be sure, dependent upon more people for the satisfactions of their life-needs than are rural people and thus are associated with a greater number of organized groups, but they are less dependent upon particular persons, and their dependence upon others is confined to a highly fractionalized aspect of the other's round of activity. This is essentially what is meant by saying that the city is characterized by secondary rather than primary contacts. The contacts of the city may indeed be face to face, but they are nevertheless impersonal, superficial, transitory, and segmental. The reserve, the indifference, and the blasé outlook which urbanites manifest in their relationships may thus be regarded as devices for immunizing themselves against the personal claims and expectations of others.

The superficiality, the anonymity, and the transitory character of urban social relations make intelligible, also, the sophistication and the rationality generally ascribed to city-dwellers. Our acquaintances tend to stand in a relationship of utility to us in the sense that the role which each one plays in our life is overwhelmingly regarded as a means for the achievement of our own ends. Whereas the individual gains, on the one hand, a certain degree of emancipation or freedom from the personal and emotional controls of intimate groups, he loses, on the other hand, the spontaneous self-expression, the morale, and the sense of participation that comes with living in an integrated society. This constitutes essentially the state of *anomie,* or the social void, to which Durkheim alludes in attempting to account for the various forms of social disorganization in technological society.

The segmental character and utilitarian accent of interpersonal relations in the city find their institutional expression in the proliferation of specialized tasks which we see in their most developed form in the professions. The operations of the pecuniary nexus lead to predatory relationships, which tend to obstruct the efficient functioning of the social order unless checked by professional codes and occupational etiquette. The premium put upon utility and efficiency suggests the adaptability of the corporate device for the organization of enterprises in which individuals can engage only in groups. The advantage that the corporation has over the individual entrepreneur and the partnership in the urban-industrial world derives not only from the possibility it affords of centralizing the resources of thousands of individuals or from the legal privilege of limited liability and perpetual succession, but from the fact that the corporation has no soul.

The specialization of individuals, particularly in their occupations, can proceed only, as Adam Smith pointed out, upon the basis of an enlarged market, which in turn accentuates the division of labor. This enlarged market is only in part supplied by the city's hinterland; in large measure it is found among the large numbers that the city itself contains. The dominance of the city over the surrounding hinterland becomes explicable in terms of the division of labor which urban life occasions and promotes. The extreme degree of interdependence and the unstable equilibrium of urban life are closely associated with the division of labor and the specialization of occupations. This interdependence and this instability are increased by the tendency of each city to specialize in those functions in which it has the greatest advantage.

In a community composed of a larger number of individuals than can know one another intimately and can be assembled in one spot, it becomes necessary to communicate through indirect media and to articulate individual interests by a process of delegation. Typically in the city, interests are made effective through representation. The individual counts for little, but the voice of the representative is heard with a deference roughly proportional to the numbers for whom he speaks.

While this characterization of urbanism, in so far as it derives from large numbers, does not by any means exhaust the sociological inferences that might be drawn from our knowledge of the relationship of the size of a group to the characteristic behavior of the members, for the sake of brevity the assertions made may serve to exemplify the sort of propositions that might be developed.

Density

As in the case of numbers, so in the case of concentration in limited space certain consequences of relevance in sociological analysis of the city emerge. Of these only a few can be indicated.

As Darwin pointed out for flora and fauna and as Durkheim noted in the case of human societies, an increase in numbers when area is held constant (i.e., an increase in density) tends to produce differentiation and specialization, since only in this way can the area support increased numbers. Density thus reinforces the effect of numbers in diversifying men and their activities and in increasing the complexity of the social structure.

On the subjective side, as Simmel has suggested, the close physical contact of numerous individuals necessarily produces a shift in the media through which we orient ourselves to the urban milieu, especially to our fellow-men. Typically, our physical contacts are close but our social contacts are distant. The urban world puts a premium on visual recognition. We see the uniform which denotes the role of the functionaries, and are oblivious to the personal eccentricities hidden behind the uniform. We tend to acquire and develop a sensitivity to a world of artifacts, and become progressively farther removed from the world of nature.

We are exposed to glaring contrasts between splendor and squalor, between riches and poverty, intelligence and ignorance, order and chaos. The competition for space is great, so that each area generally tends to be put to the use which yields the greatest economic return. Place of work tends to become dissociated from place of residence, for the proximity of industrial and commercial establishments makes an area both economically and socially undesirable for residential purposes.

Density, land values, rentals, accessibility, healthfulness, prestige, aesthetic consideration, absence of nuisances such as noise, smoke, and dirt determine the desirability of various areas of the city as places of settlement for different sections of the population. Place and nature of work income, racial and ethnic characteristics, social status, custom, habit, taste, preference, and prejudice are among the significant factors in accordance with which the urban population is selected and distributed into more or less distinct settlements. Diverse population elements inhabiting a compact settlement thus become segregated from one another in the degree in which their requirements and modes of life are incompatible and in the measure in which they are antagonistic. Similarly, persons of homogeneous status and needs unwittingly drift into, consciously select, or are forced by circumstances into the same area. The different parts of the city acquire specialized functions, and the city consequently comes to resemble a mosaic of social worlds in which the transition from one to the other is abrupt. The juxtaposition of divergent personalities and modes of life tends to produce a relativistic perspective and a sense of toleration of differences which may be regarded as prerequisites for rationality and which lead toward the secularization of life.

The close living together and working together of individuals who have no sentimental and emotional ties foster a spirit of competition, aggrandizement, and mutual exploitation. Formal controls are instituted to counteract irresponsibility and potential disorder. Without rigid adherence to predictable routines a large compact society would scarcely be able to maintain itself. The clock and the traffic signal are symbolic of the basis of our social order in the urban world. Frequent close physical contact, coupled with great social distance, accentuates the reserve of unattached individuals toward one another and, unless compensated by other opportunities for response, gives rise to loneliness. The necessary frequent movement of great numbers of individuals in a congested habitat causes friction and irritation. Nervous tensions which derive from such personal frustrations are increased by the rapid tempo and the complicated technology under which life in dense areas must be lived.

Heterogeneity

The social interaction among such a variety of personality types in the urban milieu tends to break down the rigidity of caste lines and to complicate the class structure; it thus induces a more ramified and differentiated framework of social stratification than is found in more integrated societies. The heightened mobility of the individual, which brings him within the range of stimulation by a great number of diverse individuals and subjects him to fluctuating status in the differentiated social groups that compose the social structure of the city, brings him toward the acceptance of instability and insecurity in the world at large as a norm. This fact helps to account, too, for the sophistication and cosmopolitanism of the urbanite. No single group has the undivided allegiance of the individual. The groups with which he is affiliated do not lend themselves readily to a simple hierarchical arrangement. By virtue of his different interests arising out of different aspects of social life, the individual acquires membership in widely divergent groups, each of which functions only with reference to a single segment of his personality. Nor do these groups easily permit of a concentric arrangement so that the narrower ones fall within the circumference of the more inclusive ones, as is more likely to be the case in the rural community or in primitive societies. Rather the groups with which the person typically is affiliated are tangential to each other or intersect in highly variable fashion.

Partly as a result of the physical footlooseness of the population and partly as a result of their social mobility, the turnover in group membership generally is rapid. Place of residence, place and character of employment, income, and interests fluctuate, and the task of holding organizations together and maintaining and promoting intimate and lasting acquaintanceship between the members is difficult. This applies strikingly to the local areas within the city into which persons become segregated more by virtue of differences in race, language, income, and social status than through choice or positive attraction to people like themselves. Overwhelmingly the city-dweller is not a home-owner, and since a transitory habitat does not generate binding traditions and sentiments, only rarely is he a true neighbor.

There is little opportunity for the individual to obtain a conception of the city as a whole or to survey his place in the total scheme. Consequently he finds it difficult to determine what is to his own "best interests" and to decide between the issues and leaders presented to him by the agencies of mass suggestion. Individuals who are thus detached from the organized bodies which integrate society comprise the fluid masses that make collective behavior in the urban community so unpredictable and hence so problematical.

Although the city, through the recruitment of variant types to perform its diverse tasks and the accentuation of their uniqueness through competition and the premium upon eccentricity, novelty, efficient performance, and inventiveness, produces a highly differentiated population, it also exercises a leveling influence. Wherever large numbers of differently constituted individuals congregate, the process of depersonalization also enters. This leveling tendency inheres in part in the economic basis of the city. The development of large cities, at least in the modern age, was largely dependent upon the concentrative force of steam. The rise of the factory made possible mass production for an impersonal market. The fullest exploitation of the possibilities of the division of labor and mass production, however, is possible only with standardization of processes and products. A money economy goes hand in hand with such a system of production. Progressively as cities have developed upon a background of this system of production, the pecuniary nexus which implies the purchasability of services and things has displaced personal relations as the basis of association. Individuality under these circumstances must be replaced by categories. When large numbers have to make common use of facilities and institutions, those facilities and institutions must serve the needs of the average person rather than those of particular individuals. The services of the public utilities, of the recreational, educational, and cultural institutions, must be adjusted to mass requirements. Similarly, the cultural institutions, such as the schools, the movies, the radio, and the newspapers, by virtue of their mass clientele, must necessarily operate as leveling influences. The political process as it appears in urban life could not be understood unless one examined the mass appeals made through modern propaganda techniques. If the individual would participate at all in the social, political, and economic life of the city, he must subordinate some of his individuality to the demands of the larger community and in that measure immerse himself in mass movements. . . .

On the basis of the three variables, number, density of settlement, and degree of heterogeneity, of the urban population, it appears possible to explain the characteristics of urban life and to account for the differences between cities of various sizes and types. . . .

DISCUSSION QUESTIONS

1. According to Wirth, simply increasing the number of people in a specific area (namely, increasing population density) has many consequences for social interaction and social structure. List and talk about several of these.

2. Are you from an urban area—a city, rather than a suburb or rural area? In your opinion, does Wirth's argument apply to the area where you live?

INFOTRAC COLLEGE EDITION

You can use your access to InfoTrac College Edition to learn more about the subjects covered in this essay. Some suggested search terms include:

anomie   population heterogeneity
population density   urbanism

76

American Apartheid

DOUGLAS S. MASSEY AND NANCY A. DENTON

Douglas S. Massey and Nancy A. Denton argue that segregation, particularly residential segregation, is a fundamental dimension of race relations in the United States, and is all too often ignored by policymakers and even scholars. It is a major cause of many of the ills of race relations in this country. They argue that it is the "missing link" in past attempts to understand the urban poor.

It is quite simple. As soon as there is a group area then all your uncertainties are removed and that is, after all, the primary purpose of this Bill [requiring racial segregation in housing].

Minister of the Interior,
Union of South Africa
legislative debate on the
the Group Areas Act of 1950

During the 1970s and 1980s a word disappeared from the American vocabulary. It was not in the speeches of politicians decrying the multiple ills besetting American cities. It was not spoken by government officials

From: Douglas S. Massey and Nancy A Denton. 1993. *American Apartheid: Segregation and the Making of the Underclass.* Cambridge MA: Harvard University Press, pp. 1–7. Reprinted with permission.

responsible for administering the nation's social programs. It was not mentioned by journalists reporting on the rising tide of homelessness, drugs, and violence in urban America. It was not discussed by foundation executives and think-tank experts proposing new programs for unemployed parents and unwed mothers. It was not articulated by civil rights leaders speaking out against the persistence of racial inequality; and it was nowhere to be found in the thousands of pages written by social scientists on the urban underclass. The word was segregation.

Most Americans vaguely realize that urban America is still a residentially segregated society, but few appreciate the depth of black segregation or the degree to which it is maintained by ongoing institutional arrangements and contemporary individual actions. They view segregation as an unfortunate holdover from a racist past, one that is fading progressively over time. If racial residential segregation persists, they reason, it is only because civil rights laws passed during the 1960s have not had enough time to work or because many blacks still prefer to live in black neighborhoods. The residential segregation of blacks is viewed charitably as a "natural" outcome of impersonal social and economic forces, the same forces that produced Italian and Polish neighborhoods in the past and that yield Mexican and Korean areas today.

But black segregation is not comparable to the limited and transient segregation experienced by other racial and ethnic groups, now or in the past. No group in the history of the United States has ever experienced the sustained high level of residential segregation that has been imposed on blacks in large American cities for the past fifty years. This extreme racial isolation did not just happen; it was manufactured by whites through a series of self-conscious actions and purposeful institutional arrangements that continue today. Not only is the depth of black segregation unprecedented and utterly unique compared with that of other groups, but it shows little sign of change with the passage of time or improvements in socioeconomic status.

If policymakers, scholars, and the public have been reluctant to acknowledge segregation's persistence, they have likewise been blind to its consequences for American blacks. Residential segregation is not a neutral fact; it systematically undermines the social and economic well-being of blacks in the United States. Because of racial segregation, a significant share of black America is condemned to experience a social environment where poverty and joblessness are the norm, where a majority of children are born out of wedlock, where most families are on welfare, where educational failure prevails, and where social and physical deterioration abound. Through prolonged exposure to such an environment, black chances for social and economic success are drastically reduced.

Deleterious neighborhood conditions are built into the structure of the black community. They occur because segregation concentrates poverty to build a set of mutually reinforcing and self-feeding spirals of decline into black neighborhoods. When economic dislocations deprive a segregated group of employment and increase its rate of poverty, socioeconomic deprivation inevitably becomes more concentrated in neighborhoods where that group lives. The damaging social consequences that follow from increased poverty are spatially concentrated as well, creating uniquely disadvantaged environments that become progressively isolated—geographically, socially, and economically—from the rest of society.

The effect of segregation on black well-being is structural, not individual. Residential segregation lies beyond the ability of any individual to change; it constrains black life chances irrespective of personal traits, individual motivations, or private achievements. For the past twenty years this fundamental fact has been swept under the rug by policymakers, scholars, and theorists of the urban underclass. Segregation is the missing link in prior attempts to understand the plight of the urban poor. As long as blacks continue to be segregated in American cities, the United States cannot be called a race-blind society.

THE FORGOTTEN FACTOR

The present myopia regarding segregation is all the more startling because it once figured prominently in theories of racial inequality. Indeed, the ghetto was once seen as central to black subjugation in the United States. In 1944 Gunnar Myrdal wrote in *An American Dilemma* that residential segregation "is basic in a mechanical sense. It exerts its influence in an indirect and impersonal way: because Negro people do not live near white people, they cannot . . . associate with each other in the many activities founded on common neighborhood. Residential segregation. . . becomes reflected in uni-racial schools, hospitals, and other institutions" and creates "an artificial city . . . that permits any prejudice on the part of public officials to be freely vented on Negroes without hurting whites."

Kenneth B. Clark, who worked with Gunnar Myrdal as a student and later applied his research skills in the landmark *Brown v. Topeka* school integration case, placed residential segregation at the heart of the U.S. system of racial oppression. In *Dark Ghetto*, written in 1965, he argued that "the dark ghetto's invisible walls have been erected by the white society, by those who have power, both to confine those who have no power and to perpetuate their powerlessness. The dark ghettos are social, political, educational, and—above all—economic colonies. Their inhabitants are subject peoples, victims of the greed, cruelty, insensitivity, guilt, and fear of their masters."

Public recognition of segregation's role in perpetuating racial inequality was galvanized in the late 1960s by the riots that erupted in the nation's ghettos. In their aftermath, President Lyndon B. Johnson appointed a commission chaired by Governor Otto Kerner of Illinois to identify the causes of the violence and to propose policies to prevent its recurrence. The Kerner Commission released its report in March 1968 with the shocking admonition that the United States was "moving toward two societies, one black, one white—separate and unequal." Prominent among the causes that the commission identified for this growing racial inequality was residential segregation.

In stark, blunt language, the Kerner Commission informed white Americans that "discrimination and segregation have long permeated much of American life; they now threaten the future of every American." "Segregation and poverty have created in the racial ghetto a destructive environment totally unknown to most white Americans. What white Americans have never fully understood—but what the Negro can never forget—is that white society is deeply implicated in the ghetto. White institutions created it, white institutions maintain it, and white society condones it."

The report argued that to continue present policies was "to make permanent the division of our country into two societies; one, largely Negro and poor, located in the central cities; the other, predominantly white and affluent, located in the suburbs." Commission members rejected a strategy of ghetto enrichment coupled with abandonment of efforts to integrate, an approach they saw "as another way of choosing a permanently divided country." Rather, they insisted that the only reasonable choice for America was "a policy which combines ghetto enrichment with programs designed to encourage integration of substantial numbers of Negroes into the society outside the ghetto."

America chose differently. Following the passage of the Fair Housing Act in 1968, the problem of housing discrimination was declared solved, and residential segregation dropped off the national agenda. Civil rights leaders stopped pressing for the enforcement of open housing, political leaders increasingly debated employment and educational policies rather than housing integration, and academicians focused their theoretical scrutiny on everything from culture to family structure, to institutional racism, to federal welfare systems. Few people spoke of racial segregation as a problem or acknowledged its persisting consequences. By the end of the 1970s residential segregation became the forgotten factor in American race relations.

While public discourse on race and poverty became more acrimonious and more focused on divisive issues such as school busing, racial quotas, welfare, and affirmative action, conditions in the nation's ghettos steadily deteriorated. By the end of the 1970s, the image of poor minority families mired in an endless cycle of unemployment, unwed childbearing, illiteracy, and dependency had coalesced into a compelling and powerful concept: the urban underclass. In the view of many middle-class whiles, inner cities had come to house a large population of poorly educated single mothers and jobless men—mostly black and Puerto Rican—who were unlikely to exit poverty and become self-sufficient. In the ensuing national debate on the causes for this persistent poverty, four theoretical explanations gradually emerged: culture, racism, economics, and welfare.

Cultural explanations for the underclass can be traced to the work of Oscar Lewis, who identified a "culture of poverty" that he felt promoted patterns of behavior inconsistent with socioeconomic advancement. According to Lewis, this culture originated in endemic unemployment and chronic social immobility, and provided an ideology that allowed poor people to cope with feelings of hopelessness and despair that arose because their chances for socioeconomic success were remote. In individuals, this culture was typified by a lack of impulse control, a strong present-time orientation, and little ability to defer gratification. Among families, it yielded an absence of childhood, an early initiation into sex, a prevalence of free marital unions, and a high incidence of abandonment of mothers and children.

Although Lewis explicitly connected the emergence of these cultural patterns to structural conditions in society, he argued that once the culture of poverty was established, it became an independent cause of persistent poverty. This idea was further elaborated in 1965 by the Harvard sociologist and then Assistant Secretary of Labor Daniel Patrick Moynihan, who in a confidential report to the President focused on the relationship between male unemployment, family instability, and the intergenerational transmission of poverty, a process he labeled a "tangle of pathology." He warned that because of the structural absence of employment in the ghetto, the black family was disintegrating in a way that threatened the fabric of community life.

When these ideas were transmitted through the press, both popular and scholarly, the connection between culture and economic structure was somehow lost, and the argument was popularly perceived to be that "people were poor because they had a defective culture." This position was later explicitly adopted by the conservative theorist Edward Banfield, who argued that lower-class culture—with its limited time horizon, impulsive need for gratification, and psychological self-doubt—was primarily responsible for persistent urban poverty. He believed that these cultural traits were largely imported, arising primarily because cities attracted lower-class migrants.

The culture-of-poverty argument was strongly criticized by liberal theorists as a self-serving ideology that "blamed the victim." In the ensuing wave of reaction, black families were viewed not as weak but, on the contrary, as resilient and well adapted survivors in an oppressive and racially prejudiced society. Black disadvantages were attributed not to a defective culture but to the persistence of institutional racism in the United States. According to theorists of the underclass such as Douglas Glasgow and Alphonso Pinkney, the black urban underclass came about because deeply imbedded racist practices within American institutions—particularly schools and the economy—effectively kept blacks poor and dependent.

As the debate on culture versus racism ground to a halt during the late 1970s, conservative theorists increasingly captured public attention by focusing on a third possible cause of poverty: government welfare policy. According to Charles Murray, the creation of the underclass was rooted in the liberal welfare state. Federal antipoverty programs altered the incentives governing the behavior of poor men and women, reducing the desirability of marriage, increasing the benefits of unwed childbearing, lowering the attractiveness of menial labor, and ultimately resulted in greater poverty.

A slightly different attack on the welfare state was launched by Lawrence Mead, who argued that it was not the generosity but the permissiveness of the U.S. welfare system that was at fault. Jobless men and unwed mothers should be required to display "good citizenship" before being supported by the state. By not requiring anything of the poor, Mead argued, the welfare state undermined their independence and competence, thereby perpetuating their poverty.

This conservative reasoning was subsequently attacked by liberal social scientists, led principally by the sociologist William Julius Wilson, who had long been arguing for the increasing importance of class over race in understanding the social and economic problems facing blacks. In his 1987 book *The Truly Disadvantaged*, Wilson argued that persistent urban poverty stemmed primarily from the structural transformation of the inner-city economy. The decline of manufacturing, the suburbanization of employment, and the rise of a low-wage service sector dramatically reduced the number of city jobs that paid wages sufficient to support a family, which led to high rates of joblessness among minorities and a shrinking pool of "marriageable" men (those financially able to support a family). Marriage thus became less attractive to poor women, unwed childbearing increased, and female-headed families proliferated. Blacks suffered disproportionately from these trends because, owing to past discrimination, they were concentrated in locations and occupations particularly affected by economic restructuring.

Wilson argued that these economic changes were accompanied by an increase in the spatial concentration of poverty within black neighborhoods. This new geography of poverty, he felt, was enabled by the civil rights revolution of the 1960s, which provided middle-class blacks with new opportunities outside the ghetto. The out-migration of middle-class families from ghetto areas left behind a destitute community lacking the institutions, resources, and values necessary for success in postindustrial society. The urban underclass thus arose from a complex interplay of civil rights policy, economic restructuring, and a historical legacy of discrimination.

Theoretical concepts such as the culture of poverty, institutional racism, welfare disincentives, and structural economic change have all been widely debated. None of these explanations, however, considers residential segregation to be an important contributing cause of urban poverty and the underclass. In their principal works, Murray and Mead do not mention segregation at all and Wilson refers to racial segregation only as a historical legacy from the past, not as an outcome that is institutionally supported and actively created today. Although Lewis mentions segregation sporadically in his writings, it is not assigned a central role in the set of structural factors responsible for the culture of poverty, and Banfield ignores it entirely. Glasgow, Pinkney, and other theorists of institutional racism mention the ghetto frequently, but generally call not for residential desegregation but for race-specific policies to combat the effects of discrimination in the schools and labor markets. In general, then, contemporary theorists of urban poverty do not see high levels of black-white segregation as particularly relevant to understanding the underclass or alleviating urban poverty.

The purpose of this [argument] is to redirect the focus of public debate back to issues of race and racial segregation and to suggest that they should be fundamental to thinking about the status of black Americans and the origins of the urban underclass. Our quarrel is less with any of the prevailing theories of urban poverty than with their systematic failure to consider the important role that segregation has played in mediating, exacerbating, and ultimately amplifying the harmful social and economic processes they treat.

We join earlier scholars in rejecting the view that poor urban blacks have an autonomous "culture of poverty" that explains their failure to achieve socioeconomic success in American society. We argue instead that residential segregation has been instrumental in creating a structural niche within which a deleterious set of attitudes and behaviors—a culture of segregation—has arisen and flourished. Segregation created the structural conditions for the emergence of an oppositional culture that devalues work, schooling, and marriage and that stresses attitudes and behaviors that are antithetical and often hostile to success in the larger economy. Although poor black neighborhoods still contain many people who lead conventional, productive lives, their example has been overshadowed in recent years by a growing concentration of poor, welfare-dependent families that is an inevitable result of residential segregation.

We readily agree with Douglas, Pinkney, and others that racial discrimination is widespread and may even be institutionalized within large sectors of American society, including the labor market, the educational system, and the welfare bureaucracy. We argue, however, that this view of black subjugation is incomplete without understanding the special role that residential segregation plays in enabling all other forms of racial oppression. Residential segregation is the institutional apparatus that supports other racially discriminatory processes and binds them together into a coherent and uniquely effective system of racial subordination. Until the black ghetto is dismantled as a basic institution of American urban life, progress ameliorating racial inequality in other arenas will be slow, fitful, and incomplete.

We also agree with William Wilson's basic argument that the structural transformation of the urban economy undermined economic supports for the black community during the 1970s and 1980s. We argue, however, that in the absence of segregation, these structural changes would not have produced the disastrous social and economic outcomes observed in inner cities during these decades. Although rates of black poverty were driven up by the economic dislocations Wilson identifies, it was segregation that confined the increased deprivation to a small number of densely settled, tightly packed, and geographically isolated areas.

Wilson also argues that concentrated poverty arose because the civil rights revolution allowed middle-class blacks to move out of the ghetto. Although we remain open to the possibility that class-selective migration did occur, we argue that concentrated poverty would have happened during the 1970s with or without black middle-class migration. Our principal objection to Wilson's focus on middle-class out-migration is not that it did not occur, but that it is misdirected: focusing on the flight of the black middle class deflects attention from the real issue, which is the limitation of black residential options through segregation.

Middle-class households—whether they are black, Mexican, Italian, Jewish, or Polish—always try to escape the poor. But only blacks must attempt their escape within a highly segregated, racially segmented housing market. Because of segregation, middle-class blacks are less able to escape than other groups, and as a result are exposed to more poverty. At the same time, because of segregation no one will move into a poor black neighborhood except other poor blacks. Thus both middle-class blacks and poor blacks lose compared with the poor and middle class of other groups: poor blacks live under unrivaled concentrations of poverty and affluent blacks live in neighborhoods that are far less advantageous than those experienced by the middle class of other groups.

Finally, we concede Murray's general point that federal welfare policies are linked to the rise of the urban underclass, but we disagree with his specific hypothesis that generous welfare payments, by themselves, discouraged employment, encouraged unwed childbearing, undermined the strength of the family, and thereby caused persistent poverty. We argue instead that welfare payments were only harmful to the socioeconomic well-being of groups that were residentially segregated. As poverty rates rose among blacks in response to the economic dislocations of the 1970s and 1980s, so did the use of welfare programs. Because of racial segregation, however, the higher levels of welfare receipt were confined to a small number of isolated, all-black neighborhoods. By promoting the spatial concentration of welfare use, therefore, segregation created a residential environment within which welfare dependency was the norm, leading to the intergenerational transmission and broader perpetuation of urban poverty. . . .

Our fundamental argument is that racial segregation—and its characteristic institutional form, the black ghetto—are the key structural factors responsible for the perpetuation of black poverty in the United States. Residential segregation is the principal organizational feature of American society that is responsible for the creation of the urban underclass. . . .

DISCUSSION QUESTIONS

1. Regardless of your race or ethnicity, did you grow up in a racially segregated environment? How central in your life was this fact? What consequences did it have? If you know anyone who did, what in your estimation were the effects on their life?

2. What is the "culture of poverty" view? Do you agree with it? What do Massey and Denton have to say about it?

INFOTRAC COLLEGE EDITION

You can use your access to InfoTrac College Edition to learn more about the subjects covered in this essay. Some suggested search terms include:

"culture of poverty"
hypersegregation

residential segregation
institutional racism

77

Black, Brown, Red, and Poisoned

REGINA AUSTIN AND MICHAEL SCHILL

The principle of environmental racism states that race, more so than class (socioeconomic status), explains the unfortunate residential closeness of people of color to toxic waste dumps, chemical plants, oil refineries, incinerators, and other toxic sources. Here Regina Austin and Michael Schill review the issue of environmental racism as well as activist strategies to combat it.

People of color throughout the United States are receiving more than their fair share of the poisonous fruits of industrial production. They live cheek by jowl with waste dumps, incinerators, landfills, smelters, factories, chemical plants, and oil refineries whose operations make them sick and kill them young. They are poisoned by the air they breathe, the water they drink, the fish they catch, the vegetables they grow, and, in the case of children, the very ground they play on. Even the residents of some of the most remote rural hamlets of the South and Southwest suffer from the ill effects of toxins.

This [essay] examines some of the reasons why communities of color bear a disparate burden of pollution. It also brings into focus the commonality of their struggles and some strategies that are useful in overcoming environmental injustice.

THE PATH OF LEAST RESISTANCE

The disproportionate location of sources of toxic pollution in communities of color is the result of various development patterns. In some cases, the residential communities where people of color now live were originally the homes of whites who worked in the facilities that generate toxic emissions. The housing and the industry sprang up roughly simultaneously. Whites vacated the housing (but not necessarily the jobs) for better shelter as their socioeconomic status improved, and poorer black and brown folks who enjoy much less residential mobility took their place. In other cases, housing for African Americans and Latino Americans was built in the vicinity of existing industrial operations because the land was cheap and the people were poor. For example, Richmond, California, was developed downwind from a Chevron oil refinery when African Americans migrated to the area to work in shipyards during World War II.

In yet a third pattern, sources of toxic pollution were placed in existing minority communities. The explanations for such sitings are numerous; some reflect the

From: Robert D. Bullard, ed. 1994. *Unequal Protection: Environmental Justice and Communities of Color.* San Francisco: Sierra Club Books, pp. 53–73. Reprinted with permission.

impact of racial and ethnic discrimination. The impact, of course, may be attenuated and less than obvious. The most neutral basis for a siting choice is probably the natural characteristics of the land, such as mineral content of the soil. . . . Low population density would appear to be a similar criterion. It has been argued, however, that in the South, a sparse concentration of inhabitants is correlated with poverty, which is in turn correlated with race. "It follows that criteria for siting hazardous waste facilities which include density of population will have the effect of targeting rural black communities that have high rates of poverty."

Likewise, the compatibility of pollution with preexisting uses might conceivably make some sites more suitable than others for polluting operations. Pollution tends to attract other sources of pollutants, particularly those associated with toxic disposal. For example, Chemical Waste Management, Inc. (Chem Waste) has proposed the construction of a toxic waste incinerator outside of Kettleman City, California, a community composed largely of Latino farm workers. Chem Waste also has proposed to build a hazardous waste incinerator in Emelle, a predominantly African American community located in the heart of Alabama's "black belt." The company already has hazardous waste landfills in Emelle and Kettleman City.

According to the company's spokeswoman, Chem Waste placed the landfill in Kettleman City "because of the area's geological features. Because the landfill handles toxic waste, . . . it is an ideal spot for the incinerator"; the tons of toxic ash that the incinerator will generate can be "contained and disposed of at the installation's landfill." Residents of Kettleman City face a "triple whammy" of threats from pesticides in the fields, the nearby hazardous waste landfill, and a proposed hazardous waste incinerator. This case is not unique.

After reviewing the literature on hazardous waste incineration, one commentator has concluded that "[m]inority communities represent a 'least cost' option for waste incineration . . . because much of the waste to be incinerated is already in these communities." Despite its apparent neutrality, then, siting based on compatibility may be related to racial and ethnic discrimination, particularly if such discrimination influenced the siting of preexisting sources of pollution.

Polluters know that communities of low-income and working-class people with no more than a high school education are not as effective at marshaling opposition as communities of middle- or upper-income people. People of color in the United States have traditionally had less clout with which to check legislative and executive abuse or to challenge regulatory laxity. Private corporations, moreover, can have a powerful effect on the behavior of public officials. Poor minority people wind up the losers to them both.

People of color are more likely than whites to be economically impoverished, and economic vulnerability makes impoverished communities of color prime targets for "risky" technologies. Historically, these communities are more likely than others to tolerate pollution-generating commercial development in the hope that economic benefits will inure to the community in the form of jobs, increased taxes, and civic improvements. Once the benefits start to flow, the community may be reluctant to forgo them even when they are accompanied by poisonous spills or emissions. This was said to be the case in Emelle, in Sumter County, Alabama, site of the nation's largest hazardous waste landfill.

Sumter County's population is roughly 70 percent African American, and 30 percent of its inhabitants fall below the poverty line. Although the landfill was apparently leaking, it was difficult to rally support against the plant among African American politicians because its operations contributed an estimated $15.9 million to the local economy in the form of wages, local purchases of goods and services, and per-ton landfill user fees.

Of course, benefits do not always materialize after the polluter begins operations. For example, West Harlem was supposed to receive, as a trade-off for accepting New York City's largest sewage treatment plant, an elaborate state park to be built on the roof of the facility. The plant is functioning, fouling the air with emissions of hydrogen sulfide and promoting an infestation of rats and mosquitoes. The park, however, has yet to be completed, the tennis courts have been removed from the plan completely, and the "first-rate" restaurant has been scaled down to a pizza parlor.

In other cases, there is no net profit to distribute among the people. New jobs created by the poisonous enterprises are "filled by highly skilled labor from outside the community," while the increased tax revenues go not to "social services or other community development projects, but . . . toward expanding the infrastructure to better serve the industry."

Once a polluter has begun operations, the victims' options are limited. Mobilizing a community against an existing polluter is more difficult than organizing opposition to a proposed toxic waste-producing activity. Resignation sets in, and the resources for attacking ongoing pollution are not as numerous, and the tactics not as potent, as those available during the proposal stage. Furthermore, though some individuals are able to escape toxic poisoning by moving out of the area, the flight of others will be blocked by limited incomes, housing discrimination, and restrictive land use regulations.

THREAT TO BARRIOS, GHETTOS, AND RESERVATIONS

Pollution is no longer accepted as an unalterable consequence of living in the "bottom" (the least pleasant, poorest area minorities can occupy) by those on the bottom of the status hierarchy. Like anybody else, people of color are distressed by accidental toxic spills, explosions, and inexplicable patterns of miscarriages and cancers, and they are beginning to fight back, from Maine to Alaska.

To be sure, people of color face some fairly high barriers to effective mobilization against toxic threats, such as limited time and money; lack of access to technical, medical, and legal expertise; relatively weak influence in political and media circles; and ideological conflicts that pit jobs against the environment. Limited fluency in English and fear of immigration authorities will keep some of those affected, especially Latinos, quiescent. Yet despite the odds, poor minority people are responding to their poisoning with a grass-roots movement of their own.

Activist groups of color are waging grass-roots environmental campaigns all over the country. Although they are only informally connected, these campaigns reflect certain shared characteristics and goals. The activity of activists of color is indicative of a grassroots movement that occupies a distinctive position relative to both the mainstream movement and the white grass-roots environmental movement. The environmental justice movement is antielitist and antiracist. It capitalizes on the social and cultural differences of people of color as it cautiously builds alliances with whites and persons of the middle class. It is both fiercely environmental *and* conscious of the need for economic development in economically disenfranchised communities. Most distinctive of all, this movement has been extremely outspoken in challenging the integrity and bona fides of mainstream establishment environmental organizations.

 People of color have not been mobilized to join grass-roots environmental campaigns because of their general concern for the environment. Characterizing a problem as being "environmental" may carry weight in some circles, but it has much less impact among poor minority people. It is not that people of color are uninterested in the environment—a suggestion the grass-roots activists find insulting. In fact they are more likely to be concerned about pollution than are people who are wealthier and white. Rather, in the view of many people of color, environmentalism is associated with the preservation of wildlife and wilderness, which simply is not more important than the survival of people and the communities in which they live; thus, the mainstream movement has its priorities skewed. . . .

CAPITALIZING ON THE RESOURCES
OF COMMON CULTURE

For people of color, social and cultural differences such as language are not handicaps but the communal resources that facilitate mobilization around issues like toxic poisoning. As members of the same race, ethnicity, gender, and even age cadre, would-be participants share cultural traditions, modes, and mores that encourage cooperation and unity. People of color may be more responsive to organizing efforts than whites because they already have experience with collective action through community groups and institutions such as churches, parent-teacher associations, and town watches or informal social networks. Shared criticisms of racism, a distrust of corporate power, and little expectation that government will be responsive to their complaints are common sentiments in communities of color and support the call to action around environmental concerns.

 Grass-roots environmentalism is also fostered by notions that might be considered feminist or womanist. Acting on a realization that toxic poisoning is a threat to home and family, poor minority women have moved into the public realm to confront corporate and government officials whose modes of analysis reflect patriarchy, white supremacy, and class and scientific elitism. There are numerous examples of women of color whose strengths and talents have made them leaders of grass-roots environmental efforts.

 The organization Mothers of East Los Angeles (MELA) illustrates the link between group culture and mobilization in the people of color grass-roots environmental movement. Persistent efforts by MELA defeated proposals for constructing a state prison and a toxic waste incinerator in the group's mostly Latino American neighborhood in East Los Angeles.

Similarly, the Lumbee Indians of Robeson County, North Carolina, who attach spiritual significance to a river that would have been polluted by a hazardous waste facility proposed by the GSX Corporation, waged a campaign against the facility on the ground of cultural genocide. Throughout the campaign, "Native American dance, music, and regalia were used at every major public hearing. Local Lumbee churches provided convenient meeting locations for GSX planning sessions. Leaflet distribution at these churches reached significant minority populations in every pocket of the county's nearly 1,000 square miles."

Concerned Citizens of Choctaw defeated a plan to locate a hazardous waste facility on their lands in Philadelphia, Mississippi. The Good Road Coalition, a grassroots Native American group based on the Rosebud Reservation in South Dakota, defeated plans by a Connecticut-based company to build a 6,000-acre garbage landfill on the Rosebud. Local residents initiated a recall election, defeating several tribal council leaders and the landfill proposal. The project, dubbed "dances with garbage," typifies the lengths that the Lakota people and other Native Americans will go to preserve their land—which is an essential part of their religion and culture.

Consider, finally, the Toxic Avengers of El Puente, a group of environmental organizers based in the Williamsburg section of Brooklyn, New York. The name is taken from the title of a horror movie. The group attacks not only environmental racism but also adultism and adult superiority and privilege. The members, whose ages range from nine to twenty-eight, combine their activism with programs to educate themselves and others about the science of toxic hazards.

The importance of culture in the environmental justice movement seems not to have produced the kind of distrust and misgivings that might impede interaction with white working-class and middle-class groups engaged in grass-roots environmental activism. There are numerous examples of ethnic-based associations working in coalitions with one another, with majority group associations, and with organizations from the mainstream. There are also localities in which the antagonism and suspicion that are the legacy of white racism have kept whites and African Americans from uniting against a common toxic enemy. The link between the minority groups and the majority groups seems grounded in material exchange, not ideological fellowship. The white groups attacking toxins at the grass-roots level have been useful sources of financial assistance and information about tactics and goals. . . .

BRIDGING THE JUSTICE-ENVIRONMENT GAP

At the same time that environmental justice activists are battling polluters, some are engaged on another front in a struggle against elitism and racism that exist within the mainstream environmental movement. There are several substantive points of disagreement between grass-roots groups of color and mainstream environmental organizations. First, communities of color are tired of shouldering the fallout from environmental regulation. A letter sent to ten of the establishment environmental organizations by the Southwest Organizing Project and numerous activists of color engaged in the grass-roots environmental struggle illustrates the level of exasperation:

> Your organizations continue to support and promote policies which emphasize the clean-up and preservation of the environment on the backs of working people in general and people of color in particular. In the name of eliminating environmental hazards at any cost, across the country industrial and other economic activities which employ us are being shut down, curtailed or prevented while our survival needs and cultures are ignored. We suffer the end results of these actions, but are never full participants in the decision-making which leads to them. . . .

Another threat to communities of color is the growing popularity of NIMBY (not in my backyard) groups. People of color have much to fear from these groups because their communities are the ones most likely to lose the contests to keep the toxins out. The grass-roots environmentalists argue that rather than trying to bar polluters, who will simply locate elsewhere, energies should be directed at bringing the amount of pollution down to zero. In lieu of NIMBY, mainstream environmentalists should be preaching NIABY (not in anyone's backyard).

Finally, conservation organizations are making "debt-for-nature" swaps throughout the so-called Third World. Through swaps, conservation organizations procure ownership of foreign indebtedness (either by gift or by purchase at a reduced rate) and negotiate with foreign governments for reduction of the debt in exchange for land. Grass-roots environmental activists of color complain that these deals, which turn conservation organizations into creditors of so-called Third World peoples, legitimate the debt and the exploitation on which it is based. . . .

DISCUSSION QUESTIONS

1. In your opinion, is environmental racism (the closeness of people of color to toxic waste sites) mostly because of race, class (socioeconomic status), or both, or some other reason all together? Discuss.

2. Assume that you are asked to organize a group to combat environmental racism in or near your own hometown or city. What kinds of issues would you focus on and how might you begin to go about organizing such a group?

INFOTRAC COLLEGE EDITION

You can use your access to InfoTrac College Edition to learn more about the subjects covered in this essay. Some suggested search terms include:

environmental racism/environmental justice   pollution
hazardous waste   toxic waste dumps

78

The New Feminist Movement

VERTA TAYLOR AND NANCY WHITTIER

Reviewing the development of feminism as a social movement, Verta Taylor and Nancy Whittier identify the different ideologies and structures of the women's movement. They argue that the women's movement has entered a period of abeyance at the same time that it is profoundly affecting the consciousness, culture, and structure of U.S. society.

From a social movement perspective, women in the United States have always had sufficient grievances to create the context for feminist activity. Indeed, instances of collective action on the part of women abound in history, especially if one includes female reform societies, women's church groups, alternative religious societies, and women's clubs. However, collective activity on the part of women directed specifically toward improving their own status has flourished primarily in periods of generalized social upheaval, when sensitivity to moral injustice, discrimination, and social inequality has been widespread in the society as a whole (Chafe 1977). The first wave of feminism in this country grew out of the abolitionist struggle of the 1830s and peaked during an era of social reform in the 1890s, and the contemporary movement emerged out of the general social discontent of the 1960s. Although the women's movement did not die between these periods of heightened activism, it varied greatly in form and intensity.

Structural conditions underlie the emergence of protest (Oppenheimer 1973; Huber & Spitze 1983; Chafetz & Dworkin 1986). Chafetz and Dworkin (1986), for example, propose that as industrialization and urbanization bring greater education for women, expanding public roles create role and status conflicts for middle class women, who then develop the discontent and gender consciousness necessary for a women's movement. The changing shape and size of women's movements depend on the opportunities for women to organize on their own behalf, the resources available to them, and their collective identity and interpretation of their grievances. All of these vary for different groups of women and at different times and places (West & Blumberg 1990). Thus, it is to be expected that the ideology, structure, and strategies adopted by the women's movement are quite different depending on when, where, and by what groups of women it is organized. . . .

From: Verta Taylor and Nancy Whittier. "The New Feminist Movement."

IDEOLOGY

While ideas do not necessarily cause social movements, ideology is a central component in the life of any social movement. The new feminist movement, like most social movements, is not ideologically monolithic. In the 1970s, it was commonplace to characterize the women's movement as consisting of three ideological strands: liberal feminism, socialist feminism, and radical feminism (Ryan 1989; Buechler 1990). Feminist ideology today continues to be a mix of several orientations that differ in the scope of change sought, the extent to which gender inequality is linked to other systems of domination, especially class, race/ethnicity, and sexuality, and the significance attributed to gender differences. Our analysis here explores the diversity of feminism by focusing on the evolution of the dominant ideologies that have motivated participants in the two major branches of the new feminist movement from its inception, liberal feminism and radical feminism.

The first wave of the women's movement was by and large a liberal feminist reform movement. It asked for equality within the existing social structure and, indeed, in many ways functioned like other reform movements to reaffirm existing values within the society (Ferree & Hess 1985). Nineteenth-century feminists believed that if they obtained the right to an education, the right to own property, the right to vote, employment rights—in other words, equal civil rights under the law—they would attain equality with men.

The basic ideas identified with contemporary liberal or "mainstream" feminism have changed little since their formulation in the nineteenth century when they seemed progressive, even radical (Eisenstein 1981). Contemporary liberal feminist ideology holds that women lack power simply because we are not, as women, allowed equal opportunity to compete and succeed in the male-dominated economic and political arenas but, instead, are relegated to the subordinate world of home, domestic labor, motherhood, and family. Its major strategy for change is to gain legal and economic equalities and to obtain access to elite positions in the workplace and in politics while, at the same time, making up for the fact that women's starting place in the "race of life" is unequal to men's (Eisenstein 1981). Thus, liberal feminists tend to place as much emphasis on changing individual women as they do on changing society. . . .

In sum, the liberal concept of equality involves equality under the law, in the workplace, and in the public arena, and fails to recognize how deeply women's inequality is rooted in their responsibility for the care of children and the home (Hartmann 1981; Huber & Spitze 1983), their dependence on men in the context of traditional heterosexual marriage (Rich 1980), and in men's use of their dominant status to preserve male advantage by establishing the rules that distribute rewards (Reskin 1988). . . .

Radical feminist ideology dates to Simone de Beauvoir's early 1950s theory of "sex class," which was developed further in the late 1960s among small groups of radical women who fought the subordination of women's liberation within the New Left, and which flourished in the feminist movement in the 1970s (Beauvior 1952; Firestone 1970; Millett 1971; Atkinson 1974; Rubin 1975; Rich 1976, 1980; Griffin 1978; Daly 1978; Eisenstein 1981; Hartmann 1981; Frye 1983;

Hartsock 1983; MacKinnon 1983). The radical approach recognizes women's identity and subordination as a "sex class," views gender as the primary contradiction and foundation for the unequal distribution of a society's rewards and privileges, and recasts relations between women and men in political terms (Echols 1989). Defining women as a "sex class" means no longer treating patriarchy in individual terms but acknowledging the social and structural nature of women's subordination. Radical feminists hold that in all societies, institutions and social patterns are structured to maintain and perpetuate gender inequality and that female disadvantage permeates virtually all aspects of sociocultural and personal life. Further, through the gender division of labor, social institutions are linked so that male superiority depends upon female subordination (Acker 1980; Hartmann 1981; Chafetz 1990). In the United States, as in most industrialized societies, power, prestige, and wealth accrue to those who control the distribution of resources outside the home in the economic and political spheres. The sexual division of labor that assigns childcare and domestic responsibilities to women not only ensures gender inequality in the family system but perpetuates male advantage in political and economic institutions as well.

In contrast to the liberal position, radical feminists do not deny that men are privileged as men and that they benefit as a group from their privilege—not just in the public arena but also in relation to housework, reproduction, sexuality, and marriage. Rather, they view patriarchy as a system of power that structures and sustains male advantage in every sphere of life, in economic, political, and family institutions, in the realms of religion, law, science and medicine, and in the interactions of everyday life. To unravel the complex structure on which gender inequality rests requires, from a radical feminist perspective, a fundamental transformation of all institutions in society and the existing relations among them. To meet this challenge, radical feminists have formulated influential critiques of the family, marriage, love, motherhood, heterosexuality, rape, battering and other forms of sexual violence, capitalism, the medicalization of childbirth, reproductive policies, the media, science, language and culture, the beauty industry, politics and the law, and technology and its impact on the environment. Thus, radical feminism is a transformational politics engaged in a fight against female disadvantage and the masculinization of culture. Its ultimate vision is revolutionary in scope: a fundamentally new social order that eliminates the sex-class system and replaces it with new ways of defining and structuring experience. . . .

STRUCTURE

Social movements do not generally have a single, central organization or unified direction. Rather, the structure of any general and broad-based social movement is more diffuse—composed of a number of relatively independent organizations that differ in ideology, structure, goals, and tactics—is characterized by decentralized leadership, and is loosely connected by multiple and overlapping memberships, friendship networks, and cooperation in working toward common goals (Gerlach & Hine 1970). The organizational structure of the new feminist movement has conformed to this model from its beginnings (Freeman 1975; Cassell 1977). While the movement as a whole is characterized by a decentralized structure, the various organizations that comprise it vary widely in structure. The diversity of feminist organizational forms reflects both ideological differences and, as Freeman (1979) points out, the movement's diverse membership base (for example, differences in members' prior organizational expertise, experience in other movements, expectations, social status, age or generation, and relations with different target groups).

There have been two main types of organizational structure in the new feminist movement since its resurgence, reflecting the two main sources of feminist organizing in the late 1960s: bureaucratically structured movement organizations with hierarchical leadership and democratic decision-making procedures, such as the National Organization for Women (NOW); and smaller collectively structured groups that formed a more diffuse social movement community held together by a feminist political culture. It is important to recognize, however, that while these two strands emerged separately, they have not remained distinct and opposed to each other. On the contrary, the two structures have converged as bureaucratic organizations adopted some of the innovations of collectivism, and feminist collectives became more formally structured (Staggenborg 1988, 1989; Martin 1990; Ryan 1992; Whittier 1995). In addition, many individual activists are involved in a variety of organizations with differing structures. . . .

FROM HEYDAY TO ABEYANCE

The early 1980s saw a rapid decrease in the number of feminist organizations and a transformation in the form and activities of the women's movement. In part, this was a response to the successes of the New Right: so powerful were antifeminist sentiments and forces that members of a major political party, the Republican party, were elected in 1980 on a platform developed explicitly to "put women back in their place." After forty years of faithful support of the ERA, the Republican party dropped it from its platform, called for a constitutional amendment to ban abortion, and aligned itself with the economic and social policies of the New Right. After the election of the conservative Reagan administration in 1980, federal funds and grants were rarely available to feminist service organizations, and because other social service organizations were also hard hit by budget cuts, competition increased for relatively scarce money from private foundations. As a result, many feminist programs such as rape crisis centers, shelters for battered women, and job training programs were forced to close or limit their services. . . .

The women's movement has suffered not only from opposition, but also from its apparent success. Overt opposition to the feminist movement had been muted in the mid-to late 1970s. Elites in politics, education, and industry gave the appearance of supporting feminist aims through largely ineffectual affirmative action programs and the appointment of a few token women to high positions in their respective areas. Meanwhile, the popular image of feminism advanced by the mass media suggested that the women's movement had won its goals, making feminism an anachronism. Despite the real-life difficulties women encountered trying to balance paid employment and a "second shift" of housework and child-care (Hochschild 1989), the image of the working woman became the feminine ideal. The public discourse implied that since women had already achieved equality with men, they no longer needed a protest movement, unless they happened to be lesbians and man haters. Both popular and scholarly writers, in short, declared the 1980s and 90s a "post-feminist" era.

We are suggesting here that the women's movement of the late 1980s and 90s is in abeyance. Verta Taylor (1989), building on research on the women's movement of the 1950s (Rupp & Taylor 1987), suggests that movements adopt abeyance structures in order to survive in hostile political climates. Movements in abeyance are in a "holding pattern," during which activists from an earlier period maintain the ideology and structural base of the movement, but few new recruits join. A movement in abeyance is primarily oriented toward maintaining itself rather than confronting the established order directly. Focusing on building an alternative culture, for example, is a means of surviving when external resources are not available and the political structure is not amenable to challenge. The structure of the women's movement has changed as mass mobilization and confrontation of the social system have declined. Nevertheless, feminist resistance continues in different forms. Patricia Hill-Collins suggests that resistance can occur at three levels: the individual level of consciousness, the cultural level, and the social structural level (1990:227). This conceptualization allows us to recognize that protest takes many forms and to acknowledge the role of social movements in changing consciousness and culture.

The women's movement has sought to make profound changes in the lives of women. At the level of consciousness and individual actions, women who were active in the women's movement of the 1960s and 70s have continued to shape their lives around their feminist beliefs in the 1980s and 90s, even when they are not involved in organized feminist activity. For example, many feminists hold jobs in government, social service organizations, or Women's Studies and other academic programs that allow them to incorporate their political goals into their work, and continue to choose leisure activities, significant relationships, and dress and presentation of self that are consistent with feminist ideology (Whittier 1995). The consciousness and lives of women who do not identify as feminist have also been altered by the women's movement. . . .

At the cultural level, the feminist social movement community has continued to thrive into the 1990s, with such events as, for example, an "annual multicultural multiracial conference on aging" for lesbians (*Off Our Backs* 1991:12), feminist cruises, several annual women's music festivals and a women's comedy festival in different parts of the country. Gatherings and conferences in 1990 included groups such as Jewish lesbian daughters of holocaust survivors, women motorcyclists, fat dykes, practitioners of Diannic Wicca, Asian lesbians, practitioners of herbal medicine, and survivors of incest. Newsletters and publications exist for groups including women recovering from addictions, women's music profession-

als and fans, lesbian separatists, disabled lesbians, lesbian couples, feminists interested in sadomasochism, feminists opposed to pornography and a multitude of
others. The growth of the feminist community underscores the flowering of lesbian feminism in the late 1980s and 90s. A wide variety of lesbian and lesbian
feminist books and anthologies have been published on topics ranging from lesbian feminist ethics, to separatism, to sexuality, to commitment ceremonies for
lesbian couples (see, for example, Hoagland 1988, Hoagland & Penelope 1988,
Loulan 1990, Butler 1991), reflecting diverse perspectives that have been hotly
debated in the pages of lesbian publications and at conferences and festivals. . . .

   At the social structural level, the feminist movement has not been unrespon-
sive to the conservative backlash. In fact, the gains of the New Right in the late
1980s sparked some of the largest feminist demonstrations and actions in years. In
April, 1989, NOW and abortion rights groups organized a national demonstration in Washington, DC, that drew between 300,000 and 600,000 women and
men to protest restrictions on abortion. Additional national and local demonstrations followed, and pro-choice activists organized electoral lobbying, defense of
abortion clinics, and conferences, and attempted to form coalitions across racial
and ethnic lines and among women of different ages (Staggenborg 1991; Ryan
1992). The National Abortion Rights Action League (NARAL) experienced a
growth in membership from 200,000 in 1989 to 400,000 in 1990 (Staggenborg
1991: 138), and membership in NOW also continued to grow in the late 1980s
and 90s after a decline in the early 1980s, with a membership of 250,000 in 1989.
In addition, a wide variety of feminist organizations continue to pursue social
change at state and local levels (Ferree & Martin 1995).

   The women's movement has also had a substantial impact on other social
movements of the 1980s and 90s. Movements such as the gay and lesbian movement, AIDS movement, recovery from addictions, New Age spirituality, and the
animal rights movement have been profoundly influenced by feminist values and
ideology, including the emphasis on collective structure and consensus, the notion of the personal as political, goddess-worship, and the critique of patriarchal
mistreatment of animals and ecological resources. The women's movement also
trained a large number of feminist activists in the 1970s, particularly lesbians, who
have participated in new social movements and integrated feminism into them
(Cavin 1990; Whittier 1995). For example, the gay and lesbian movement has
begun expanding its health concerns to include breast cancer as well as AIDS and
has adopted strategies of the feminist anti-rape movement to confront violence
against gays and lesbians. . . .

   Given the scope and size of women's movement activity in the late 1980s and
90s, why do we argue that the movement is in abeyance? We think that, although
the movement may resurge in the mid 1990s, the level of mass mobilization and
confrontation of the social structural system clearly declined following 1982. Because feminism in the late 80s and 90s is focused more on consciousness and culture and has established roots in other social movements of the period, feminist
protest is less visible than it was during the heyday of the women's movement.
Notably, in keeping with the patterns that characterize movements in abeyance
(Taylor 1989), the most active feminists in the late 1980s and 90s have been

women who became involved with the movement during the late 1960s and 1970s, were transformed by their involvement, and formed a lasting commitment to feminist goals. Despite support for feminist goals, many young women do not identify themselves as feminists, apparently because the identity of feminist is stigmatized. A feminist is seen as someone who deviates from gender norms by being unattractive, aggressive, hostile to men, opposed to marriage and motherhood, lesbian, and seeking to imitate men (Schneider 1988, Dill unpublished). Despite the gains made by women in some areas, gender norms are still so rigid and deeply internalized that they successfully deter many women who otherwise support the feminist agenda from participating in the movement.

Yet, some younger women have joined the women's movement in the late 1980s and 90s despite the risks entailed in identifying with a stigmatized and unpopular cause. In a study of young feminist activists in the 1990s, Kim Dill has found that a new generation of women has been recruited to feminism primarily through Women's Studies courses and through the transmission of feminism from mothers to daughters (unpublished). In short, the institutionalized gains of the heyday of feminist activism in the 1970s are enabling the women's movement to survive and to spread its ideology to new recruits.

The history of the women's movement, and its present survival despite the challenges it has confronted from within its own ranks and from a conservative political climate, suggest that because feminism is a response to the fundamental social cleavage of gender it will continue to exist (Taylor & Rupp 1993). As one generation of feminists fades from the scene with its ultimate goals unrealized, another takes up the challenge (Rossi 1982). But each new generation of feminists does not simply carry on where the previous generation left off. Rather, it speaks for itself and defines its own objectives and strategies, often to the dismay and disapproval of feminists from earlier generations. . . . As Myra Ferree and Beth Hess (1985:182) point out, "feminism is not simply a form of received wisdom" but something that evolves with each new cycle of feminist activism. Both continuity and change, then, will characterize the feminism of the twenty-first century.

REFERENCES

Acker, J. 1980. "Women and stratification: A review of recent literature." *Contemporary Sociology* 9:25–35.

Atkinson, T. G. 1974. *Amazon odyssey.* New York: Links.

Beauvoir, S. de. 1952. *The second sex.* New York: Bantam.

Buechler, Steven M. 1990. *Women's movements in the United States.* New Brunswick, NJ: Rutgers.

Butler, Becky. 1991. *Ceremonies of the heart: Celebrating lesbian unions.* Seattle, WA: The Seal Press.

Cassell, J. 1977. *A group called women: Sisterhood and symbolism in the feminist movement.* New York: David McKay.

Cavin, Susan. 1990. "The invisible army of women: Lesbian social protests, 1969–1988." Pp. 321–332 in *Women and social protest,* edited by Guida West and Rhoda Blumberg. New York: Oxford University Press.

Chafe, W. H. 1977. *Women and equality: Changing patterns in American culture.* New York: Oxford University Press.

Chafetz, Janet. 1990. *Gender equity: An integrated theory of stability and change.* Newbury Park, CA: Sage.

Chafetz, Janet and Gary Dworkin. 1986. *Female revolt.* Totowa, NJ: Rowman and Allenheld.

Daly, Mary. 1978. *Gyn/ecology.* Boston: Beacon.

Dill, Kim. Unpublished. "Feminism in the nineties: The influence of collective identity and community on young feminist activists." Master's thesis. The Ohio State University, 1991.

Echols, Alice. 1989. *Daring to be bad: Radical feminism in America 1967–1975.* Minneapolis: University of Minnesota Press.

Eisenstein, Z. 1981. *The radical future of liberal feminism.* New York: Longman.

Ferree, Myra Marx, and Beth B. Hess. 1985. *Controversy and coalition: The new feminist movement.* Boston: Twayne.

————and Patricia Yancey Martin. 1995. *Feminist Organizations: Harvest of the New Women's Movement.* Philadelphia: Temple University Press.

Firestone, S. 1970. *The dialectic of sex.* New York: William Morrow.

Freeman, Jo. 1975. *The politics of women's liberation.* New York: David McKay.

————. 1979. "Resource mobilization and strategy: A model for analyzing social movement organization actions." Pp. 167–89. in *The dynamics of social movements,* edited by M. N. Zald and J. D. McCarthy, Cambridge, MA: Winthrop.

Frye, Marilyn. 1983. *The politics of reality: Essays in feminist theory.* Trumansburg, NY: Crossing Press.

Gerlach, L. P., and V. H. Hine. 1970. *People, power, change: Movements of social transformation.* Indianapolis: Bobbs-Merrill.

Griffin, S. 1978. *Women and nature.* New York: Harper & Row.

Hartmann, H. 1981. "The family as the locus of gender, class, and political struggle: The example of housework." *Signs* 6 (Spring):366–94.

Hartsock, N.C.M. 1983. *Money, sex, and power: Toward a feminist historical materialism.* New York: Longman.

Hill-Collins. Patricia. 1990. *Black feminist thought.* Boston: Unwin Hyman.

Hoagland, Sarah Lucia. 1988. *Lesbian ethics: Toward new value.* Palo Alto, CA: Institute of Lesbian Studies.

————and Julia Penelope, eds. 1988. *For lesbians only.* London: Onlywomen Press.

Hochschild, Arlie. 1989. *The second shift.* New York: Avon.

Huber, J. 1973. "From sugar and spice to professor." In *Academic women on the move,* edited by A. S. Rossi and A. Calderwood. New York: Russell Sage Foundation.

Loulan, JoAnn. 1990. *The lesbian erotic dance.* San Francisco: Spinsters Book Company.

MacKinnon, C. A. 1983. "Feminism, Marxism, method, and the state: Toward feminist jurisprudence." *Signs* 8(4):635–58.

Martin, Patricia Yancey. 1990. "Rethinking feminist organizations." *Gender and Society* 4(2):182–206.

Millett, K. 1971. *Sexual politics.* New York: Avon.

Off Our Backs 1991. "Passages 7—Beyond the barriers." Vol. 21(6):12.

Oppenheimer, Valerie Kincade, 1973. "Demographic influence on female employment and the status of women." Pp. 184–199. in *Changing women in a changing society,* edited by Joan Huber, Chicago: University of Chicago Press.

Reskin, Barbara. 1988. "Bringing the men back in: Sex differentiation and the devaluation of women's work." *Gender and Society* 2:58–81.

Rich, Adrienne. 1976. *Of woman born.* New York: Norton.

————. 1980. "Compulsory heterosexuality and lesbian existence." *Signs* 5:631–60.

Rossi, A. S. 1982. *Feminists in politics.* New York: Academic Press.

Rubin, G. 1975. "The traffic in women: Notes on the 'political economy' of sex. In *Toward an anthropology of women,* edited by Rayne Reiter. New York: Monthly Review Press.

Rupp, Leila J. and Verta Taylor, 1987. *Survival in the doldrums: The American women's rights movement, 1945 to 1960s.* New York: Oxford University Press.

Ryan, Barbara, 1989. "Ideological purity and feminism: The U.S. women's movement from 1966 to 1975." *Gender and Society* 3:239–257.

———. 1992. *Feminism and the women's movement.* New York: Routledge.

Schneider, Beth. 1988. "Political generations in the contemporary women's movement." *Sociological Inquiry* 58:4–21.

Staggenborg, Suzanne. 1988. "The consequences of professionalization and formalization in the pro-choice movement." *American Sociological Review* 53:585–606.

———. 1989. "Stability and innovation in the women's movement: A comparison of two movement organizations." *Social Problems* 36:75–92.

Taylor, John. 1991. "Are you politically correct?" *New York Magazine.* January 21:33–40.

——— and Leila Rupp. 1993. "Women's culture and lesbian feminist activism: A reconsideration of cultural feminism." *Signs* 19:32–61.

Taylor, Verta. 1989. "Social movement continuity: The women's movement in abeyance. *American Sociological Review* 54:761–775.

West, Guida and Rhoda Lois Blumberg. 1990. *Women and social protest.* New York: Oxford University Press.

Whittier, Nancy E. 1995. *Feminist Generations: The Persistence of the Radical Women's Movement.* Philadelphia: Temple University Press.

DISCUSSION QUESTIONS

1. Compare and contrast the ideologies of liberal and radical feminism. What strategies does each use in the quest for women's liberation?

2. What evidence do Taylor and Whittier provide for the influence of the feminist movement in each of these areas: women's consciousness, cultural phenomena, and social structural change?

INFOTRAC COLLEGE EDITION

You can use your access to InfoTrac College Edition to learn more about the subjects covered in this essay. Some suggested search terms include:

feminism
lesbian feminism
liberal feminism

radical feminism
second wave feminism
women's movement

79

The Evolution of the U.S. Environmental Movement from 1970 to 1990:

An Overview

RILEY E. DUNLAP AND ANGELA G. MERTIG

Like all social movements, the environmental movement has particular organizational and ideological roots. Riley Dunlap and Angela Mertig identify the social and historical forces that have fostered the environmental movement and point to several contemporary developments that are shaping this movement in the current period.

The U.S. environmental movement has proven to be exceptionally successful and enduring, as demonstrated by the twentieth anniversary of Earth Day on 22 April 1990. A national poll conducted a week earlier found nearly two-thirds (63%) of Americans to be in support of Earth Day and only a tiny fraction (3%) opposed to it (Hart/Teeter, 1990). It is not surprising, therefore, that Earth Day 1990 turned out to be an enormously successful event, both in the United States and worldwide, leading to the claim that "it united more people concerned about a single cause than any other global event in history" (Cahn and Cahn, 1990, p. 17).

The success of Earth Day 1990 indicates that the environmental movement is not only alive and well after two decades but that it may be stronger than ever. Few social movements achieve such widespread acceptance, and fewer still are able to celebrate a twentieth anniversary. Why has environmentalism been able to avoid the fate of most short-lived movements, and how has it changed since the first Earth Day? Such questions are the focus of this volume, which is intended to provide an overview of the evolution of U.S. environmentalism during the past two decades.

From: Riley E. Dunlap and Angela G. Mertig. 1992. *American Environmentalism: The U.S. Environmental Movement*. 1970–1990. Philadelphia: Taylor and Francis, Inc., pp. 1–10. Reprinted with permission.

THE EMERGENCE OF THE U.S. ENVIRONMENTAL
MOVEMENT: A BRIEF HISTORY

The organizational and ideological roots of contemporary environmentalism are commonly traced to the progressive conservation movement that emerged in the late nineteenth century in reaction to reckless exploitation of our nation's natural resources. Early conservationists led by Gifford Pinchot (with support from Theodore Roosevelt) emphasized the wise management of natural resources for continued human use, but a few individuals such as John Muir argued for the preservation of nature for its own sake. Although the two factions eventually came into conflict, their joint efforts led to legislation establishing early national parks and agencies such as the U.S. Forest Service. They also spawned conservation organizations such as the Sierra Club and the National Audubon Society.

World War I deflected the nation's attention from conservation (O'Riordan, 1971), but after the war the United States was confronted by massive environmental calamities such as flooding and the Dust Bowl, as well as by the Great Depression. A second wave of conservationism arose during the Franklin Roosevelt administration and emphasized the mitigation of resource problems (e.g., flood control and soil conservation) as well as the development of resources (e.g., energy through the Tennessee Valley Authority) to stimulate economic recovery. Although these efforts were sidetracked by World War II, the 1950s saw a third wave of conservationism. More emphasis was placed on preservation of areas of natural beauty and wilderness for public enjoyment in this era, spearheaded by older organizations such as the Sierra Club. Widely publicized efforts to save the Grand Canyon and Dinosaur National Monument provided these organizations with considerable momentum. The resulting "wilderness movement" (McCloskey, 1972) was accompanied by continued concern about the future availability of natural resources as well as by growing concerns about overpopulation and air and water quality (Paehlke, 1989).

These old and new issues began to coalesce in the 1960s and gradually evolved into environmental concerns. Epitomized by Rachel Carsons' (1962) analysis of the subtle and wide-ranging impacts of pesticides on the natural environment and human beings in *Silent Spring,* these newer concerns were much broader than those of conservation. Environmental problems tended to (a) be more complex in origin, often stemming from new technologies; (b) have delayed, complex, and difficult-to-detect effects; and (c) have consequences for human health and well-being as well as for the natural environment (Mitchell, 1989). Encompassing both pollution and loss of recreational and aesthetic resources, such problems were increasingly viewed as threats to our quality of life (Hays, 1987).

Thus, by the late 1960s the third wave of conservationism had evolved into modern environmentalism with the transformation formalized by the national celebration of Earth Day 1970. With its reported 20 million participants (Dunlap and Gale, 1972), Earth Day not only marked the replacement of conservation with the full panoply of environmental issues but it mobilized (albeit temporarily) a far broader base of support than had any of the prior waves of conservationism (Mitchell, 1989). What accounted for this transformation?

Analyses of the emergence of environmentalism (e.g., Hays, 1987) have emphasized one or more of the following:

(1) The 1960s had given rise to an activist culture that encouraged people, especially youths, to take direct action to solve society's ills.

(2) Scientific knowledge about environmental problems such as smog began to grow, as did media coverage of such problems and major accidents such as the 1969 Santa Barbara oil spill.

(3) A rapid increase in outdoor recreation brought many people into direct contact with environmental degradation and heightened their commitment to preservation.

(4) Perhaps most fundamentally, tremendous post–World War II economic growth created widespread affluence, eventually lowering concern with materialism and generating concern over the quality of life.

(5) Many of the existing conservation organizations broadened their focus to encompass a wide range of environmental issues and attracted substantial support from foundations, enabling them to mobilize increased support for environmental causes. In the process they transformed themselves into environmental organizations (Mitchell, 1989).

Although its origins are undoubtedly complex, the environmental movement had clearly "arrived" by 1970, as shown by the tremendous growth in the size of the conservation-era organizations, the development of newer organizations, and in wide-spread public support. The emergence of the environmental movement in the 1960s and early 1970s was accompanied by creation of new federal agencies, such as the Environmental Protection Agency and the Council on Environmental Quality, and by legislation aimed at combating air and water pollution and requiring "environmental impact statements." In short, by the early 1970s, society had accepted environmentalists' view of environmental quality as a social problem.

THE NATURAL HISTORY OF SOCIAL MOVEMENTS

The importance of social movements in generating "social problems" has received considerable attention from sociologists, and some have argued that the two are analytically inseparable (Mauss, 1975, 1989). In this view, proponents of problems coalesce into one or more social movement organizations that attempt to mobilize others to work to ameliorate the problematic conditions (environmental degradation, racism, sexism, etc.). These organizations must obtain the support of the media, funding sources, the public, and ultimately of policymakers. Although many such "social-problem movements" fail to get off the ground, some are successful in mobilizing enough support to generate societal action aimed at solving the problem.

Solutions generally take the form of new government regulations and agencies, which signify the institutionalization of the movement. In the process of achieving such success a movement typically loses momentum: Its organizations evolve into formalized interest groups staffed by activists-turned-bureaucrats, many of its leaders are co-opted by government to staff the new agencies or simply tire of battle, and support dwindles as the media turn to newer issues and the public assumes the problematic conditions are being taken care of by government. Efforts to revitalize the movement and avoid stagnation and co-option may lead to rancorous infighting and fragmentation, with "die-hard activists" disavowing those co-opted by government or seduced into working "within the system." These trends may result in the demise of the movement, as it disappears with little if any improvement in the problematic conditions that generated it. This "natural history" model seems to fit the rise-and-fall pattern of many social-problem movements (Mauss, 1975).

A similar analysis of the problem-solving cycle has been offered by political scientists, who focus on the policy development and implementation stages. They argue that policies designed to solve social problems seldom succeed for two major reasons: First, interest groups (which successful social movements become at institutionalization) often achieve only symbolic victories, with government passing reassuring but essentially meaningless legislation. Second, even well-intended agencies are likely to fail, typically because they are captured by the very interests they were designed to regulate. This pattern of regulatory agency failure has been termed the "natural decay" model of government problem-solving efforts (see Sabatier and Mazmanian, 1980).

In short, social scientists posit a pattern in which social problems—such as "the environment"—are regularly discovered or created by activists, who are occasionally successful in getting the larger society to accept their definition of conditions as problematic and in need of amelioration. Such efforts are generally transitory and seldom fully successful, however, and generally experience a natural decline, often with little, if any, improvement in the problematic conditions. Has the environmental movement avoided this fate, as suggested by the success of the 1990 Earth Day anniversary, and, if so, why?

THE STAGES OF MODERN ENVIRONMENTALISM

The environmental movement was clearly institutionalized in the late 1960s and early 1970s, as signified by a flood of new groups at the national and especially the local levels, formalized media attention, and far-reaching legislation (Fessler, 1990). However, it appeared to lose steam fairly quickly. Within two or three years most organizations experienced slowed growth rates and several (especially at the local level) disappeared, public awareness and concern declined significantly, and explicitly anti-environmental counter-movements were being launched (Hays, 1987). It appeared that the movement had passed its peak and was experiencing a natural decline by the mid-1970s (see Albrecht, 1976). These trends accelerated during the pro-environmental Carter administration (1976–1980), as much of the public was lulled into thinking that environmental problems were being solved and many leading environmentalists were coopted into the administration (Manes, 1990). Nevertheless, the strong organizational base did not fade away, and environmental protection continued to receive substantial if not consensual support from the public (Mitchell, 1989).

Ironically, the election of Ronald Reagan reversed these trends by stimulating a resurgence of environmental concern and activism. The anti-environmental orientation of his administration, highlighted by Department of Interior Secretary Watt and Environmental Protection Agency Director Gorsuch, provided environmental organizations with reason—and ammunition—for mobilizing opposition to his policies. Several of the national organizations were remarkably successful in recruiting new members, and their increased visibility and intensified lobbying helped stimulate some degree of Congressional opposition to administration policies.

In the process, the environmental movement not only avoided a demise but it experienced a major revitalization, gaining increased membership, new organizations (especially at the local levels), and renewed support from the public and policymakers. This suggests that efforts by government to overtly repress or "de-institutionalize" a well-entrenched movement (as opposed to using more subtle "capture and co-optation" techniques) may backfire. By threatening environmentalists' hard-won "interest group" status, the Reagan administration rekindled the movement's zeal and activism. However, although the movement successfully used the administration's hostility to its own benefit, it was largely unsuccessful in preventing Watt, Gorsuch, and their successors from halting nearly two decades of progress in federal environmental reform (Hays, 1987).

This clearly highlighted the weakness of relying on the governmental regulatory apparatus that the national organizations had worked so hard to create (Manes, 1990; Scarce, 1990).

The revitalization of environmentalism in the 1980s was sparked by opposition to the Reagan agenda and facilitated by the relatively high level of resources available to the environmental movement. However, it was also stimulated by several underlying phenomena (Mitchell, 1989):

(1) The inherent and widespread appeal of environmental protection, stemming from the highly visible and increasingly threatening nature of environmental problems to virtually all segments of the population;

(2) the fact that progress in areas such as urban air quality was quickly offset by the emergence of new problems, often of a wider scale (e.g., acid rain) and more ominous nature (e.g., ozone depletion) than the older ones;

(3) the increasing societal recognition of continual, unanticipated environmental deterioration, resulting from the institutionalization of "environmental science" in government, academia, and especially environmental organizations themselves (and obviously stimulated by media attention); and

(4) the fact that environmental awareness had been institutionalized not only within the movement per se, but within many government agencies, scholarly associations, educational institutions, and even churches. Such trends have led Oates to argue that "a substantial degree of ecological consciousness has

become a permanent part of the American value system" (Oates, 1989, p. 186; see also Milbrath, 1984; Paehlke, 1989).

INCREASED DIVERSITY: STRENGTH OR WEAKNESS?

Although environmentalism has clearly endured over the past two decades, with unintentional aid from its opposition, it nonetheless has changed substantially. The major change appears to be its vastly increased diversity. As Gottlieb noted, "By the end of the 1980s . . . environmentalism meant many different things to different groups and movements" (Gottlieb, 1990, p. 42). Although this diversity may lead to fragmentation, which Mauss (1975) sees as a precursor to the demise of a movement, we believe that it may prove to be an important strength of contemporary environmentalism. . . .

A fundamental change in environmentalism since 1970 has been the rapid increase in the number and prominence of local grassroots organizations. These loosely coordinated volunteer groups typically develop in response to local environmental hazards that pose a threat to human health. Although epitomized by Love Canal, such groups have emerged in reaction to a wide range of problems besides toxic wastes, for example, existing or proposed landfills and waste incinerators. . . .

The recent emergence of grassroots environmentalism within minority communities is especially important, since racial and ethnic minorities traditionally have been wary of the environmental movement, fearing that it deflects attention from social justice concerns. . . . Growing opposition to "environmental racism," the practice of locating "LULUs" (locally unwanted land uses) within minority communities, has led to the blending of environmental and equity concerns into an "environmental justice" movement. . . .

Another recent development is the emergence of a radical wing of environmentalism (although, admittedly, NIMBY ("Not in my backyard") campaigns sometimes use radical tactics). Fueled by a sense that the national organizations have failed in their efforts to reform the system (especially during the Reagan era), radical environmentalism calls for direct action (e.g., sit-ins) and illegal actions (e.g., "monkey-wrenching") if necessary (Manes, 1990; Scarce, 1990). Concerned primarily with wilderness and wildlife, many radical environmentalists draw inspiration from the philosophy of *deep ecology*. They see mainstream or reformist organizations (and most NIMBY groups) as shallow in their primary concern with human welfare, and they endorse a "biocentric" perspective that denies the favored status of *Homo sapiens*. . . .

It is clear that environmentalism has been globalized within the past decade. Because environmental problems are increasingly transnational and often global in scale and therefore require international action, because such problems are in turn related to conditions within other nations (e.g., rainforest destruction in Brazil), and because U.S. environmentalists are often called on to assist environmental movements abroad (especially in the Third World), the major U.S. environmental organizations have adopted an increasingly global orientation

(McCormick, 1989). In addition, international organizations such as Greenpeace have found considerable support in the United States. . . .

CONCLUSION: WHEN IS A MOVEMENT SUCCESSFUL?

When environmentalism is judged against the typical social movement, we believe it is appropriate to argue that environmentalism has been a resounding success. By the first Earth Day in 1970 the environmental movement had achieved enormous visibility within our society and—albeit with some ups and downs—it has remained a viable sociopolitical force for more than two decades. However, social-problem movements develop for the purpose of solving problematic social conditions and, in this respect, environmentalism has not been atypical.

Many leading environmentalists. . . have acknowledged that the movement has largely failed in its goal of protecting the quality of the environment. As Denis Hayes, key organizer for both the first and twentieth Earth Days, stated, "The world is in worse shape today than it was twenty years ago" (Hayes, 1990, p. 56). Of course, others are quick to point out that the situation would be far worse had the movement not been around. . . .

REFERENCES

Albrecht. S. L. 1976. Legacy of the environmental movement. *Environment and Behavior,* 8:147–168.

Cahn, R., and P. Cahn. 1990. Did Earth Day change the world? *Environment,* September:16–20, 36–43.

Carson, R. 1962. *Silent spring.* New York: Houghton-Mifflin.

Dunlap, R. E., and R. P. Gale. 1972. Politics and ecology: A political profile of student ecoactivists. *Youth and Society,* 3:379–397.

Fessler, P. 1990, January 20. A quarter-century of activism erected a bulwark of laws. *Congressional Quarterly Weekly Report,* 48:154–156.

Gottlieb, R. 1990. An odd assortment of allies: American environmentalism in the 1990s. *Gannett Center Journal,* 4:37–47.

Hart/Teeter. 1990. NBC News/*Wall Street Journal*: National Survey No. 6. Washington, DC: Author.

Hayes, D. 1990. Earth Day 1990: Threshold of the green decade. *Natural History,* April:55–58, 67–70.

Hays, S. P. 1987. *Beauty, health, and permanence: Environmental politics in the United States, 1955–1985.* New York: Cambridge University Press.

Manes, C. 1990. *Green rage: Radical environmentalism and the unmaking of civilization.* Boston: Little, Brown.

Mauss, A. L. 1975. *Social problems as social movements.* Philadelphia: J. B. Lippincott.

————. 1989. Beyond the illusion of social problems theory. *Perspectives on Social Problems,* 1:19–39.

McCloskey. M. 1972. Wilderness movement at the crossroads, 1945–1970. *Pacific Historical Review,* 41:346–364.

McCormick, J. 1989. *Reclaiming Paradise: The global environmental movement.* Bloomington: Indiana University Press.

Milbrath. L. W. 1984. *Environmentalists: Vanguard for a new society.* Albany: State University of New York Press.

Mitchell, R. C. 1989. From conservation to environmental movement: The development of the modern environmental lobbies. In *Government and Environmental Politics,* ed. M. J. Lacey, pp. 81–113. Washington. DC: Wilson Center Press.

Oates, D. 1989. *Earth rising: Ecological belief in an age of science.* Corvallis: Oregon State University Press.

O'Riordan, T. 1971. The third American conservation movement: New

implications for public policy. *Journal of American Studies,* 5:155–171.

Paehlke, R. C. 1989. *Environmentalism and the future of progressive politics.* New Haven, CT: Yale University Press.

Sabatier, P., and D. Mazmanian. 1980. The implementation of public policy: A framework of analysis. *Policy Studies Journal,* 8:538–560.

Scarce, R. 1990. *Eco-warriors: Understanding the radical environmental movement.* Chicago: Noble Press.

DISCUSSION QUESTIONS

1. What do Dunlap and Mertig mean by the "natural history of social movements"? What events and conditions have shaped the natural history of the environmental movement?

2. What specific changes do Dunlap and Mertig identify as characteristic of the environmental movement now? How do these developments influence the strategies used by environmentalists to achieve their goals?

INFOTRAC COLLEGE EDITION

You can use your access to InfoTrac College Edition to learn more about the subjects covered in this essay. Some suggested search terms include:

Earth Day
environmental movement
Environmental Protection Agency

environmentalism
social movement theory

80

Sexual Politics,
Sexual Communities

JOHN D'EMILIO

The gay and lesbian liberation movement has its origins in other historical movements that challenged the oppression of gays, as well as in the civil rights movement, women's movement, and student movement of the 1960s. Reviewing the emergence of the gay and lesbian liberation movement, John D'Emilio reviews the contributions of this movement and its connections to other forms of activism.

Since June 1969, when a police raid of a Greenwich Village gay bar sparked several nights of rioting by male homosexuals, gay men and women in the United States have enlisted in ever growing numbers in a movement to emancipate themselves from the laws, the public policies, and the attitudes that have consigned them to an inferior position in society. In ways pioneered by other groups that have suffered a caste-like status, homosexuals and lesbians have formed organizations, conducted educational campaigns, lobbied inside legislative halls, picketed outside them, rioted in the streets, sustained self-help efforts, and constructed alternative separatist institutions on their road to liberation. They have worked to repeal statutes that criminalize their sexual behavior and to eliminate discriminatory practices. They have labored to unravel the ideological web that supports degrading stereotypes. Like other minorities, gay women and men have struggled to discard the self-hatred they have internalized. Many of them have rejected the negative definitions that American society has affixed to their sexuality and, instead, have begun to embrace their identity with pride.

From the beginning a curious inconsistency appeared between the rhetoric of the gay liberation movement and the reality of its achievements. On the one hand, activists in the early 1970s repeatedly stressed, in their writing and their public comments, the intertwining themes of silence, invisibility, and isolation. Gay men and lesbians, the argument ran, were invisible to society and to each other, and they lived isolated from their own kind. A vast silence surrounded the topic of homosexuality, perpetuating both invisibility and isolation. On the other hand, gay liberationists exhibited a remarkable capacity to mobilize their allegedly hidden, isolated constituency, and the movement grew with amazing rapidity. By

From: John D'Emilio. 1998. *Sexual Politics, Sexual Communities,* second ed. Chicago: The University of Chicago Press, pp. 1–5, 223–239.

the mid-1970s, homosexuals and lesbians had formed more than 1,000 organizations scattered throughout the country. Many of these groups directed their energy outward, exerting pressure on legislatures, schools, the media, churches, and the professions. Activists proved capable of turning out tens of thousands of individuals for demonstrations, and they won impressive victories in relatively quick order. Many lesbian and gay male organizations also looked inward, toward their constituency. Activists created newspapers, magazines, health clinics, churches, multipurpose social centers, and specialized businesses—in short, a range of institutions that implied the existence of a separate, cohesive gay community.

Clearly, what the movement achieved and how lesbians and gay men responded to it belied the rhetoric of isolation and invisibility. Isolated men and women do not create, almost overnight, a mass movement premised upon a shared group identity. In combating prejudicial attitudes and discriminatory practices, moreover, gay liberationists encountered a quite clearly articulated body of thought about homosexuality. And, if lesbians and homosexuals were indeed invisible, the movement's leaders displayed an uncanny ability to find them in large numbers. . . .

Although the homophile movement never managed to fire the imagination of its constituency during the 1960s, it owed whatever dynamism it did possess to the example set by other discontented groups.

By the late 1960s, . . . a distinctively new culture of protest had taken shape in the United States, with which the reform orientation of the gay movement contrasted oddly. At Columbia University, for instance, the student homophile league peacefully picketed a forum on homosexuality on the same day that black and white student radicals initiated a week-long occupation of campus buildings. . . . The gay movement continued to pursue equality through the courts at a time when militant civil rights workers were shifting toward a strategy of community organizing and an ideology of black power. As homophile activists marched with neatly lettered placards in front of federal buildings, urban ghettos burst into flame, bombs exploded in banks and military-connected university facilities, and Black Panthers and Weathermen called for "armed struggle" against American imperialism. A generation of blacks and whites, men and women, was rising in revolt, but for the most part homophile activists remained curiously detached from the rebellions that were rocking the nation.

Despite the distance between gay activists and young radicals, the protests of the 1960s had more than superficial relevance to the situation of homosexuals and lesbians. Each of the separate strands of the "Movement"—black power, the student New Left, the counterculture, and women's liberation—spoke in a special way to gay women and men. Taken together and appropriated by those stigmatized for their sexuality, the ideology and tactics of the mass movements of the 1960s had the power to transform not only the organized struggle of homosexuals and lesbians for freedom but also the everyday quality and condition of gay life in the United States.

Young black militants in organizations like the Student Nonviolent Coordinating Committee and the Congress of Racial Equality led the move from the exuberant, optimistic reform spirit of the early 1960s to the angry, confrontational politics of the second half of the decade. By 1965 the successes of the civil rights movement had provoked a "crisis of victory," as racial inequality proved more intractable than the Southern legal codes that buttressed it. The combination of violent attacks by Southern whites, a sense of betrayal by Northern liberals and Democratic politicians, the explosive anger of ghetto residents, the growing appeal of the nationalist Black Muslims, and the war in Vietnam aroused impatience among many civil rights veterans and bred a cynicism toward the integrationist goals and nonviolent tactics that initially characterized the struggle for racial justice.

When SNCC leader Stokely Carmichael raised the black power cry in June 1966, he voiced more than a slogan. Black power quickly came to embody a distinctive form of politics and culture. In many ways it represented a reversal of the dominant, long-term trend in American society, the assimilation of ethnic and cultural minorities in the American melting pot. Black power advocates began speaking about structural racism and systematic oppression rather than prejudice and discrimination. As their goals, liberation replaced equality, and self-determination superseded integration. They talked of organizing the black community, fashioning an independent power base, and preserving their autonomy and separateness from white society. Instead of minimizing the differences between the races, these new militants celebrated them. Black became beautiful, as young radicals took the stigma out of skin color and made it a source of pride.

. . . . While many of the young channeled their energy into political protest and directed their fury toward institutions, others adopted a cultural radicalism. In some ways the hippie counterculture and the New Left worked at cross purposes, but in other ways they complemented and reinforced one another. In rebelling against the hypocrisy and alienation of modern American life, counterculture enthusiasts rejected the detached, objective mode of scientific inquiry that seemed to lead inexorably to the conflagration in Vietnam and pursued instead a politics of experience that celebrated the subjective. The counterculture sought a revolution in consciousness, a transformation of self that would create a personality, ethics, and style of living consistent with the political and social criticism of the New Left. Young hippies refused to conform to the expectations and values of white middle-class America. They constructed alternative living arrangements, adopted new styles of dress, ingested mind-expanding drugs, and embraced a sexual morality that often shocked ordinary Americans. Embedded in this experimentation was the conviction that one *could* remake one's self, that the quest for authenticity and fulfillment in a corrupt society *was* attainable—in short, that one could live the revolution now.

Eventually, white women in the New Left took many of the key concepts that underlay the radicalism of the decade, applied them to their own situation, and, out of the tension between rhetoric and reality, created a women's liberation movement. The stress within the civil rights movement and among early SDS members on equality, self-determination, and participatory democracy almost demanded a resurgence of feminism, as young radical women found their contributions to the New Left devalued, their leadership capacity inhibited, and their identities submerged in those of the men with whom they were intimate. Moreover, the peculiar combination of politics and culture that characterized protest in the 1960s gave a boost to the women's movement. By rejecting traditional forms of family life, the counterculture delegitimized the normative female role of housewife and mother. Yet, paradoxically, its pursuit of a free-flowing, unrestricted sexuality also intensified for young women the experience of sexual objectification. Under the guise of liberation, politicized women endured new forms of alienating human relationships.

Women's liberation added another dimension to political protest. Radical feminists questioned the very categories of male and female upon which most individuals' sense of self could rest securely. Women placed gender alongside race and class as a systematically enforced, socially constructed form of inequality. The injustices committed by fathers, husbands, and lovers were no more excusable than those of generals at the Pentagon. Intimate relationships became arenas of struggle, the bedroom and the kitchen battlegrounds, as women's liberationists fashioned a sexual politics that encompassed every aspect of personal life.

Though much of the decade's rhetoric was overblown and never delivered what it promised, it nonetheless proved rousing enough to prod millions of the young and the not so young into action against social injustice or into revamping their own lives. In fact, the radicalism of the 1960s derived its force in no small part because it bound the personal and the political so tightly together that the two could no longer be distinguished. Even the vocabulary of the Movement had dual meanings. The right of self-determination could be applied to individuals as well as to a people; revolutions occurred in both consciousness and society; the future shape of the country could be found not only in alternative institutions but in alternative styles of living. The affirmation of subjective personal experience as a primary source of knowledge and a reliable guide to action encouraged many Americans to make decisions they might otherwise not have risked and to see those choices as part of a vast movement for social change.

. . . On Friday, June 27, 1969, shortly before midnight, two detectives from Manhattan's Sixth Precinct set off with a few other officers to raid the Stonewall Inn, a gay bar on Christopher Street in the heart of Greenwich Village. They must have expected it to be a routine raid. New York was in the midst of a mayoral campaign—always a bad time for the city's homosexuals—and John Lindsay, the incumbent who had recently lost his party's primary, had reason to agree to a police cleanup. Moreover, a few weeks earlier the Sixth Precinct had received a new commanding officer who marked his entry into the position by initiating a series of raids on gay bars. The Stonewall Inn was an especially inviting target. Operating without a liquor license, reputed to have ties with organized crime, and offering scantily clad go-go boys as entertainment, it brought an "unruly" element to Sheridan Square, a busy Village intersection. Patrons of the Stonewall tended to be young and nonwhite. Many were drag queens, and many came from the burgeoning ghetto of runaways living across town in the East Village.

However, the customers at the Stonewall that night responded in any but the usual fashion. As the police released them one by one from inside the bar, a crowd accumulated on the street. Jeers and catcalls arose from the onlookers when a paddy wagon departed with the bartender, the Stonewall's bouncer, and three drag queens. A few minutes later, an officer attempted to steer the last of the patrons, a lesbian, through the bystanders to a nearby patrol car. "She put up a struggle," the *Village Voice* reported, "from car to door to car again." At that moment,

> the scene became explosive. Limp wrists were forgotten. Beer cans and bottles were heaved at the windows and a rain of coins descended on the cops. . . . Almost by signal the crowd erupted into cobblestone and bottle heaving. . . . From nowhere came an uprooted parking meter—used as a battering ram on the Stonewall door. I heard several cries of "let's get some gas," but the blaze of flame which soon appeared in the window of the Stonewall was still a shock.[1]

Reinforcements rescued the shaken officers from the torched bar, but their work had barely started. Rioting continued far into the night, with Puerto Rican transvestites and young street people leading charges against rows of uniformed police officers and then withdrawing to regroup in Village alleys and side streets.

By the following night, graffiti calling for "Gay Power" had appeared along Christopher Street. Knots of young gays—effeminate, according to most reports—gathered on corners, angry and restless. Someone heaved a sack of wet garbage through the window of a patrol car. On nearby Waverly Place, a concrete block landed on the hood of another police car that was quickly surrounded by dozens of men, pounding on its doors and dancing on its hood. Helmeted officers from the tactical patrol force arrived on the scene and dispersed with swinging clubs an impromptu chorus line of gay men in the middle of a full kick. At the intersection of Greenwich Avenue and Christopher Street, several dozen queens screaming "Save Our Sister!" rushed a group of officers who were clubbing a young man and dragged him to safety. For the next few hours, trash fires blazed, bottles and stones flew through the air, and cries of "Gay Power!" rang in the streets as the police, numbering over 400, did battle with a crowd estimated at more than 2,000.

After the second night of disturbances, the anger that had erupted into street fighting was channeled into intense discussion of what many had begun to memorialize as the first gay riot in history. Allen Ginsberg's stature in the 1960s had risen almost to that of guru for many counterculture youth. When he arrived at the Stonewall on Sunday evening, he commented on the change that had already taken place. "You know, the guys there were so beautiful," he told a reporter. "They've lost that wounded look that fags all had ten years ago."[2] The New York Mattachine Society hastily assembled a special riot edition of its newsletter that characterized the events, with camp humor, as "The Hairpin Drop Heard Round the World." It scarcely exaggerated. Before the end of July, women and men in New York had formed the Gay Liberation Front, a self-proclaimed revolutionary organization in the style of the New Left. Word of the Stonewall riot and GLF spread rapidly among the networks of young radicals scattered across the country, and within a year gay liberation groups had sprung into existence on college campuses and in cities around the nation.

The Stonewall riot was able to spark a nationwide grassroots "liberation" effort among gay men and women in large part because of the radical movements that had so inflamed much of American youth during the 1960s. Gay liberation used the demonstrations of the New Left as recruiting grounds and appropriated the tactics of confrontational politics for its own ends. The ideas that suffused youth protest found their way into gay liberation, where they were modified and adapted to describe the oppression of homosexuals and lesbians. The apocalyptic rhetoric and the sense of impending revolution that surrounded the Movement by the end of the decade gave to its newest participants an audacious daring that made the dangers of a public avowal of their sexuality seem insignificant.

The first gay liberationists attracted so many other young radicals not only because of a common sexual identity but because they shared a similar political perspective. Gay liberationists spoke in the hyperbolic phrases of the New Left. They talked of liberation from oppression, resisting genocide, and making a revolution against "imperialist Amerika.". . .

Gay liberationists targeted the same institutions as homophile militants, but their disaffection from American society impelled them to use tactics that their predecessors would never have adopted. Bar raids and street arrests of gay men in New York City during August 1970 provoked a march by several thousand men and women from Times Square to Greenwich Village, where rioting broke out. Articles hostile to gays in the *Village Voice* and in *Harper's* led to the occupation of publishers' offices. In San Francisco a demonstration against the *Examiner* erupted into a bloody confrontation with the police. Chicago Gay Liberation invaded the 1970 convention of the American Medical Association, while its counterpart in San Francisco disrupted the annual meeting of the American Psychiatric Association. At a session there on homosexuality a young bearded gay man danced around the auditorium in a red dress, while other homosexuals and lesbians scattered in the audience shouted "Genocide!" and "Torture!" during the reading of a paper on aversion therapy. Politicians campaigning for office found themselves hounded by scruffy gay militants who at any moment might race across the stage where they were speaking or jump in front of a television camera to demand that they speak out against the oppression of homosexuals. The confrontational tactics and flamboyant behavior thrust gay liberationists into the public spotlight. Although their actions may have alienated some homosexuals and lesbians, they inspired many others to join the movement's ranks.

From its beginning, gay liberation transformed the meaning of "coming out." Previously coming out had signified the private decision to accept one's homosexual desires and to acknowledge one's sexual identity to other gay men and women. Throughout the 1950s and 1960s, leaders of the homophile cause had in effect extended their coming out to the public sphere through their work in the movement. But only rarely did they counsel lesbians and homosexuals at large to follow their example, and when they did, homophile activists presented it as a selfless step taken for the benefit of others. Gay liberationists, on the other hand, recast coming out as a profoundly political act that could offer enormous personal benefits to an individual. The open avowal of one's sexual identity, whether at work, at school, at home, or before television cameras, symbolized the shedding of the self-hatred that gay men and women internalized, and consequently it promised an immediate improvement in one's life. To come out of the "closet" quintessentially expressed the fusion of the personal and the political that the radicalism of the late 1960s exalted.

Coming out also posed as the key strategy for building a movement. Its impact on an individual was often cathartic. The exhilaration and anger that surfaced when men and women stepped through the fear of discovery propelled them into political activity. Moreover, when lesbians and homosexuals came out, they crossed a critical dividing line. They relinquished their invisibility, made themselves vulnerable to attack, and acquired an investment in the success of the movement in a way that mere adherence to a political line could never accomplish. Visible lesbians and gay men also served as magnets that drew others to them. Furthermore, once out of the closet, they could not easily fade back in. Coming out provided gay liberation with an army of permanent enlistees.

A second critical feature of the post-Stonewall era was the appearance of a strong lesbian liberation movement. Lesbians had always been a tiny fraction of the homophile movement. But the almost simultaneous birth of women's liberation and gay liberation propelled large numbers of them into radical sexual politics. Lesbians were active in both early gay liberation groups and feminist organizations. Frustrated and angered by the chauvinism they experienced in gay groups and the hostility they found in the women's movement, many lesbians opted to create their own separatist organizations. . . .

Although gay liberation and women's liberation both contributed to the growth of a lesbian-feminist movement, the latter exerted a greater influence. The feminist movement offered the psychic space for many women to come to a self-definition as lesbian. Women's liberation was in its origins a separatist movement, with an ideology that defined men as the problem and with organizational forms from consciousness-raising groups to action-oriented collectives that placed a premium on female solidarity. As women explored their oppression together, it became easier to acknowledge their love for other women. The seeming contradiction between an ideology that focused criticism on men per se and the ties of heterosexual feminists to males often provoked a crisis of identity. Lesbian-feminists played upon this contradiction. "A lesbian is the rage of all women condensed to the point of explosion," wrote New York Radicalesbians in "The Woman-Identified Woman," one of the most influential essays of the sexual liberation movements. . . .

Besides the encouragement it provided for women to come out, women's liberation served lesbians—and gay men—in another way. The feminist movement continued to thrive during the 1970s. Its ideas permeated the country, its agenda worked itself into the political process, and it effected deep-seated changes in the lives of tens of millions of women and men. Feminism's attack upon traditional sex roles and the affirmation of a nonreproductive sexuality that was implicit in such demands as unrestricted access to abortion paved a smoother road for lesbians and homosexuals who were also challenging rigid male and female stereotypes and championing an eroticism that by its nature did not lead to procreation. Moreover, lesbians served as a bridge between the women's movement and gay liberation, at the very least guaranteeing that sectors of each remained amenable to the goals and perspectives of the other. Feminism helped to remove gay life and gay politics from the margins of American society.

. . . Stonewall thus marked a critical divide in the politics and consciousness of homosexuals and lesbians. A small, thinly spread reform effort suddenly grew into a large, grassroots movement for liberation. The quality of gay life in America was permanently altered as a furtive subculture moved aggressively into the open.

NOTES

1. *Village Voice,* July 3, 1969, p. 18.
2. Ibid.

DISCUSSION QUESTIONS

1. What are the historical precursors to the gay and lesbian liberation movement? How does understanding this history change your concept of the
   contemporary character of this movement?
2. What historical conditions were necessary for the development of the gay
   and lesbian movement? How have these changed and what influence will
   this likely have on the future of this movement?

INFOTRAC COLLEGE EDITION

You can use your access to InfoTrac College Edition to learn more about the subjects covered in this essay. Some suggested search terms include:

gay liberation   sexual revolution
gay rights   social construction of homosexuality
homophile movement   Stonewall riot
National Gay and Lesbian Task Force

81

The Genius of the Civil Rights Movement:

Can It Happen Again?

ALDON MORRIS

The Civil Rights Movement is arguably the most influential movement in the United States during the twentieth century. Aldon Morris reviews the development of the civil rights movements and notes the products of this movement, including the mobilization of other national and international movements and the transformations in academic scholarship that the movement generated. By identifying the particular historical and social circumstances in which the civil rights movement developed, he also asks whether such a movement is possible again.

It is important for African Americans, as well as all Americans, to take a look backward and forward as we approach the turn of a new century, indeed a new millennium. When a panoramic view of the entire history of African Americans is taken into account, it becomes crystal clear that African American social protest has been crucial to Black liberation. In fact, African American protest has been critical to the freedom struggles of people of color around the globe and to progressive people throughout the world.

The purpose of this essay is: 1) to revisit the profound changes that the modern Black freedom struggle has achieved in terms of American race relations; 2) to assess how this movement has affected the rise of other liberation movements both nationally and internationally; 3) to focus on how this movement has transformed how scholars think about social movements; 4) to discuss the lessons that can be learned from this groundbreaking movement pertaining to future African American struggles for freedom in the next century.

It is hard to imagine how pervasive Black inequality would be today in America if it had not been constantly challenged by Black protests throughout each century since the beginning of slavery. The historical record is clear that slave resistance and slave rebellions and protest in the context of the Abolitionist movement were crucial to the overthrow of the powerful slave regime.

The establishment of the Jim Crow regime was one of the great tragedies of the late nineteenth and early twentieth centuries. The overthrow of slavery represented one of those rare historical moments where a nation had the opportunity

From: Aldon Morris, Northwestern University. Reprinted by permission of author.

to embrace a democratic future or to do business as usual by reinstalling undemocratic practices. In terms of African Americans, the White North and South chose to embark along undemocratic lines.

For Black people, the emergence of the Jim Crow regime was on of the greatest betrayals that could be visited upon a people who had hungered for freedom so long; what made it even worse for them is that the betrayal emerged from the bosom of a nation declaring to all the world that it was the beacon of democracy.

The triumph of Jim Crow ensured that African Americans would live in a modern form of slavery that would endure well into the second half of the twentieth century. The nature and consequences of the Jim Crow system are well known. It was successful in politically disenfranchising the Black population and in creating economic relationships that ensured Black economic subordination. Work on wealth by sociologists Melvin Oliver and Thomas Shapiro (1995), as well as Dalton Conley (1999), are making clear that wealth inequality is the most drastic form of inequality between Blacks and Whites. It was the slave and Jim Crow regimes that prevented Blacks from acquiring wealth that could have been passed down to succeeding generations. Finally, the Jim Crow regime consisted of a comprehensive set of laws that stamped a badge of inferiority on Black people and denied them basic citizenship rights.

The Jim Crow regime was backed by the iron fist of southern state power, the United States Supreme Court, and white terrorist organizations. Jim Crow was also held in place by white racist attitudes. As Larry Bobo has pointed out, "The available survey data suggests that anti-Black attitudes associated with Jim Crow were once widely accepted . . . [such attitudes were] expressly premised on the notion that Blacks were the innate intellectual, cultural, and temperamental inferior to Whites (Bobo, 1997:35)." Thus, as the twentieth century opened, African Americans were confronted with a powerful social order designed to keep them subordinate. As long as the Jim Crow order remained intact, the Black masses could breathe neither freely nor safely. Thus, nothing less than the overthrow of a social order was the daunting task that faced African Americans during the early decades of the twentieth century.

The voluminous research on the modern civil rights movement has reached a consensus: That movement was the central force that toppled the Jim Crow regime. To be sure, there were other factors that assisted in the overthrow including the advent of the television age, the competition for Northern Black votes between the two major parties, and the independence movement in Africa which sought to overthrow European domination. Yet it was the Civil Rights movement itself that targeted the Jim Crow regime and generated the great mass mobilizations that would bring it down.

What was the genius of the Civil Rights movement that made it so effective in fighting a powerful and vicious opposition? The genius of the Civil Rights movement was that its leaders and participants recognized that change could occur if they were able to generate massive crises within the Jim Crow order—crises of such magnitude that the authorities of oppression must yield to the demands of the movement to restore social order. Max Weber defined power as the ability to realize one's will despite resistance. Mass disruption generated power. That was the strategy of nonviolent direct action. By utilizing tactics of disruption, implemented by thousands of disciplined demonstrators who had been mobilized through their churches, schools, and voluntary associations, the Civil Rights movement was able to generate the necessary power to overcome the Jim Crow regime. The famous crises created in places like Birmingham and Selma, Alabama, coupled with the important less visible crises that mushroomed throughout the nation, caused social breakdown in Southern business and commerce, created unpredictability in all spheres of social life, and strained the resources and credibility of Southern state governments while forcing white terrorist groups to act on a visible stage where the whole world could watch. At the national level, the demonstrations and repressive measures used against them generated foreign policy nightmares because they were covered by foreign media in Europe, the Soviet Union, and Africa. Therefore what gave the mass-based sit-ins, boycotts, marches, and jailing their power was their ability to generate disorder.

As a result, within ten years—1955 to 1965—the Civil Rights movement had toppled the Jim Crow order. The 1964 Civil Rights Bill and the 1965 Voting Rights Act brought the regime of formal Jim Crow to a close.

The Civil Rights movement unleashed an important social product. It taught that a mass-based grass roots social movement that is sufficiently organized, sustained, and disruptive is capable of generating fundamental social change. In other words, it showed that human agency could flow from a relatively powerless and despised group that was thought to be backward, incapable of producing great leaders.

Other oppressed groups in America and around the world took notice. They reasoned that if American Blacks could generate such agency they should be able to do likewise. Thus the Civil Rights movement exposed the agency available to oppressed groups. By agency I refer to the empowering beliefs and action of individuals and groups that enable them to make a difference in their own lives and in the social structures in which they are embedded.

Because such agency was made visible by the Civil Rights movement, disadvantaged groups in America sought to discover and interject their agency into their own movements for social change. Indeed, movements as diverse as the Student movement, the Women's movement, the Farm Worker's movement, the Native American movement, the Gay and Lesbian movement, the Environmental movement, and the Disability Rights movement all drew important lessons and inspiration from the Civil Rights movement. From that movement other groups discovered how to organize, how to build social movement organizations, how to mobilize large numbers of people, how to devise appropriate tactics and strategies, how to infuse their movement activities with cultural creativity, how to confront and defeat authorities, and how to unleash the kind of agency that generates social change.

For similar reasons, the Black freedom struggle was able to effect freedom struggles internationally. For example, nonviolent direct action has inspired oppressed groups as diverse as Black South Africans, Arabs of the Middle East, and pro-democracy demonstrators in China to engage in collective actions. The sit-in tactic made famous by the Civil Rights movement, has been used in liberation movements throughout the third world, in Europe, and in many other foreign countries. The Civil Rights movement's national anthem "We Shall Overcome" has been interjected into hundreds of liberation movements both nationally and internationally. Because the Civil Rights movement has been so important to international struggles, activists from around the world have invited civil rights participants to travel abroad. Thus early in Poland's Solidarity movement Bayard Rustin was summoned to Poland by that movement. As he taught the lessons of the Civil Rights movement, he explained that "I am struck by the complete attentiveness of the predominantly young audience, which sits patiently, awaiting the translations of my words." (Rustin, undated)

Therefore, as we seek to understand the importance of the Black Freedom Struggle, we must conclude the following: the Black Freedom Struggle had provided a model and impetus for social movements that have exploded on the American and international landscapes. This impact has been especially pronounced in the second half of the twentieth century.

What is less obvious is the tremendous impact that the Black Freedom Struggle has had on the scholarly study of social movements. Indeed, the Black freedom struggle has helped trigger a shift in the study of social movements and collective action. The Black movement has provided scholars with profound empirical and theoretical puzzles because it has been so rich organizationally and tactically and because it has generated unprecedented levels of mobilization. Moreover, this movement has been characterized by a complex leadership base, diverse gender roles, and it has revealed the tremendous amount of human agency that usually lies dormant within oppressed groups. The empirical realities of the Civil Rights movement did not square with the theories used by scholars to explain social movements prior to the 1960s.

Previous theories did not focus on the organized nature of social movements, the social movement organizations that mobilize them, the tactical and strategic choices that make them effective, nor the rationally planned action of leaders and participants who guide them. In the final analysis, theories of social movements lacked a theory that incorporated human agency at the core of their conceptual apparatuses. Those theories conceptualized social movements as spontaneous, largely unstructured, and discontinuous with institutional and organizational behavior. Movement participants were viewed as reacting to various forms of strain and doing so in a non-rational manner. In these frameworks, human agency was conceptualized as reactive, created by uprooted individuals seeking to reestablish a modicum of personal and social stability. In short, social movement theories prior to the Civil Rights movement operated with a vague, weak vision of agency to explain phenomena that are driven by human action.

The predictions and analytical focus of social movement theories prior to the 1970s stood in sharp contrast to the kind of theories that would be needed to capture the basic dynamics that drove the Civil Rights movement. It became apparent to social movement scholars that if they were to understand the Civil Rights movement and the multiple movements it spun, the existing theoretical landscape would have to undergo a radical process of reconceptualization.

As a result, the field of social movements has been reconceptualized and this retheoritization will effect research well into the new millennium. To be credible in the current period any theory of social movements must grapple conceptually with the role of rational planning and strategic action, the role of movement leadership, and the nature of the mobilization process. How movements are gendered, how movement dynamics are bathed in cultural creativity, and how the interactions between movements and their opposition determine movement outcomes are important questions. At the center of this entire matrix of factors must be an analysis of the central role that human agency plays in social movements and in the generation of social change.

Thanks, in large part, to the Black freedom struggle, theories of social movements that grapple with real dynamics in concrete social movements are being elaborated. Intellectual work in the next century will determine how successful scholars will be in unraveling the new empirical and theoretical puzzles thrust forth by the Black freedom movement. Although it was not their goal, Black demonstrators of the Civil Rights movement changed an academic discipline.

A remaining question is: Will Black protest continue to be vigorous in the twenty-first century, capable of pushing forward the Black freedom agenda? It is not obvious that Black protest will be as sustainable and as paramount as it has been in previous centuries. To address this issue we need to examine the factors important to past protests and examine how they are situated in the current context.

Social movements are more effective when they can identify a clear-cut enemy. Who or what is the clear-cut enemy of African Americans of the twenty-first century? Is it racism, and if so, who embodies it? Is it capitalism, and if so, how is this enemy to be loosened from its abstract perch and concretized? In fact, we do not currently have a robust concept that grasps the modern form of domination that Blacks currently face. Because the modern enemy has become opaque, slippery, illusive, and covert, the launching of Black protest has become more difficult because of conceptual fuzziness.

Second, during the closing decades of the twentieth century the Black class structure has become more highly differentiated and it is no longer firmly anchored in the Black community. There is some danger, therefore, that the cross fertilization between different strata within the Black class structure so important to previous protest movements may have become eroded to the extent that it is no longer fully capable of launching and sustaining future Black protest movements.

Third, will the Black community of the twenty-first century possess the institutional strength required for sustaining Black protest? Black colleges have been weakened because of the racial integration of previously all white institutions of higher learning and because many Black colleges are being forced to integrate. The degree of institutional strength of the church has eroded because some of them have migrated to the suburbs in an attempt to attract affluent Blacks. In other instances, the Black Church has been unable to attract young people of the inner city who find more affinity with gangs and the underground economy. Moreover, a great potential power of the Black church is not being realized because its male clergy refuse to empower Black women as preachers and pastors. The key question is whether the Black church remains as close to the Black masses—especially to poor and working classes—as it once was. That closeness determines its strength to facilitate Black protest.

In short, research has shown conclusively that the Black church, Black colleges and other Black community organizations were critical vehicles through which social protest was organized, mobilized and sustained. A truncated class structure was also instrumental to Black protest. It is unclear whether during the twenty-first century these vehicles will continue to be effective tools of Black protest or whether new forces capable of generating protest will step into the vacuum.

In conclusion, I foresee no reason why Black protest should play a lesser role for Black people in the twenty-first century. Social inequality between the races will continue and may even worsen especially for poorer segments of the Black communities. Racism will continue to effect the lives of all people of color. If future changes are to materialize, protest will be required. In 1898 as Du Bois glanced toward the dawn of the twentieth century, he declared that in order for Blacks to achieve freedom they would have to protest continuously and energetically. This will become increasingly true for the twenty-first century. The question is whether organizationally, institutionally, and intellectually the Black community will have the wherewithal to engage in the kind of widespread and effective social protest that African Americans have utilized so magnificently. If previous centuries are our guide, then major surprises on the protest front should be expected early in the new millennium.

REFERENCES

Bobo, L. 1997. "The Color Line, the Dilemma, and the Dream: Race Relations in America at the Close of the Twentieth Century." In *Civil Rights and Social Wrongs: Black-White Relations since World War II,* edited by J. Higham, pp. 31–55. University Park, PA: Penn State University Press.

Conley, Dalton. 1999. *Being Black, Living in the Red: Race, Wealth, and Social Policy in America.* Berkeley: University of California Press.

Rustin, Bayard. no date. *Report on Poland.* New York: A. Philip Randolph Institute.

Oliver, Melvin, and Thomas E. Shapiro. 1995. *Black Wealth/White Wealth: A New Perspective on Racial Inequality.* New York: Routledge.

DISCUSSION QUESTIONS

1. What does Morris mean by "the genius of the Civil Rights movement?" Can you imagine such a strategy being an effective means of combating the oppression of racial groups today? If so, how; if not, why not?

2. What does Morris identify as the products of the Civil Rights movement? What does this teach you about the connections between contemporary social movements and the Civil Rights movement?

INFOTRAC COLLEGE EDITION

You can use your access to InfoTrac College Edition to learn more about the subjects covered in this essay. Some suggested search terms include:

Black Panther Party
Black power movement
Black protest
Civil Rights movement

disability rights
Jim Crow segregation
Montgomery Bus Boycott

82

Gemeinschaft and Gesellschaft

FERDINAND TÖNNIES

In this classic essay distinguishing gemeinschaft *and* gesellschaft, *Ferdinand Tönnies describes the differing social relationships that emerge in these two forms of society. His theory of social change shows society increasingly moving from gemeinschaft to gesellschaft, particularly as cities and nations develop more elaborate systems of commerce.*

All intimate, private, and exclusive living together, so we discover, is understood as life in Gemeinschaft (community). Gesellschaft (society) is public life—it is the world itself. In Gemeinschaft with one's family, one lives from birth on, bound to it in weal and woe. One goes into Gesellschaft as one goes into a strange country. A young man is warned against bad Gesellschaft, but the expression bad Gemeinschaft violates the meaning of the word. Lawyers may speak of domestic (*häusliche*) Gesellschaft, thinking only of the legalistic concept of social association; but the domestic Gemeinschaft, or home life with its immeasurable influence upon the human soul, has been felt by everyone who ever shared it. Likewise, a bride or groom knows that he or she goes into marriage as a complete Gemeinschaft of life *(communio totius vitae)*. A Gesellschaft of life would be a contradiction in and of itself. One keeps or enjoys another's Gesellschaft, but not his Gemeinschaft in this sense. One becomes a part of a religious Gemeinschaft; religious Gesellschaften (associations or societies), like any other groups formed for given purposes, exist only in so far as they, viewed from without, take their places among the institutions of a political body or as they represent conceptual elements of a theory; they do not touch upon the religious Gemeinschaft as such. There exists a Gemeinschaft of language, of folkways or mores, or of beliefs; but, by way of contrast, Gesellschaft exists in the realm of business, travel, or sciences. So of special importance are the commercial Gesellschaften; whereas, even though a certain familiarity and Gemeinschaft may exist among business partners, one could indeed hardly speak of commercial Gemeinschaft. To make the word combination "joint-stock Gemeinschaft" would be abominable. On the other hand, there exists a Gemeinschaft of ownership in fields, forest, and pasture. The Gemeinschaft of property between man and wife cannot be called Gesellschaft of property. Thus many differences become apparent. . . .

The Gemeinschaft by blood, denoting unity of being, is developed and differentiated into Gemeinschaft of locality, which is based on a common habitat. A further differentiation leads to the Gemeinschaft of mind, which implies only

From: Ferdinand Tönnies. 1957. [1887]. *Community & Society,* translated and edited by Charles P. Loomis. East Lansing, MI: The Michigan State University Press, pp. 33–43, 64–65, 226–233.

co-operation and co-ordinated action for a common goal. Gemeinschaft of lo-cality may be conceived as a community of physical life, just as Gemeinschaft of mind expresses the community of mental life. In conjunction with the others, this last type of Gemeinschaft represents the truly human and supreme form of community. . . .

The theory of the Gesellschaft deals with the artificial construction of an ag-gregate of human beings which superficially resembles the Gemeinschaft in so far as the individuals live and dwell together peacefully. However, in the Gemein-schaft they remain essentially united in spite of all separating factors, whereas in the Gesellschaft they are essentially separated in spite of all uniting factors. In the Gesellschaft, as contrasted with the Gemeinschaft, we find no action that can be derived from an a priori and necessarily existing unity; no actions, therefore, which manifest the will and the spirit of the unity even if performed by the indi-vidual; no actions which, in so far as they are performed by the individual, take place on behalf of those united with him. In the Gesellschaft such actions do not exist. On the contrary, here everybody is by himself and isolated, and there exists a condition of tension against all others. Their spheres of activity and power are sharply separated, so that everybody refuses to everyone else contact with and ad-mittance to his sphere; i.e., intrusions are regarded as hostile acts. Such a negative attitude toward one another becomes the normal and always underlying relation of these power-endowed individuals, and it characterizes the Gesellschaft in the condition of rest; nobody wants to grant and produce anything for another indi-vidual, nor will he be inclined to give ungrudgingly to another individual, if it be not in exchange for a gift or labor equivalent that he considers at least equal to what he has given. . . .

The exterior forms of community life as represented by natural will and Gemeinschaft were distinguished as house, village, and town. These are the last-ing types of real and historical life. In a developed Gesellschaft, as in the earlier and middle stages, people live together in these different ways. The town is the highest, viz., the most complex, form of social life. Its local character, in com-mon with that of the village, contrasts with the family character of the house. Both village and town retain many characteristics of the family; the village re-tains more, the town less. Only when the town develops into the city are these characteristics almost entirely lost. Individuals or families are separate identities, and their common locale is only an accidental or deliberately chosen place in which to live. But as the town lives on within the city, elements of life in the Gemeinschaft, as the only real form of life, persist within the Gesellschaft, al-though lingering and decaying. On the other hand, the more general the con-dition of Gesellschaft becomes in the nation or a group of nations, the more this entire "country" or the entire "world" begins to resemble one large city. . . .

The city is typical of Gesellschaft in general. It is essentially a commercial town and, in so far as commerce dominates its productive labor, a factory town. Its wealth is capital wealth which, in the form of trade, usury, or industrial capital, is used and multiplies. Capital is the means for the appropriation of products of labor or for the exploitation of workers. The city is also the center of science and culture, which always go hand in hand with commerce and industry. Here the arts must make a living; they are exploited in a capitalistic way. Thoughts spread and change with astonishing rapidity. Speeches and books through mass distribution become stimuli of far-reaching importance. . . .

To conclude our theory, two periods stand thus contrasted with each other in the history of the great systems of culture: a period of Gesellschaft follows a period of Gemeinschaft. The Gemeinschaft is characterized by the social will as concord, folkways, mores, and religion; the Gesellschaft by the social will as convention, legislation, and public opinion. The concepts correspond to the types of external social organization, which may be classed as follows:

A. Gemeinschaft

1. Family life = concord. Man participates in this with all his sentiments. Its real controlling agent is the people (Volk).

2. Rural village life = folkways and mores. Into this, man enters with all his mind and heart. Its real controlling agent is the commonwealth.

3. Town life = religion. In this, the human being takes part with his entire conscience. Its real controlling agent is the church.

B. Gesellschaft

1. City life = convention. This is determined by man's intentions. Its real controlling agent is Gesellschaft per se.

2. National life = legislation. This is determined by man's calculations. Its real controlling agent is the state.

3. Cosmopolitan life = public opinion. This is evolved by man's consciousness. Its real controlling agent is the republic of scholars.

With each of these categories a predominant occupation and a dominating tendency in intellectual life are related in the following manner:

(A) 1. Home (or household) economy, based upon liking or preference, viz., the joy and delight of creating and conserving. Understanding develops the norms for such an economy.

    2. Agriculture, based upon habits, i.e., regularly repeated tasks. Co-operation is guided by custom.

    3. Art, based upon memories, i.e., of instruction, of rules followed, and of ideas conceived in one's own mind. Belief in the work and the task unites the artistic wills.

(B) 1. Trade based upon deliberation; namely, attention, comparison, calculation are the basis of all business. Commerce is deliberate action per se. Contracts are the custom and creed of business.

    2. Industry based upon decisions; namely, of intelligent productive use of capital and sale of labor. Regulations rule the factory.

    3. Science, based upon concepts, as is self-evident. Its truths and opinions then pass into literature and the press and thus become part of public opinion. . . .

DISCUSSION QUESTIONS

1. What forms of gemeinschaft remain in contemporary society and do you agree with Tönnies that life is increasingly characterized by the less personal gesellschaft forms of association?

2. Some might argue that Tönnies' conception of social change underestimates the more personal forms of interaction that occur even within a gesellschaft setting, such as work. What forms of gemeinschaft do you observe even in settings that might be described as primarily a gesellschaft?

INFOTRAC COLLEGE EDITION

You can use your access to InfoTrac College Edition to learn more about the subjects covered in this essay. Some suggested search terms include:

depersonalization	gesellschaft
gemeinschaft	urbanization

83

Goods Move. People Move. Ideas Move. And Cultures Change.

ERLA ZWINGLE

Erla Zwingle's essay describes the impact of globalization on various world cultures. His travels reveal the Westernization of non-Western countries with both positive and negative consequences. He challenges critics who believe this globalization change leads to all cultures becoming the same. Instead, he argues that cultures transform each other and human connections can be made across the globe.

Once I started looking for them, these moments were everywhere: That I should be sitting in a coffee shop in London drinking Italian espresso served by an Algerian waiter to the strains of the Beach Boys singing "I wish they all could be California girls. . . ." Or hanging around a pub in New

From: Erla Zwingle. "Goods Move. People Move. Ideas Move. And Cultures Change." *National Geographic* (August 1999): 12–33. Reprinted with permission.

Delhi that serves Lebanese cuisine to the music of a Filipino band in rooms decorated with barrels of Irish stout, a stuffed hippo head, and a vintage poster announcing the Grand Ole Opry concert to be given at the high school in Douglas, Georgia. Some Japanese are fanatics for flamenco. Denmark imports five times as much Italian pasta as it did ten years ago. The classic American blond Barbie doll now comes in some 30 national varieties—and this year emerged as Austrian and Moroccan.

Today we are in the throes of a worldwide reformation of cultures, a tectonic shift of habits and dreams called, in the curious argot of social scientists, "globalization." It's an inexact term for a wild assortment of changes in politics, business, health, entertainment. "Modern industry has established the world market. . . . All old-established national industries. . . . are dislodged by new industries whose . . . products are consumed, not only at home, but in every quarter of the globe. In place of the old wants . . . We find new wants, requiring for their satisfaction the products of distant lands and climes." Karl Marx and Friedrich Engels wrote this 150 years ago in *The Communist Manifesto*. Their statement now describes an ordinary fact of life.

How people feel about this depends a great deal on where they live and how much money they have. Yet globalization, as one report stated, "is a reality, not a choice." Humans have been weaving commercial and cultural connections since before the first camel caravan ventured afield. In the 19th century the postal service, newspapers, transcontinental railroads, and great steam-powered ships wrought fundamental changes. Telegraph, telephone, radio, and television tied tighter and more intricate knots between individuals and the wider world. Now computers, the Internet, cellular phones, cable TV, and cheaper jet transportation have accelerated and complicated these connections.

Still, the basic dynamic remains the same: Goods move. People move. Ideas move. And cultures change. The difference now is the speed and scope of these changes. It took television 13 years to acquire 50 million users; the Internet took only five. . . .

Westernization, I discovered over months of study and travel, is a phenomenon shot through with inconsistencies and populated by very strange bedfellows. Critics of Western culture blast Coke and Hollywood but not organ transplants and computers. Boosters of Western culture can point to increased efforts to preserve and protect the environment. Yet they make no mention of some less salubrious aspects of Western culture, such as cigarettes and automobiles, which, even as they are being eagerly adopted in the developing world, are having disastrous effects. Apparently westernization is not a straight road to hell, or to paradise either.

But I also discovered that cultures are as resourceful, resilient, and unpredictable as the people who compose them. In Los Angeles, the ostensible fountainhead of world cultural degradation, I saw more diversity than I could ever have supposed—at Hollywood High School the student body represents 32 different languages. In Shanghai I found that the television show *Sesame Street* has been redesigned by Chinese educators to teach Chinese values and traditions. "We borrowed an American box," one told me, "and put Chinese content into it." In India, where there are more than 400 languages and several very strict religions, McDonald's serves mutton instead of beef and offers a vegetarian menu acceptable to even the most orthodox Hindu. . . .

Los Angeles is fusion central, where cultures mix and morph. Take Tom Sloper and mah-jongg. Tom is a computer geek who is also a mah-jongg fanatic. This being America, he has found a way to marry these two passions and sell the result. He has designed a software program, *Shanghai: Dynasty,* that enables you to play mah-jongg on the Internet. This ancient Chinese game involves both strategy and luck, and it is still played all over Asia in small rooms that are full of smoke and the ceaseless click of the chunky plastic tiles and the fierce concentration of the players. It is also played by rich society women at country clubs in Beverly Hills and in apartments on Manhattan's Upper West Side. But Tom, 50, was playing it at his desk in Los Angeles one evening in the silence of a nearly empty office building.

Actually, he only appeared to be alone. His glowing computer screen showed a game already in progress with several habitual partners: "Blue Whale," a man in Cologne, Germany, where the local time was 4:30 a.m.; Russ, from Dayton, Ohio; and "yobydderf" (or Freddyboy spelled backward), a Chinese-American who lives in Edina, Minnesota. (According to one study 64 percent of Asian-American families are linked to the Internet, compared with 33 percent of all U. S. families.) Tom played effortlessly as we talked.

"I've learned about 11 different styles of mah-jongg," he told me with that detached friendliness of those whose true connection is with machines. "There are a couple of different ways of playing it in America. We usually play Chinese mah-jongg. The Japanese style is the most challenging—more hoops to jump through."

I watched the little tiles, like the cards in solitaire, bounce around the screen. From what I gathered, it has to do with collecting similar or sequential tiles of dots, characters, bamboos, winds, or dragons into groupings called pungs, kongs, or, in some cases, chows. As Tom played, he and his partners conversed by typing short comments to each other.

"I'm trying to coax Fred to be a better sport—he kind of gloats too much when he wins, and he really likes the Spice Girls too," Tom remarked as his fingers gripped the mouse. "Oh, he got a pung of dragons, I hate him. . . ." The mouse clicked, some tiles shifted.

Does he ever play with real people? "Oh yeah," Tom replied. "Once a week at the office in the evening, and Thursday at lunch." A new name appeared on the screen. "There's Fred's mother. Can't be, they're in Vegas. Oh, it must be his sister. There's my eight dot. TJ's online too, she's the one from Wales—a real night owl. She's getting married soon, and she lives with her fiancé, and sometimes he gets up and says, 'Get off that damn computer!' "

Tom played on into the night. At least it was night where I was. He, an American playing a Chinese game with people in Germany, Wales, Ohio, and Minnesota, was up in the cybersphere far above the level of time zones. It is a realm populated by individuals he's never met who may be more real to him than the people who live next door. "Can't be, they're in Vegas." The global village gets a fiber-optic party line. . . .

Early on I realized that I was going to need some type of compass to guide me through the wilds of global culture. So when I was in Los Angeles, I sought out Alvin Toffler, whose book *Future Shock* was published in 1970. In the nearly three decades since, he has developed and refined a number of interesting ideas, explained in *The Third Wave,* written with his wife, Heidi.

What do we know about the future now, I asked, that we didn't know before? "We now know that order grows out of chaos," he answered immediately. "You cannot have significant change, especially on the scale of Russia or China, without conflict. Not conflicts between East and West, or North and South, but 'wave' conflicts between industrially dominant countries and predominantly agrarian countries, or conflicts within countries making a transition from one to the other."

Waves, he explained, are major changes in civilization. The first wave came with the development of agriculture, the second with industry. Today we are in the midst of the third, which is based on information. "In 1956 something new began to happen, which amounts to the emergence of a new civilization," Toffler said. "It was in that year that U.S. service and knowledge workers outnumbered blue-collar factory workers. In 1957 Sputnik went up. Then jet aviation became commercial, television became universal, and computers began to be widely used. And with all these changes came changes in culture."

"What's happening now is the trisection of world power," he continued. "Agrarian nations on the bottom, smokestack countries in between, and knowledge-based economies on top." There are a number of countries—Brazil, for example—where all three civilizations coexist and collide.

"Culturally we'll see big changes," Toffler said. "You're going to turn on your TV and get Nigerian TV and Fijian TV in your own language." Also, some experts predict that the TV of the future, with 500 cable channels, may be used by smaller groups to foster their separate, distinctive cultures and languages.

"People ask, 'Can we become third wave and still remain Chinese?' Yes," Toffler says. "You can have a unique culture made of your core culture. But you'll be the Chinese of the future, not of the past." . . .

Change: It's a reality, not a choice. But what will be its true driving force? Cultures don't become more uniform; instead, both old and new tend to transform each other. The late philosopher Isaiah Berlin believed that, rather than aspire to some utopian ideal, a society should strive for something else: "not that we agree with each other," his biographer explained, "but that we can understand each other."

In Shanghai one October evening I joined a group gathered in a small, sterile hotel meeting room. It was the eve of Yom Kippur, the Jewish Day of Atonement, and there were diplomats, teachers, and businessmen from many Western countries. Elegant women with lively children, single men, young fathers. Shalom Greenberg, a young Jew from Israel married to an American, was presiding over his first High Holy Days as rabbi of the infant congregation.

"It's part of the Jewish history that Jews went all over the world," Rabbi Greenberg reflected. "They received a lot from local cultures, but they also kept their own identity."

The solemn liturgy proceeded, unchanged over thousands of years and hundreds of alien cultures: "Create in me a clean heart, O God, and renew a right spirit within me," he intoned. I'm neither Jewish nor Chinese, but sitting there I didn't feel foreign—I felt at home. The penitence may have been Jewish, but the aspiration was universal.

Global culture doesn't mean just more TV sets and Nike shoes. Linking is humanity's natural impulse, its common destiny. But the ties that bind people around the world are not merely technological or commercial. They are the powerful cords of the heart.

DISCUSSION QUESTIONS

1. What does the presence of U.S. culture (movies, television, fast food, music) in non-Western countries teach people about Americans? What images do they see that are inaccurate or misleading?

2. How has U.S. culture changed because of globalization? How has improved access to other cultures influenced teenagers in the United States?

INFOTRAC COLLEGE EDITION

You can use your access to InfoTrac College Edition to learn more about the subjects covered in this essay. Some suggested search terms include:

global village	technology
globalization	Westernization

84

Technological Diffusion or Cultural Imperialism?

Measuring the Information Revolution

R. ALAN HEDLEY

Alan Hedley's essay outlines the changes brought about by the technological information revolution. He examines the influence of the personal computer and access to the Internet on cultural diversity. Hedley argues that the global connections made through technology bring cultures together, but present the risk of cultural dominance.

> **The Internet is like a 20-foot tidal wave coming, and we are in kayaks.**
>
> (Grove, cited in Schlender 1996:46)

These are the words of Andy Grove, CEO of Intel, the company credited with launching the PC microelectronic revolution in 1971 (Gilder, 1989, 91-112). By most accounts, we are deeply ensconced in a technological era in which all of us are (or can be) interconnected via the information superhighway. What, then, is Grove talking about? . . .

The now overused term, "information highway," only came into being during the 1970s as a result of scientists at Corning Glass who "created a medium [optical fiber] that could transport unprecedented amounts of information on laser beams for commercially viable distances" (Diebold, 1990:132). "Today's most advanced light wave systems can relay data at 1.7 gigabits per second—fast enough to transmit the entire *Encyclopedia Britannica* in just two seconds" (Diebold, 1990:148) While this speed is impressive, should current work in developing "high temperature" superconducting material yield the results anticipated, such a resistance-free communications line could "transmit the text equivalent of one thousand *Encyclopedia Britannicas*. . . . [in just one second]. . . . Such a transmission line could transmit the entire 25 million books of the Library of Congress, the world's largest library, in two minutes" (Meredith, 1987:25). Clearly, we have come only a brief distance along the information highway. . . .

From: *International Journal of Comparative Sociology* (June 1998): 39, 198–215.

Contrary to many studies, careful analysis of the information revolution reveals that it is still very much in a nascent stage. Grove's analogy is appropriate. While the information technology (IT) market is growing quickly, it is limited primarily to the developed countries. Just five G-7[1] nations (US, Japan, Germany, France, and the UK) accounted for 80% of the IT market in 1994 (Organization for Economic Cooperation and Development, 1996:7). Furthermore, even within these countries, more than 90% of households do not have access to the Internet (OECD, 1996:106). In other words, most of us are facing Grove's tidal wave without any boats at all. Consequently, effects that we can now attribute to the information revolution are likely to intensify as IT permeates the mainstream.

If we examine how information technology is developing within the countries where it has taken hold, we can extend Grove's analogy. While some of us may be in kayaks (Americans and Canadians), others are in life boats (Japanese, and possibly Germans), and still others are in craft yet to be clearly identified. In other words, information technology is being introduced and adapted along now familiar cultural fault lines; this aspect forms a major focus of my paper.

My analysis first conceptualizes and then measures the multidimensional term "informationalization." This allows us to observe both similarities and differences in this phenomenon as it is manifested in each of the G-7 countries.

Of particular interest to the analysis is Japan because of its non-European heritage. It serves as an important theoretical case to ascertain whether characteristics thought to be inherent in the informationalization process are in fact integral to it or are instead artifacts arising out of a common (Western) cultural response to it. In other words, a comparison of Japan with the other G-7 nations allows us to address the question: "What is peculiar to the structure of . . . [an] informational society and what is specific to the history of a given country?" (Castells and Aoyama, 1994:9). . . .

CONCEPTUALIZATION AND MEASUREMENT

For the purposes of this paper, information technology refers primarily to computer hardware, components, software, and services, but also involves the new telecommunications infrastructure essential to computer networking.

A major problem in measuring the modern information revolution is that economic and labor force data are classified into traditional industrial categories. Thus, while we can identify the manufacturing component of this technological innovation in terms of computers produced and numbers employed in producing them, it is next to impossible to measure the application of this innovation within the economy. These data are "hidden" throughout all the standard industrial and occupational codes. As a result, we are limited to those applications specifically identified as "computer services."

THE INFORMATION SOCIETY
IN COMPARATIVE PERSPECTIVE

The data indicate that the computer industry has had relatively little impact to date. For example, few workers are employed directly in computing, and revenues from the industry contribute only marginally to each country's economy. Although there is a sizable IT infrastructure in place within the G-7 nations, just a small fraction of the general populace actually use computers. Rather than revealing that we are fully engaged in the information era, these data suggest a more modest conclusion.

Japan concentrates much more on the manufacture of computers and components than does the United States and the other G-7 countries. In contrast, the U.S., Canada, and European countries (with the possible exception of Germany) are more heavily involved in computer services and software (Gates, 1995:235). This different emphasis on hardware and software constitutes a major underlying cultural difference between Japan and the other G-7 nations, especially those whose native tongue is English.

Of the top twenty IT firms which together accounted for 68.7% of total revenues in 1993 (OECD, 1996:36), ten were American and seven were Japanese. Recent trend data indicate that the American lead is increasing, as industry growth is highest in software and services.

The U.S. has more than three times the computing power of Japan as measured by millions of instructions per second per capita. It appears that Japan exports much more of its computer hardware than does the U.S. Generally, the English-speaking countries occupy the top three positions on this indicator.

The data reveal relatively little penetration, certainly in relation to older technological innovations. For example, compare to 1994 per capita computer consumption throughout the G-7 countries, the number of TVs per capita was almost twice as high in 1970 (UNESCO, 1995:Table 9,2), and per capita car ownership was more than double in 1985 (*World in Figures,* 1988:23). It appears that the adoption of computers by the populace has yet to achieve critical mass.

However, within the English-speaking G-7 countries this threshold may already have been reached. Approximately one-third of American households have a personal computer, half of them further connected by modem to the Internet (OECD, 1996:106). Although personal ownership figures are not as high in Canada and the UK, these countries too appear to be well on their way.

Conversely, Japan is an anomaly among the G-7 countries. Although it leads in computer production, it lags in the business and personal application of this technology.

DISCUSSION

In 1960, Clark Kerr and his colleagues set out "the logic or imperatives" of the original industrial revolution (Kerr et al., 1964). They argued that because the results of technology can be precisely measured in terms of quantity and quality of output, it is possible to determine which technology is superior in accomplishing specific objectives, and consequently the superior technology becomes universally adopted. Their central proposition stated that the introduction of technologically superior industrial techniques leads to structural adaptations that in turn affect other aspects of society until eventually all industrialized societies, no matter how dissimilar they were initially, converge in certain patterns of social organization and behavior.

Forty years earlier, William F. Ogburn (1922) coined his now famous term "cultural lag." Ogburn theorized that changes in material culture or "the applications of scientific discovery and the material products of technology" (1956:89) occur at a faster rate than changes in the nonmaterial, adaptive culture (values, norms, patterns of social organization, etc.), thereby causing maladjustment in the nonmaterial culture, or cultural lag.

Employing these two perspectives on social change, it is useful to compare the oncoming revolution with earlier technological transformations. Concerning the introduction of computer and telecommunications technology, Kerr and his associates are correct in that incremental technical standards of power, capacity and speed (at reduced cost) have been generally acknowledged throughout the process (Tapscott, 1996:95-121). However, unlike previous technological revolutions, a significant part of the technical process itself constitutes what Ogburn would define as nonmaterial culture. Even though it commands a computer in binary code, computer software originates in words, the effective currency of culture. Furthermore, according to George Gilder (1989:328), we are now reaching the stage where "the distinction between hardware and software will all but vanish."

Although earlier technologies incorporated aspects of nonmaterial culture into their design in the form of standards and regulations, these were more limited in scope. But information technology, by its very nature, is largely nonmaterial. Consequently, to the extent that only one culture or one linguistic group produces the bulk of software, as is presently the case, and "as hardware designs increasingly embody software concepts" (Gilder, 1989:329), then certainly the possibility exists for cultural convergence (control?) on a massive scale.

Let us look at the present day Internet, the precursor of the information highway (Gates, 1995). "By January 1995, 65% of the [4.85 million] hosts connected to the Internet were in the United States, 22% in Europe, and a mere 7% were in the Asia-Oceania region (of which 2% were in Japan)" (OECD, 1996:33). This means that the Internet, both its "material" foundation and its "nonmaterial" content, is overwhelmingly American-based, English-speaking, and Western-focused. Although "the fastest-growing number of Internet hosts (percent change) in the third quarter of 1994 were [in] Argentina, Iran, Peru, Egypt, the Philippines, the Russian Federation, Slovenia, and Indonesia" (Negroponte, 1996:182), the large, affluent, Western foundation of the Internet is unlikely to be affected. It will simply have more sites to infuse.

Perhaps Grove's analogy should be rephrased:

The Internet is like an American (or Western) tidal wave coming, and the rest of the world is in kayaks.

Although it has attracted many critics in the past, the theory of cultural convergence (inundation?) will breathe new life when it enters the information era.

While Americans have enthusiastically taken to the Internet and its means of "universal" access—the PC, the Japanese have loyally remained attached to the mainframe and its sense of security (OECD, 1996:23). The personal computer permits individual initiative, the mainframe collective enterprise—hence the reference to "kayaks" and "lifeboats" in the introduction.

The Japanese have not forsaken the mainframe to the same extent as other G-7 countries for additional cultural reasons. For example, space is very scarce in Japan; it is more economical to install one large mainframe than thousands of PCs (OECD, 1996:135). Also, representing kanji characters on a standard keyboard creates another space-related problem (Gates, 1995:237).

This analysis reveals that not only are there different cultural responses to technological innovation, but in some cases the innovations themselves are culturally packaged. As evidence of this assertion, imagine for a moment the Japanese and not the Americans were pioneers of the Net. While technological advantage now appears to reside with the Americans and other English-language countries, it would be foolish to draw any firm conclusions. One has only to remember the oil shocks of the early 1970s and the strategic entry of Japanese small cars into the automobile market. However, we can conclude from this analysis that several distinct forms of an information society are likely, and that the form chosen will in all probability coincide with the cultural fault lines identified in earlier research.

Also flowing from this analysis is the potential for cultural dominance that the information revolution may foster. However, unlike previous technological revolutions, what is at stake are the very minds and thought processes of those dominated. Only powerful nations currently have the ability to choose the type of information society most compatible with their cultural institutions. However, given the high stakes involved, perhaps all countries could enter into dialogue about how information technology can be introduced in culturally harmonious ways. By these means, we could finally begin to reduce the growing disparity among nations that earlier technological revolutions have imposed.

NOTE

1. G-7 refers to seven major industrial democratic countries: France, the United States, Britain, Germany, Japan, Italy, and Canada.

REFERENCES

Castells, Mantel and Yuko Aoyama. 1994. "Paths Towards the Informational Society: Employment Structure in G-7 Countries, 1920-1990." *International Labour Review* 133(1):5-33.

Diebold, John. 1990. *The Innovators: The Discoveries, Inventions, and Breakthroughs of Our Time.* New York: Truman Talley/Plume.

Gates, Bill. 1995. *The Road Ahead.* New York: Viking.

Gilder, George. 1989. *Microcosm: The Quantum Revolution in Economics and Technology* New York: Simon & Schuster.

Kerr, Clark, et al. 1964. *Industrialism and Industrial Man.* New York: Oxford University Press [originally published in 1960].

Meredith, Dennis. 1987. "Scientists Demonstrate Use of 'High Temperature' Superconductor in Electronics and Communications." *Computers and People* 36(11–12):25–27.

Negroponte, Nicholas. 1996. *Being Digital.* New York: Vintage.

OECD. 1996. *Information Technology Outlook 1995.* Paris: Organisation for Economic Cooperation and Development.

Ogburn, William F. 1922. *Social Change with Respect to Culture and Original Nature.* New York: B. W. Huebsch.

————. 1956. "Cultural Lag as Theory." Reprinted in: Otis D. Duncan (ed.), *William F. Ogburn on Culture and Social Change.* Chicago: University of Chicago Press, pp. 86–95.

Schlender, Brent. 1996. "A Conversation with the Lords of Wintel." *Fortune* 134(1):42–58.

Tapscott, Don. 1996. *The Digital Economy: Promise and Peril in the Age of Networked Intelligence.* New York: McGraw-Hill.

1995. *Statistical Yearbook.* Paris: UNESCO.

World in Figures. 1988. Boston: G.K. Hall.

DISCUSSION QUESTIONS

1. How has the Internet influenced your education? What benefits do you get from the Internet and what disadvantages exist?

2. How does the personal computer compare to the television with regard to influencing changes in U.S. culture? Which is more transforming, the computer or the television?

INFOTRAC COLLEGE EDITION

You can use your access to InfoTrac College Edition to learn more about the subjects covered in this essay. Some suggested search terms include:

computers
cultural lag
G-7 nations

information superhighway
information technology
Internet

85

The Biotech Century:
Human Life as Intellectual Property

JEREMY RIFKIN

In this essay, Jeremy Rifkin examines the consequences of advances in biotechnology for a world economy and individual property rights. Scientific discoveries in gene therapy and cloning introduce problems of delineating ownership over parts of the human body. Rifkin outlines these issues and suggests that concerns over "genetic rights" will become the major political and economic concern of our immediate future.

In little more than a generation, our definition of life and the meaning of existence is likely to be radically altered. Long-held assumptions about nature, including our own human nature, are likely to be rethought. Many age-old practices regarding sexuality, reproduction, birth and parenthood could be partially abandoned. Ideas about equality and democracy are also likely to be redefined, as well as our vision of what is meant by terms such as "free will" and "progress." Our very sense of self and society will likely change during what I call the emerging Biotech Century, as it did when the early Renaissance spirit swept over medieval Europe more than 600 years ago.

Although Dolly the sheep and talk of cloning have gathered sensational headlines and captured the public imagination, many forces are quietly converging to create this powerful new social current. At the epicenter is a technology revolution unmatched in all of history in its power to remake ourselves, our institutions and our world: Scientists are beginning to reorganize life at the genetic level. The new biotechnologies are already reshaping a wide range of fields, including forestry, agriculture, animal husbandry, mining, energy, bioremediation, packaging and construction materials, pharmaceuticals, medicine, and food and drink. Before our eyes lies an uncharted new landscape whose contours are being shaped in thousands of biotechnology laboratories around the world. . . .

Great economic changes in history occur when a number of technological and social forces come together to create a new "operating matrix." I see seven strands composing the operating matrix of the Biotech Century. Together, they create a framework for a new economic era:

First, the ability to isolate, identify and recombine genes is making the gene pool available, for the first time, as the primary raw resource for future economic activity. Recombinant DNA techniques and other biotechnologies allow

From: *The Nation.* (April 1998). pp. 11–19.

scientists and biotech companies to locate, manipulate and exploit genetic resources for specific economic ends.

Second, the awarding of patents on genes, cell lines, genetically engineered tissue, organs and organisms, as well as the processes used to alter them, is giving the marketplace the commercial incentive to exploit the new resources.

Third, the globalization of commerce and trade make possible the wholesale reseeding of the Earth's biosphere with a laboratory-conceived Second Genesis, an artificially produced bioindustrial nature designed to replace nature's own evolutionary scheme. A global life-science industry is already beginning to wield unprecedented power over the vast biological resources of the planet. Life-science fields ranging from agriculture to medicine are being consolidated under the umbrella of giant "life" companies in the emerging biotech marketplace.

Fourth, the mapping of the approximately 100,000 genes that make up the human genome and new breakthroughs in genetic screening, including DNA chips, somatic gene therapy and the imminent prospect of genetic engineering of human eggs, sperm and embryonic cells, are paving the way for the wholesale alteration of the human species and the birth of a commercially driven eugenics civilization.

Fifth, a spate of new scientific studies on the genetic basis of human behavior and the new sociobiology that favors nature over nurture are providing a cultural context for the widespread acceptance of new biotechnologies.

Sixth, the computer is providing the communications and organizational medium to manage the genetic information that makes up the biotech economy. All over the world, researchers are using computers to decipher, download, catalogue and organize genetic information, creating a new store of genetic capital for use in the bioindustrial age. Computational technologies and genetic technologies are fusing into a powerful new technological reality. Microsoft chairman Bill Gates sums up the new collaborative efforts by saying, "This is the information age, and biological information is probably the most interesting information we are deciphering and trying to decide to change. It's all a question of how, not if."

Seventh, a new cosmological narrative about evolution is beginning to challenge the neo-Darwinian citadel with a view of nature compatible with the operating assumptions of the new technologies and the new global economy. The new ideas about nature provide the legitimizing framework for the Biotech Century by suggesting that the way we are reorganizing our economy and society are amplifications of nature's own principles and, therefore, justifiable.

In short, the Biotech Century brings with it a new resource base, a new set of transforming technologies, new forms of commercial protection, a global trading market to reseed the earth with an artificial Second Genesis, an emerging eugenics science, a new supporting sociology, a new communications tool to organize and manage economic activity at the genetic level and a new cosmological narrative. Together, genes, biotechnologies, life patents, the global life-science industry, human gene screening and surgery, the new cultural currents, computers and revised theories of evolution are beginning to remake the world.

HUMAN LIFE AS INTELLECTUAL PROPERTY

Genes are the "green gold" of the Biotech Century. The economic and political forces that control the genetic resources of the planet will exercise tremendous power over the future world economy, just as the industrial age access to and control over fossil fuels and valuable metals helped determine control over world markets. Multinational corporations are already scouting the continents, hoping to locate microbes, plants, animals and humans with rare genetic traits that might have future market potential. After locating the desired traits, biotech companies are modifying them and then seeking patent protection for their new "inventions." . . .

It seems there is no place on earth too remote for the gene hunters to go. In April 1997 the *Los Angeles Times* reported on a scientific expedition led by Dr. Noe Zamel, a University of Toronto medical geneticist, and financed by Sequana Therapeutics of La Jolla, California. Sequana is one of a handful of new biotech start-up companies dedicated to gene prospecting. The "genomics" firms are on the commercial frontier of the emerging biotech revolution.

The Sequana team traveled, by way of a South African Navy ship, to the tiny volcanic island of Tristan da Cunha, a strip of forty square miles in the middle of the Atlantic Ocean often referred to as the world's loneliest island. Its few hundred residents are the descendants of British sailors who arrived in 1817. What makes this small inbred local population interesting to Zamel and his team is that half of them suffer from asthma. Scientists hope to find the gene or genes responsible and patent them.

Company scientists took blood samples from 270 of the island's 300 residents and later reported that it had located two "candidate genes" responsible for asthma. However, thus far the company has refused to share its findings with other researchers in the field, giving rise to charges that it is putting commercial considerations above collaborative efforts to find a cure for the disease. For their part, genomics companies like Sequana acknowledge that they're in business to exploit the human genome commercially and could not hope to turn a profit if they couldn't keep their research proprietary—at least until it's patented. Sequana and other genomics firms maintain that market incentives are the best and most efficient way to advance the research. Others aren't so sure. "The issue is a nightmare," said one geneticist working for a large pharmaceutical company, who spoke off the record to *Nature* magazine's editor.

Foreign nationals aren't the only ones whose cell lines and genomes are being patented by commercial companies in the United States. In a precedent-setting case in California, an Alaska businessman named John Moore found his own body parts had been patented, without his knowledge, by U.C.L.A. and licensed to the Sandoz Pharmaceutical Corporation. Moore had been diagnosed as having a rare cancer and underwent treatment at U.C.L.A. An attending physician and university researcher discovered that Moore's spleen tissue produced a blood protein that facilitates the growth of white blood cells that are valuable anti-cancer agents. The university created a cell line from Moore's spleen tissue and obtained a patent on its "invention" in 1984. The cell line is estimated to be worth more than $3 billion. Moore subsequently sued the University of California, claiming a property right over his own tissue. In 1990 the California Supreme Court ruled against Moore, holding that he had no property right over his own body tissues. Human body parts, the court argued, could not be bartered as a commodity in the marketplace. However, the court did say that the "inventors" had a responsibility to inform Moore of the commercial potential of his tissue and, for that reason alone, had breached their fiduciary responsibility and might be liable for some kind of monetary damage. Still, the court upheld the primary claim of the university that the cell line itself, while not the property of Moore, could justifiably be claimed as the property of U.C.L.A. The irony of the decision was captured by Judge Allen Broussard in his dissenting opinion. He wrote that

> the majority's rejection of plaintiff's conversion cause of action does *not* mean that body parts may not be bought and sold for research or commercial purpose or that no private individual or entity may benefit economically from the fortuitous value of plaintiff's diseased cells. Far from elevating these biological materials above the marketplace, the majority's holding simply bars *plaintiff,* the source of the cells, from obtaining the benefit of the cell's value, but permits *defendants,* who allegedly obtained the cells from plaintiff by improper means, to retain and exploit the full economic value of their ill-gotten gains free of. . .liability.

The extraordinary implications of privatizing the human body—parceling it out in the form of intellectual property to commercial institutions—are illustrated quite poignantly in the case of a patent awarded by the European Patent Office to a U.S. company named Biocyte. The patent gives the firm ownership of all human blood cells that have come from the umbilical cord of a newborn child and are being used for any therapeutic purposes. The patent is so broad that it allows this one company to refuse the use of any blood cells from the umbilical cord to any individual or institution unwilling to pay the patent fee. Blood cells from the umbilical cord are particularly important for marrow transplants, making it a very valuable commercial asset. It should be emphasized that this patent was awarded simply because Biocyte was able to isolate the blood cells and deepfreeze them. The company made no change in the blood itself. Still, the company now possesses commercial control over this part of the human body. . . .

The entrepreneurial scramble to patent the genome of the human family has picked up substantial momentum over the past several years, in large part because of the quickened pace of mapping and sequencing the approximately 100,000 genes that make up the human genome. As soon as a gene is tagged, its "discoverer" is likely to apply for a patent, often before even knowing the function or role of the gene. . . .

The increasing consolidation of corporate control over the genetic blueprints of life as well as the technologies to exploit them is alarming, especially when we stop to consider that the biotech revolution will affect every aspect of our lives. The way we eat; the way we date and marry; the way we have our babies; the way our children are raised and educated; the way we work; the way we engage in politics; the way we express our faith; the way we perceive the world around us and our place in it—all our individual and shared realities will be deeply touched by the new technologies of the Biotech Century.

THE BACKLASH

The debate over life patents is one of the most important issues ever to face the human family. Life patents strike at the core of our beliefs about the very nature of life and whether it is to be conceived of as having intrinsic, or mere utility, value.

The last great debate of this kind occurred in the nineteenth century, over the issue of human slavery, with abolitionists arguing that every human being has intrinsic value and "God-given rights" and cannot be made the personal commercial property of another human being. The abolitionists' argument ultimately prevailed, and legally sanctioned slavery was abolished in every country in the world where it was still being practiced.

Like the anti-slavery abolitionists of the nineteenth century, a new generation of genetic activists is beginning to challenge the very concept of patenting life, arguing that life is imbued with intrinsic value and therefore can never be legitimately reduced to commercial intellectual property controlled by global life-science conglomerates and traded as mere utilities in the marketplace. Feminists, farmers, animal rights groups, consumer organizations, health advocates and social justice organizations around the world are coalescing into a new and powerful countervailing force to the growing genetic commerce that trades in the blueprints of life.

In May 1995 a coalition of more than 200 religious leaders, including the titular heads of virtually every major Protestant denomination, more than a hundred Catholic bishops and Jewish, Muslim, Buddhist and Hindu leaders, announced their opposition to the granting of patent on animal and human genes, organs, tissues and organisms. The effort was organized by the Foundation on Economic Trends, which I head.

The coalition, the largest assemblage of U.S. religious leaders to come together on an issue of mutual interest in the twentieth century, said that the patenting of life marked the most serious challenge to the notion of God's creation in history. How can life be defined as an invention to be profited from by scientists and corporations when it is freely given as a gift of God, ask the theologians? Either life is God's creation or a human invention, but it can't be both. Speaking for the coalition, Jaydee Hanson, an executive with the United Methodist Church, said, "We believe that humans and animals are creations of God, not humans, and as such should not be patented as inventions." While not all the religious leaders oppose "process" patents for the techniques used to create transgenic life forms, they are unanimous in their opposition to the patenting of the life forms and the parts themselves. They are keenly aware of the profound consequences of shifting authorship from God to scientists and transnational companies and are determined to hold the line against any attempt by "man" to stake his own claim as the prime mover and sovereign architect of life on earth.

What might it mean for subsequent generations to grow up in a world where they come to think of all life as mere invention—where the boundaries between the sacred and the profane, and between intrinsic and utility value, have all but disappeared, reducing life itself to an objectified status, devoid of any unique or essential quality that might differentiate it from the strictly mechanical? The battle to keep the earth's gene pool an open commons, free of commercial exploitation, is going to become one of the critical struggles of the Biotech Age. "Genetic rights," in turn, is likely to emerge as the seminal issue of the coming era, defining much of the political agenda of the Biotech Century.

DISCUSSION QUESTIONS

1. What negative consequences can we expect for a society that has not yet learned to regulate major scientific developments? Do we need regulation over something like cloning? Why or why not?

2. How can we catch up with the "cultural lag" created by advances in biotechnology? What legal issues do we need to consider when trying to settle issues of ownership of human body parts? Is this an issue for legislation?

INFOTRAC COLLEGE EDITION

You can use your access to InfoTrac College Edition to learn more about the subjects covered in this essay. Some suggested search terms include:

bioethics   gene therapy
biotechnology   intellectual property
cloning

Glossary

A

achieved status: a status attained by effort

affective neutrality: the emotional state where no emotional expression is allowed

age cohort: an aggregate group of people born during the same time period

alienation: the feeling of powerlessness and separation from one's group or society

anamnesis: a recalling to mind; reminiscence

anomie: a social structural condition existing when social regulations (norms) in a society breakdown

antipodal categories: categories of people that are considered directly opposite to one another in the social system

anti-Semitism: hostility toward or discrimination against Jewish people

asceticism: the practice of promoting strict self-denial as a measure of spiritual discipline

ascribed status: a status determined by birth

assimilation: process by which a minority becomes socially, economically, and culturally absorbed within the dominant society

authoritarian personality: a personality characterized by a tendency to rigidly categorize people, to submit to authority, to rigidly conform, and to be intolerant of ambiguity

authority: power that is perceived by others as legitimate

B

backstage behavior: behavior we normally do not display in public

beliefs: shared ideas held collectively by people within a given culture

bigotry: the state of mind of people intolerantly devoted to their own opinions and prejudices

biotechnology: applied biological science (such as bioengineering or recombinant DNA technology)

bipolar concept of racism: conceiving the problems of race in the United States as being about white and black relations only

block grant: a large amount of money allocated by the federal government to a state government for a specific purpose

boundary maintenance: the act(s) of preserving something that indicates or fixes a limit into group membership

bourgeoisie: term used to loosely describe either the ruling or middle class in a capitalist society

brainwashing: a forcible indoctrination to induce someone to give up basic political, social, or religious beliefs and to accept contrasting, regimented ideas

bureaucracy: a type of formal organization characterized by an authority hierarchy, with a clear division of labor, explicit rules, and impersonality

C

capital: accumulated goods devoted to the production of other goods

capitalism: an economic system based on the principles of market competition, private property, and the pursuit of profit

capitalist class: those persons who own the means of production in a society

caste system: a system of stratification (characterized by low social mobility) in which one's place in the stratification system is determined and fixed by birth

chauvinism: an attitude of superiority toward members of the opposite sex; also, behavior expressive of such an attitude

chi-square: a test statistic used to determine the statistical significance of a relationship between two variables

class: see social class

classism: prejudice or discrimination based on social class

clique: a narrow, exclusive circle or group of persons, especially one held together by common interests, views, or purposes

collective action (also collective behavior): behavior that occurs when the usual conventions are suspended and people collectively establish new norms of behavior in response to an emerging situation

collective consciousness: the body of beliefs that are common to a community or society and that give people a sense of belonging

collectivism: a political or economic theory advocating collective control over production and distribution of a system

coming out process: the process of defining oneself in terms of some prior latent or "secret" role, such as publicly defining oneself as gay or lesbian

commodity chain: the network of production and labor processes by which a product becomes a finished commodity

communism: an economic system where the state is the sole owner of the systems of production

conflict theory: a theoretical perspective that emphasizes the role of power and coercion in producing social order

consanguinity: relationship because of the same ancestry or descent

conservationism: a theoretical perspective that posits a careful preservation of a natural resource

consumerism: a preoccupation with and inclination toward the buying of consumer goods

contingent worker: a temporary worker

conversion: an experience associated with a decisive adoption of religion

corrections: a process by which convicted criminals are resocialized to be acceptable, behaving members of a society

correlation: a statistical technique that analyzes patterns of association between pairs of sociological variables

counterculture: subculture created as a reaction against the values of the dominant culture

critical modernist: a social theorist who embraces the globalization of industrialization, but expects the negative affects of this process to be solved ethically

cult: a religious group devoted to a specific cause or charismatic leader

cultural icon: an object of uncritical devotion within a society or culture

cultural imperialism: forcing all groups in a society to accept the dominant culture as their own

culture: the complex system of meaning and behavior that defines the way of life for a given group or society

culture lag: the delay in cultural adjustments to changing social conditions

culture of poverty: the argument that poverty is a way of life and, like other cultures, is passed on from generation to generation

cyberspace interaction: interaction occurring when two or more persons share a virtual reality experience via communication and interaction with each other

cyborg: a human being who is linked (as for temporary adaptation to a hostile space environment) to one or more mechanical devices upon which some vital physiological functions depend

D

data: the systematic information that is used to investigate research questions

data set: a large collection of systematic information used in sociological research

decentralization: the redistribution of population and industry from urban centers to outlying areas

deconstruction: an intellectual approach to understanding how ideas and images are socially determined by analyzing and dissecting the content of such images and ideas

deep ecology: belief system stating that saving the environment is not just for humans but also for other forms of life as well

de facto: by practice or in fact, even if not in law

deindustrialization: term referring to the structural transformation of a society from a manufacturing-based economy to a service-based economy

de jure: by rule, by policy, or by law

deterrence: a punishment intended to prevent or discourage the behavior being punished

deviance: behavior that is recognized as violating expected rules and norms

diaspora: the connection across nations of people related by common historical descent

discrimination: overt negative and unequal treatment of the members of some social group or stratum solely because of their membership in that group or stratum

discursive practice: proceeding coherently from topic to topic; marked by analytical reasoning

distribution: the position, arrangement, or frequency of occurrence (as of the members of a group) over an area or throughout a space or unit of time

division of labor: the systematic interrelation of different tasks that develops in complex societies

docent: a German term referring to a college or university lecturer

dominant culture: the culture of the most powerful group in society

double consciousness: having the realization that one is being viewed as the "other" in a social situation, and also being able to perceive and understand the dominant group

downsizing: term referring to action by companies to eliminate job positions in order to cut the firm's operating costs

dramatic metaphor: a perspective used to suggest that social interaction has a likeness to dramas presented on a stage

dual labor market: theoretical description of the occupational system defining it as divided into two major segments: the primary labor market and the secondary labor market

dyad: a group consisting of two people

E

economic restructuring: contemporary transformations in the basic structure of work that are permanently altering the workplace, including the changing composition of the workplace, deindustrialization, the use of enhanced technology, and the development of a global economy

economy: the system on which the production, distribution, and consumption of goods and services is based

egalitarian: societies or groups where men and women share power

emotional displacement theory: Freud's social psychological theory positing that emotion can be redirected from one person or object to another

emotional labor (or management): work intended to produce a desired emotional effect on a client

endogamy: the practice of selecting mates from within one's group

Enlightenment: the period in eighteenth and nineteenth century Europe characterized by faith in the ability of human reason to solve society's problems

entitlement: a government program providing benefits to members of a specified group

environmentalism: advocacy of the preservation or improvement of the natural environment, especially the movement to control pollution

environmental racism: phrase describing the disproportionate location of sources of toxic pollution in or very near communities of color

epistemology: the division of philosophy that investigates the nature and origin of knowledge

essentialist formulation of race: theory positing that race is objective, fixed, and biologically determined

ethnic enclave: a regional or neighborhood location with people of distinct cultural origins

ethnic group: a social category of people who share a common culture, such as a common language or dialect, a common religion, and common norms, practices, and customs

ethnography: a descriptive account of social life and culture in a particular social system based on observations of what people actually do

ethnomethodology: a technique for studying human interaction by deliberately disrupting social norms and observing how individuals attempt to restore normalcy

evolutionary social theory: a theory of social change predicting that societies change in a single direction over time

exchange theory: an approach that explains social interaction on the basis of rewards, punishments, and the exchange of valued resources with others

exogamy: the practice of selecting mates from outside one's group

export processing zones: an area in a developing country where little bureaucratic regulation is enforced on import or exports

expression rules: situational guidelines for the normative indicators (facial or otherwise) of emotions

F

factor: a tendency or dimension of a cluster of related variables

factor analysis: research method where a number of related variables are reduced to a smaller number of variables by determining the similarities among them

false consciousness: the idea that subordinated classes internalize the view of the ruling class

family: a primary group of people—usually related by ancestry, marriage, or adoption—who form a cooperative economic unit to care for any offspring (and each other) and who are committed to maintaining the group over time

feeling rules: situational guidelines for emotional displays appropriate to a specific situation

feminism: beliefs, actions, and theories that attempt to bring justice, fairness, and equity to all women, regardless of their race, age, class, sexual orientation, or other characteristics

feminization of poverty: the trend whereby a growing proportion of the poor are women and children

feminization of the workforce: phrase used to describe the entry of women into the workforce in large numbers beginning in the early 1970s

feudalism: the system of political organization prevailing in Europe from the ninth to about the fifteenth centuries having as its basis the relation of lord to vassal with all land held by the upper class

field research: the process of gathering data in a naturally occurring social setting

filial duty: the set of responsibilities befitting a son or daughter

flexible specialization production: a type of production system that allows firms to produce small batches of differentiated goods for diverse customers, rather than large batches of similar goods for most customers

folkways: the general standards of behavior adhered to by a group

formal rules: written regulations of civil law

free-market capitalism: see capitalism

frequency: the number of individuals in a single class or category

functionalism: a theoretical perspective that interprets each part of society in terms of how it contributes to the stability of the whole society

G

G-7: (group of 7) term referring to the 7 most industrialized nations in the world, specifically the United States, Japan, Canada, Germany, Italy, France, and the United Kingdom

game stage: the stage in childhood when children become capable of taking a multitude of roles at the same time

geisha: a Japanese woman who is trained to provide entertaining company for men

gemeinschaft: German for community; a state characterized by a sense of fellow feeling among the members of a society, including strong personal ties, sturdy primary group memberships, and a sense of personal loyalty to one another; associated with rural life

gender: socially learned expectations and behaviors associated with members of each sex

gender identity: one's definition of self as a woman or man

gender role: the learned expectations associated with being a man or a woman

gendered institution: total pattern of gender relationships embedded in social institutions

generalized other: the abstract composite of social roles and social expectations

genocide: the deliberate and systematic destruction of a racial, political, or cultural group

gerontologist: one who studies the branch of knowledge dealing with aging and the problems of the aged

gesellschaft: German for society; a form of social organization characterized by a high division of labor, less prominence of personal ties, the lack of a sense of community among the members, and the absence of a feeling of belonging; associated with urban life

global care chain: a social network where a series of personal links between mothers across the globe transfer caring (paid or unpaid) to children other than their own

global city: coordination centers for the global economy where multi-national corporations and large financial concerns house their headquarters, along with many other business service firms

global culture: diffusion of a single culture throughout the world

global economy: term used to refer to the fact that all dimensions of the economy now cross national borders

globalization: increased economic, political, and social interconnectedness and interdependence among societies in the world

gross domestic product (GDP): the total monetary value of all goods and services produced in a country in one year

grounded theory: a theory based on observations of the real world rather than solely on abstract reasoning

group: a collection of individuals who interact and communicate, share goals and norms, and who have a subjective awareness as "we"

growth coalitions: groups of elite entrepreneurs who excel at capital acquisition

H

hegemony: ascendancy or dominance of one social group or culture over another

hegira: a journey undertaken to escape from a dangerous or undesirable situation

heterogeneous: consisting of dissimilar or diverse ingredients or constituents

heterosexism: the institutionalization of heterosexuality as the only socially legitimate sexual orientation

homogamy: the pattern by which people select mates with similar social characteristics to their own

homogeneous: of the same or similar kind or composition

homophobia: the fear and hatred of homosexuality

hypothesis: a statement about what one expects to find when one does research

I

"I": the preverbal part of the mind that experiences everything without the reflection and thought that language makes possible

ideology: a belief system that tries to explain and justify the status quo

impression management: a process by which people control how others perceive them

indicator: something that represents an abstract concept

individualism: a doctrine that states that the interests of the individual are and ought to be paramount

inductive reasoning: a logical process of building general principles from specific observations

industrialization: term referring to sustained economic growth following the application of raw materials and other more intellectual resources to mechanized production

in-group: a group with which one feels a sense of solidarity or community of interests

in-group solidarity: affective group cohesion based on association with like others

initiation rite: ceremonies, ordeals, or instructions with which one is made a member of a sect or society or is invested with a particular function or status

institution: see social institution

institutional privileges: institutionalized benefits given to those of the dominant group that seem to "naturally" afford this group greater opportunity

instrumental: emotionally neutral, task-oriented (goal-oriented) behavior

intellectual capital: intangible assets or know-how used in the production of goods that can be copyrighted; for example, computer software innovations

interlocking directorate: organizational linkages created when the same people sit on the boards of directors of a number of different corporations

internalization: to incorporate (values, patterns of culture, etc.) within the self as conscious or subconscious guiding principles through learning or socialization

internalized oppression: the process by which members of a dominated social group accept the view of them held by the dominant group

internal job ladder: the structure of opportunities for job placement and promotion within firms

internal structure of cultural values: mental organization or representation of the abstract standards in a society that define ideal principles

issues: problems that affect large numbers of people and have their origins in the institutional arrangements and history of a society

K

kinship system: the pattern that defines people's family relationships to one another

L

labeling theory: a theory that interprets the responses (or "label") of others as most significant in determining the behavior of people

labor market: the available supply of jobs

language: a set of symbols and rules that, put together in a meaningful way, provide a complex communication system

La Raza: Spanish phrase meaning literally "the Race"; used to refer to a social movement promoting the rights of Latinos

leisure gap: term referring to the differences in women's and men's spare time due to the division of labor within a household

lesbian feminism: a feminist theoretical perspective and movement philosophy that posits that women's relationship with each other, sexual and otherwise, should be valued and supported

liberal feminism: a feminist theoretical perspective asserting that the origin of women's inequality is in barriers blocking women's advancement

life chances: the opportunities that people have in common by virtue of belonging to a particular class

life course perspective: sociological framework for studying aging that connects people's personal attributes, the roles they occupy, and the life events they experience to their sociohistorical context

linguistic minority: the community of sign language users

longitudinal study: a research design dealing with change within a specific group over a period of time

looking-glass self: the idea that people's conception of self arises through reflection about their relationship to others

M

macroinstitution: an established and organized broad-scale system of social behavior with a recognized purpose, such as a culture or a society

mail-order bride: the system whereby men in a developed country order future brides from catalogues of women of developing countries

mass mobilization: large scale organization of persons to address an issue

matrilineal kinship system: kinship systems in which family lineage (or ancestry) is traced through the mother

"me": the social being we see as ourself in any social situation

means of production: the system by which goods are produced and distributed

means-testing: method used to determine eligibility for governmental assistance or benefits based on income

measurement accuracy: degree of precision of a measure

measurement error: inaccuracy due to flaws in a measurement instrument

median: the midpoint in a series of values that are arranged in numerical order

mediator (statistical): a variable that helps establish the causal relationship between two other variables (variable B in a situation where A causes B which causes C)

medicalization of deviance: a social process through which a norm-violating behavior is culturally defined as a disease and is treated as a medical condition

meritocracy: a system (as an educational system) presuming that the most talented are chosen and moved ahead on the basis of their achievement

metaphysics: a branch of philosophy that examines first principles of knowledge building

methodology: term referring to the practices and techniques used to gather, process, and interpret theories about social life

microinstitution: an established and organized small-scale system of social behavior with a recognized purpose, such as a small group or dyad

minority group: any distinct group in society that shares common group characteristics and is forced to occupy low status in society because of prejudice and discrimination

miscegenation: the mixing of the races through marriage

modeling: children's imitation of adults' behaviors

model minority: a minority group used as an example to suggest that social mobility is possible for minority groups (ignoring the fact that the model minority has only achieved partial success)

modernist thinking: a belief in the power of rational thought and science to explain, predict, and manipulate the environment

modernization: term referring to the social process by which traditional societies achieve industrial capitalist economies

moral relativist: a person who believes that judgement of good and bad are relative to the individual and to the time or place in which she or he acts

mores: strict norms that control moral and ethical behavior

multiculturalism movement: the push to introduce more courses on different and diverse subcultures and groups, ethnic groups, and gender studies into the elementary, high-school, and college curricula

multidimensional theory of intelligence: a theory that there are several different kinds of intelligence that do not necessarily overlap

multivariate analysis: any of several statistical or research methods for examining the effects of more than two variables at the same time

mythopoetic men's movement: organized group of men inspired by Robert Blye's philosophies of recapturing a form of masculinity represented in Western mythology to promote a positive male identity that some believe was allegedly lost in the feminist movement

N

natal family: birth family

nativist movements: social movement that champions policies favoring native inhabitants as opposed to immigrants

natural law: a principle or body of laws considered to be derived from nature or religion and believed to be ethically binding in human society

natural selection: a process in nature resulting in the survival and perpetuation of only those forms of animal and plant life having certain favorable characteristics that best enable them to adapt to a specific environment

NIMBY: environmental movement with the slogan "not in my back yard"

nonrepresentative sample: a sample not similar to the population from which it was drawn

norms: the specific cultural expectations for how to act in a given situation

nuclear family: social unit comprised of a man and a woman living together with their children

O

objectification: process of treating a person or group of people as inanimate objects, devoid of their humanity

objectivity: the absence of bias in making or interpreting observations

occupational segregation: the pattern by which workers are separated into different occupations on the basis of social characteristics such as race and gender

oppositional culture: the complex system of meaning and behavior that is defined as contrary to the dominant culture

oppression: systematic, institutionalized mistreatment of one social group by another social group

organic metaphor: refers to the similarity early sociologists saw between society and other organic systems

organization: see social organization

out-group: a group that is distinct from one's own and is usually an object of hostility or dislike

outsourcing: hiring temporary workers outside of a firm to do certain jobs; such temporary workers seldom receive employee benefits

P

pacifist: a person opposed to war or violence of any kind

paradigm: a framework for understanding phenomenon based on a particular and identifiable set of assumptions

parochialism: quality or state of being limited in range or scope

participant observation: a method whereby the sociologist becomes both a participant in the group being studied and a scientific observer of the group

pathology: the functional manifestations of a disease

patriarchy: a society or group where men have power over women

patrimonialism: the act of deriving anything from one's father or ancestors

phenotype: the visible properties of an organism that are produced by the interaction of all or part of the genetic constitution of an individual or group and the environment

play stage: the stage in childhood when children begin to take on the roles of significant people in their environment

pluralist model: a theoretical model of power in society as coming from the representation of diverse interests of different groups in society

pogrom: systematic, organized harrassment and massacre of people; specifically, harrassment and/or massacre of Jewish people

political economy: term referring to the interdependent workings and interests of governing and material wealth production and distribution systems

politics of exclusion: methods or tactics used to bar a non-dominant social group from obtaining the privileges, resources, etc. enjoyed by the dominant social group

politics of inclusion: methods or tactics used to embrace or involve all social groups in any aspect of social life

positivism: a system of thought in which accurate observation and description is considered the highest form of knowledge

postmodernism: a theoretical perspective based on the idea that society is not an objective thing, but is found in the words and images—or discourses—that people use to represent and describe behavior and ideas

postmodern market individualism: the idea that the interests of the individual in an unfettered free market system ought to be ethically paramount

poverty line: the figure established by the government to indicate the amount of money needed to support the basic needs of a household

power: a person or group's ability to exercise influence and control over others

power elite model: a theoretical model of power positing a strong link between government and business

predictive validity: the extent to which a test accurately predicts later college grades, or some other outcome, such as likelihood of graduating

prejudice: the negative evaluation of a social group, and individuals within that group, based upon conceptions about that social group that are held despite facts that contradict it

presentation of self: Goffman's phrase referring to how people display themselves during social interaction

prestige: the subjective value with which different groups or people are judged

primary group: a group characterized by intimate, face-to-face interaction and relatively long-lasting relationships

primary source: first-hand accounts of events used as data

primordialist: a social theorist who believes national interests must be primary concerns in the new global economy not global concerns

privatization: to change (as a business or industry) from public to private control or ownership

proletarianization: (1) the process by which parts of the middle class become effectively absorbed into the working class; (2) the process by which an occupational category is downgraded in occupational status and is more closely akin to working-class jobs

proletarians: the laboring class; especially the class of industrial workers who lack their own means of production and hence sell their labor to live

proportion: the ratio of a part to the total

Protestant ethic: belief that hard work and self-denial lead to salvation and success

psychotherapy: treatment of a mental or emotional disorder based on Freud's psychoanalytic theory

Puritan: a member of a sixteenth and seventeenth century Protestant group in England and New England opposing ceremonial worship

Q

qualitative method: sociological research methodology based on interpretive and, usually, non-quantitative observation

R

race: a social category or social construction based on certain characteristics, some biological, that have been assigned social importance in the society

racial formation: process by which groups come to be defined as a "race" through social institutions such as the law and the schools

racial project: an organized effort to interpret and represent resources along particular racial lines

racism: the perception and treatment of a racial or ethnic group, or member of that group, as intellectually, socially, and culturally inferior to one's own group

radical feminism: feminist theoretical perspective that interprets patriarchy as the primary cause of women's oppression

radical movements: social movements that seek fundamental change in the structure of society

random sample: a sample that gives everyone in the population an equal chance of being selected

rate: parts per some number (e.g., per 10,000; per 100,000)

rational choice paradigm: the theoretical perspective positing that any alternative chosen is done because of a person's calculated self-interest

rationalization of society: term used by Max Weber to describe society being increasingly organized around legal, empirical, and scientific forms of thought

reactionists: those who tend toward a former and usual outmoded political or social order or policy

real wages: wage rate adjusting for the annual rate of inflation

recidivism: a tendency to relapse into a former pattern of criminal behavior

redlining: practice by which bank officials excluded racial groups from purchasing homes in particular residential areas to preserve the all-white composition of such neighborhoods

reductionism: explaining a complex system's working with a simplistic explanation

refugee: one that flees, especially one who flees to a foreign country or power to escape danger or persecution

reliability: the likelihood that a particular measure would produce the same results if the measure were repeated

religion: an institutionalized system of symbols, beliefs, values, and practices by which a group of people interprets and responds to what they feel is sacred and that provides answers to questions of ultimate meaning

religiosity: the intensity and consistency of practice of a person's (or group's) faith

religious socialization: the process by which one learns a particular religious faith

residual welfare state: term referring to an economic system that is designed to have a free market system foundation and a state government that plays a minimal role in the distribution of public benefits such as welfare

ritual: symbolic activities that express a group's spiritual convictions

role: the expected behavior associated with a given status in society

role conflict: two or more roles associated with contradictory expectations

role negotiation: working out for oneself the expectations of a specific role based on one's own beliefs and the social beliefs associated with the role

role strain: conflicting expectations within the same role

S

sample: any subset of units from a population that a researcher studies

science: a body of thought about the natural world that rests on the idea that reliable knowledge must be based on systematic, observable facts that will lead anyone who considers them to the same conclusions

scientific method: the steps in a research process, including observation, hypothesis testing, analysis of data, and generalization

scientific objectivity: see objectivity

script: a learned performance of a social role

secondary crime: deviant behavior required to maintain and sustain the original deviant behavior

secondary source: second-hand account of an event used as data

"second shift": Hochschild's term referring to women's second round of work after wage labor when they come home to manage their households

secular: the ordinary beliefs of daily life that are specifically not religious

seder: a Jewish home or community service including a ceremonial dinner held on the first evening of the Passover in commemoration of the exodus from Egypt

self: the relatively stable set of perceptions of who we are in relation to ourselves, others, and the social system

self-fulfilling prophecy: the process by which merely applying a label changes behavior and thus tends to justify the label

service sector: the part of the labor market that makes up the non-manual, non-agricultural jobs

sex: biological identity as male or female

sexism: a system of practices and beliefs through which women are controlled and exploited because of the significance given to differences between the sexes

sex segregation: the distribution of men and women in any social group or society

sexual orientation: the manner in which individuals experience sexual arousal and pleasure

shamanism: belief in an unseen world of gods, demons, and ancestral spirits responsive only to the priests who use magic for the purpose of curing the sick, divining the hidden, and controlling events

snowball sample: a subset of a population obtained by sampling one individual who recommends another to be included in the sample, and so forth

social class: the social structural position that groups hold relative to the economic, social, political, and cultural resources of society

social construction of reality: the process by which what we perceive as real is given objective meaning through a process of social interaction

social contract: an unwritten agreement between two social groups that suggests an exchange by all of certain individual rights for certain collective securities

social control: a process by which groups and individuals within those groups are brought into conformity with dominant social expectations

Social Darwinism: the idea that society evolves to allow the survival of the fittest

social exchange: transferring something between individuals, but not necessarily involving money

social facts: social patterns that are external to individuals

social identity: the role or status that distinguishes a person within a social context

social institution: an established and organized system of social behavior with a recognized purpose

social interaction: behavior between two or more people that is given meaning

socialist feminism: a feminist theoretical perspective that interprets the origins of women's oppression as lying in the system of capitalism

social mobility: a person's movement over time from one social class to another

social movement: a group that acts with some continuity and organization to promote or resist change in society

social network: a set of links of social interaction between individuals or other social units such as groups or organizations

social order: the stable social organization of the social world

social organization: the order established in social groups

social sanctions: mechanisms of social control that enforce norms

social speedup: the phenomenon of having more to do in the same amount of time than was once the case

social stratification: a relatively fixed hierarchical arrangement in society by which groups have different access to resources, power, and perceived social worth; a system of structured social inequality

social structure: the patterns of social relationships and social institutions that comprise society

society: a system of social interactions that includes both culture and social organization

socioeconomic status (SES): a measure of class standing, typically indicated by income, occupational prestige, and educational attainment

sociological imagination: the ability to see the societal patterns that influence individual and group life

sociology: the study of human behavior in society

Spanish Inquisition: a former Roman Catholic tribunal for the discovery and punishment of alleged heresy

specialization: structural adaptation of a person or group of persons to a particular function in a particular environment

spirit of capitalism: the dynamic force needed to move individuals to work hard at acquiring wealth and investment

spirituality: sensitivity or attachment to religious values

standard deviation: a statistic showing the spread or dispersion of scores in a sample

state: the organized system of power and authority in society

statistically significant: term describing the relationship between variables that is larger than would be expected by chance alone

status: an established position in a social structure that carries with it a degree of prestige

status hierarchy: the power and prestige order of the group

steering: the practice of some real estate agents by which ethnic and racial minorities were influenced to buy houses that are not located in predominantly white communities in order to preserve the racial makeup of white communities

stereotype: an oversimplified set of beliefs about the members of a social group or social stratum that is used to categorize individuals of that group

stigma: an attribute that is socially devalued and discredited

structural functionalism: see functionalism

subculture: the culture of groups whose values and norms of behavior are somewhat different from those of the dominant culture

subjectivity: belonging to the thinking subject rather than outside of the subject

sunshine modernist: a social theorist who uncritically accepts the globalization of industrialization as good, no matter what the consequences

survey: to query (someone or a group or a sample) in order to collect data for the analysis of some aspect of a group or area

symbolic communication: interaction using abstract concepts to represent physical and social objects

symbolic interaction theory: a theoretical perspective claiming that people act toward things because of the meaning things have for them

T

taboo: a prohibition imposed by social custom

teleological: exhibiting or relating to design or purpose especially in nature

troubles: privately felt problems that come from events or feelings in one individual's life

U

underemployment: a term used to describe being employed at a level below what would be expected, given a person's level of training or education

unemployment rate: the percentage of those not working, but officially defined as looking for work

urban underclass: a grouping of people, largely minority and poor, who live at the absolute bottom of the socioeconomic ladder in urban areas

V

validity: the degree to which an indicator accurately measures or reflects a concept

values: the abstract standards in a society or group that define ideal principles

W

wealth: the monetary value of what someone actually owns, calculated by adding all financial assets (stocks, bonds, property, insurance, value of investments, etc.) and subtracting debts; also called net worth

welfare system: the public benefit system of a society designed to provide for the needs of those who cannot fully provide for themselves or their families

Westernization: the process by which non-Western peoples convert to or adopt Western culture

working class: those persons who do not own the means of production and therefore must sell their labor in order to earn a living

working poor: employed people whose wages are too low to bring their standard of living above the poverty level

Index